CONTINUOUS AND DISCRETE SIGNALS AND SYSTEMS

PRENTICE HALL INFORMATION AND SYSTEM SCIENCES SERIES
Thomas Kailath, Editor

ÅSTRÖM & WITTENMARK	*Computer-Controlled Systems: Theory and Design, 3/E*
BASSEVILLE & NIKIROV	*Detection of Abrupt Changes: Theory & Application*
FRIEDLAND	*Advanced Control System Design*
GREEN & LIMEBEER	*Robust Linear Control*
HAYKIN	*Adaptive Filter Theory 3/E*
HAYKIN	*Blind Deconvolution*
JAIN	*Fundamentals of Digital Image Processing*
JOHANSSON	*System Modeling and Identification*
KAILATH	*Linear Systems*
KWAKERNAAK & SIVAN	*Modern Signals and Systems*
LANDAU	*System Identification and Control Design Using P.I.M. + Software*
LJUNG	*System Identification: Theory for the User*
LJUNG & GLAD	*Modeling of Dynamic Systems*
MACOVSKI	*Medical Imaging Systems*
MOSCA	*Optimal, Predictive and Adaptive Control*
RUGH	*Linear System Theory, 2/E*
SOLIMAN & SRINATH	*Continuous and Discrete Signals and Systems, 2/E*
SOLO & KONG	*Adaptive Signal Processing Algorithms: Stability & Performance*
SRINATH, RAJASEKARAN, & VISWANATHAN	*Introduction to Statistical Signal Processing with Applications*
VISWANADHAM & NARAHARI	*Performance Modeling of Automated Manufacturing Systems*

CONTINUOUS AND DISCRETE SIGNALS AND SYSTEMS

SECOND EDITION

SAMIR S. SOLIMAN

QUALCOMM Incorporated
San Diego, California

MANDYAM D. SRINATH

Southern Methodist University
Dallas, Texas

PRENTICE-HALL INTERNATIONAL, INC.

Original edition, entitled CONTINUOUS AND DISCRETE SIGNALS AND SYSTEMS, 2nd Edition, by SOLIMAN, SAMIR S.; MANDYAM D. SRINATH, published by Pearson Education, Inc, publishing as Prentice Hall, Copyright © 1998.

ISBN 978-89-450-0633-2

Contents

Preface

The second edition of *Continuous and Discrete Signals and Systems* is a modified version of the first edition based on our experience in using it as a textbook in the introductory course on signals and systems at Southern Methodist University, as well as the comments of numerous colleagues who have used the book at other universities. The result, we hope, is a book that provides an introductory, but comprehensive treatment of the subject of continuous and discrete-time signals and systems. Some changes that we have made to enhance the quality of the book is to move the section on orthogonal representations of signals from Chapter 1 to the beginning of Chapter 3 on Fourier series, which permits us to treat Fourier series as a special case of more general representations. Other features are the addition of sections on practical reconstruction filters, sampling-rate conversion, and A/D and D/A converters to Chapter 7. We have also added several problems in various chapters, emphasizing computer usage. However, we have not suggested or required the use of any specific mathematical software packages as we feel that this choice should be left to the preference of the instructor. Overall, about a third of the problems and about a fifth of the examples in the book have been changed.

As noted in the first edition, the aim of building complex systems that perform sophisticated tasks imposes on engineering students a need to enhance their knowledge of signals and systems, so that they are able to use effectively the rich variety of analysis and synthesis techniques that are available. Thus signals and systems is a core course in the Electrical Engineering curriculum in most schools. In writing this book we have tried to present the most widely used techniques of signal and system analysis in an appropriate fashion for instruction at the junior or senior level in electrical engineering. The concepts and techniques that form the core of the book are of fundamental importance and should prove useful also to engineers wishing to update or extend their understanding of signals and systems through self-study.

The book is divided into two major parts. In the first part, a comprehensive treatment of continuous-time signals and systems is presented. In the second part, the results are extended to discrete-time signals and systems. In our experience, we have found that covering both continuous-time and discrete-time systems together, frequently confuses students and they often are not clear as to whether a particular concept or technique applies to continuous-time or discrete-time systems, or both. The result is that they often use solution techniques that simply do not apply to particular problems. Since most students are familiar with continuous-time signals and systems in the basic courses leading up to this course, they are able to follow the development of the theory and analysis of continuous-time systems without difficulty. Once they have become familiar with this material which is covered in the first five chapters, students should be ready to handle discrete-time signals and systems.

The book is organized such that all the chapters are distinct but closely related with smooth transitions between chapters, thereby providing considerable flexibility in course design. By appropriate choice of material, the book can be used as a text in several courses such as transform theory (Chapters 1, 3, 4, 5, 7, and 8), continuous-time signals and systems (1, 2, 3, 4, and 5), discrete-time signals and systems (Chapters 6, 7, 8, and 9), and signals and systems: continuous and discrete (Chapters 1, 2, 3, 4, 6, 7, and 8). We have been using the book at Southern Methodist University for a one-semester course covering both continuous-time and discrete-time systems and it has proved successful.

Normally, a signals and systems course is taught in the third year of a four-year undergraduate curriculum. Although the book is designed to be self-contained, a knowledge of calculus through integration of trigonometric functions, as well as some knowledge of differential equations, is presumed. A prior exposure to matrix algebra as well as a course in circuit analysis is preferable but not necessary. These prerequisite skills should be mastered by all electrical engineering students by their junior year. No prior experience with system analysis is required. While we use mathematics extensively, we have done so, not rigorously, but in an engineering context. We use examples extensively to illustrate the theoretical material in an intuitive manner.

As with all subjects involving problem solving, we feel that it is imperative that a student sees many solved problems related to the material covered. We have included a large number of examples that are worked out in detail to illustrate concepts and to show the student the application of the theory developed in the text. In order to make the student aware of the wide range of applications of the principles that are covered, applications with practical significance are mentioned. These applications are selected to illustrate key concepts, stimulate interest, and bring out connections with other branches of electrical engineering.

It is well recognized that the student does not fully understand a subject of this nature unless he or she is given the opportunity to work out problems in using and applying the basic tools that are developed in each chapter. This not only reinforces the understanding of the subject matter, but, in some cases, allows for the extension of various concepts discussed in the text. In certain cases, even new material is introduced via the problem sets. Consequently, over 260 end-of-chapter problems have been included. These problems are of various types, some being straightforward applications of the basic ideas presented in the chapters, and are included to ensure that the student understands the material fully. Some are moderately difficult, and other problems require that the student apply the theory he or she learned in the chapter to problems of practical importance.

The relative amount of "Design" work in various courses is always a concern for the electrical engineering faculty. The inclusion in this text of analog- and digital-filter design as well as other design-related material is in direct response to that concern.

At the end of each chapter, we have included an item-by-item summary of all the important concepts and formulas covered in that chapter as well as a checklist of all important terms discussed. This list serves as a reminder to the student of material that deserves special attention.

Throughout the book, the emphasis is on linear time-invariant systems. The focus in Chapter 1 is on signals. This material, which is basic to the remainder of the book, considers the mathematical representation of signals. In this chapter, we cover a variety of subjects such as periodic signals, energy and power signals, transformations of the independent variable, and elementary signals.

Chapter 2 is devoted to the time-domain characterization of continuous-time (CT) linear time-invariant (LTIV) systems. The chapter starts with the classification of continuous-time systems and then introduces the impulse-response characterization of LTIV systems and the convolution integral. This is followed by a discussion of systems characterized by linear constant-coefficient differential equations. Simulation diagrams for such systems are presented and used as a stepping stone to introduce the state variable concept. The chapter concludes with a discussion of stability.

To this point the focus is on the time-domain description of signals and systems. Starting with Chapter 3, we consider frequency-domain descriptions. We begin the chapter with a consideration of the orthogonal representation of arbitrary signals. The Fourier series are then introduced as a special case of the orthogonal representation for periodic signals. Properties of the Fourier series are presented. The concept of line spectra for describing the frequency content of such signals is given. The response of linear systems to periodic inputs is discussed. The chapter concludes with a discussion of the Gibbs phenomenon.

Chapter 4 begins with the development of the Fourier transform. Conditions under which the Fourier transform exists are presented and its properties discussed. Applications of the Fourier transform in areas such as amplitude modulation, multiplexing, sampling, and signal filtering are considered. The use of the transfer function in determining the response of LTIV systems is discussed. The Nyquist sampling theorem is derived from the impulse-modulation model for sampling. The several definitions of bandwidth are introduced and duration-bandwidth relationships discussed.

Chapter 5 deals with the Laplace transform. Both unilateral and bilateral Laplace transforms are defined. Properties of the Laplace transform are derived and examples are given to demonstrate how these properties are used to evaluate new Laplace transform pairs or to find the inverse Laplace transform. The concept of the transfer function is introduced and other applications of the Laplace transform such as for the solution of differential equations, circuit analysis, and control systems are presented. The state-variable representation of systems in the frequency domain and the solution of the state equations using Laplace transforms are discussed.

The treatment of continuous-time signals and systems ends with Chapter 5, and a course emphasizing only CT material can be ended at this point. By the end of this chapter, the reader should have acquired a good understanding of continuous-time signals and systems and should be ready for the second half of the book in which discrete-time signals and systems analysis are covered.

We start our consideration of discrete-time systems in Chapter 6 with a discussion of elementary discrete-time signals. The impulse-response characterization of discrete-time systems is presented and the convolution sum for determining the response to arbitrary inputs is derived. The difference equation representation of discrete-time systems and their solution is given. As in CT systems, simulation diagrams are discussed as a means of obtaining the state-variable representation of discrete-time systems.

Chapter 7 considers the Fourier analysis of discrete-time signals. The Fourier series for periodic sequences and the Fourier transform for arbitrary signals are derived. The similarities and differences between these and their continuous-time counterparts are brought out and their properties and applications discussed. The relation between the continuous-time and discrete-time Fourier transforms of sampled analog signals is derived and used to obtain the impulse-modulation model for sampling that is considered in Chapter 4. Reconstruction of sampled analog signals using practical reconstruction devices such as the zero-order hold is considered. Sampling rate conversion by decimation and interpolation of sampled signals is discussed. The chapter concludes with a brief description of A/D and D/A conversion.

Chapter 8 discusses the *Z*-transform of discrete-time signals. The development follows closely that of Chapter 5 for the Laplace transform. Properties of the *Z*-transform are derived and their application in the analysis of discrete-time systems developed. The solution of difference equations and the analysis of state-variable systems using the *Z*-transform are also discussed. Finally, the relation between the Laplace and the *Z*-transforms of sampled signals is derived and the mapping of the s-plane into the z-plane is discussed.

Chapter 9 introduces the discrete Fourier transform (DFT) for analyzing finite-length sequences. The properties of the DFT are derived and the differences with the other transforms discussed in the book are noted. The interpretation of the DFT as a matrix operation on a data vector is used to briefly note its relation to other orthogonal transforms. The application of the DFT to linear system analysis and to spectral estimation of analog signals is discussed. Two popular fast Fourier transform (FFT) algorithms for the efficient computation of the DFT are presented.

The final chapter, Chapter 10, considers some techniques for the design of analog and digital filters. Techniques for the design of two low-pass analog filters, namely, the Butterworth and the Chebyshev filters, are given. The impulse invariance and bilinear techniques for designing digital IIR filters are derived. Design of FIR digital filters using window functions is also discussed. An example to illustrate the application of FIR filters to approximate nonconventional filters is presented. The chapter concludes with a very brief overview of computer-aided techniques.

In addition, four appendices are included. They should prove useful as a readily available source for some of the background material in complex variables and matrix algebra necessary for the course. A somewhat extensive list of frequently-used formulas is also included.

We wish to acknowledge the many people who have helped us in writing this book, especially the students on whom much of this material was classroom tested, and the reviewers whose comments were very useful. We have tried to incorporate most of their comments in preparing this second edition of the book. We wish to thank Dyan Muratalla, who typed a substantial part of the manuscript. Finally, we would like to thank our wives and families for their patience during the completion of this book.

S. Soliman
M.D. Srinath

Chapter 1

Representing Signals

1.1 INTRODUCTION

Signals are detectable physical quantities or variables by means of which messages or information can be transmitted. A wide variety of signals are of practical importance in describing physical phenomena. Examples include the human voice, television pictures, teletype data, and atmospheric temperature. Electrical signals are the most easily measured and the most simply represented type of signals. Therefore, many engineers prefer to transform physical variables to electrical signals. For example, many physical quantities, such as temperature, humidity, speech, wind speed, and light intensity, can be transformed, using transducers, to time-varying current or voltage signals. Electrical engineers deal with signals that have a broad range of shapes, amplitudes, durations, and perhaps other physical properties. For example, a radar-system designer analyzes high-energy microwave pulses, a communication-system engineer who is concerned with signal detection and signal design analyzes information-carrying signals, a power engineer deals with high-voltage signals, and a computer engineer deals with millions of pulses per second.

Mathematically, signals are represented as functions of one or more independent variables. For example, time-varying current or voltage signals are functions of one variable (time), the vibration of a rectangular membrane can be represented as a function of two spatial variables (x and y coordinates), the electrical field intensity can be looked upon as a function of two variables (time and space), and finally, an image signal can be regarded as a function of two variables (x and y coordinates). In this introductory course of signals and systems, we focus attention on signals involving one independent variable, which we take to be time, although it can be different in some specific applications.

We begin this chapter with an introduction to two classes of signals that we are concerned with throughout the text, namely, continuous-time and discrete-time signals. Then, in Section 1.3, we define periodic signals. Section 1.4 deals with the issue of power and energy signals. A number of transformations of the independent variable are discussed in Section 1.5. In Section 1.6, we introduce several important elementary signals that not only occur frequently in applications, but also serve as a basis for representing other signals. Other types of signals that are of importance to engineers are mentioned in Section 1.7.

1.2 CONTINUOUS-TIME VS. DISCRETE-TIME SIGNALS

One way to classify signals is according to the nature of the independent variable. If the independent variable is continuous, the corresponding signal is called a continuous-time signal and is defined for a continuum of values of the independent variable. A telephone or radio signal as a function of time and an atmospheric pressure as a function of altitude are examples of continuous-time signals. (See Figure 1.2.1.)

Corresponding to any instant t_1 and an infinitesimally small positive real number ε, let us denote the instants $t_1 - \varepsilon$ and $t_1 + \varepsilon$ by t_1^- and t_1^+, respectively. If $x(t_1^-) = x(t_1^+) = x(t_1)$, we say that $x(t)$ is continuous at $t = t_1$. Otherwise it is discontinuous at t_1, and the amplitude of signal $x(t)$ has a jump at that point. Signal $x(t)$ is said to be continuous if it is continuous for all t. A signal that has only a finite or a countably infinite number of discontinuities is said to be piecewise continuous if the jump in amplitude at each discontinuity is finite.

There are many continuous-time signals of interest that are not continuous. An example is the rectangular pulse function $\mathrm{rect}(t/\tau)$ (see Figure 1.2.2), which is defined as

$$\mathrm{rect}(t/\tau) = \begin{cases} 1, & |t| < \dfrac{\tau}{2} \\ 0, & |t| > \dfrac{\tau}{2} \end{cases} \tag{1.2.1}$$

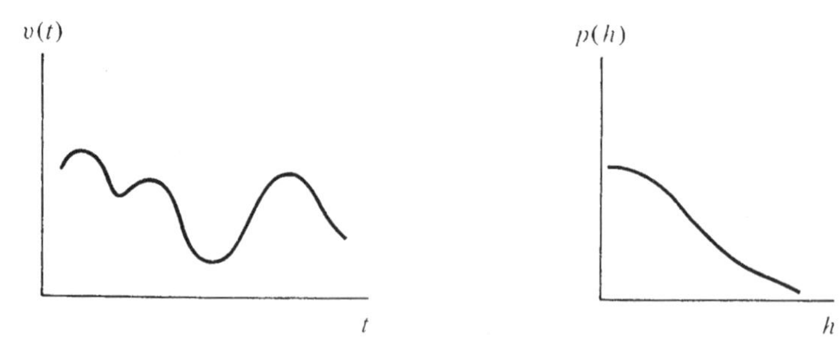

(a) (b)

Figure 1.2.1 Examples of continuous-time signals.

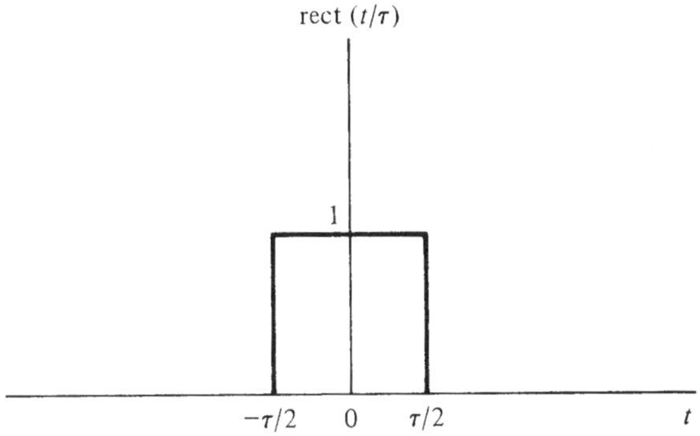

Figure 1.2.2 A rectangular pulse signal.

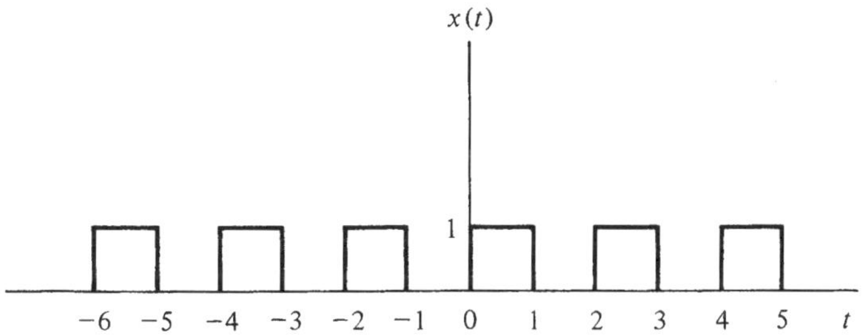

Figure 1.2.3 A pulse train.

This signal is piecewise continuous, since it is continuous everywhere except at $t = \pm\tau/2$ and the magnitude of the jump at these points is 1. Another example is the pulse train shown in Figure 1.2.3. This signal is continuous at all t except $t = 0, \pm 1, \pm 2, \ldots$.

At a point of discontinuity t_1, the value of the signal $x(t)$ is usually considered to be undefined. However, in order to be able to consider both continuous and piecewise continuous signals in a similar manner, we will assign the value

$$x(t_1) = \frac{1}{2}[x(t_1^+) + x(t_1^-)] \tag{1.2.2}$$

to $x(t)$ at the point of discontinuity $t = t_1$.

If the independent variable takes on only discrete values $t = kT_s$, where T_s is a fixed positive real number and k ranges over the set of integers (i.e., $k = 0, \pm 1, \pm 2$, etc.), the corresponding signal $x(kT_s)$ is called a discrete-time signal. Discrete-time signals arise naturally in many areas of business, economics, science, and engineering. Examples are the amount of a loan payment in the *kth* month, the weekly Dow Jones stock index, and the output of an information source that produces one of the digits 1, 2, ..., M every T_s seconds. We consider discrete-time signals in more detail in Chapter 6.

1.3 PERIODIC VS. APERIODIC SIGNALS

Any continuous-time signal that satisfies the condition

$$x(t) = x(t + nT), \qquad n = 1, 2, 3, \ldots \tag{1.3.1}$$

where $T > 0$ is a constant known as the fundamental period, is classified as a periodic signal. A signal $x(t)$ that is not periodic is referred to as an aperiodic signal. Familiar examples of periodic signals are the sinusoidal functions. A real-valued sinusoidal signal can be expressed mathematically by a time-varying function of the form

$$x(t) = A \sin(\omega_0 t + \phi) \tag{1.3.2}$$

where

A = amplitude
ω_0 = radian frequency in rad/s
ϕ = initial phase angle with respect to the time origin in rad

This sinusoidal signal is periodic with fundamental period $T = 2\pi/\omega_0$ for all values of ω_0.

The sinusoidal time function described in Equation (1.3.2) is usually referred to as a sine wave. Examples of physical phenomena that approximately produce sinusoidal signals are the voltage output of an electrical alternator and the vertical displacement of a mass attached to a spring under the assumption that the spring has negligible mass and no damping. The pulse train shown in Figure 1.2.3 is another example of a periodic signal, with fundamental period $T = 2$. Notice that if $x(t)$ is periodic with fundamental period T, then $x(t)$ is also periodic with period $2T, 3T, 4T, \ldots$. The fundamental frequency, in radians, (radian frequency) of the periodic signal $x(t)$ is related to the fundamental period by the relationship

$$\omega_0 = \frac{2\pi}{T} \tag{1.3.3}$$

Engineers and most mathematicians refer to the sinusoidal signal with radian frequency $\omega_k = k\omega_0$ as the kth harmonic. For example, the signal shown in Figure 1.2.3 has a fundamental radian frequency $\omega_0 = \pi$, a second harmonic radian frequency $\omega_2 = 2\pi$, and a third harmonic radian frequency $\omega_3 = 3\pi$. Figure 1.3.1 shows the first, second, and third harmonics of signal $x(t)$ in Eq. (1.3.2) for specific values of A, ω_0, and ϕ. Note that the waveforms corresponding to each harmonic are distinct. In theory, we

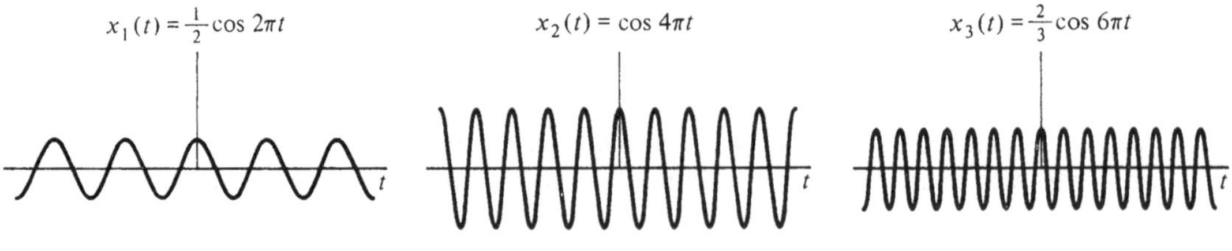

Figure 1.3.1 Harmonically related sinusoids.

can associate an infinite number of distinct harmonic signals with a given sinusoidal waveform.

Periodic signals occur frequently in physical problems. In this section, we discuss the mathematical representation of such signals. In Chapter 3, we show how to represent any periodic signal in terms of simple ones, such as sine and cosine.

Example 1.3.1

Harmonically related continuous-time exponentials are sets of complex exponentials with fundamental frequencies that are all multiples of a single positive frequency ω_0. Mathematically,

$$\phi_k(t) = \exp[jk\omega_0 t], \quad k = 0, \pm 1, \pm 2, \ldots \tag{1.3.4}$$

We show that for $k \neq 0$, $\phi_k(t)$ is periodic with fundamental period $2\pi/|k\omega_0|$ or fundamental frequency $|k\omega_0|$.

In order for signal $\phi_k(t)$ to be periodic with period $T > 0$, we must have

$$\exp[jk\omega_0(t + T)] = \exp[jk\omega_0 t]$$

or, equivalently,

$$T = \frac{2\pi}{|k\omega_0|} \tag{1.3.5}$$

Note that since a signal that is periodic with period T is also periodic with period lT for any positive integer l, then all signals $\phi_k(t)$ have a common period of $2\pi/\omega_0$.

The sum of two periodic signals may or may not be periodic. Consider the two periodic signals $x(t)$ and $y(t)$ with fundamental periods T_1 and T_2, respectively. We investigate under what conditions the sum

$$z(t) = ax(t) + by(t)$$

is periodic and what the fundamental period of this signal is if the signal is periodic. Since $x(t)$ is periodic with period T_1, it follows that

$$x(t) = x(t + kT_1)$$

Similarly,

$$y(t) = y(t + lT_2)$$

where k and l are integers such that

$$z(t) = ax(t + kT_1) + by(t + lT_2)$$

In order for $z(t)$ to be periodic with period T, one needs

$$ax(t + T) + by(t + T) = ax(t + kT_1) + by(t + lT_2)$$

We therefore must have

$$T = kT_1 = lT_2$$

or, equivalently,

$$\frac{T_1}{T_2} = \frac{l}{k}$$

In other words, the sum of two periodic signals is periodic only if the ratio of their respective periods can be expressed as a rational number.

Example 1.3.2

We wish to determine which of the following signals are periodic.

(a) $x_1(t) = \sin\frac{2\pi}{3}t$

(b) $x_2(t) = \sin\frac{2\pi}{5}t\cos 4\frac{\pi}{3}t$

(c) $x_3(t) = \sin 3t$

(d) $x_4(t) = x_1(t) - 2x_3(t)$

For these signals, $x_1(t)$ is periodic with period $T_1 = 3$. We can write $x_2(t)$ as the sum of two sinusoids with periods $T_{21} = 15/13$ and $T_{22} = 15/7$. Since $13T_{21} = 7T_{22}$, it follows that $x_2(t)$ is periodic with period $T_2 = 15$. $x_3(t)$ is periodic with period $T_3 = 2\pi/13$. Since we cannot find integers k and l such that $kT_1 = lT_3$, it follows that $x_4(t)$ is not periodic.

Note that if $x(t)$ and $y(t)$ have the same period T, then $z(t) = x(t) + y(t)$ is periodic with period T; i.e., linear operations (addition in this case) do not affect the periodicity of the resulting signal. Nonlinear operations on periodic signals (such as multiplication) produce periodic signals with different fundamental periods. The following example demonstrates this fact.

Example 1.3.3

Let $x(t) = \cos\omega_1 t$ and $y(t) = \cos\omega_2 t$. Consider the signal $z(t) = x(t)y(t)$. Signal $x(t)$ is periodic with periodic $2\pi/\omega_1$, and signal $y(t)$ is periodic with period $2\pi/\omega_2$. The fact that $z(t) = x(t)y(t)$ has two components, one with radian frequency $\omega_2 - \omega_1$ and the other with radian frequency $\omega_2 + \omega_1$, can be seen by rewriting the product $x(t)y(t)$ as

$$\cos\omega_1 t\cos\omega_2 t = \frac{1}{2}[\cos(\omega_2 - \omega_1)t + \cos(\omega_2 + \omega_1)t]$$

If $\omega_1 = \omega_2 = \omega$, then $z(t)$ will have a constant term (1/2) and a second-harmonic term ($1/2\cos 2\omega t$). In general, nonlinear operations on periodic signals can produce higher order harmonics.

Since a periodic signal is a signal of infinite duration that should start at $t = -\infty$ and go on to $t = \infty$, it follows that all practical signals are aperiodic. Nevertheless, the study of the system response to periodic inputs is essential (as we shall see in Chapter 4) in the process of developing the system response to all practical inputs.

1.4 ENERGY AND POWER SIGNALS

Let $x(t)$ be a real-valued signal. If $x(t)$ represents the voltage across a resistance R, it produces a current $i(t) = x(t)/R$. The instantaneous power of the signal is $Ri^2(t) = x^2(t)/R$, and the energy expended during the incremental interval dt is $x^2(t)/R\,dt$. In general, we do not know whether $x(t)$ is a voltage or a current signal, and in order to normalize power, we assume that $R = 1$ ohm. Hence, the instantaneous power associated with signal $x(t)$ is $x^2(t)$. The signal energy over a time interval of length $2L$ is defined as

$$E_{2L} = \int_{-L}^{L} |x(t)|^2\, dt \tag{1.4.1}$$

and the total energy in the signal over the range $t \in (-\infty, \infty)$ can be defined as

$$E = \lim_{L\to\infty} \int_{-L}^{L} |x(t)|^2\, dt \tag{1.4.2}$$

The average power can then be defined as

$$P = \lim_{L\to\infty} \left[\frac{1}{2L} \int_{-L}^{L} |x(t)|^2\, dt\right] \tag{1.4.3}$$

Although we have used electrical signals to develop Equations (1.4.2) and (1.4.3), these equations define the energy and power, respectively, of any arbitrary signal $x(t)$.

When the limit in Equation (1.4.2) exists and yields $0 < E < \infty$, signal $x(t)$ is said to be an energy signal. Inspection of Equation (1.4.3) reveals that energy signals have zero power. On the other hand, if the limit in Equation (1.4.3) exists and yields $0 < P < \infty$, then $x(t)$ is a power signal. Power signals have infinite energy.

As stated earlier, periodic signals are assumed to exist for all time from $-\infty$ to $+\infty$ and, therefore, have infinite energy. If it happens that these periodic signals have finite average power (which they do in most cases), then they are power signals. In contrast, bounded finite-duration signals are energy signals.

Example 1.4.1

In this example, we show that for a periodic signal with period T, the average power is

$$P = \frac{1}{T}\int_{0}^{T} |x(t)|^2\, dt \tag{1.4.4}$$

If $x(t)$ is periodic with period T, then the integral in Equation (1.4.3) is the same over any interval of length T. Allowing the limit to be taken in a manner such that $2L$ is an integral multiple of the period (i.e., $2L = mT$), we find that the total energy of $x(t)$ over an interval of length $2L$ is m times the energy over one period. The average power is then

$$\begin{aligned} P &= \lim_{m\to\infty}\left[\frac{1}{mT}\, m \int_{0}^{T} |x(t)|^2\, dt\right] \\ &= \frac{1}{T}\int_{0}^{T} |x(t)|^2\, dt \end{aligned}$$

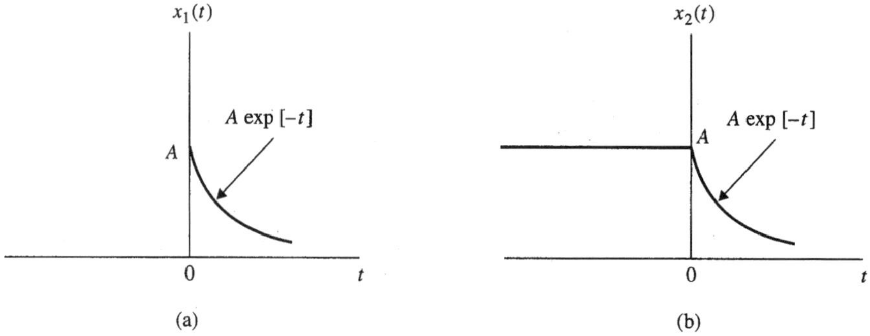

Figure 1.4.1 Signals for Example 1.4.2.

Example 1.4.2

Consider the signals in Figure 1.4.1. We wish to determine whether these signals are energy or power signals. The signal in Figure 1.4.1(a) is aperiodic with total energy

$$E = \int_0^\infty A^2 \exp[-2t]\, dt = \frac{A^2}{2}$$

which is finite. Therefore this signal is an energy signal with energy $A^2/2$. The average power is

$$P = \lim_{L\to\infty} \left(\frac{1}{2L}\int_{-L}^{L} A^2 \exp[-2t]\, dt\right)$$

$$= \lim_{L\to\infty} \frac{A^2}{4L} = 0$$

and is zero as expected.

The energy in the signal in Figure 1.4.2(b) is found as

$$E = \lim_{L\to\infty} \left[\int_{-L}^{0} A^2\, dt + \int_0^L A^2 \exp[-2t]\, dt\right] = \lim_{L\to\infty} A^2\left[L + \frac{1}{2}(1 - \exp[-2L])\right]$$

which is clearly unbounded. Thus this signal is not an energy signal. Its power can be found as

$$E = \lim_{L\to\infty} \frac{1}{2L}\left[\int_{-L}^{0} A^2\, dt + \int_0^L A^2 \exp[-2t]\, dt\right] = \frac{A^2}{2}$$

so that this is a power signal with average power $A^2/2$.

Example 1.4.3

Consider the sinusoidal signal

$$x(t) = A \sin(\omega_0 t + \phi)$$

This signal is periodic with period

$$T = \frac{2\pi}{\omega_0}$$

The average power of the signal is

$$P = \frac{1}{T}\int_0^T A^2 \sin^2(\omega_0 t + \phi)\, dt$$
$$= \frac{A^2\omega_0}{2\pi}\int_0^{2\pi/\omega_0}\left[\frac{1}{2} - \frac{1}{2}\cos(2\omega_0 t + 2\phi)\right] dt$$
$$= \frac{A^2}{2}$$

The last step follows because the signal $\cos(2\omega_0 t + 2\phi)$ is periodic with period $T/2$ and the area under a cosine signal over any interval of length lT, where l is a positive integer, is always zero. (You should have no trouble confirming this result if you draw two complete periods of $\cos(2\omega_0 t + 2\phi)$).

Example 1.4.4

Consider the two aperiodic signals shown in Figure 1.4.2. These two signals are examples of energy signals. The rectangular pulse shown in Figure 1.4.2(a) is strictly time limited, since $x_1(t)$ is identically zero outside the duration of the pulse. The other signal is asymptotically time limited in the sense that $x_2(t) \to 0$ as $t \to \pm\infty$. Such signals may also be described loosely as "pulses." In either case, the average power equals zero. The energy for signal $x_1(t)$ is

$$E_1 = \lim_{L\to\infty}\int_{-L}^{L} x_1^2(t)\, dt$$
$$= \int_{-\tau/2}^{\tau/2} A^2\, dt = A^2\tau$$

For $x_2(t)$,

$$E_2 = \lim_{L\to\infty}\int_{-L}^{L} A^2 \exp[-2a|t|]\, dt$$
$$= \lim_{L\to\infty}\frac{A^2}{a}(1 - \exp[-2aL]) = \frac{A^2}{a}$$

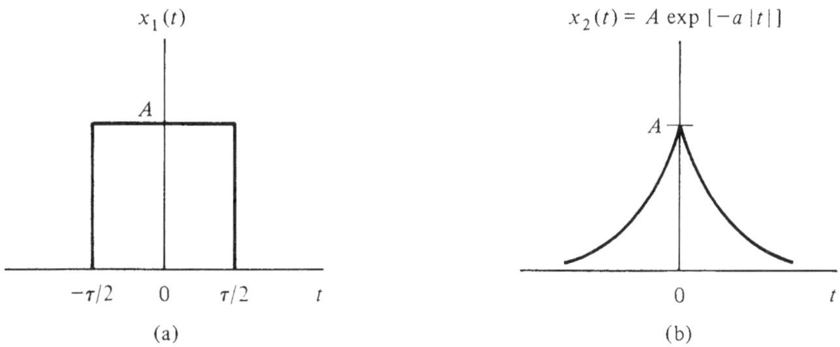

Figure 1.4.2 Signals for Example 1.4.4.

Since E_1 and E_2 are finite, $x_1(t)$ and $x_2(t)$ are energy signals. Almost all time-limited signals of practical interest are energy signals.

1.5 TRANSFORMATIONS OF THE INDEPENDENT VARIABLE

A number of important operations are often performed on signals. Most of these operations involve transformations of the independent variable. It is important that the reader know how to perform such operations and understand the physical meaning of each one. The three operations we discuss in this section are shifting, reflecting, and time scaling.

1.5.1 The Shifting Operation

Signal $x(t - t_0)$ represents a time-shifted version of $x(t)$; see Figure 1.5.1. The shift in time is t_0. If $t_0 > 0$, then the signal is delayed by t_0 seconds. Physically, t_0 cannot take on negative values, but from the analytical viewpoint, $x(t - t_0)$, $t_0 < 0$, represents an advanced replica of $x(t)$. Signals that are related in this fashion arise in applications such as radar, sonar, communication systems, and seismic signal processing.

Example 1.5.1

Consider the signal $x(t)$ shown in Figure 1.5.2. We want to plot $x(t - 2)$ and $x(t + 3)$. It can easily be seen that

$$x(t) = \begin{cases} t + 1, & -1 \le t \le 0 \\ 1, & 0 < t \le 2 \\ -t + 3, & 2 < t \le 3 \\ 0, & \text{otherwise} \end{cases}$$

To perform the time-shifting operation, replace t by $t - 2$ in the expression for $x(t)$:

$$x(t-2) = \begin{cases} (t-2) + 1, & -1 \le t - 2 \le 0 \\ 1, & 0 \le t - 2 \le 2 \\ -(t-2) + 3, & 2 < t - 2 \le 3 \\ 0, & \text{otherwise} \end{cases}$$

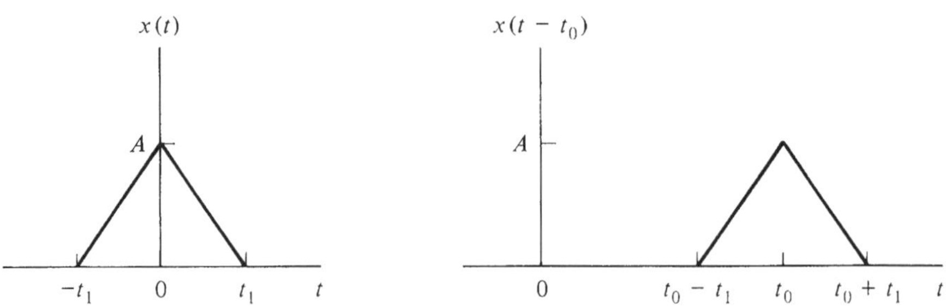

Figure 1.5.1 The shifting operation.

or, equivalently,

$$x(t-2) = \begin{cases} t-1, & 1 \le t \le 2 \\ 1, & 2 < t \le 4 \\ -t+5, & 4 < t \le 5 \\ 0, & \text{otherwise} \end{cases}$$

The signal $x(t-2)$ is plotted in Figure 1.5.3(a) and can be described as $x(t)$ shifted two units to the right on the time axis. Similarly, it can be shown that

$$x(t+3) = \begin{cases} t+4, & -4 \le t \le -3 \\ 1, & -3 < t \le -1 \\ -t, & -1 < t \le 0 \\ 0, & \text{otherwise} \end{cases}$$

The signal $x(t+3)$ is plotted in Figure 1.5.3(b) and represents a shifted version of $x(t)$, shifted three units to the left.

Example 1.5.2

Vibration sensors are mounted on the front and rear axles of a moving vehicle to pick up vibrations due to the roughness of the road surface. The signal from the front sensor is $x(t)$ and is shown in Figure 1.5.4. The signal from the rear axle sensor is modeled as $x(t-120)$. If the sensors are placed 6 ft apart, it is possible to determine the speed of the vehicle by comparing the signal from the rear axle sensor with the signal from the front axle sensor.

Figure 1.5.5 illustrates the time-delayed version of $x(t)$ where the delay is 120 ms, or

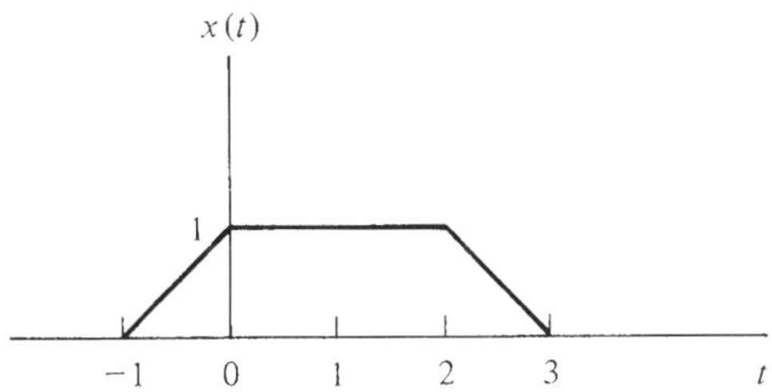

Figure 1.5.2 Plot of $x(t)$ for Example 1.5.1.

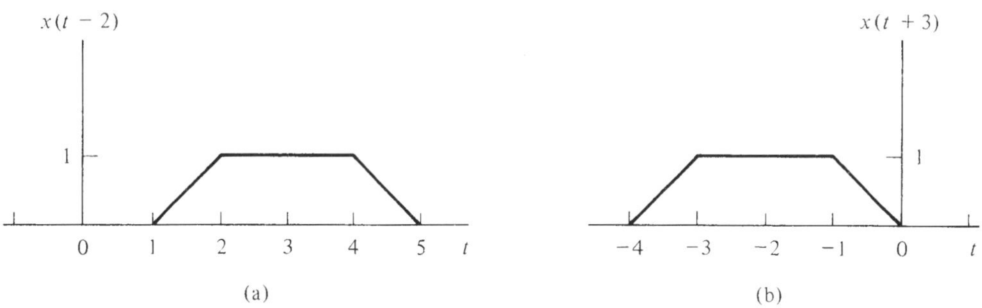

Figure 1.5.3 The shifting of $x(t)$ of Example 1.5.1.

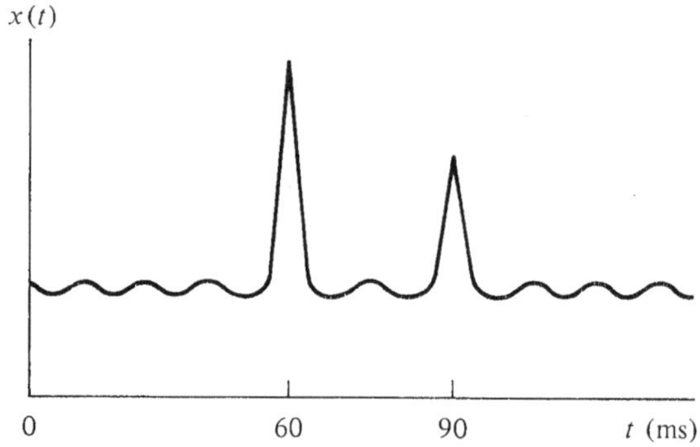

Figure 1.5.4 Front axle sensor signal for Example 1.5.2.

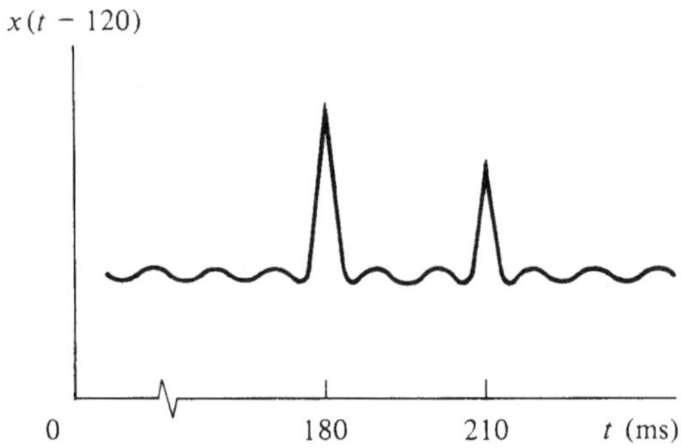

Figure 1.5.5 Rear axle sensor signal for Example 1.5.2.

0.12 s. The delay τ between the sensor signals from the front and rear axles is related to the distance d between the two axles and the speed v of the vehicle by

$$d = v\tau$$

so that

$$\begin{aligned} v &= \frac{d}{\tau} \\ &= \frac{6\,ft}{0.12\,s} = 50\,ft/s \end{aligned}$$

Example 1.5.3

A radar placed to detect aircraft at a range R of 45 nautical miles (nmi) (1 nautical mile = 6076.115 ft) transmits the pulse-train signal shown in Figure 1.5.6. If there is a target, the transmitted signal is reflected back to the radar's receiver. The radar operates by mea-

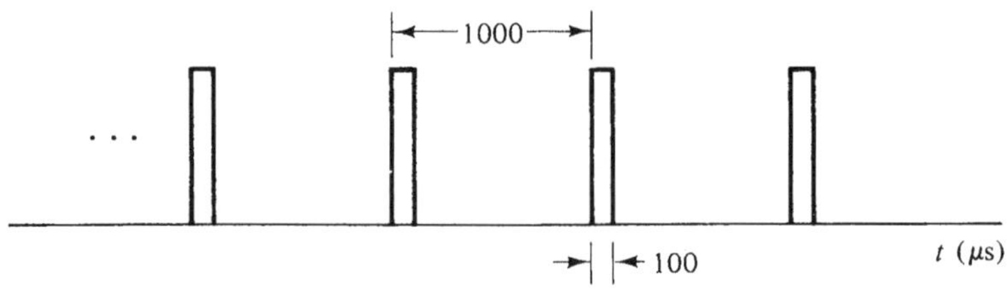

Figure 1.5.6 Radar-transmitted signal for Example 1.5.3.

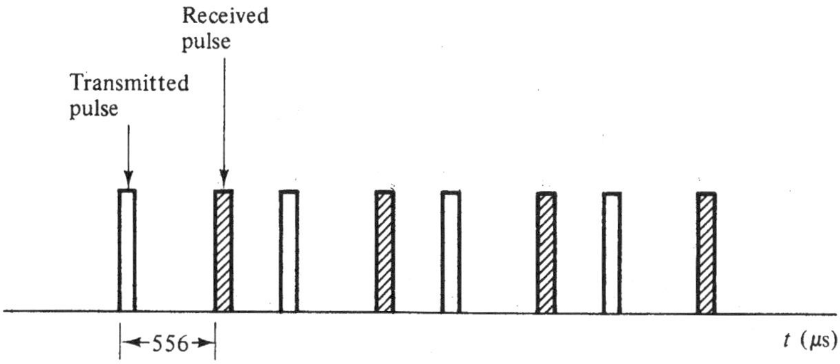

Figure 1.5.7 Transmitted and received pulse train of Example 1.5.3.

suring the time delay between each transmitted pulse and the corresponding return, or echo. The velocity of propagation of the radar signal, C, is equal to 161,875 nmi/s.

The round-trip delay is

$$\tau = \frac{2R}{C} = \frac{2 \times 45}{161{,}875} = 0.556 \text{ ms}$$

Therefore, the received pulse train is the same as the transmitted pulse train, but shifted to the right by 0.556 ms; see Figure 1.5.7.

1.5.2 The Reflection Operation

The signal $x(-t)$ is obtained from the signal $x(t)$ by a reflection about $t = 0$ (i.e., by reversing $x(t)$), as shown in Figure 1.5.8. Thus, if $x(t)$ represents a signal out of a video recorder, then $x(-t)$ is the signal out of a video player when the rewind switch is pushed on (assuming that the rewind and play speeds are the same).

Example 1.5.4

We want to draw $x(-t)$ and $x(3 - t)$ if $x(t)$ is as shown in Figure 1.5.9(a). The signal $x(t)$ can be written as

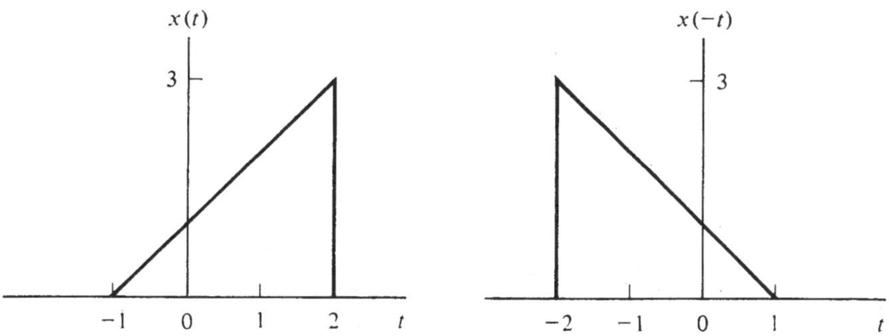

Figure 1.5.8 The reflection operation.

$$x(t) = \begin{cases} t+1, & -1 \le t \le 0 \\ 1, & 0 < t \le 2 \\ 0, & \text{otherwise} \end{cases}$$

The signal $x(-t)$ is obtained by replacing t by $-t$ in the last equation so that

$$x(-t) = \begin{cases} -t+1, & -1 \le -t \le 0 \\ 1, & 0 < -t \le 2 \\ 0, & \text{otherwise} \end{cases}$$

or, equivalently,

$$x(-t) = \begin{cases} -t+1, & 0 \le t \le 1 \\ 1, & -2 < t \le 0 \\ 0, & \text{otherwise} \end{cases}$$

The signal $x(-t)$ is illustrated in Figure 1.5.9(b) and can be described as $x(t)$ reflected about the vertical axis. Similarly, it can be shown that

$$x(3-t) = \begin{cases} 4-t, & 3 \le t \le 4 \\ 1, & 1 < t \le 3 \\ 0, & \text{otherwise} \end{cases}$$

The signal $x(3-t)$ is shown in Figure 1.5.9(c) and can be viewed as $x(t)$ reflected and then shifted three units to the right. This result is obtained as follows:

$$x(3-t) = x(-(t-3))$$

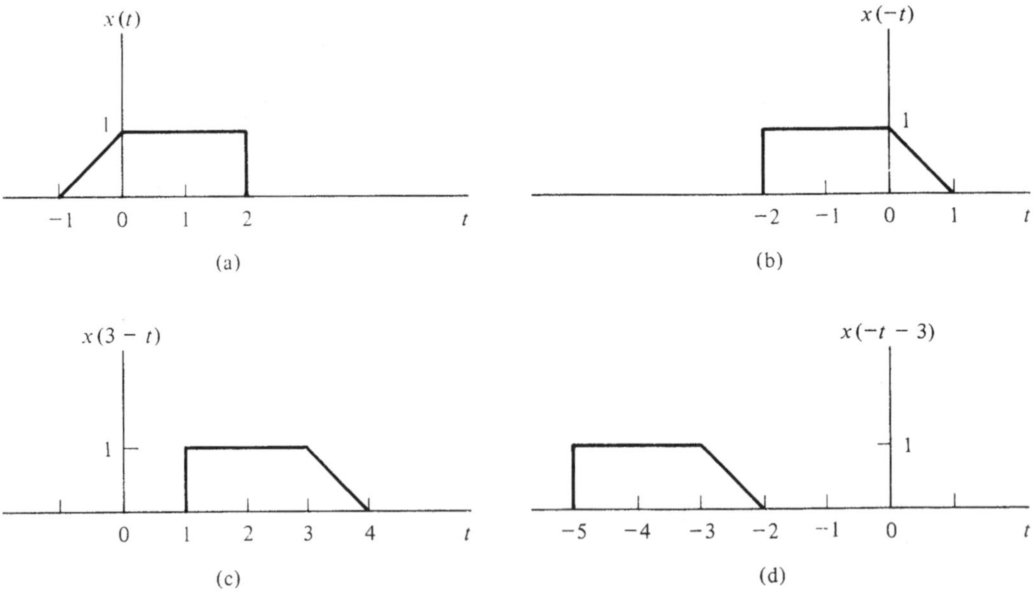

Figure 1.5.9 Plots of $x(-t)$ and $x(3-t)$ for Example 1.5.4.

Note that if we first shift $x(t)$ by three units and then reflect the shifted signal, the result is $x(-t-3)$, which is shown in Figure 1.5.9(d). Therefore, the operations of shifting and reflecting are not commutative.

In addition to its use in representing physical phenomena such as that in the video recorder example, reflection is extremely useful in examining the symmetry properties that the signal may possess. A signal $x(t)$ is referred to as an even signal, or is said to be even symmetric, if it is identical to its reflection about the origin—that is, if

$$x(-t) = x(t) \tag{1.5.1}$$

A signal is referred to as odd symmetric if

$$x(-t) = -x(t) \tag{1.5.2}$$

An arbitrary signal $x(t)$ can always be expressed as a sum of even and odd signals as

$$x(t) = x_e(t) + x_o(t) \tag{1.5.3}$$

where $x_e(t)$ is called the even part of $x(t)$ and is given by (see Problem 1.14)

$$x_e(t) = \frac{1}{2}[x(t) + x(-t)] \tag{1.5.4}$$

and $x_o(t)$ is called the odd part of $x(t)$ and is expressed as

$$x_o(t) = \frac{1}{2}[x(t) - x(-t)] \tag{1.5.5}$$

Example 1.5.5

Consider the signal $x(t)$ defined by

$$x(t) = \begin{cases} 1, & t > 0 \\ 0, & t < 0 \end{cases}$$

The even and odd parts of this signal are, respectively,

$$x_e(t) = \frac{1}{2}, \quad \text{all } t \text{ except } t = 0$$

$$x_o(t) = \begin{cases} -\dfrac{1}{2}, & t < 0 \\ \dfrac{1}{2}, & t > 0 \end{cases}$$

The only problem here is the value of these functions at $t = 0$. If we define $x(0) = 1/2$ (the definition here is consistent with our definition of the signal at a point of discontinuity), then

$$x_e(0) = \frac{1}{2} \quad \text{and} \quad x_o(0) = 0$$

Signals $x_e(t)$ and $x_o(t)$ are plotted in Figure 1.5.10.

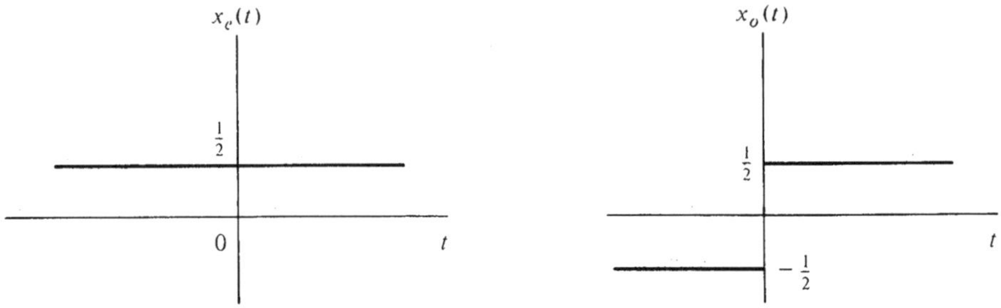

Figure 1.5.10 Plots of $x_e(t)$ and $x_o(t)$ for $x(t)$ in Example 1.5.5.

Example 1.5.6

Consider the signal

$$x(t) = \begin{cases} A\exp[-\alpha t], & t > 0 \\ 0, & t < 0 \end{cases}$$

The even part of the signal is

$$x_e(t) = \begin{cases} \frac{1}{2}A\exp[-\alpha t], & t > 0 \\ \frac{1}{2}A\exp[\alpha t], & t < 0 \end{cases}$$

$$= \frac{1}{2}A\exp[-\alpha|t|]$$

The odd part of $x(t)$ is

$$x_o(t) = \begin{cases} \frac{1}{2}A\exp[-\alpha t], & t > 0 \\ -\frac{1}{2}A\exp[\alpha t], & t < 0 \end{cases}$$

Signals $x_e(t)$ and $x_o(t)$ are as shown in Figure 1.5.11.

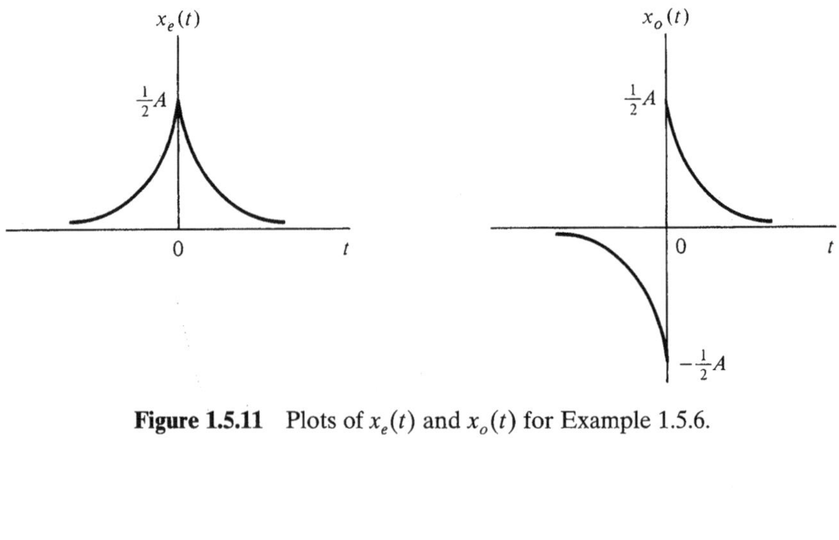

Figure 1.5.11 Plots of $x_e(t)$ and $x_o(t)$ for Example 1.5.6.

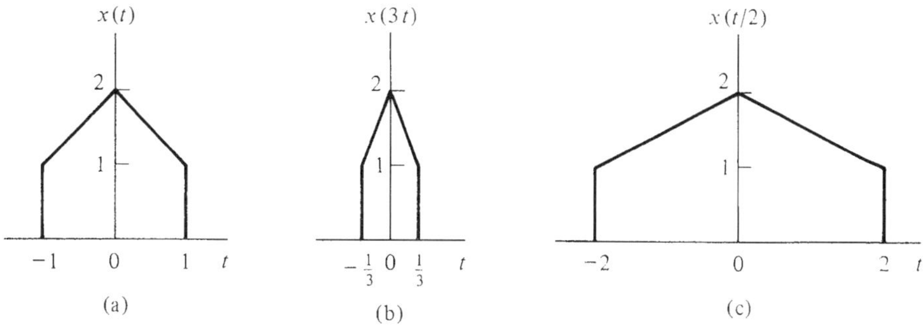

Figure 1.5.12 The time-scaling operation.

1.5.3 The Time-Scaling Operation

Consider the signals $x(t)$, $x(3t)$, and $x(t/2)$, as shown in Figure 1.5.12. As is seen in the figure, $x(3t)$ can be described as $x(t)$ contracted by a factor of 3. Similarly, $x(t/2)$ can be described as $x(t)$ expanded by a factor of 2. Both $x(3t)$ and $x(t/2)$ are said to be time-scaled versions of $x(t)$. In general, if the independent variable is scaled by a parameter η, then $x(\eta t)$ is a compressed version of $x(t)$ if $|\eta| > 1$ (the signal exists in a smaller time interval) and is an expanded version of $x(t)$ if $|\eta| < 1$ (the signal exists in a larger time interval). If we think of $x(t)$ as the output of a videotape recorder, then $x(3t)$ is the signal obtained when the recording is played back at three times the speed at which it was recorded, and $x(t/2)$ is the signal obtained when the recording is played back at half speed.

Example 1.5.7

Suppose we want to plot the signal $x(3t - 6)$, where $x(t)$ is the signal shown in Figure 1.5.2. Using the definition of $x(t)$ in Example 1.5.1 we obtain

$$x(3t-6) = \begin{cases} 3t-5, & \frac{5}{3} \leq t \leq 2 \\ 1, & 2 < t \leq \frac{8}{3} \\ -3t+9, & \frac{8}{3} < t \leq 3 \\ 0, & \text{otherwise} \end{cases}$$

A plot of $x(3t - 6)$ versus t is illustrated in Figure 1.5.13 and can be viewed as $x(t)$ compressed by a factor of 3 (or time scaled by a factor of 1/3) and then shifted two units of time to the right. Note that if $x(t)$ is shifted first and then time scaled by a factor of 1/3, we will obtain a different signal; therefore, shifting and time scaling are not commutative. The result we did get can be justified as follows:

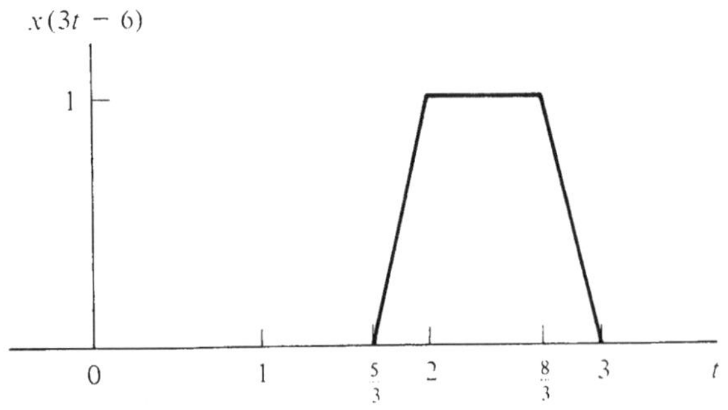

Figure 1.5.13 Plot of $x(3t - 6)$ of Example 1.5.7.

$$x(3t - 6) = x(3(t - 2))$$

This equation indicates that we perform the scaling operation first and then the shifting operation.

Example 1.5.8

We often encounter signals of the type

$$x(t) = 1 - A\exp[-\alpha t]\cos(\omega_0 t + \phi)$$

Figure 1.5.14 shows $x(t)$ for typical values of A, α and ω_0. As can be seen, this signal eventually goes to a steady state value of 1 as t becomes infinite. In practice, it is assumed that the signal has settled down to a final value when it stays within a specified percentage of its final theoretical value. This percentage is usually chosen to be 5% and the time t_s after which the signal stays within this range is defined as the settling time t_s. As can be seen from Figure 1.5.14, t_s can be determined by solving

$$1 + A\exp[-\alpha t_s] = 1.05$$

so that

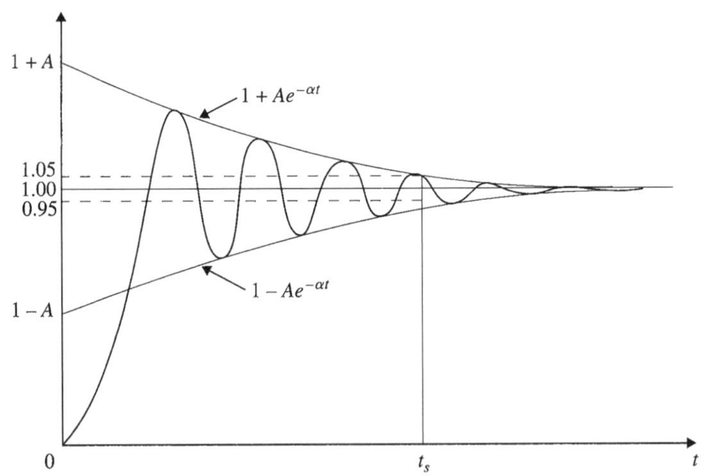

Figure 1.5.14 Signal $x(t)$ for Example 1.5.8.

$$t_s = -\frac{1}{\alpha}\ln\left[\frac{0.05}{A}\right]$$

Let

$$x(t) = 1 - 2.3\exp[-10.356t]\cos[5t]$$

We will find t_s for $x(t)$, $x(t/2)$ and $x(2t)$.

For $x(t)$, since $A = 2.3$ and $\alpha = 10.356$, we get $t_s = 0.3697$ s.

Since

$$x(t/2) = 1 - 2.3\exp[-5.178t]\cos[2.5t]$$

and

$$x(2t) = 1 - 2.3\exp[-20.712t]\cos[10t]$$

we get $t_s = 0.7394$ s and $t_s = 0.1849$ s for $x(t/2)$ and $x(2t)$ respectively. These results are expected since $x(t)$ is compressed by a factor of 2 in the first case and is expanded by the same factor in the second case.

In conclusion, for any general signal $x(t)$, the transformation $\alpha t + \beta$ on the independent variable can be performed as follows:

$$x(\alpha t + \beta) = x(\alpha(t + \beta/\alpha)) \tag{1.5.6}$$

where α and β are assumed to be real numbers. The operations should be performed in the following order:

1. Scale by α. If α is negative, reflect about the vertical axis.
2. Shift to the right by β/α if β and α have different signs, and to the left by β/α if β and α have the same sign.

Note that the operation of reflecting and time scaling is commutative, whereas the operation of shifting and reflecting or shifting and time scaling is not.

1.6 ELEMENTARY SIGNALS

Several important elementary signals that occur frequently in applications also serve as a basis for representing other signals. Throughout the book, we will find that representing signals in terms of these elementary signals allows us to better understand the properties of both signals and systems. Furthermore, many of these signals have features that make them particularly useful in the solution of engineering problems and, therefore, of importance in our subsequent studies.

1.6.1 The Unit Step Function

The continuous-time unit step function is defined as

$$u(t) = \begin{cases} 1, & t > 0 \\ 0, & t < 0 \end{cases} \tag{1.6.1}$$

and is shown in Figure 1.6.1.

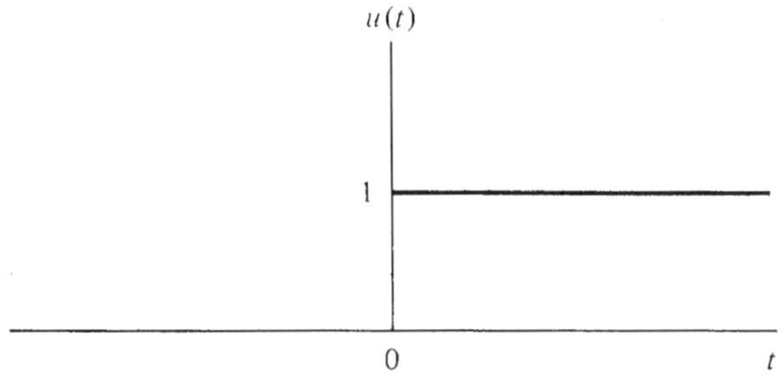

Figure 1.6.1 Continuous-time unit step function.

This signal is an important signal for analytic studies, and it also has many practical applications. Note that the unit step function is continuous for all t except at $t = 0$, where there is a discontinuity. According to our earlier discussion, we define $u(0) = 1/2$. An example of a unit step function is the output of a 1-V dc voltage source in series with a switch that is turned on at time $t = 0$.

Example 1.6.1

The rectangular pulse signal shown in Figure 1.6.2 is the result of an on-off switching operation of a constant voltage source in an electric circuit.

In general, a rectangular pulse that extends from $-a$ to $+a$ and has an amplitude A can be written as a difference between appropriately shifted step functions, i.e.,

$$A \operatorname{rect}(t/2a) = A[u(t + a) - u(t - a)] \tag{1.6.2}$$

In our specific example,

$$2 \operatorname{rect}(t/2) = 2[u(t + 1) - u(t - 1)]$$

Example 1.6.2

Consider the signum function (written sgn) shown in Figure 1.6.3. The unit sgn function is defined by

$$\operatorname{sgn} t = \begin{cases} 1, & t > 0 \\ 0, & t = 0 \\ -1, & t < 0 \end{cases} \tag{1.6.3}$$

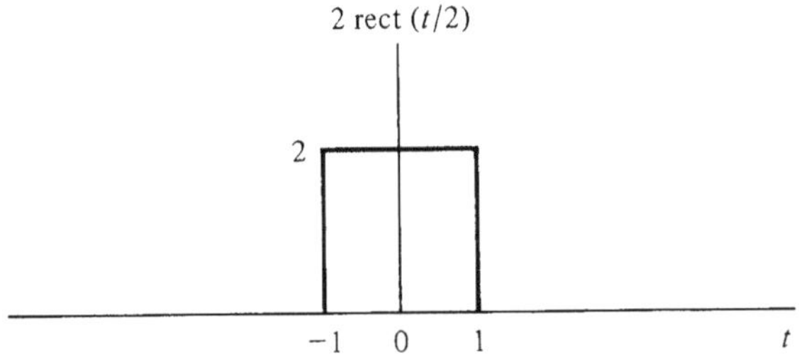

Figure 1.6.2 Rectangular pulse signal of Example 1.6.1.

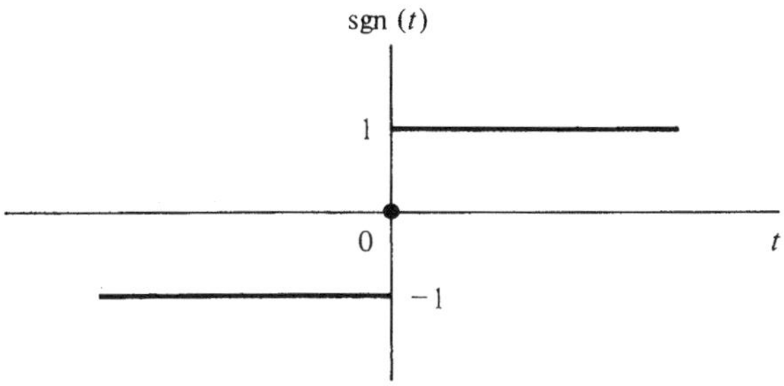

Figure 1.6.3 The signum function.

The signum function can be expressed in terms of the unit step function as

$$\operatorname{sgn} t = -1 + 2u(t)$$

The signum function is one of the most often used signals in communication and in control theory.

1.6.2 The Ramp Function

The ramp function shown in Figure 1.6.4 is defined by

$$r(t) = \begin{cases} t, & t \geq 0 \\ 0, & t < 0 \end{cases} \tag{1.6.4}$$

The ramp function is obtained by integrating the unit step function:

$$\int_{-\infty}^{t} u(\tau)d\tau = r(t)$$

The device that accomplishes this operation is called an integrator. In contrast to both the unit step and the signum functions, the ramp function is continuous at $t = 0$. Time scaling a unit ramp by a factor α corresponds to a ramp function with slope α. (A unit ramp function has a slope of unity.) An example of a ramp function is the linear-sweep waveform of a cathode-ray tube.

Example 1.6.3

Let $x(t) = u(t + 2) - 2u(t + 1) + 2u(t) - u(t - 2) - 2u(t - 3) + 2u(t - 4)$. Let $y(t)$ denote its integral. Then

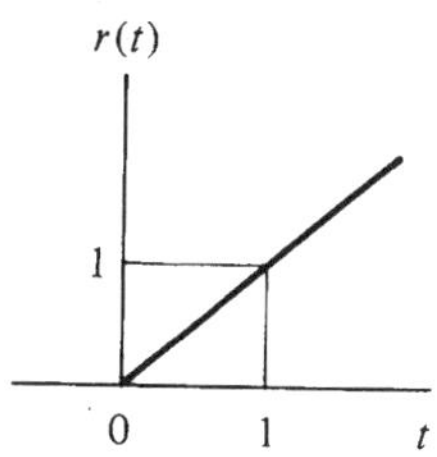

Figure 1.6.4 The ramp function.

$$y(t) = r(t+2) - 2r(t+1) + 2r(t) - r(t-2) - 2r(t-3) + 2r(t-4)$$

Signal $y(t)$ is sketched in Figure 1.6.5.

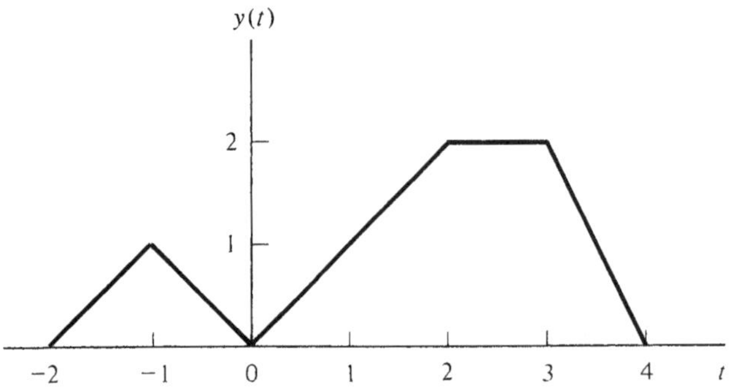

Figure 1.6.5 The signal used in Example 1.6.3.

1.6.3 The Sampling Function

A function frequently encountered in spectral analysis is the sampling function $\mathrm{Sa}(x)$, defined by

$$\mathrm{Sa}(x) = \frac{\sin x}{x} \tag{1.6.5}$$

Since the denominator is an increasing function of x and the numerator is bounded ($|\sin x| \le 1$), Sa (x) is simply a damped sine wave. Figure 1.6.6(a) shows that Sa (x) is an even function of x having its peak at $x = 0$ and zero-crossings at $x = \pm n\pi$. The value of the function at $x = 0$ is established by using l'Hôpital's rule. A closely related function is sinc x, which is defined by

$$\operatorname{sinc} x = \frac{\sin \pi x}{\pi x} = \mathrm{Sa}(\pi x) \tag{1.6.6}$$

and is shown in Figure 1.6.6(b). Note that sinc x is a compressed version of Sa (x); the compression factor is π.

1.6.4 The Unit Impulse Function

The unit impulse signal $\delta(t)$, often called the Dirac delta function or, simply, the delta function, occupies a central place in signal analysis. Many physical phenomena such as point sources, point charges, concentrated loads on structures, and voltage or current sources acting for very short times can be modeled as delta functions. Mathematically, the Dirac delta function is defined by

$$\int_{t_1}^{t_2} x(t)\delta(t)\,dt = x(0), \quad t_1 < 0 < t_2 \tag{1.6.7}$$

provided that $x(t)$ is continuous at $t = 0$. The function $\delta(t)$ is depicted graphically by a spike at the origin, as shown in Figure 1.6.7, and possesses the following properties:

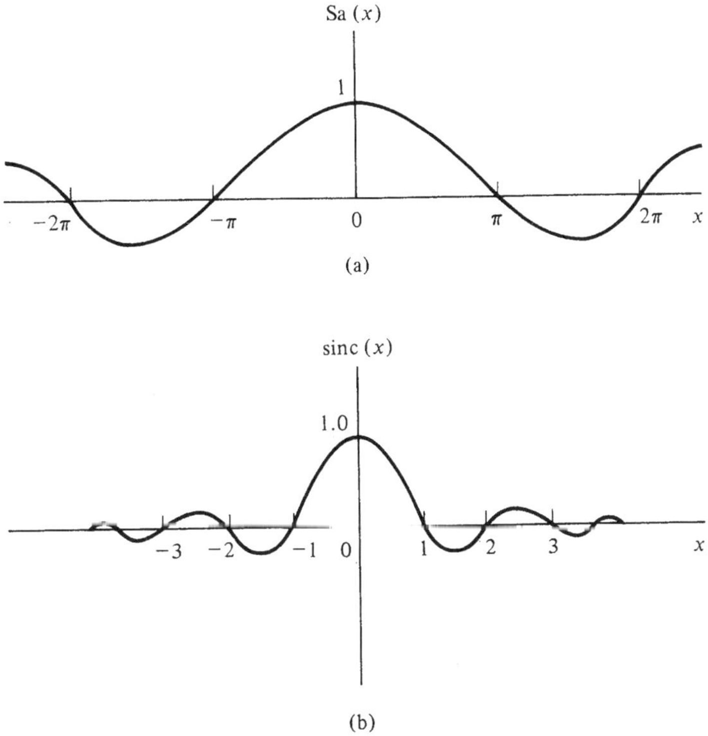

Figure 1.6.6 The sampling function.

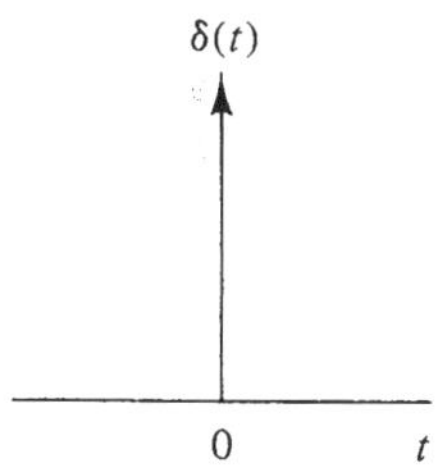

Figure 1.6.7 Representation of the unit impulse function $\delta(t)$.

1. $\delta(0) \to \infty$
2. $\delta(t) = 0, t \neq 0$
3. $\int_{-\infty}^{\infty} \delta(t)\, dt = 1$
4. $\delta(t)$ is an even function; i.e., $\delta(t) = \delta(-t)$

As just defined, the δ function does not conform to the usual definition of a function. However, it is sometimes convenient to consider it as the limit of a conventional function as some parameter ε approaches zero. Several examples are shown in Figure 1.6.8; all such functions have the following properties for "small" ε:

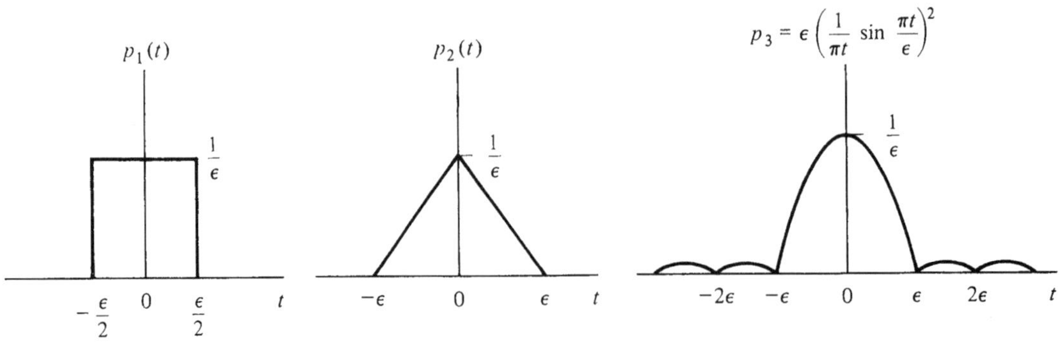

Figure 1.6.8 Engineering models for $\delta(t)$.

1. The value at $t = 0$ is very large and becomes infinity as ε approaches zero.
2. The duration is relatively very short and becomes zero as ε becomes zero.
3. The total area under the function is constant and equal to 1.
4. The functions are all even.

Example 1.6.4

Consider the function defined as

$$p(t) = \lim_{\varepsilon \to 0^+} \varepsilon\left(\frac{1}{\pi t}\sin\frac{\pi t}{\varepsilon}\right)^2$$

This function satisfies all the properties of a delta function, as can be shown by rewriting it as

$$p(t) = \lim_{\varepsilon \to 0^+} \frac{1}{\varepsilon}\left(\frac{\sin \pi t/\varepsilon}{\pi t/\varepsilon}\right)^2$$

so that

1. $p(0) = \lim_{\varepsilon \to 0^+} (1/\varepsilon) = \infty$. Here we used the well-known limit $\lim_{\tau \to 0} (\sin\tau)/\tau = 1$.
2. For values of $t \neq 0$,

$$\begin{aligned} p(t) &= \lim_{\varepsilon \to 0^+} \varepsilon\left(\frac{1}{\pi t}\sin\frac{\pi t}{\varepsilon}\right)^2 \\ &= (\lim_{\varepsilon \to 0^+} \varepsilon)\left[\lim_{\varepsilon \to 0^+}\left(\frac{1}{\pi t}\sin\frac{\pi t}{\varepsilon}\right)^2\right] \end{aligned}$$

The second limit is bounded by 1, but the first limit vanishes as $\varepsilon \to 0^+$; therefore,

$$p(t) = 0, \quad t \neq 0$$

3. To show that the area under $p(t)$ is unity, we note that

$$\begin{aligned} \int_{-\infty}^{\infty} p(t)\,dt &= \lim_{\varepsilon \to 0} \frac{1}{\epsilon}\int_{-\infty}^{\infty}\left(\frac{\sin(\pi t/\varepsilon)}{\pi t/\varepsilon}\right)^2 dt \\ &= \frac{1}{\pi}\int_{-\infty}^{\infty}\frac{\sin^2(\tau)}{\tau^2}\,d\tau \end{aligned}$$

where the last step follows after using the substitution

$$\tau = \frac{\pi t}{\varepsilon}$$

Since (see Appendix B)

$$\int_{-\infty}^{\infty} \frac{\sin^2 \tau}{\tau^2}\, d\tau = \pi$$

it follows that

$$\int_{-\infty}^{\infty} p(t)\, dt = 1$$

4. It is clear that $p(t) = p(-t)$; therefore, $p(t)$ is an even function.

Three important properties repeatedly used when operating with delta functions are the sifting property, the sampling property, and the scaling property.

Sifting Property. The sifting property is expressed in the equation

$$\int_{t_1}^{t_2} x(t)\delta(t - t_0)\, dt = \begin{cases} x(t_0), & t_1 < t_0 < t_2 \\ 0, & \text{otherwise} \end{cases} \tag{1.6.8}$$

This can be seen by using the change of variables $\tau = t - t_0$ to obtain

$$\begin{aligned} \int_{t_1}^{t_2} x(t)\delta(t - t_0)\, dt &= \int_{t_1 - t_0}^{t_2 - t_0} x(\tau + t_0)\delta(\tau)\, d\tau \\ &= x(t_0), \quad t_1 < t_0 < t_2 \end{aligned}$$

by Equation (1.6.7). Notice that the right-hand side of Equation (1.6.8) can be looked at as a function of t_0. This function is discontinuous at $t_0 = t_1$ and $t_0 = t_2$. Following our notation, the value of the function at t_1 or t_2 should be given by

$$\int_{t_1}^{t_2} x(t)\delta(t - t_0)\, dt = \frac{1}{2}x(t_0), \quad t_0 = t_1 \quad \text{or} \quad t_0 = t_2 \tag{1.6.9}$$

The sifting property is usually used in lieu of Equation (1.6.7) as the definition of a delta function located at t_0. In general, the property can be written as

$$x(t) = \int_{-\infty}^{\infty} x(t)\delta(t - \tau)\, d\tau \tag{1.6.10}$$

which implies that the signal $x(t)$ can be expressed as a continuous sum of weighted impulses. This result can be interpreted graphically if we approximate $x(t)$ by a sum of rectangular pulses, each of width Δ seconds and of varying heights, as shown in Figure 1.6.9. That is,

$$\hat{x}(t) = \sum_{k=-\infty}^{\infty} x(k\Delta)\, \text{rect}\,((t - k\Delta)/\Delta)$$

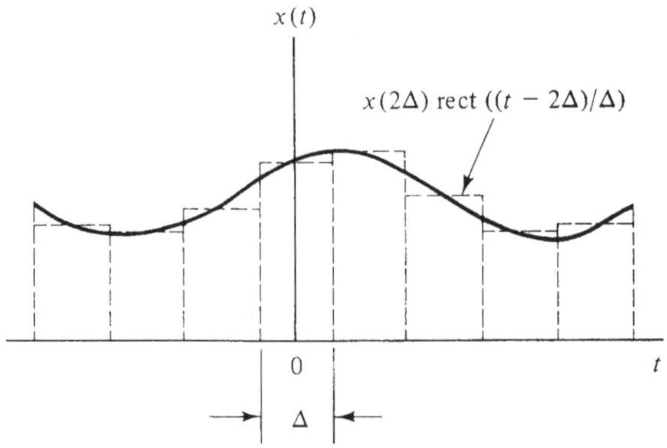

Figure 1.6.9 Approximation of signal $x(t)$.

which can be written as

$$\hat{x}(t) = \sum_{k=-\infty}^{\infty} x(k\Delta)\left[\frac{1}{\Delta}\text{rect}\,((t - k\Delta)/\Delta)\right][k\Delta - (k - 1)\Delta]$$

Now, each term in the sum represents the area under the kth pulse in the approximation $\hat{x}(t)$. We thus let $\Delta \to 0$ and replace $k\Delta$ by τ, so that $k\Delta - (k - 1)\Delta = d\tau$, and the summation becomes an integral. Also, as $\Delta \to 0$, $1/\Delta\ \text{rect}((t - k\Delta)/\Delta)$ approaches $\delta(t - \tau)$, and Equation (1.6.10) follows. The representation of Equation (1.6.10), along with the superposition principle, is used in Chapter 2 to study the behavior of a special and important class of systems known as linear time-invariant systems.

Sampling Property. If $x(t)$ is continuous at t_0, then

$$x(t)\delta(t - t_0) = x(t_0)\delta(t - t_0) \tag{1.6.11}$$

Graphically, this property can be illustrated by approximating the impulse signal by a rectangular pulse of width Δ and height $1/\Delta$, as shown in Figure 1.6.10, and then allowing Δ to approach zero, to obtain

$$\lim_{\Delta \to 0} x(t)\frac{1}{\Delta}\text{rect}((t - t_0)/\Delta) = x(t_0)\delta(t - t_0)$$

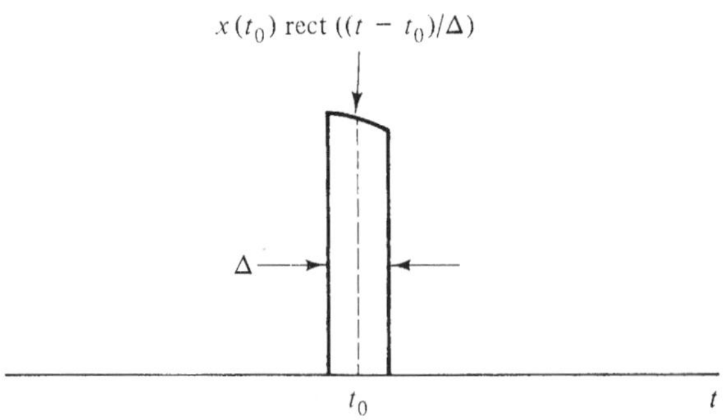

Figure 1.6.10 The sampling property of the impulse function.

Mathematically, two functions $f_1(\delta(t))$ and $f_2(\delta(t))$ are equivalent over the interval (t_1, t_2) if, for any continuous function $y(t)$,

$$\int_{t_1}^{t_2} y(t) f_1(\delta(t))\, dt = \int_{t_1}^{t_2} y(t) f_2(\delta(t))\, dt$$

Therefore, $x(t)\delta(t - t_0)$ and $x(t_0)\delta(t - t_0)$ are equivalent, since

$$\int_{t_1}^{t_2} y(t)x(t)\delta(t - t_0)\, dt = y(t_0)x(t_0) = \int_{t_1}^{t_2} y(t)x(t_0)\delta(t - t_0)\, dt$$

Note the difference between the sifting property and the sampling property: The right-hand side of Equation (1.6.8) is the value of the function evaluated at some point, whereas the right-hand side of Equation (1.6.11) is still a delta function with strength equal to the value of $x(t)$ evaluated at $t = t_0$.

Scaling Property. The scaling property is given by

$$\delta(at + b) = \frac{1}{|a|}\delta\left(t + \frac{b}{a}\right) \tag{1.6.12}$$

This result is interpreted by considering $\delta(t)$ as the limit of a unit area pulse $p(t)$ as some parameter ε tends to zero. The pulse $p(at)$ is a compressed (expanded) version of $p(t)$ if $a > 1 (a < 1)$, and its area is $1/|a|$. (Note that the area is always positive.) By taking the limit as $\varepsilon \to 0$, the result is a delta function with strength $1/|a|$. We show this by considering the two cases $a > 0$ and $a < 0$ separately. For $a > 0$, we have to show that

$$\int_{t_1}^{t_2} x(t)\delta(at + b)\, dt = \int_{t_1}^{t_2} \frac{1}{a} x(t)\delta\left(t + \frac{b}{a}\right) dt, \quad t_1 < \frac{-b}{a} < t_2$$

Applying the sifting property to the right-hand side yields

$$\frac{1}{a} x\left(\frac{-b}{a}\right)$$

To evaluate the left-hand side, we use the transformation of variables

$$\tau = at + b$$

Then $dt = (1/a)\, d\tau$, and the range $t_1 < t < t_2$ becomes $at_1 + b < \tau < at_2 + b$. The left-hand side now becomes

$$\int_{t_1}^{t_2} x(t)\delta(at + b)\, dt = \int_{at_1+b}^{at_2+b} x\left(\frac{\tau - b}{a}\right)\delta(\tau)\frac{1}{a}\, d\tau$$
$$= \frac{1}{a} x\left(\frac{-b}{a}\right)$$

which is the same as the right-hand side.

When $a < 0$, we have to show that

$$\int_{t_1}^{t_2} x(t)\delta(at+b)\,dt = \int_{t_1}^{t_2} \frac{1}{|a|} x(t)\delta\left(t+\frac{b}{a}\right)dt, \quad t_1 < \frac{-b}{a} < t_2$$

Using the sifting property, we evaluate the right-hand side. The result is

$$\frac{1}{|a|} x\left(\frac{-b}{a}\right)$$

For the left-hand side, we use the transformation $\tau = at + b$, so that

$$dt = \frac{1}{a}\,d\tau = -\frac{1}{|a|}\,d\tau$$

and the range of τ becomes $-|a|t_2 + b < \tau < -|a|t_1 + b$, resulting in

$$\begin{aligned}\int_{t_1}^{t_2} x(t)\delta(at+b)\,dt &= \int_{-|a|t_1+b}^{-|a|t_2+b} x\left(\frac{\tau-b}{a}\right)\delta(\tau)\frac{-1}{|a|}\,d\tau \\ &= \frac{1}{|a|}\int_{-|a|t_2+b}^{-|a|t_1+b} x\left(\frac{\tau-b}{a}\right)\delta(\tau)\,d\tau \\ &= \frac{1}{|a|} x\left(\frac{-b}{a}\right)\end{aligned}$$

Notice that before using the sifting property in the last step, we interchanged the limits of integration and changed the sign of the integrand, since

$$-|a|t_2 + b < -|a|t_1 + b$$

Example 1.6.5

Consider the Gaussian pulse

$$p(t) = \frac{1}{\sqrt{2\pi\varepsilon^2}} \exp\left[\frac{-t^2}{2\varepsilon^2}\right]$$

The area under this pulse is always 1; that is,

$$\int_{-\infty}^{\infty} \frac{1}{\sqrt{2\pi\varepsilon^2}} \exp\left[\frac{-t^2}{2\varepsilon^2}\right] dt = 1$$

It can be shown that $p(t)$ approaches $\delta(t)$ as $\varepsilon \to 0$. (See Problem 1.19.) Let $a > 1$ be any constant. Then $p(at)$ is a compressed version of $p(t)$. It can be shown that the area under $p(at)$ is $1/a$, and as ε approaches 0, $p(at)$ approaches $\delta(at)$.

Example 1.6.6

Suppose we want to evaluate the following integrals:

a. $\displaystyle\int_{-2}^{1} (t+t^2)\delta(t-3)\,dt$

b. $\displaystyle\int_{-2}^{4} (t+t^2)\delta(t-3)\,dt$

c. $\int_0^3 \exp[t-2]\delta(2t-4)\,dt$

d. $\int_{-\infty}^{t} \delta(\tau)\,d\tau$

a. Using the sifting property yields

$$\int_{-2}^{1} (t+t^2)\delta(t-3)\,dt = 0$$

since $t = 3$ is not in the interval $-2 < t < 1$.

b. Using the sifting property yields

$$\int_{-2}^{4} (t+t^2)\delta(t-3)\,dt = 3 + 3^2 = 12$$

since $t = 3$ is within the interval $-2 < t < 4$.

c. Using the scaling property and then the sifting property yields

$$\int_0^3 \exp[t-2]\delta(2t-4)\,dt = \int_0^3 \exp[t-2]\frac{1}{2}\delta(t-2)\,dt$$

$$= \frac{1}{2}\exp[0] = \frac{1}{2}$$

d. Consider the following two cases:

Case 1: $t < 0$

In this case, point $\tau = 0$ is not within the interval $-\infty < \tau < t$, and the result of the integral is zero.

Case 2: $t > 0$

In this case, $\tau = 0$ lies within the interval $-\infty < \tau < t$, and the value of the integral is 1. Summarizing, we obtain

$$\int_{-\infty}^{t} \delta(\tau)\,d\tau = \begin{cases} 1, & t > 0 \\ 0, & t < 0 \end{cases}$$

But this is by definition a unit step function; therefore, the functions $\delta(t)$ and $u(t)$ form an integral-derivative pair. That is,

$$\frac{d}{dt}u(t) = \delta(t) \tag{1.6.13}$$

$$\int_{-\infty}^{t} \delta(\tau)\,d\tau = u(t) \tag{1.6.14}$$

The unit impulse is one of a class of functions known as singularity functions. Note that the definition of $\delta(t)$ in Equation (1.6.7) does not make sense if $\delta(t)$ is an ordinary function. It is meaningful only if $\delta(t)$ is interpreted as a functional, i.e., as a process of assigning the value $x(0)$ to the signal $x(t)$. The integral notation is used merely as a convenient way of describing the properties of this functional, such as linearity, shifting, and scaling. We now consider how to represent the derivatives of the impulse function.

1.6.5 Derivatives of the Impulse Function

The (first) derivative of the impulse function, or unit doublet, denoted by $\delta'(t)$, is defined by

$$\int_{t_1}^{t_2} x(t)\delta'(t - t_0)\,dt = -x'(t_0), \quad t_1 < t_0 < t_2 \tag{1.6.15}$$

provided that $x(t)$ possesses a derivative $x'(t_0)$ at $t = t_0$. This result can be demonstrated using integration by parts as follows:

$$\begin{aligned}\int_{t_1}^{t_2} x(t)\delta'(t - t_0)\,dt &= \int_{t_1}^{t_2} x(t)d[\delta(t - t_0)] \\ &= x(t)\,\delta(t - t_0)\Big|_{t_1}^{t_2} - \int_{t_1}^{t_2} x'(t)\,\delta(t - t_0)\,dt \\ &= 0 - 0 - x'(t_0)\end{aligned}$$

since $\delta(t) = 0$ for $t \neq 0$. It can be shown that $\delta'(t)$ possesses the following properties:

1. $x(t)\delta'(t - t_0) = x(t_0)\delta'(t - t_0) - x'(t_0)\delta(t - t_0)$
2. $\int_{-\infty}^{t} \delta'(\tau - t_0)\,d\tau = \delta(t - t_0)$
3. $\delta'(at + b) = \frac{1}{|a|}\delta'\left(t + \frac{b}{a}\right)$

Higher order derivatives of $\delta(t)$ can be defined by extending the definition of $\delta'(t)$. For example, the nth-order derivative of $\delta(t)$ is defined by

$$\int_{t_1}^{t_2} x(t)\delta^{(n)}(t - t_0)\,dt = (-1)^n x^{(n)}(t_0), \quad t_1 < t_0 < t_2 \tag{1.6.16}$$

provided that such derivative exists at $t = t_0$. The graphical representation of $\delta'(t)$ is shown in Figure 1.6.11.

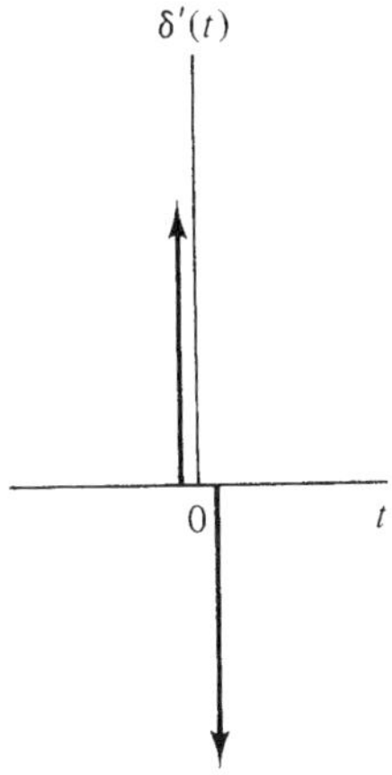

Figure 1.6.11 Representation of $\delta'(t)$.

Example 1.6.7

The current through an inductor of 1-mH is $i(t) = 10\exp[-2t]u(t) - \delta(t)$ amperes. The voltage drop across the inductor is given by

$$\begin{aligned} v(t) &= 10^{-3}\frac{d}{dt}[10\exp[-2t]u(t) - \delta(t)] \\ &= -2 \times 10^{-2}\exp[-2t]u(t) + 10^{-2}\exp[-2t]\delta(t) - 10^{-2}\delta'(t) \text{ volts} \\ &= -2 \times 10^{-2}\exp[-2t]u(t) + 10^{-2}\delta(t) - 10^{-2}\delta'(t) \text{ volts} \end{aligned}$$

where the last step follows from Equation (1.6.11).

Figures 1.6.12(a) and (b) demonstrate the behavior of the inductor current $i(t)$ and the voltage $v(t)$ respectively.

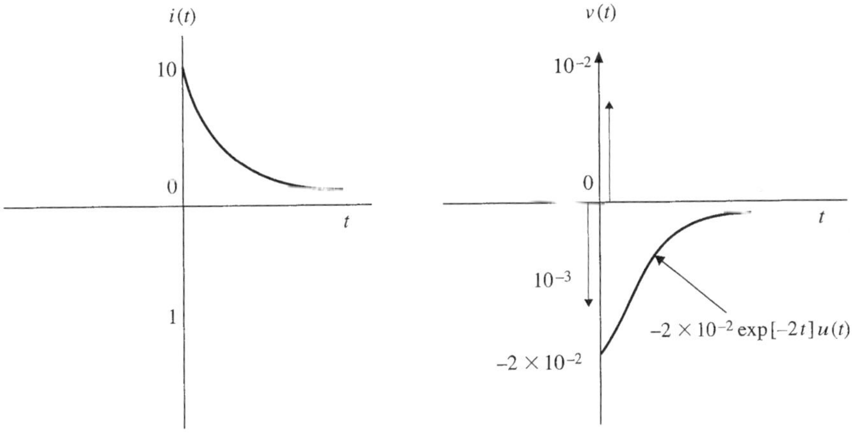

Figure 1.6.12 $v(t)$ and $dv(t)/dt$ for Example 1.6.7.

Note that the derivative of $x(t)u(t)$ is obtained using the product rule of differentiation, i.e.

$$\frac{d}{dt}[x(t)u(t)] = x(t)\delta(t) + x'(t)u(t)$$

whereas the derivative of $x(t)\delta(t)$ is

$$\frac{d}{dt}[x(t)\delta(t)] = \frac{d}{dt}[x(0)\delta(t)]$$

This result cannot be obtained by direct differentiation of the product, because $\delta(t)$ is interpreted as a functional rather than an ordinary function.

Example 1.6.8

We will evaluate the following integrals:

(a) $\int_{-4}^{4} (t-2)^2\delta'\left(-\frac{1}{3}t + \frac{1}{2}\right)dt$

(b) $\int_{-4}^{1} t\exp[-2t]\delta''(t-1)\,dt$

For (a), we have

$$\int_{-4}^{4} (t-2)^2\delta'\left(-\frac{1}{3}t + \frac{1}{2}\right)dt = \int_{-4}^{4} 3(t-2)^2\delta'\left(t - \frac{3}{2}\right)dt$$

$$= \int_{-4}^{4} 3(t-2)^2\delta'\left(t - \frac{3}{2}\right)dt$$

$$= \int_{-4}^{4}\left[\frac{3}{4}\delta'\left(t-\frac{3}{2}\right) + 9\delta\left(t-\frac{3}{2}\right)\right]dt = \frac{3}{4}\delta\left(t-\frac{3}{2}\right) + 9$$

For (b), we have

$$\int_{-4}^{1} t\exp[-2t]\delta''(t-1)\,dt = (4t-4)\exp[-2t]\big|_{t=1} = 0$$

1.7 OTHER TYPES OF SIGNALS

There are many other types of signals that electrical engineers work with very often. Signals can be classified broadly as random and nonrandom, or deterministic. The study of random signals is well beyond the scope of this text, but some of the ideas and techniques that we discuss are basic to more advanced topics. Random signals do not have the kind of totally predictable behavior that deterministic signals do. Voice, music, computer output, TV, and radar signals are neither pure sinusoids nor pure periodic waveforms. If they were, then by knowing one period of the signal, we could predict what the signal would look like for all future time. Any signal that is capable of carrying meaningful information is in some way random. In other words, in order to contain information, a signal must, in some manner change in a nondeterministic fashion.

Signals can also be classified as analog or digital signals. In science and engineering, the word "analog" means to act similarly, but in a different domain. For example, the electric voltage at the output terminals of a stereo amplifier varies in exactly the same way as does the sound that activated the microphone that is feeding the amplifier. In other words, the electric voltage $v(t)$ at every instant of time is proportional (analog) to the air pressure that is rapidly varying with time. Simply, an analog signal is a physical quantity that varies with time, usually in a smooth and continuous fashion.

The values of a discrete-time signal can be continuous or discrete. If a discrete-time signal takes on all possible values on a finite or infinite range, it is said to be a continuous-amplitude discrete-time signal. Alternatively, if the discrete-time signal takes on values from a finite set of possible values, it is said to be a discrete-amplitude discrete-time signal, or, simply, a digital signal. Examples of digital signals are digitized images, computer input, and signals associated with digital information sources.

Most of the signals that we encounter in nature are analog signals. A basic reason for this is that physical systems cannot respond instantaneously to changing inputs. Moreover, in many cases, the signal is not available in electrical form, thus requiring the use of a transducer (mechanical, electrical, thermal, optical, and so on) to provide an electrical signal that is representative of the system signal. Transducers generally cannot respond instantaneously to changes and tend to smooth out the signals.

Digital signal processing has developed rapidly over the past two decades, chiefly because of significant advances in digital computer technology and integrated-circuit fabrication. In order to process signals digitally, the signals must be in digital form (discrete in time and discrete in amplitude). If the signal to be processed is in analog form, it is first converted to a discrete-time signal by sampling at discrete instants in time. The discrete-time signal is then converted to a digital signal by a process called quantization.

Quantization is the process of converting a continuous-amplitude signal into a discrete-amplitude signal and is basically an approximation procedure. The whole procedure is called analog-to-digital (A/D) conversion, and the corresponding device is called an A/D converter.

1.8 SUMMARY

- Signals can be classified as continuous-time or discrete-time signals.
- Continuous-time signals that satisfy the condition $x(t) = x(t + T)$ are periodic with fundamental period T.
- The fundamental radian frequency of the periodic signal is related to the fundamental period T by the relationship

$$\omega_0 = \frac{2\pi}{T}$$

- The complex exponential $x(t) = \exp[j\omega_0 t]$ is periodic with period $T = 2\pi/\omega_0$ for all ω_0.

- Harmonically related continuous-time exponentials

$$x_k(t) = \exp[jk\omega_0 t]$$

are periodic with common period $T = 2\pi/\omega_0$.
- The energy E of the signal $x(t)$ is defined by

$$E = \lim_{L\to\infty} \int_{-L}^{L} |x(t)|^2\,dt$$

- The power P of the signal $x(t)$ is defined by

$$P = \lim_{L\to\infty} \frac{1}{2L} \int_{-L}^{L} |x(t)|^2\,dt$$

- The signal $x(t)$ is an energy signal if $0 < E < \infty$.
- The signal $x(t)$ is a power signal if $0 < P < \infty$.
- The signal $x(t - t_0)$ is a time-shifted version of $x(t)$. If $t_0 > 0$, then the signal is delayed by t_0 seconds. If $t_0 < 0$, then $x(t - t_0)$ represents an advanced replica of $x(t)$.
- The signal $x(-t)$ is obtained by reflecting $x(t)$ about $t = 0$.
- The signal $x(\alpha t)$ is a scaled version of $x(t)$. If $\alpha > 1$, then $x(\alpha t)$ is a compressed version of $x(t)$, whereas if $0 < \alpha < 1$, then $x(\alpha t)$ is an expanded version of $x(t)$.
- The signal $x(t)$ is even symmetric if

$$x(t) = x(-t)$$

- The signal $x(t)$ is odd symmetric if

$$x(t) = -x(-t)$$

- Unit impulse, unit step, and unit ramp functions are related by

$$u(t) = \int_{-\infty}^{t} \delta(\tau)\,d\tau$$

$$r(t) = \int_{-\infty}^{t} u(\tau)\,d\tau$$

- The sifting property of the δ function is

$$\int_{t_1}^{t_2} x(t)\delta(t - t_0)\,dt = \begin{cases} x(t_0), & t_1 < t_0 < t_2 \\ 0, & \text{otherwise} \end{cases}$$

- The sampling property of the δ function is

$$x(t)\delta(t - t_0) = x(t_0)\delta(t - t_0)$$

1.9 CHECKLIST OF IMPORTANT TERMS

Aperiodic signals
Continuous-time signals
Discrete-time signals
Elementary signals
Energy signals
Periodic signals
Power signals
Rectangular pulse
Reflection operation
Sampling function
Scaling operation
Shifting operation
Signum function
Sinc function
Unit impulse function
Unit ramp function
Unit step function

1.10 PROBLEMS

1.1. Find the fundamental period T of each of the following signals:

$$\cos(\pi t),\ \sin(2\pi t),\ \cos(3\pi t),\ \sin(4\pi t),\ \cos\left(\frac{\pi}{2}t\right),\ \sin\left(\frac{\pi}{3}t\right),$$

$$\cos\left(\frac{5\pi}{2}t\right),\ \sin\left(\frac{4\pi}{3}t\right),\ \cos\left(\frac{\pi}{4}t\right),\ \sin\left(\frac{2\pi}{3}t\right),\ \cos\left(\frac{3\pi}{5}t\right)$$

1.2. Sketch the following signals:

(a) $x(t) = \sin\left(\frac{\pi}{4}t + 20°\right)$

(b) $x(t) = t + e^{3t}, \quad 0 \le t \le 2$

(c) $x(t) = \begin{cases} t+2 & t \le -2 \\ 0 & -2 \le t \le 2 \\ t-2 & 2 \le t \end{cases}$

(d) $x(t) = 2\exp[-t], \quad 0 \le t \le 1, \quad$ and $\quad x(t+1) = x(t)$ for all t

1.3. Show that if $x(t)$ is periodic with period T, then it is also periodic with period nT, $n = 2, 3, \ldots$.

1.4. Show that if $x_1(t)$ and $x_2(t)$ have period T, then $x_3(t) = ax_1(t) + bx_2(t)$ (a, b constant) has the same period T.

1.5. Use Euler's form, $\exp[j\omega t] = \cos\omega t + j\sin\omega t$, to show that $\exp[j\omega t]$ is periodic with period $T = 2\pi/\omega$.

1.6. Are the following signals periodic? If so, find their periods.

(a) $x(t) = \sin\left(\frac{\pi}{3}t\right) + 2\cos\left(\frac{8\pi}{3}t\right)$

(b) $x(t) = \exp\left[j\frac{7\pi}{6}t\right] + \exp\left[j\frac{5\pi}{6}t\right]$

(c) $x(t) = \exp\left[j\frac{7\pi}{6}t\right] + \exp\left[\frac{5}{6}t\right]$

(d) $x(t) = \exp\left[j\frac{5\pi}{6}t\right] + \exp\left[\frac{\pi}{6}t\right]$

(e) $x(t) = 2\sin\left(\frac{3\pi}{8}t\right) + \cos\left(\frac{3}{4}t\right)$

1.7. If $x(t)$ is a periodic signal with period T, show that $x(at)$, $a > 0$, is a periodic signal with period T/a, and $x(t/b)$, $b > 0$, is a periodic signal with period bT. Verify these results for $x(t) = \sin t$, $a = b = 2$.

1.8. Determine whether the following signals are power or energy signals or neither. Justify your answers.
(a) $x(t) = A \sin t, \quad -\infty < t < \infty$
(b) $x(t) = A[u(t - a) - u(t + a)]$
(c) $x(t) = r(t) - r(t - 1)$
(d) $x(t) = \exp[-at]u(t), \quad a > 0$
(e) $x(t) = tu(t)$
(f) $x(t) = u(t)$
(g) $x(t) = A \exp[bt], \quad b > 0$

1.9. Repeat Problem 1.8 for the following signals:
(a) $x(t) = t \sin\left(\frac{\pi}{3} t\right)$
(b) $x(t) = \exp[-2|t|] \sin(\pi t)$
(c) $x(t) = \exp[4|t|]$
(d) $x(t) = \exp\left[j \frac{5\pi}{6}\right] t$
(e) $x(t) = 2 \sin\left(\frac{3\pi}{8} t\right) \cos\left(\frac{2\pi}{3} t\right)$
(f) $x(t) = \begin{cases} 1, & t < 0 \\ \exp[3t], & 0 \le t \end{cases}$

1.10. Show that if $x(t)$ is periodic with period T, then

$$\left| \int_0^T x(t)\, dt \right| \le \sqrt{PT}$$

where P is the average power of the signal.

1.11. Let

$$x(t) = \begin{cases} -t + 1, & -1 \le t < 0 \\ t, & 0 \le t < 2 \\ 2, & 2 \le t < 3 \\ 0, & \text{otherwise} \end{cases}$$

(a) Sketch $x(t)$.
(b) Sketch $x(t - 2)$, $x(t + 3)$, $x(-3t - 2)$, and $x\left(\frac{2}{3} t + \frac{1}{2}\right)$ and find the analytical expressions for these functions.

1.12. Repeat Problem 1.11 for

$$x(t) = \begin{cases} 2t + 2, & -1 \le t < 0 \\ 2t - 2, & 0 \le t < 1 \end{cases}$$

1.13. Sketch the following signals:
(a) $x_1(t) = u(t) + 5u(t - 1) - 2u(t - 2)$
(b) $x_2(t) = r(t) - r(t - 1) - u(t - 2)$
(c) $x_3(t) = \exp[-t]u(t)$
(d) $x_4(t) = 2u(t) + \delta(t - 1)$

(e) $x_5(t) = u(t)u(t - a), \quad a > 0$
(f) $x_6(t) = u(t)u(a - t), \quad a > 0$
(g) $x_7(t) = u(\cos t)$
(h) $x_1(t)x_2\left(t + \frac{1}{2}\right)$
(i) $x_1\left(-\frac{t}{3} + \frac{1}{2}\right)x_3(t - 2)$
(j) $x_2(t)x_3(2 - t)$

1.14. **(a)** Show that

$$x_e(t) = \frac{1}{2}[x(t) + x(-t)]$$

is an even signal.

(b) Show that

$$x_o(t) = \frac{1}{2}[x(t) - x(-t)]$$

is an odd signal.

1.15. Consider the simple FM stereo transmitter shown in Figure P1.15.
(a) Sketch the signals $L + R$ and $L - R$.
(b) If the outputs of the two adders are added, sketch the resulting waveform.
(c) If signal $L - R$ is inverted and added to signal $L + R$, sketch the resulting waveform.

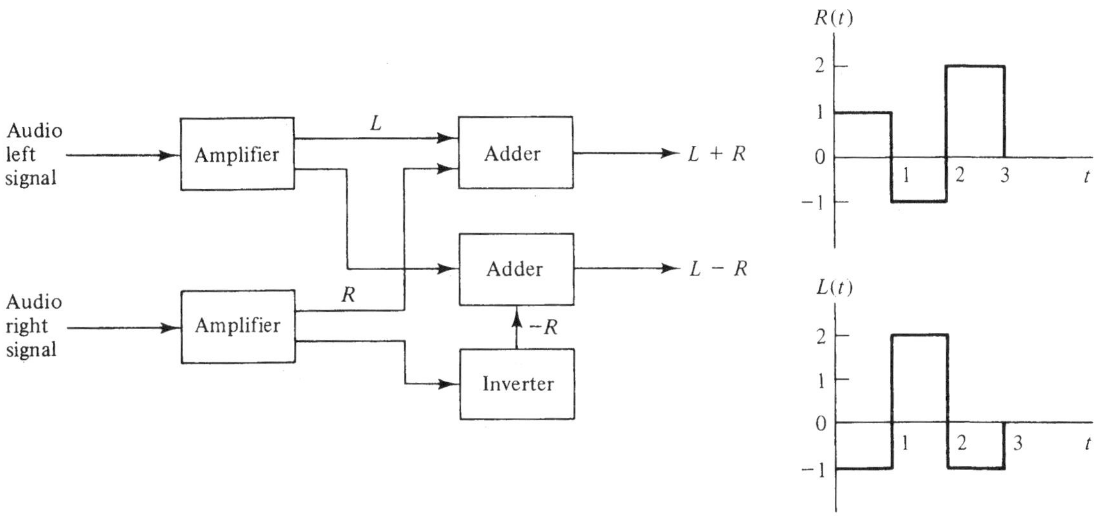

Figure P1.15

1.16. For each of the signals shown in Figure P1.16, write an expression in terms of unit step and unit ramp basic functions.

1.17. If the duration of $x(t)$ is defined as the time at which $x(t)$ drops to $1/e$ of the value at the origin, find the duration of the following signals:
(a) $x_1(t) = A\exp[-t/T]u(t)$
(b) $x_2(t) = x_1(3t)$

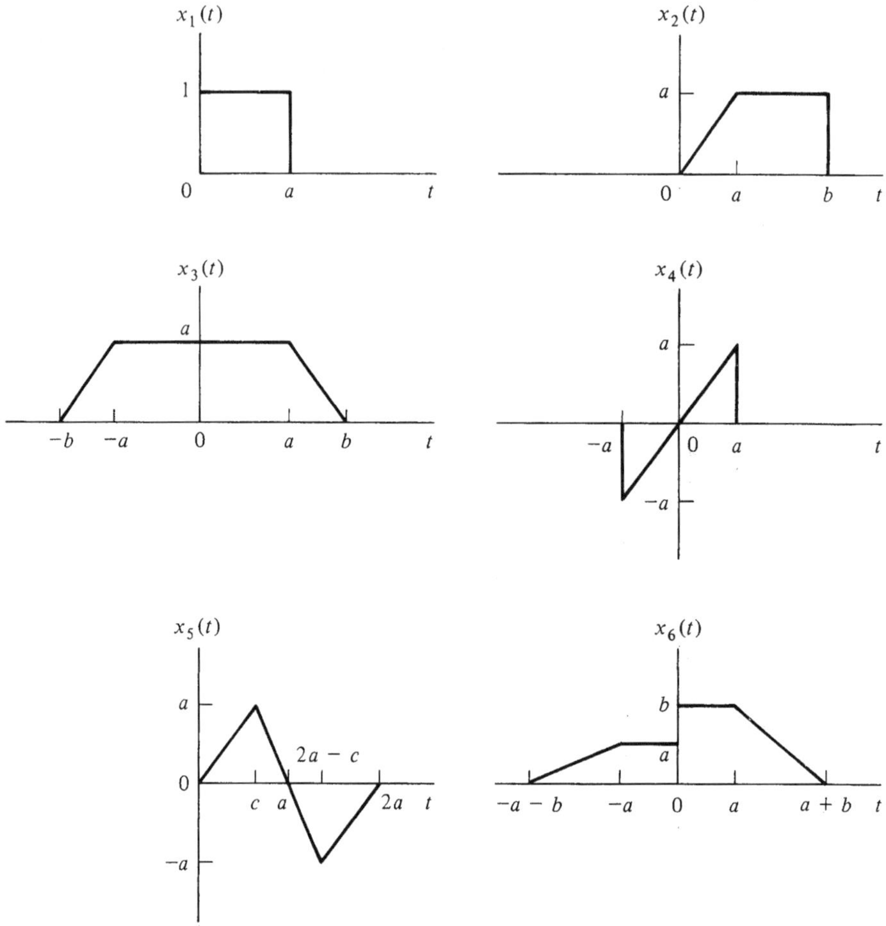

Figure P1.16

(c) $x_3(t) = x_1(t/2)$

(d) $x_4(t) = 2x_1(t)$

1.18. The signal $x(t) = \text{rect}(t/2)$ is transmitted through the atmosphere and is reflected by different objects located at different distances. The received signal is

$$y(t) = x(t) + 0.5x\left(t - \frac{T}{2}\right) + 0.25x(t - T), \quad T \gg 2$$

Signal $y(t)$ is processed as shown in Fig. P1.18.

(a) Sketch $y(t)$ for $T = 10$.

(b) Sketch $z(t)$ for $T = 10$.

1.19. Check whether each of the following can be used as a mathematical model of a delta function:

(a) $p_1(t) = \lim_{\varepsilon \to 0} \dfrac{1}{2\varepsilon\pi \cosh(t/\varepsilon)}$

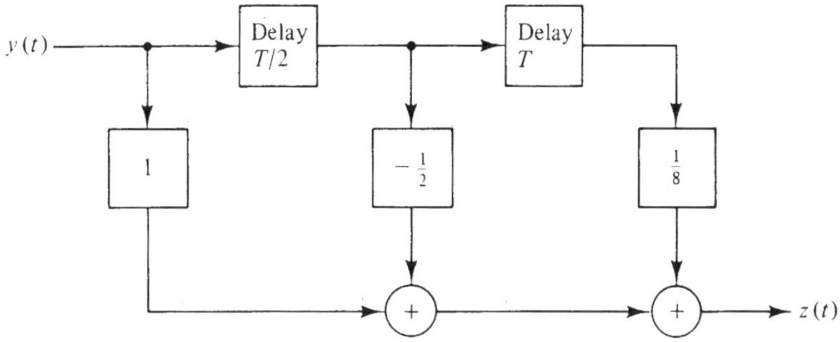

Figure P1.18

(b) $p_2(t) = \lim_{\varepsilon \to 0} \sqrt{\frac{1}{2\pi\varepsilon^2}} \exp\left[\frac{-t^2}{2\varepsilon^2}\right]$

(c) $p_3(t) = \lim_{\varepsilon \to 0} \frac{2\varepsilon}{4\pi^2 t^2 + \varepsilon^2}$

(d) $p_4(t) = \lim_{\varepsilon \to 0} \frac{1}{\pi} \frac{\varepsilon}{t^2 + \varepsilon^2}$

(e) $p_5(t) = \lim_{\varepsilon \to 0} \varepsilon \exp[-\varepsilon |t|]$

(f) $p_6(t) = \lim_{\varepsilon \to 0} \frac{1}{\pi} \frac{\sin \varepsilon t}{t}$

1.20. Evaluate the following integrals:

(a) $\int_{-\infty}^{\infty} \left(\frac{2}{3}t - \frac{3}{2}\right) \delta(t-1)\, dt$

(b) $\int_{-\infty}^{\infty} (t-1)\delta\left(\frac{2}{3}t - \frac{3}{2}\right) dt$

(c) $\int_{-3}^{-2} \left[\exp(-t+1) + \sin\left(\frac{2\pi}{3}t\right)\right] \delta\left(t - \frac{3}{2}\right) dt$

(d) $\int_{-3}^{2} \left[\exp(-t+1) + \sin\left(\frac{2\pi}{3}t\right)\right] \delta\left(t - \frac{3}{2}\right) dt$

(e) $\int_{-\infty}^{\infty} \exp[-5t+1]\delta'(t-5)\, dt$

1.21. The probability that a random variable x is less than α is found by integrating the probability density function $f(x)$ to obtain

$$P(x \le \alpha) = \int_{-\infty}^{\alpha+} f(x)\, dx$$

Given that

$$f(x) = 0.2\delta(x+2) + 0.3\delta(x) + 0.2\delta(x-1) + 0.1[u(x-3) - u(x-6)]$$

find

(a) $P(x \le -3)$

(b) $P(x \le 1.5)$

(c) $P(x \le 4)$
(d) $P(x \le 6)$

1.22. The velocity of 1 g of mass is

$$v(t) = \exp[-(t+1)]u(t+1) + \delta(t-1)$$

(a) Plot $v(t)$
(b) Evaluate the force

$$f(t) = m\frac{d}{dt}[v(t)]$$

(c) If there is a spring connected to the mass with constant $k = 1$ N/m, find the force

$$f_k(t) = k\int_{-\infty}^{t} v(\tau)\,d\tau$$

1.23. Sketch the first and second derivatives of the following signals:
(a) $x(t) = u(t) + 5u(t-1) - 2u(t-2)$
(b) $x(t) = r(t) - r(t-1) + 2u(t-2)$
(c) $x(t) = \begin{cases} 2t+2 & -1 \le t < 0 \\ 2t-2 & 0 \le t < 1 \end{cases}$

1.11 COMPUTER PROBLEMS

1.24. The integral

$$\int_{t_1}^{t_2} x(t)y(t)\,dt$$

can be approximated by a summation of rectangular strips, each of width Δt, as follows:

$$\int_{t_1}^{t_2} x(t)y(t)\,dt \simeq \sum_{n=1}^{N} x(n\Delta t)y(n\Delta t)\Delta t$$

Here, $\Delta t = (t_2 - t_1)/N$. Write a program to verify that

$$\sqrt{\frac{1}{2\pi\varepsilon^2}}\exp\left[\frac{-t^2}{2\varepsilon^2}\right]$$

can be used as a mathematical model for the delta function by approximating the following integrals by a summation:

(a) $\displaystyle\int_{-1}^{1}(t+1)\sqrt{\frac{1}{2\pi\varepsilon^2}}\exp\left[-\frac{t^2}{2}\varepsilon^2\right]dt$

(b) $\displaystyle\int_{-2}^{1}(t+1)\sqrt{\frac{1}{2\pi\varepsilon^2}}\exp\left[-\frac{(t+1)^2}{2\varepsilon^2}\right]dt$

(c) $\displaystyle\int_{-1}^{2}(t+1)\sqrt{\frac{1}{2\pi\varepsilon^2}}\exp\left[-\frac{(t-1)^2}{2\varepsilon^2}\right]dt$

1.25. Repeat Problem 1.24 for the following integrals:

(a) $\displaystyle\int_{-1}^{1}\exp[-t]\frac{2\varepsilon}{\varepsilon^2 + 4\pi^2 t^2}\,dt$

(b) $\displaystyle\int_{-2}^{1}\exp[-t]\frac{2\varepsilon}{\varepsilon^2 + 4\pi^2 (t+1)^2}\,dt$

(c) $\displaystyle\int_{-1}^{2}\exp[-t]\frac{2\varepsilon}{\varepsilon^2 + 4\pi^2 (t-1)^2}\,dt$

Chapter 2

Continuous-Time Systems

2.1 INTRODUCTION

Every physical system is broadly characterized by its ability to accept an input such as voltage, current, force, pressure, displacement, etc., and to produce an output in response to this input. For example, a radar receiver is an electronic system whose input is the reflection of an electromagnetic signal from the target and whose output is a video signal displayed on the radar screen. Similarly, a robot is a system whose input is an electric control signal and whose output is a motion or action on the part of the robot. A third example is a filter, whose input is a signal corrupted by noise and interference and whose output is the desired signal. In brief, a system can be viewed as a process that results in transforming input signals into output signals.

We are interested in both continuous-time and discrete-time systems. A continuous-time system is a system in which continuous-time input signals are transformed into continuous-time output signals. Such a system is represented pictorially as shown in Figure 2.1.1(a), where $x(t)$ is the input and $y(t)$ is the output. A discrete-time system is a system that transforms discrete-time inputs into discrete-time outputs. (See Figure 2.1.1(b)). Continuous-time systems are treated in this chapter, and discrete-time systems are discussed in Chapter 6.

In studying the behavior of systems, the procedure is to model mathematically each element that comprises the system and then to consider the interconnection of elements. The result is described mathematically either in the time domain, as in this chapter, or in the frequency domain, as in Chapters 3 and 4.

In this chapter, we show that the analysis of linear systems can be reduced to the study of the response of the system to basic input signals.

(a) (b)

Figure 2.1.1 Examples of continuous-time and discrete-time systems.

2.2 CLASSIFICATION OF CONTINUOUS-TIME SYSTEMS

Our intent in this section is to lend additional substance to the concept of systems by discussing their classification according to the way the system interacts with the input signal. This interaction, which defines the model for the system, can be linear or nonlinear, time invariant or time varying, memoryless or with memory, causal or noncausal, stable or unstable, and deterministic or nondeterministic. For the most part, we are concerned with linear, time-invariant, deterministic systems. In this section, we briefly examine the properties of each of these classes.

2.2.1 Linear and Nonlinear Systems

When the system is linear, the superposition principle can be applied. This important fact is precisely the reason that the techniques of linear-system analysis have been so well developed. Superposition simply implies that the response resulting from several input signals can be computed as the sum of the responses resulting from each input signal acting alone. Mathematically, the superposition principle can be stated as follows: Let $y_1(t)$ be the response of a continuous-time system to an input $x_1(t)$ and $y_2(t)$ be the response corresponding to the input $x_2(t)$. Then the system is linear (follows the principle of superposition) if

1. the response to $x_1(t) + x_2(t)$ is $y_1(t) + y_2(t)$; and
2. the response to $\alpha\, x_1(t)$ is $\alpha\, y_1(t)$, where α is any arbitrary constant.

The first property is referred to as the additivity property; the second is referred to as the homogeneity property. These two properties defining a linear system can be combined into a single statement as

$$\alpha x_1(t) + \beta x_2(t) \rightarrow \alpha y_1(t) + \beta y_2(t) \tag{2.2.1}$$

where the notation $x(t) \rightarrow y(t)$ represents the input/output relation of a continuous-time system. A system is said to be nonlinear if Equation (2.2.1) is not valid for at least one set of $x_1(t)$, $x_2(t)$, α, and β.

Example 2.2.1

Consider the voltage divider shown in Figure 2.2.1 with $R_1 = R_2$. For input $x(t)$ and output $y(t)$, this is a linear system. The input/output relation can be explicitly written as

$$y(t) = \frac{R_2}{R_1 + R_2}x(t) = \frac{1}{2}x(t)$$

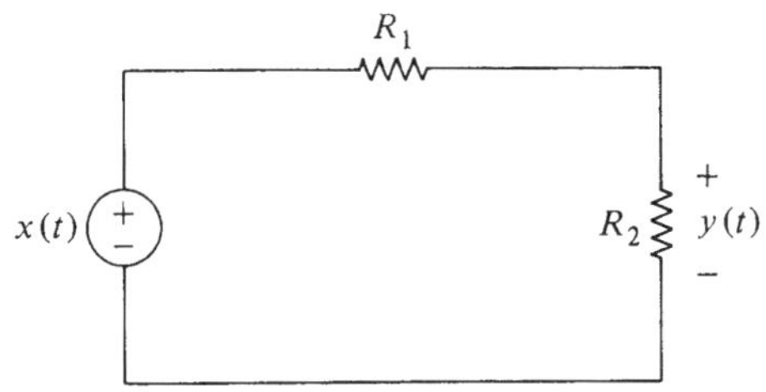

Figure 2.2.1 System for Example 2.2.1.

i.e., the transformation involves only multiplication by a constant. To prove that the system is indeed linear, one has to show that Equation (2.2.1) is satisfied. Consider the input $x(t) = ax_1(t) + bx_2(t)$. The corresponding output is

$$\begin{aligned} y(t) &= \tfrac{1}{2}x(t) \\ &= \tfrac{1}{2}[ax_1(t) + bx_2(t)] \\ &= a\tfrac{1}{2}x_1(t) + b\tfrac{1}{2}x_2(t) \\ &= ay_1(t) + by_2(t) \end{aligned}$$

where

$$y_1(t) = \tfrac{1}{2}x_1(t) \quad \text{and} \quad y_2(t) = \tfrac{1}{2}x_2(t)$$

On the other hand, if R_1 is a voltage-dependent resistor such that $R_1 = R_2x(t)$, then the system is nonlinear. The input/output relation can then be written as

$$\begin{aligned} y(t) &= \frac{R_2}{R_2x(t) + R_2}x(t) \\ &= \frac{x(t)}{x(t) + 1} \end{aligned}$$

For an input of the form

$$x(t) = ax_1(t) + bx_2(t)$$

the output is

$$y(t) = \frac{ax_1(t) + bx_2(t)}{ax_1(t) + bx_2(t) + 1}$$

This system is nonlinear because

$$\frac{ax_1(t) + bx_2(t)}{ax_1(t) + bx_2(t) + 1} \neq a\frac{x_1(t)}{x_1(t) + 1} + b\frac{x_2(t)}{x_2(t) + 1}$$

for some $x_1(t)$, $x_2(t)$, a, and b (try $x_1(t) = x_2(t)$ and $a = 1, b = 2$).

Example 2.2.2

Suppose we want to determine which of the following systems is linear:

(a)
$$y(t) = K\frac{dx(t)}{dt} \tag{2.2.2}$$

(b)
$$y(t) = \exp[x(t)] \tag{2.2.3}$$

For part (a), consider the input

$$x(t) = ax_1(t) + bx_2(t) \tag{2.2.4}$$

The corresponding output is

$$y(t) = K\frac{d}{dt}[ax_1(t) + bx_2(t)]$$

which can be written as

$$\begin{aligned} y(t) &= Ka\frac{d}{dt}x_1(t) + Kb\frac{d}{dt}x_2(t) \\ &= ay_1(t) + by_2(t) \end{aligned}$$

where

$$y_1(t) = K\frac{d}{dt}x_1(t)$$

and

$$y_2(t) = K\frac{d}{dt}x_2(t)$$

so that the system described by Equation (2.2.2) is linear.

Comparing Equation (2.2.2) with

$$v(t) = L\frac{di(t)}{dt}$$

we conclude that an ideal inductor with input $i(t)$ (current through the inductor) and output $v(t)$ (voltage across the inductor) is a linear system (element). Similarly, we can show that a system that performs integration is a linear system. (See Problem 2.1(f).) Hence, an ideal capacitor is a linear system (element).

For part (b), we investigate the response of the system to the input in Equation (2.2.4):

$$\begin{aligned} y(t) &= \exp[ax_1(t) + bx_2(t)] \\ &= \exp[ax_1(t)]\exp[bx_2(t)] \\ &\neq ay_1(t) + by_2(t) \end{aligned}$$

Therefore, the system characterized by Equation (2.2.3) is nonlinear.

Example 2.2.3

Consider the RL circuit shown in Figure 2.2.2. This circuit can be viewed as a continuous-time system with input $x(t)$ equal to voltage source $e(t)$ and with output $y(t)$ equal to the current in the inductor. Assume that at time t_0, $i_L(t_0) = y(t_0) = y_0$.

Applying Kirchhoff's current law at node a, we obtain

$$\frac{v_a(t) - e(t)}{R_1} + \frac{v_a(t)}{R_2} + i_L(t) = 0$$

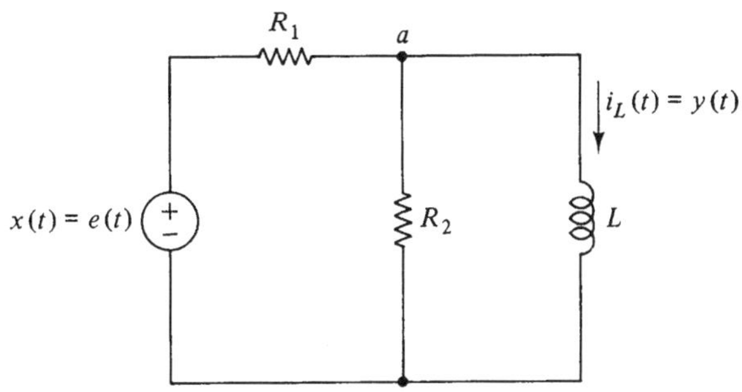

Figure 2.2.2 RL circuit for Example 2.2.3.

Since

$$v_a(t) = L\frac{di_L(t)}{dt}$$

it follows that

$$L\frac{R_1 + R_2}{R_1R_2}\frac{di_L(t)}{dt} + i_L(t) = \frac{e(t)}{R_1}$$

so that

$$\frac{di_L(t)}{dt} + \frac{R_1R_2}{L(R_1 + R_2)}i_L(t) = \frac{R_2}{L(R_1 + R_2)}e(t)$$

or

$$\frac{dy(t)}{dt} + \frac{R_1R_2}{L(R_1 + R_2)}y(t) = \frac{R_2}{L(R_1 + R_2)}x(t) \tag{2.2.5}$$

The differential equation, Equation (2.2.5), is called the input/output differential equation describing the system. To compute an explicit expression for $y(t)$ in terms of $x(t)$, we must solve the differential equation for an arbitrary input $x(t)$ applied for $t \geq t_0$. The complete solution is of the form

$$\begin{aligned} y(t) = {} & y(t_0)\exp\left[-\frac{R_1R_2}{L(R_1 + R_2)}(t - t_0)\right] \\ & + \frac{R_2}{L(R_1 + R_2)}\int_{t_0}^{t}\exp\left[-\frac{R_1R_2}{L(R_1 + R_2)}(t - \tau)\right]x(\tau)\,d\tau; \quad t \geq t_0 \end{aligned} \tag{2.2.6}$$

According to Equation (2.2.1), this system is nonlinear unless $y(t_0) = 0$. To prove this, consider the input $x(t) = \alpha x_1(t) + \beta x_2(t)$. The corresponding output is

$$\begin{aligned} y(t) = {} & y(t_0)\exp\left[-\frac{R_1R_2}{L(R_1 + R_2)}(t - t_0)\right] \\ & + \frac{\alpha R_2}{L(R_1 + R_2)}\int_{t_0}^{t}\exp\left[-\frac{R_1R_2}{L(R_1 + R_2)}(t - \tau)\right]x_1(\tau)\,d\tau \\ & + \frac{\beta R_2}{L(R_1 + R_2)}\int_{t_0}^{t}\exp\left[-\frac{R_1R_2}{R_1 + R_2}(t - \tau)\right]x_2(\tau)\,d\tau \\ \neq {} & \alpha y_1(t) + \beta y_2(t) \end{aligned}$$

This may seem surprising, since inductors and resistors are linear elements. However, the system in Figure 2.2.2 violates a very important property of linear systems, namely, that zero input should yield zero output. Therefore, if $y_0 = 0$, then the system is linear.

The concept of linearity is very important in systems theory. The principle of superposition can be invoked to determine the response of a linear system to an arbitrary input if that input can be decomposed into the (possibly infinite) sum of several basic signals. The response to each basic signal can be computed separately and added to obtain the overall system response. This technique is used repeatedly throughout the text and in most cases yields closed-form mathematical results, which is not possible for nonlinear systems.

Many physical systems, when analyzed in detail, demonstrate nonlinear behavior. In such situations, a solution for a given set of initial conditions and excitation can be found either analytically or with the aid of a computer. Frequently, it is required to determine the behavior of the system in the neighborhood of this solution. A common technique of treating such problems is to approximate the system by a linear model that is valid in the neighborhood of the operating point. This technique is referred to as linearization. Some important examples are the small-signal analysis technique applied to transistor circuits and the small-signal model of a simple pendulum.

2.2.2 Time-Varying and Time-Invariant Systems

A system is said to be time invariant if a time shift in the input signal causes an identical time shift in the output signal. Specifically, if $y(t)$ is the output corresponding to input $x(t)$, a time-invariant system will have $y(t - t_0)$ as the output when $x(t - t_0)$ is the input. That is, the rule used to compute the system output does not depend on the time at which the input is applied.

The procedure for testing whether a system is time invariant is summarized in the following steps:

1. Let $y_1(t)$ be the output corresponding to $x_1(t)$.
2. Consider a second input, $x_2(t)$, obtained by shifting $x_1(t)$,
$$x_2(t) = x_1(t - t_0)$$
and find the output $y_2(t)$ corresponding to the input $x_2(t)$.
3. From step 1, find $y_1(t - t_0)$ and compare with $y_2(t)$.
4. If $y_2(t) = y_1(t - t_0)$, then the system is time invariant; otherwise it is a time-varying system.

Example 2.2.4

We wish to determine whether the systems described by the following equations are time invariant:

(a) $y(t) = \cos x(t)$

(b) $\dfrac{dy(t)}{dt} = -ty(t) + x(t), \quad t \geq 0, \quad y(0) = 0$

Consider the system in part (a), $y(t) = \cos x(t)$. From the steps listed before:

1. For input $x_1(t)$, the output is

$$y_1(t) = \cos x_1(t) \tag{2.2.7}$$

2. Consider the second input, $x_2(t) = x_1(t - t_0)$. The corresponding output is

$$\begin{aligned} y_2(t) &= \cos x_2(t) \\ &= \cos x_1(t - t_0) \end{aligned} \tag{2.2.8}$$

3. From Equation (2.2.7)

$$y_1(t - t_0) = \cos x_1(t - t_0) \tag{2.2.9}$$

4. Comparison of Equations (2.2.8) and (2.2.9) shows that the system $y(t) = \cos x(t)$ is time invariant.

Now consider the system in Part (b).

1. If the input is $x_1(t)$, it can be easily verified by direct substitution in the differential equation that the output $y_1(t)$ is given by

$$y_1(t) = \int_0^t \exp\left[-\frac{t^2}{2} + \frac{\tau^2}{2}\right] x_1(\tau)\,d\tau \tag{2.2.10}$$

2. Consider the input $x_2(t) = x_1(t - t_0)$. The corresponding output is

$$\begin{aligned} y_2(t) &= \int_0^t \exp\left[-\frac{t^2}{2} + \frac{\tau^2}{2}\right] x_2(\tau)\,d\tau \\ &= \int_0^t \exp\left[-\frac{t^2}{2} + \frac{\tau^2}{2}\right] x_1(\tau - t_0)\,d\tau \\ &= \int_{-t_0}^{t-t_0} \exp\left[-\frac{t^2}{2} + \frac{(\tau + t_0)^2}{2}\right] x_1(\tau)\,d\tau \end{aligned} \tag{2.2.11}$$

3. From Equation (2.2.10),

$$y_1(t - t_0) = \int_0^{t-t_0} \exp\left[-\frac{(t - t_0)^2}{2} + \frac{\tau^2}{2}\right] x_1(\tau)\,d\tau \neq y_2(t) \tag{2.2.12}$$

4. Comparison of Equations (2.2.11) and (2.2.12) leads to the conclusion that the system is not time invariant.

2.2.3 Systems with and without Memory

For most systems, the inputs and outputs are functions of the independent variable. A system is said to be memoryless, or instantaneous, if the present value of the output depends only on the present value of the input. For example, a resistor is a memoryless system, since with input $x(t)$ taken as the current and output $y(t)$ taken as the voltage, the input/output relationship is

$$y(t) = Rx(t)$$

where R is the resistance. Thus, the value of $y(t)$ at any instant depends only on the value of $x(t)$ at that instant. On the other hand, a capacitor is an example of a system with memory. With input taken as the current and output as the voltage, the input/output relationship in the case of the capacitor is

$$y(t) = \frac{1}{C}\int_{-\infty}^{t} x(\tau)\,d\tau$$

where C is the capacitance. It is obvious that the output at any time t depends on the entire past history of the input.

If a system is memoryless, or instantaneous, then the input/output relationship can be written in the form

$$y(t) = F(x(t)) \tag{2.2.13}$$

For linear systems, this relation reduces to

$$y(t) = k(t)x(t)$$

and if the system is also time invariant, we have

$$y(t) = kx(t)$$

where k is a constant.

An example of a linear, time-invariant, memoryless system is the mechanical damper. The linear dependence between force $f(t)$ and velocity $v(t)$ is

$$v(t) = \frac{1}{D}f(t)$$

where D is the damping constant.

A system whose response at the instant t is completely determined by the input signals over the past T seconds (the interval from $t - T$ to t) is a finite-memory system having a memory of length T units of time.

Example 2.2.5

The output of a communication channel $y(t)$ is related to its input $x(t)$ by

$$y(t) = \sum_{i=0}^{N} a_i x(t - T_i)$$

It is clear that the output $y(t)$ of the channel at time t depends not only on the input at time t, but also on the past history of $x(t)$, e.g.,

$$y(0) = a_0x(0) + a_1x(-T_1) + \cdots + a_nx(-T_n)$$

Therefore, this system has a finite memory of $T = \max_i(T_i)$.

2.2.4 Causal Systems

A system is causal, or nonanticipatory (also known as physically realizable), if the output at any time t_0 depends only on values of the input for $t < t_0$. Equivalently, if two inputs to a causal system are identical up to some time t_0, the corresponding outputs must also be equal up to this same time since a causal system cannot predict if the two inputs will be different after t_0 (in the future). Mathematically, if

$$x_1(t) = x_2(t); \qquad t < t_0$$

and the system is causal, then

$$y_1(t) = y_2(t);\ t < t_0$$

A system is said to be noncausal or anticipatory if it is not causal.

Causal systems are also referred to as physically realizable systems.

Example 2.2.6

In several applications, we are interested in the value of a signal $x(t)$, not at present time t, but at some time in the future, $t + \alpha$, or at some time in the past, $t - \beta$. The signal $y(t) = x(t + \alpha)$ is called a prediction of $x(t)$ while the signal $y(t - \beta)$ is the delayed version of $x(t)$. The first system is called an *ideal predictor* while the second system is an *ideal delay*.

Clearly the predictor is noncausal since the output depends on future values of the input. We can also verify this mathematically as follows. Consider the inputs

$$x_1(t) = \begin{cases} 1 & t \le 5 \\ \exp(-t) & t > 5 \end{cases}$$

and

$$x_2(t) = \begin{cases} 1 & t \le 5 \\ 0 & t > 5 \end{cases}$$

so that $x_1(t)$ and $x_2(t)$ are identical up to $t_0 = 5$.

Suppose $\alpha = 3$. The corresponding outputs are

$$y_1(t) = \begin{cases} 1 & t \le 2 \\ \exp[-(t + 3)] & t > 2 \end{cases}$$

and

$$y_2(t) = \begin{cases} 1 & t \le 2 \\ 0 & t > 2 \end{cases}$$

If the system is causal, $y_1(t) = y_2(t)$ for all $t < 5$. But $y_1(3) = \exp(-6)$ while $y_2(3) = 0$. Thus the system is noncausal.

The ideal delay is causal since its output depends only on past values of the input signal.

Example 2.2.7

We are often required to determine the average value of a signal at each time instant t. We do this by defining the *running average*, $x^{\mathrm{av}}(t)$ of signal $x(t)$. $x^{\mathrm{av}}(t)$ can be computed in several ways, for example,

$$x^{\mathrm{av}}(t) = \frac{1}{T}\int_{t-T}^{t} x(\tau)\,d\tau \tag{2.2.14}$$

or

$$x^{\mathrm{av}}(t) = \frac{1}{T}\int_{t-\frac{T}{2}}^{t+\frac{T}{2}} x(\tau)\,d\tau \tag{2.2.15}$$

Let $x_1(t)$ and $x_2(t)$ be two signals which are identical for $t \le t_0$ but are different from each other for $t > t_0$. Then, for the system of Equation (2.2.14),

$$\begin{aligned} x_1^{\mathrm{av}}(t_0) &= \frac{1}{T}\int_{t_0-T}^{t_0} x_1(\tau)\,d\tau \\ &= \frac{1}{T}\int_{t_0-T}^{t_0} x_2(\tau)\,d\tau = x_2^{\mathrm{av}}(t_0) \end{aligned} \tag{2.2.16(a)}$$

Thus this system is causal.

For the system of Equation (2.2.15)

$$x_1^{\mathrm{av}}(t_0) = \frac{1}{T}\int_{t_0-\frac{T}{2}}^{t_0+\frac{T}{2}} x_1(\tau)\,d\tau \tag{2.2.16(b)}$$

which is not equal to

$$x_2^{\mathrm{av}}(t_0) = \frac{1}{T}\int_{t_0-\frac{T}{2}}^{t_0+\frac{T}{2}} x_2(\tau)\,d\tau \tag{2.2.16(c)}$$

since $x_1(t)$ and $x_2(t)$ are not the same for $t > t_0$. This system is, therefore, noncausal.

2.2.5 Invertibility and Inverse Systems

A system is invertible if by observing the output, we can determine its input. That is, we can construct an inverse system that when cascaded with the given system, as illustrated in Figure 2.2.3, yields an output equal to the original input to the given system. In other words, the inverse system "undoes" what the given system does to input $x(t)$. So the effect of the given system can be eliminated by cascading it with its inverse system. Note that if two different inputs result in the same output, then the system is not invertible. The inverse of a causal system is not necessarily causal, in fact, it may not exist at all in any conventional sense. The use of the concept of a system inverse in the following chapters is primarily for mathematical convenience and does not require that such a system be physically realizable.

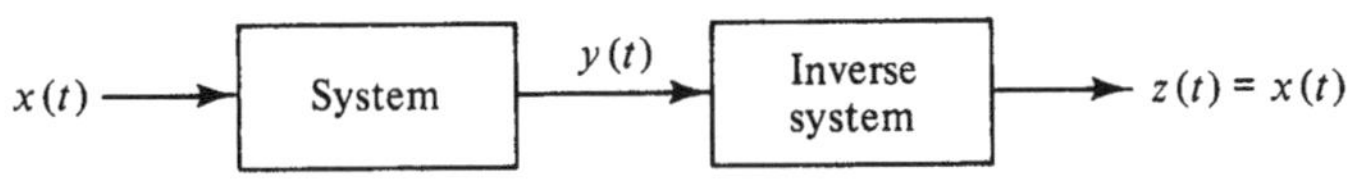

Figure 2.2.3 Concept of an inverse system.

Example 2.2.8

We want to determine if each of the following systems is invertible. If it is, we will construct the inverse system. If it is not, we will find two input signals to the system that have the same output.

(a) $y(t) = 2x(t)$

(b) $y(t) = \cos x(t)$

(c) $y(t) = \int_{-\infty}^{t} x(\tau)d\tau; \quad y(-\infty) = 0$

(d) $y(t) = x(t+1)$

For part (a), system $y(t) = 2x(t)$ is invertible with the inverse

$$z(t) = \tfrac{1}{2}y(t)$$

This idea is demonstrated in Figure 2.2.4.

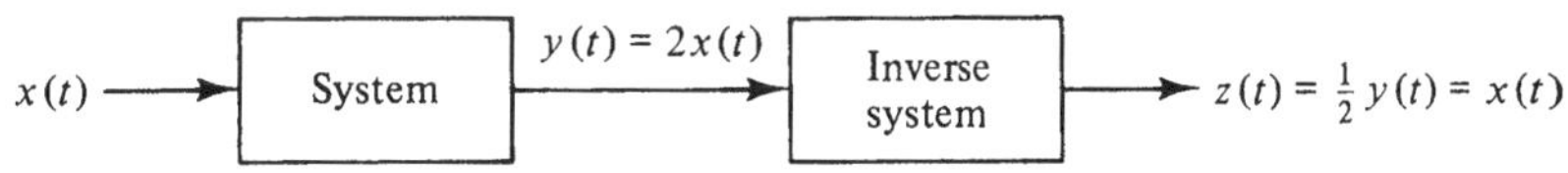

Figure 2.2.4 Inverse system for part (a) of Example 2.2.8.

For part (b), system $y(t) = \cos x(t)$ is noninvertible since $x(t)$ and $x(t) + 2\pi$ give the same output.

For part (c), system $y(t) = \int_{-\infty}^{t} x(\tau)d\tau$, $y(-\infty) = 0$, is invertible and the inverse system is the differentiator

$$z(t) = \frac{d}{dt}y(t)$$

For part (d), system $y(t) = x(t+1)$ is invertible and the inverse system is the one-unit delay

$$z(t) = y(t-1)$$

In some applications, it is necessary to perform preliminary processing on the received signal to transform it into a signal that is easy to work with. If the preliminary processing is invertible, it can have no effect on the performance of the overall system (see Problem 2.13).

2.2.6 Stable Systems

One of the most important concepts in the study of systems is the notion of stability. Whereas many different types of stability can be defined, in this section, we consider only one type, namely, bounded-input bounded-output (BIBO) stability. BIBO stability involves the behavior of the output response resulting from the application of a bounded input.

Signal $x(t)$ is said to be bounded if its magnitude does not grow without bound, i.e.,

$$|x(t)| < B < \infty, \quad \text{for all } t$$

A system is BIBO stable if, for any bounded input $x(t)$, the response $y(t)$ is also bounded. That is,

$$|x(t)| < B_1 < \infty \quad \text{implies} \quad |y(t)| < B_2 < \infty$$

Example 2.2.9

We want to determine which of these systems is stable:

(a) $y(t) = \exp[x(t)]$

(b) $y(t) = \int_{-\infty}^{t} x(\tau)d\tau$

For the system of part (a), a bounded input $x(t)$ such that $|x(t)| < B$, results in an output $y(t)$ with magnitude

$$|y(t)| = |\exp[x(t)]| = \exp[x(t)] \leq \exp[B] < \infty$$

Therefore, the output is also bounded and the system is stable.

For part (b), consider as input $x(t)$, the unit step function $u(t)$. The output $y(t)$ is then equal to

$$y(t) = \int_{-\infty}^{t} u(\tau)d\tau = r(t)$$

Thus the bounded input $u(t)$ produces an unbounded output $r(t)$ and the system is not stable.

This example serves to emphasize that for a system to be stable, *all* bounded inputs must give rise to bounded outputs. If we can find even one bounded input for which the output is not bounded, the system is unstable.

2.3 LINEAR TIME-INVARIANT SYSTEMS

In the previous section we have discussed a number of basic system properties. Two of these, linearity and time invariance, play a fundamental role in signal and system analysis because of the many physical phenomena that can be modeled by linear time-invariant systems and because a mathematical analysis of the behavior of such systems can be carried out in a fairly straightforward manner. In this section, we develop an important and useful representation for linear time-invariant (LTI) systems. This forms the foundation for linear-system theory and different transforms encountered throughout the text.

A fundamental problem in system analysis is determining the response to some specified input. Analytically, this can be answered in many different ways. One obvious way is to solve the differential equation describing the system, subject to the specified input and initial conditions. In the following section, we introduce a second method that exploits the linearity and time invariance of the system. This development results in an important integral known as the convolution integral. In Chapters 3 and 4, we consider frequency-domain techniques to analyze LTI systems.

2.3.1 The Convolution Integral

Linear systems are governed by the superposition principle. Let the responses of the system to two inputs $x_1(t)$ and $x_2(t)$ be $y_1(t)$ and $y_2(t)$ respectively. The system is linear if the response to the input $x(t) = a_1x_1(t) + a_2x_2(t)$ is equal to $y(t) = a_1y_1(t) + a_2y_2(t)$.

More generally, if the input $x(t)$ is the weighted sum of any set of signals $x_i(t)$, and if the response to $x_i(t)$ is $y_i(t)$, if the system is linear, the output $y(t)$ will be the weighted sum of the responses $y_i(t)$. That is, if

$$x(t) = a_1x_1(t) + a_2x_2(t) + \cdots + a_Nx_N(t) = \sum_{i=1}^{N} a_ix_i(t)$$

we will have

$$y(t) = a_1y_1(t) + a_2y_2(t) + \cdots + a_Ny_N(t) = \sum_{i=1}^{N} a_iy_i(t)$$

In Section 1.6, we demonstrated that the unit-step and unit-impulse functions can be used as building blocks to represent arbitrary signals. In fact, the sifting property of the δ function,

$$x(t) = \int_{-\infty}^{\infty} x(\tau)\,\delta(t - \tau)\,d\tau \tag{2.3.1}$$

shows that any signal $x(t)$ can be expressed as a continuum of weighted impulses.

Now consider a continuous-time system with input $x(t)$. Using the superposition property of linear systems (Equation 2.2.1), we can express output $y(t)$ as a linear combination of the responses of the system to shifted impulse signals; that is,

$$y(t) = \int_{-\infty}^{\infty} x(\tau)\,h(t, \tau)\,d\tau \tag{2.3.2}$$

where $h(t, \tau)$ denotes the response of a linear system to the shifted impulse $\delta(t - \tau)$. In other words, $h(t, \tau)$ is the output of the system at time t in response to input $\delta(t - \tau)$ applied at time τ. If, in addition to being linear, the system is also time invariant, then $h(t, \tau)$ should depend not on τ, but rather on $t - \tau$; i.e., $h(t, \tau) = h(t - \tau)$. This is because the time-invariance property implies that if $h(t)$ is the response to $\delta(t)$, then the response to $\delta(t - \tau)$ is simply $h(t - \tau)$. Thus, Equation (2.3.2) becomes

$$y(t) = \int_{-\infty}^{\infty} x(\tau)\,h(t - \tau)\,d\tau \tag{2.3.3}$$

The function $h(t)$ is called the impulse response of the LTI system and represents the output of the system at time t due to a unit-impulse input occurring at $t = 0$ when the system is relaxed (zero initial conditions).

The integral relationship expressed in Equation (2.3.3) is called the convolution integral of signals $x(t)$ and $h(t)$ and relates the input and output of the system by means of the system impulse response. This operation is represented symbolically as

$$y(t) = x(t) * h(t) \tag{2.3.4}$$

One consequence of this representation is that the LTI system is completely characterized by its impulse response. It is important to know that the convolution

$$y(t) = x(t) * h(t)$$

does not exist for all possible signals. The sufficient conditions for the convolution of two signals $x(t)$ and $h(t)$ to exist are:

1. Both $x(t)$ and $h(t)$ must be absolutely integrable over the interval $(-\infty, 0]$.
2. Both $x(t)$ and $h(t)$ must be absolutely integrable over the interval $[0, \infty)$.
3. Either $x(t)$ or $h(t)$ or both must be absolutely integrable over the interval $(-\infty, \infty)$.

The signal $x(t)$ is called absolutely integrable over the interval $[a, b]$ if

$$\int_a^b |x(t)|\, dt < \infty \tag{2.3.5}$$

For example, the convolutions $\sin \omega t * \cos \omega t$, $\exp[t] * \exp[t]$, and $\exp[t] * \exp[-t]$ do not exist.

Continuous-time convolution satisfies the following important properties:

Commutativity.

$$x(t) * h(t) = h(t) * x(t)$$

This property is proved by substitution of variables. The property implies that the roles of the input signal and the impulse response are interchangeable.

Associativity.

$$\begin{aligned} x(t) * h_1(t) * h_2(t) &= [x(t) * h_1(t)] * h_2(t) \\ &= x(t) * [h_1(t) * h_2(t)] \end{aligned}$$

This property is proved by changing the orders of integration. Associativity implies that a cascade combination of LTI systems can be replaced by a single system whose impulse response is the convolution of the individual impulse responses.

Distributivity.

$$x(t) * [h_1(t) + h_2(t)] = [x(t) * h_1(t)] + [x(t) * h_2(t)]$$

This property follows directly as a result of the linear property of integration. Distributivity states that a parallel combination of LTI systems is equivalent to a single system whose impulse response is the sum of the individual impulse responses in the parallel configuration. All three properties are illustrated in Figure 2.3.1.

Some interesting and useful additional properties of convolution integrals can be obtained by considering convolution with singularity signals, particularly the unit step, unit impulse, and unit doublet. From the defining relationships given in Chapter 1, it can be shown that

$$x(t) * \delta(t) = \int_{-\infty}^{\infty} x(\tau)\, \delta(t - \tau) d\tau = x(t) \tag{2.3.6}$$

Therefore, an LTI system with impulse response $h(t) = \delta(t)$ is the identity system. Now

$$x(t) * u(t) = \int_{-\infty}^{\infty} x(\tau) u(t - \tau) d\tau = \int_{-\infty}^{t} x(\tau) d\tau \tag{2.3.7}$$

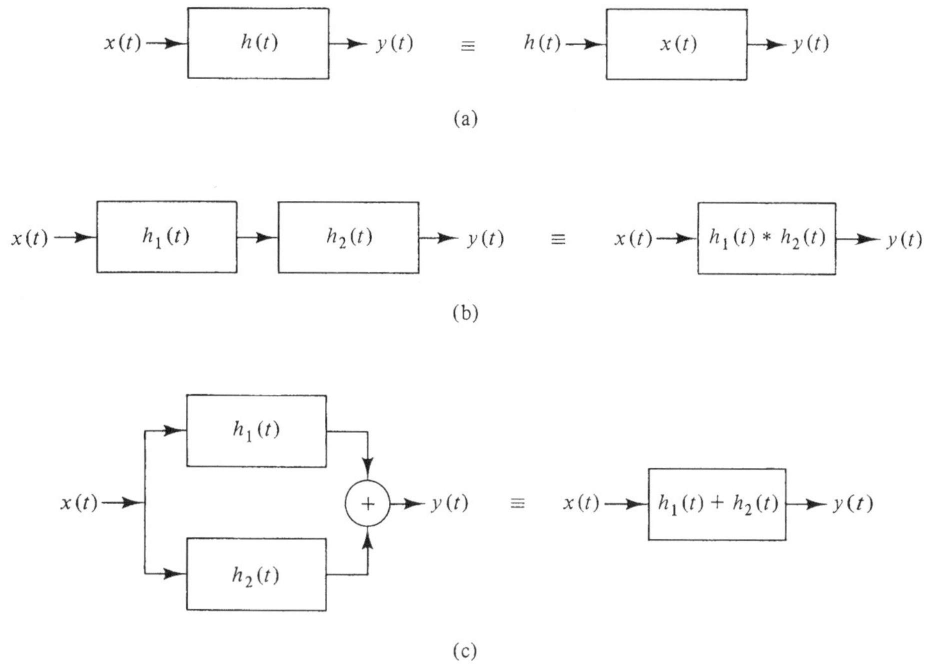

Figure 2.3.1 Properties of continuous-time convolution.

Consequently, an LTI system with impulse response $h(t) = u(t)$ is a perfect integrator. Also,

$$x(t) * \delta'(t) = \int_{-\infty}^{\infty} x(\tau)\delta'(t-\tau)d\tau = x'(t) \tag{2.3.8}$$

so that an LTI system with impulse response $h(t) = \delta'(t)$ is a perfect differentiator. The previous discussions and the discussions in Chapter 1 point out the differences between the following three operations:

$$x(t)\delta(t-a) = x(a)\delta(t-a)$$

$$\int_{-\infty}^{\infty} x(t)\delta(t-a)\,dt = x(a)$$

$$x(t) * \delta(t-a) = x(t-a)$$

The result of the first (sampling property of the delta function) is a δ-function with strength $x(a)$. The result of the second (sifting property of the delta function) is the value of the signal $x(t)$ at $t = a$, and the result of the third (convolution property of the delta function) is a shifted version of $x(t)$.

Example 2.3.1

Let the signal $x(t) = a\delta(t) + b\delta(t - t_0)$ be input to an LTI system with impulse response $h(t) = K\exp[-ct]u(t)$. The input is thus the weighted sum of two shifted δ-functions.

Since the system is linear and time invariant, it follows that the output, $y(t)$, can be expressed as the weighted sum of the responses to these δ-functions. By definition, the response of the system to a unit impulse input is equal to $h(t)$ so that

$$\begin{aligned} y(t) &= ah(t) + bh(t - t_0) \\ &= aK\exp[-ct]u(t) + bK\exp[-c(t - t_0)]u(t - t_0)) \end{aligned}$$

Example 2.3.2

The output $y(t)$ of an optimum receiver in a communication system is related to its input $x(t)$ by

$$y(t) = \int_{t-T}^{t} x(\tau)s(T - t + \tau)d\tau, \qquad 0 \le t \le T \tag{2.3.9}$$

where $s(t)$ is a known signal with duration T. Comparison of Equation (2.3.9) with Equation (2.3.3) yields

$$\begin{aligned} h(t - \tau) &= s(T - t + \tau), && 0 < t - \tau < T \\ &= 0, && \text{elsewhere} \end{aligned}$$

or

$$\begin{aligned} h(t) &= s(T - t), && 0 < t < T \\ &= 0, && \text{elsewhere} \end{aligned}$$

Such a system is called a matched filter. The system impulse response is $s(t)$ reflected and shifted by T (system is matched to $s(t)$).

Example 2.3.3

Consider the system described by

$$y(t) = \frac{1}{T}\int_{t-\frac{T}{2}}^{t+\frac{T}{2}} x(\tau)d\tau$$

As noted erlier, this system computes the running average of signal $x(t)$ over the interval $[t - T/2, t + T/2]$.

We now let $x(t) = \delta(t)$ to find the impulse response of this system as

$$\begin{aligned} h(t) &= \frac{1}{T}\int_{t-\frac{T}{2}}^{t+\frac{T}{2}} \delta(\tau)d\tau \\ &= \begin{cases} \dfrac{1}{T} & -\dfrac{T}{2} \le t \le \dfrac{T}{2} \\ 0 & \text{otherwise} \end{cases} \end{aligned}$$

where the last step follows from the sifting property, Equation (1.6.8), of the impulse function.

Example 2.3.4

Consider the LTI system with impulse response

$$h(t) = \exp[-at]u(t), \qquad a > 0$$

If the input to the system is

$$x(t) = \exp[-bt]u(t), \qquad b \neq a$$

then the output $y(t)$ is

$$y(t) = \int_{-\infty}^{\infty} \exp[-b\tau]u(\tau)\exp[-a(t-\tau)]u(t-\tau)d\tau$$

Note that

$$\begin{aligned} u(\tau)u(t-\tau) &= 1, \qquad 0 < t < \tau \\ &= 0, \qquad \text{otherwise} \end{aligned}$$

Therefore

$$\begin{aligned} y(t) &= \int_0^t \exp[-at]\exp[(a-b)\tau]d\tau \\ &= \frac{1}{a-b}[\exp(-bt) - \exp(-at)]u(t) \end{aligned}$$

Example 2.3.5

Let us find the impulse response of the system shown in Figure 2.3.2 if

$$\begin{aligned} h_1(t) &= \exp[-2t]u(t) \\ h_2(t) &= 2\exp[-t]u(t) \\ h_3(t) &= \exp[-3t]u(t) \\ h_4(t) &= 4\delta(t) \end{aligned}$$

By using the associative and distributive properties of the impulse response it follows that $h(t)$ for the system of Figure 2.3.2 is

$$\begin{aligned} h(t) &= h_1(t) * h_2(t) + h_3(t) * h_4(t) \\ &= [\exp(-t) - \exp(-2t)]u(t) + 12\exp(-3t)u(t) \end{aligned}$$

where the last step follows from Example 2.3.4 and the fact that $x(t) * \delta(t) = x(t)$.

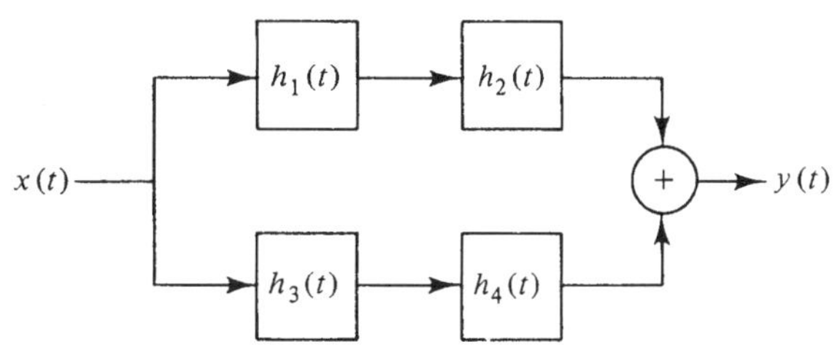

Figure 2.3.2 System of Example 2.3.5.

Example 2.3.6

The convolution has the property that the area of the convolution integral is equal to the product of the areas of the two signals entering into the convolution. The area can be computed by integrating Equation (2.3.3) over the interval $-\infty < t < \infty$, giving

$$\int_{-\infty}^{\infty} y(t)dt = \int_{-\infty}^{\infty}\int_{-\infty}^{\infty} x(\tau)h(t-\tau)d\tau\, dt$$

Interchanging the orders of integration results in

$$\int_{-\infty}^{\infty} y(t)dt = \int_{-\infty}^{\infty} x(\tau)\left[\int_{-\infty}^{\infty} h(t-\tau)dt\right]d\tau$$

$$= \int_{-\infty}^{\infty} x(\tau)[\text{area under } h(t)]d\tau$$

$$= \text{area under } x(t) \times \text{area under } h(t)$$

This result is generalized later when we discuss Fourier and Laplace transforms, but for the moment, we can use it as a tool to quickly check the answer of a convolution integral.

2.3.2 Graphical Interpretation of Convolution

Calculating $x(t) * h(t)$ is conceptually no more difficult than ordinary integration when the two signals are continous for all t. Often, however, one or both of the signals is defined in a piecewise fashion, and the graphical interpretation of convolution becomes especially helpful. We list in what follows the steps of this graphical aid to computing the convolution integration. These steps demonstrate how the convolution is computed graphically in the interval $t_{i-1} \le t \le t_i$, where the interval $[t_{i-1}, t_i]$ is chosen such that the product $x(\tau)h(t-\tau)$ is described by the same mathematical expression over the interval.

Step 1. For an arbitrary, but fixed value of t in the interval $[t_{i-1}, t_i]$, plot $x(\tau)$, $h(t-\tau)$, and the product $g(t, \tau) = x(\tau)h(t-\tau)$ as a function of τ. Note that $h(t-\tau)$ is a folded and shifted version of $h(\tau)$ and is equal to $h(-\tau)$ shifted to the right by t seconds.

Step 2. Integrate the product $g(t, \tau)$ as a function of τ. Note that the integrand $g(t, \tau)$ depends on t and τ, the latter being the variable of integration, which disappears after the integration is completed and the limits are imposed on the result. The integration can be viewed as the area under the curve represented by the integrand.

This procedure is illustrated by the following four examples.

Example 2.3.7

Consider the signals in Figure 2.3.3(a), where

$$x(t) = A\exp[-t], \qquad 0 \le t < \infty$$

$$h(t) = \frac{t}{T}, \qquad 0 \le t < T$$

Figure 2.3.3(b) shows $x(\tau)$, $h(t-\tau)$, and $x(\tau)h(t-\tau)$ with $t < 0$. The value of t always

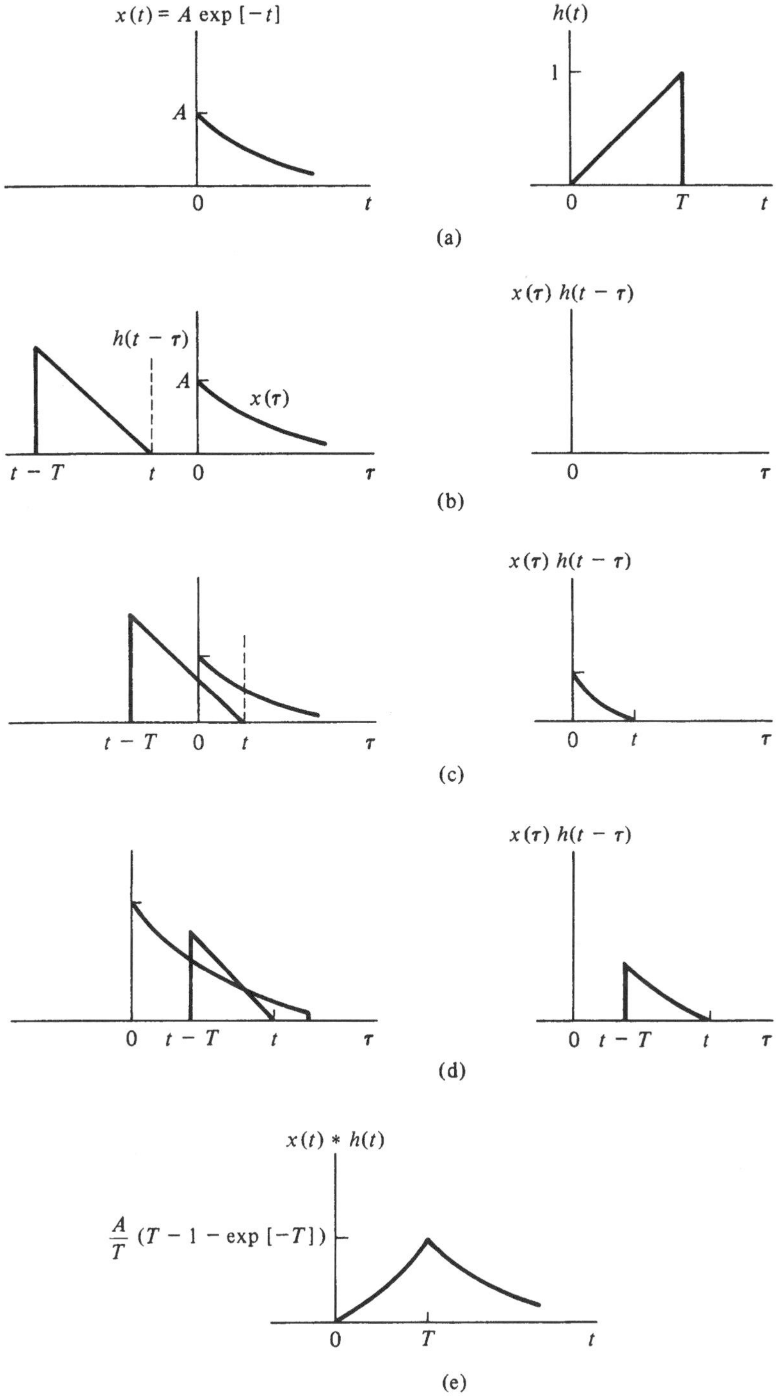

Figure 2.3.3 Graphical interpretation of the convolution for Example 2.3.7.

equals the distance from the origin of $x(\tau)$ to the shifted origin of $h(-\tau)$ indicated by the dashed line. We see that the signals do not overlap; hence, the integrand equals zero, and

$$x(t) * h(t) = 0, \qquad t \leq 0$$

When $0 \leq t \leq T$, as shown in Figure 2.3.3(c), the signals overlap for $0 \leq \tau \leq t$, so t becomes the upper limit of integration, and

$$\begin{aligned} x(t) * h(t) &= \int_0^t A \exp[-\tau] \frac{t-\tau}{T} d\tau \\ &= \frac{A}{T}\left[t - 1 + \exp[-t]\right] \qquad 0 < t \leq T \end{aligned}$$

Finally, when $t > T$, as shown in Figure 2.3.3(d), the signals overlap for $t - T \leq \tau \leq t$, and

$$\begin{aligned} x(t) * h(t) &= \int_{t-T}^t A \exp[-\tau] \frac{t-\tau}{T} d\tau \\ &= \frac{A}{T}\left[T - 1 + \exp[-T]\right]\exp[-(t-T)], \qquad t > T \end{aligned}$$

The complete result is plotted in Figure 2.3.3(e). For this example, the plot shows that convolution is a smoothing operation in the sense that $x(t) * h(t)$ is smoother than either of the original signals.

Example 2.3.8

Let us determine the convolution

$$y(t) = \text{rect}(t/2a) * \text{rect}(t/2a)$$

Figure 2.3.4 illustrates the overlapping of the two rectangular pulses for different values of t and the resulting signal $y(t)$. The result is expressed analytically as

$$y(t) = \begin{cases} 0, & t < -2a \\ t + 2a, & -2a \leq t < 0 \\ 2a - t, & 0 \leq t < 2a \\ 0, & t \geq 2a \end{cases}$$

or, in more compact form,

$$\begin{aligned} y(t) &= \begin{cases} 2a - |t|, & |t| \leq 2a \\ 0, & |t| \geq 2a \end{cases} \\ &= 2a\,\Delta(t/2a) \end{aligned}$$

This signal appears frequently in our discussion and is called the triangular signal. We use the notation $\Delta(t/2a)$ to denote the triangular signal that is of unit height, centered around $t = 0$, and has base of length $4a$.

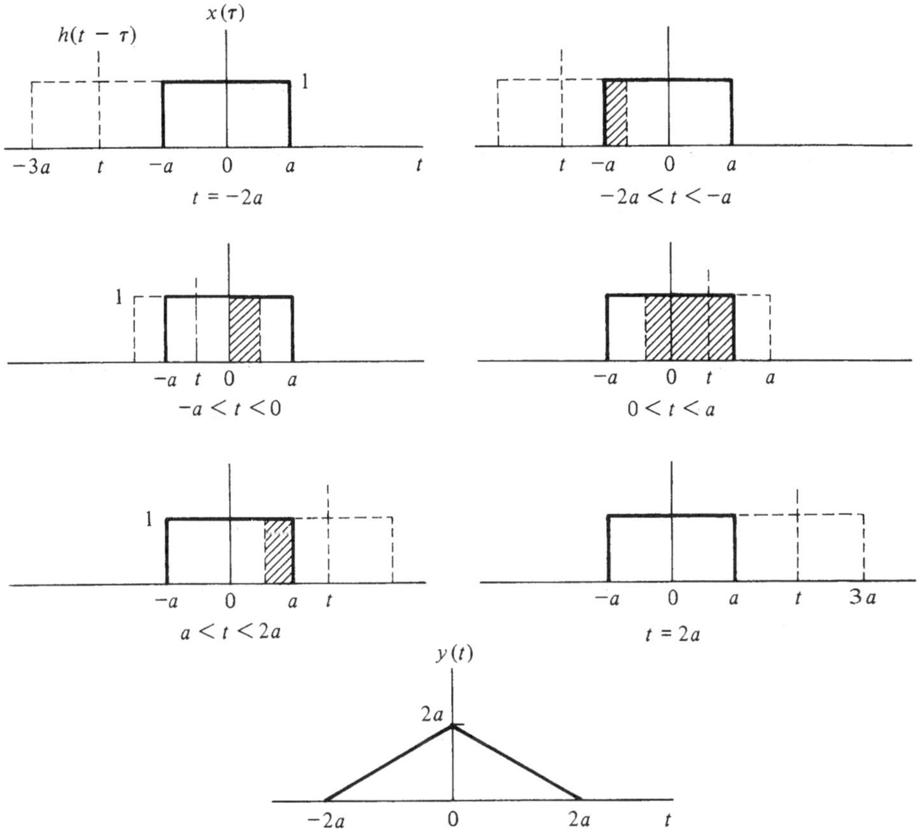

Figure 2.3.4 Graphical solution of Example 2.3.8.

Example 2.3.9

Let us compute the convolution $x(t) * h(t)$, where $x(t)$ and $h(t)$ are as follows:

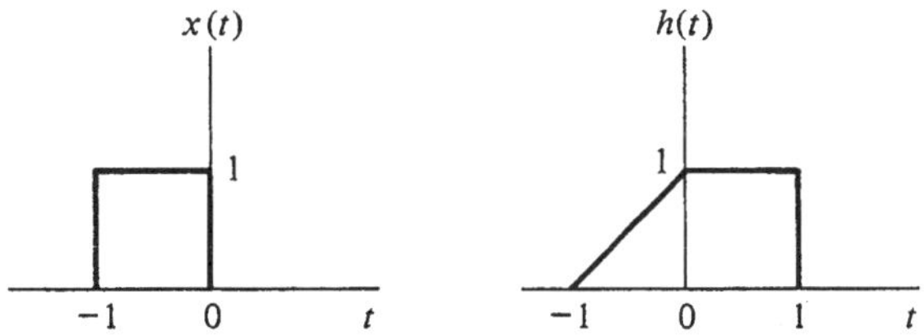

Figure 2.3.5 demonstrates the overlapping of the two signals $x(\tau)$ and $h(t-\tau)$. We can see that for $t < -2$, the product $x(\tau)h(t-\tau)$ is always zero. For $-2 \le t < -1$, the product is a triangle with base $t + 2$ and height $t + 2$; therefore, the area is

$$y(t) = \frac{1}{2}(t+2)^2 \qquad -2 \le t < -1$$

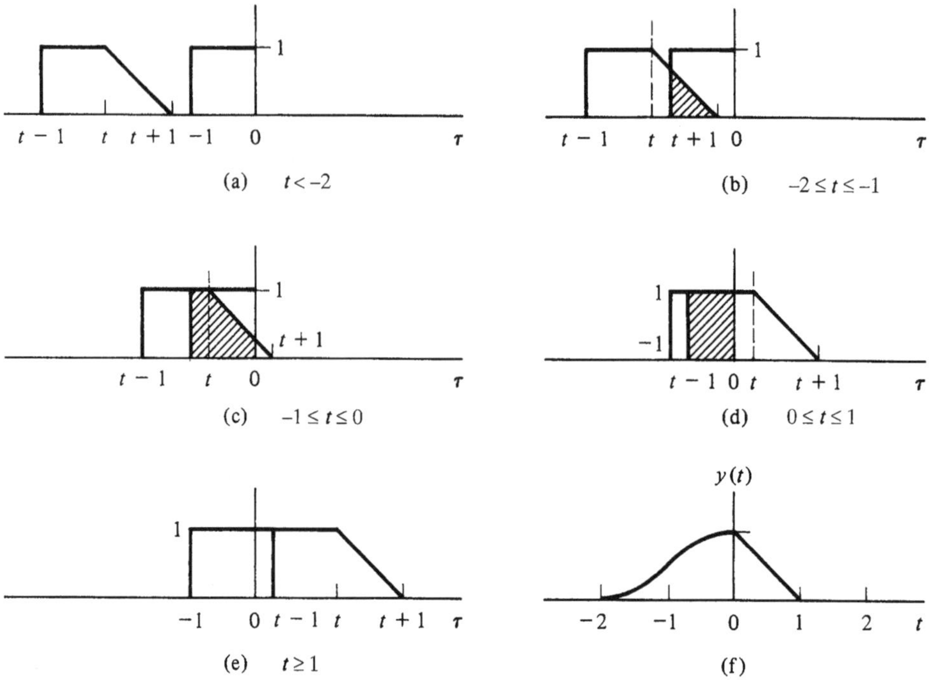

Figure 2.3.5 Graphical interpretation of $x(t) * h(t)$ for Example 2.3.9.

For $-1 \le t < 0$, the product is shown in Figure 2.3.5(c), and the area is

$$y(t) = 1 - \tfrac{1}{2}t^2, \qquad -1 \le t < 0$$

For $0 \le t < 1$, the product is a rectangle with base $1 - t$ and height 1; therefore, the area is

$$y(t) = 1 - t, \qquad 0 \le t < 1$$

For $t > 1$, the product is always zero. Summarizing, we have

$$y(t) = \begin{cases} 0, & t < -2 \\ \dfrac{(t+2)^2}{2}, & -2 \le t < -1 \\ 1 - \dfrac{t^2}{2}, & -1 \le t < 0 \\ 1 - t, & 0 \le t < 1 \\ 0, & t \ge 1 \end{cases}$$

Example 2.3.10

The convolution of the two signals shown in the following figure is evaluated using graphical interpretation.

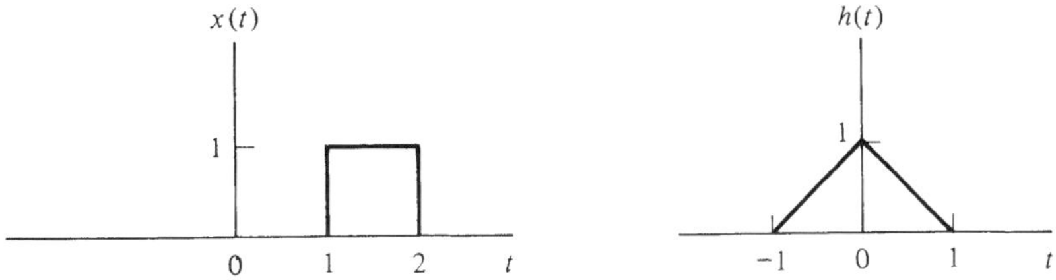

From Figure 2.3.6, we can see that for $t < 0$, the product $x(\tau)h(t-\tau)$ is always zero for all t; therefore, $y(t) = 0$. For $0 \le t < 1$, the product is a triangle with base t and height t; therefore, $y(t) = t^2/2$. For $1 \le t < 2$, the area under the product is equal to

$$y(t) = 1 - [\tfrac{1}{2}(t-1)^2 + \tfrac{1}{2}(2-t)^2], \qquad 1 \le t < 2$$

For $2 \le t < 3$, the product is a triangle with base $3 - t$ and height $3 - t$; therefore, $y(t) = (3-t)^2/2$. For $t \ge 3$, the product $x(\tau)h(t-\tau)$ is always zero. Summarizing, we have

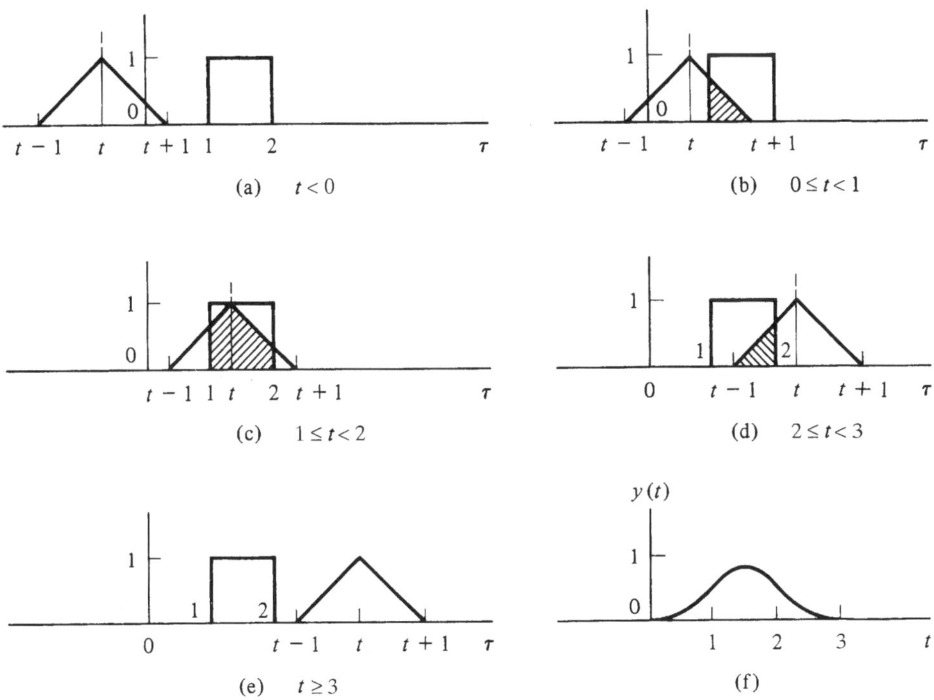

Figure 2.3.6 Convolution of $x(t)$ and $h(t)$ in Example 2.3.10.

$$y(t) = \begin{cases} 0, & t < 0 \\ \dfrac{t^2}{2}, & 0 \le t < 1 \\ 3t - t^2 - \dfrac{3}{2}, & 1 \le t < 2 \\ \dfrac{(3-t)^2}{2}, & 2 \le t < 3 \\ 0, & t \ge 3 \end{cases}$$

2.4 PROPERTIES OF LINEAR, TIME-INVARIANT SYSTEMS

The impulse response of an LTI system represents a complete description of the characteristics of the system. In this section, we examine the system properties discussed in Section 2.2 in terms of the system impulse response.

2.4.1 Memoryless LTI Systems

In Section 2.2.3, we defined a system to be memoryless if its output at any time depends only on the value of the input at the same time. There we saw that a memoryless, time-invariant system obeys an input/output relation of the form

$$y(t) = K\,x(t) \tag{2.4.1}$$

for some constant K. By setting $x(t) = \delta(t)$ in Equation (2.4.1), we see that this system has the impulse response

$$h(t) = K\,\delta(t) \tag{2.4.2}$$

2.4.2 Causal LTI Systems

As was mentioned in Section 2.2.4, the output of a causal system depends only on the present and past values of the input. Using the convolution integral, we can relate this property to a corresponding property of the impulse response of an LTI system. Specifically, for a continuous-time system to be causal, $y(t)$ must not depend on $x(\tau)$ for $\tau > t$. From Equation (2.3.3), we can see that this will be so if

$$h(t) = 0 \quad \text{for} \quad t < 0 \tag{2.4.3}$$

In this case, the convolution integral becomes

$$\begin{aligned} y(t) &= \int_{-\infty}^{t} x(\tau)h(t-\tau)\,d\tau \\ &= \int_{0}^{\infty} h(\tau)x(t-\tau)\,d\tau \end{aligned} \tag{2.4.4}$$

As an example, the system $h(t) = u(t)$ is causal, but the system $h(t) = \delta(t + t_0)$, $t_0 > 0$, is noncausal.

In general, $x(t)$ is called a causal signal if

$$x(t) = 0, \qquad t < 0$$

The signal is anticausal if $x(t) = 0$ for $t \geq 0$. Any signal that does not contain any singularities (a delta function or its derivatives) at $t = 0$ can be written as the sum of a causal part $x^+(t)$ and anticausal part $x^-(t)$, i.e.,

$$x(t) = x^+(t) + x^-(t)$$

For example, the exponential $x(t) = \exp[-t]$ can be written as

$$x(t) = \exp[-t]u(t) + \exp[-t]u(-t)$$

where the first term represents the causal part of $x(t)$ and the second term represents the anticausal part of $x(t)$. Note that multiplying the signal by the unit step ensures that the resulting signal is causal.

2.4.3 Invertible LTI Systems

Consider a continuous-time LTI system with impulse response $h(t)$. In Section 2.2.5, we mentioned that a system is invertible only if we can design an inverse system that, when connected in cascade with the original system, yields an output equal to the system input. If $h_1(t)$ represents the impulse response of the inverse system, then in terms of the convolution integral, we must, therefore, have

$$y(t) = h_1(t) * h(t) * x(t) = x(t)$$

From Equation (2.3.6), we conclude that $h_1(t)$ must satisfy

$$h_1(t) * h(t) = h(t) * h_1(t) = \delta(t) \tag{2.4.5}$$

As an example, the LTI system $h_1(t) = \delta(t + t_0)$ is the inverse of the system $h(t) = \delta(t - t_0)$.

2.4.4 Stable LTI Systems

A continuous-time system is stable if and only if every bounded input produces a bounded output. In order to relate this property to the impulse response of LTI systems, consider a bounded input $x(t)$, i.e., $|x(t)| < B$ for all t. Suppose that this input is applied to an LTI system with impulse response $h(t)$. Using Equation (2.3.3), we find that the magnitude of the output is

$$\begin{aligned} |y(t)| &= \left| \int_{-\infty}^{\infty} h(\tau)x(t-\tau)\,d\tau \right| \\ &\leq \int_{-\infty}^{\infty} |h(\tau)|\,|x(t-\tau)|\,d\tau \\ &\leq B \int_{-\infty}^{\infty} |h(\tau)|\,d\tau \end{aligned} \tag{2.4.6}$$

Therefore, the system is stable if

$$\int_{-\infty}^{\infty} |h(\tau)|\, d\tau < \infty \tag{2.4.7}$$

i.e., a sufficient condition for bounded-input, bounded-output stability of an LTI system is that its impulse response be absolutely integrable.

This is also a necessary condition for stability. That is, if the condition is violated, we can find bounded inputs for which the corresponding outputs are unbounded. For instance, let us fix t and choose as input the bounded signal $x(\tau) = \text{sgn}[h(t - \tau)]$ or, equivalently, $x(t - \tau) = \text{sgn}[h(\tau)]$. Then

$$\begin{aligned} y(t) &= \int_{-\infty}^{\infty} h(\tau)\, \text{sgn}[h(\tau)]\, d\tau \\ &= \int_{-\infty}^{\infty} |h(\tau)|\, d\tau \end{aligned}$$

Clearly, if $h(t)$ is not absolutely integrable, $y(t)$ will be unbounded.

As an example, the system with $h(t) = \exp[-t]u(t)$ is stable, whereas the system with $h(t) = u(t)$ is unstable.

Example 2.4.1

We will determine if the system with impulse responses as shown are causal or noncausal, with or without memory, and stable or unstable:

(i) $h_1(t) = t\exp[-2t]u(\text{t}) + \exp[3t]u(-t) + \delta(t - 1)$

(ii) $h_2(t) = -3\exp[2t]u(t)$

(iii) $h_3(t) = 5\delta(t + 5)$

(iv) $h_4(t) = 10\dfrac{\sin 5\pi t}{\pi t}$

Systems (i), (iii) and (iv) are noncausal since for $t < 0$, $h_i(t) \neq 0$, $i = 1, 3, 4$. Thus only System (ii) is causal.

Since $h(t)$ is not of the form $K\,\delta(t)$ for any of the systems, it follows that all the systems have memory.

To determine which of the systems are stable, we note that

$$\int_{-\infty}^{\infty} |h_1(t)|\, dt = \int_{0}^{\infty} t\exp[-2t]dt + \int_{-\infty}^{0} \exp[3t] + 1 = \frac{19}{12}$$

$$\int_{-\infty}^{\infty} |h_2(t)|\, dt = \int_{0}^{\infty} 3\exp[2t]dt \text{ is unbounded.}$$

$$\int_{-\infty}^{\infty} |h_3(t)|\, dt = 5$$

and

$$\int_{-\infty}^{\infty} |h_4(t)|\, dt = \int_{-\infty}^{\infty} 10\frac{\sin 5\pi t}{\pi t} = 20$$

Thus Systems (i), (ii) and (iv) are stable, while System (ii) is unstable.

2.5 SYSTEMS DESCRIBED BY DIFFERENTIAL EQUATIONS

The response of the RC circuit in Example 2.2.7 was described in terms of a differential equation. In fact, the response of many physical systems can be described by differential equations. Examples of such systems are electric networks comprising ideal resistors, capacitors, and inductors, and mechanical systems made of small springs, dampers, and the like. In Section 2.5.1, we consider systems with linear input/output differential equations with constant coefficients and show that such systems can be realized (or simulated) using adders, multipliers, and integrators. We shall give also some examples to demonstrate how to determine the impulse response of LTI systems described by linear, constant-coefficient differential equations. In Chapters 4 and 5, we shall present an easier, more straightforward method of determining the impulse response of LTI systems, namely, transforms.

2.5.1 Linear, Constant-Coefficient Differential Equations

Consider the continuous-time system described by the input/output differential equation

$$\frac{d^N y(t)}{dt^N} + \sum_{i=0}^{N-1} a_i \frac{d^i y(t)}{dt^i} = \sum_{i=0}^{M} b_i \frac{d^i x(t)}{dt^i} \tag{2.5.1}$$

where the coefficients a_i, $i = 1, 2, \ldots, N-1$, b_j, $j = 1, \ldots, M$ are real constants and $N > M$. In operator form, the last equation can be written as

$$\left(D^N + \sum_{i=0}^{N-1} a_i D^i\right) y(t) = \left(\sum_{i=0}^{M} b_i D^i\right) x(t) \tag{2.5.2}$$

where D represents the differentiation operator that transforms $y(t)$ into its derivative $y'(t)$. To solve Equation (2.5.2), one needs the N initial conditions

$$y(t_0), y'(t_0), \ldots, y^{(N-1)}(t_0)$$

where t_0 is some instant at which the input $x(t)$ is applied to the system and $y^i(t)$ is the ith derivative of $y(t)$.

The integer N is the order or dimension of the system. Note that if the ith derivative of the input $x(t)$ contains an impulse or a derivative of an impulse, then, to solve Equation (2.5.2) for $t > t_0$, it is necessary to know the initial conditions at time $t = t_0^-$. The reason is that the output $y(t)$ and its derivatives up to order $N-1$ can change instantaneously at time $t = t_0$. So initial conditions must be taken just prior to time t_0.

Although we assume that the reader has some exposure to solution techniques for ordinary linear differential equations, we work out a first-order case ($N = 1$) to review the usual method of solving linear, constant-coefficient differential equations.

Example 2.5.1

Consider the first-order LTI system that is described by the first-order differential equation

$$\frac{dy(t)}{dt} + ay(t) = bx(t) \tag{2.5.3}$$

where a and b are arbitrary constants. The complete solution of Equation (2.5.3) consists of the sum of the particular solution, $y_p(t)$, and the homogeneous solution, $y_h(t)$:

$$y(t) = y_p(t) + y_h(t) \tag{2.5.4}$$

The homogeneous differential equation

$$\frac{dy(t)}{dt} + ay(t) = 0$$

has a solution in the form

$$y_h(t) = C\exp[-at]$$

Using the integrating factor method, we find that the particular solution is

$$y_p(t) = \int_{t_0}^{t} \exp[-a(t-\tau)]\, bx(\tau)d\tau, \qquad t \geq t_0$$

Therefore, the general solution is

$$y(t) = C\exp[-at] + \int_{t_0}^{t} \exp[-a(t-\tau)]\, bx(\tau)d\tau, \qquad t \geq t_0 \tag{2.5.5}$$

Note that in Equation (2.5.5), the constant C has not been determined yet. In order to have the output completely determined, we have to know the initial condition $y(t_0)$. Let

$$y(t_0) = y_0$$

Then, from Equation (2.5.5),

$$y_0 = C\exp[-at_0]$$

Therefore, for $t \geq t_0$,

$$y(t) = y_0\exp[-a(t-t_0)] + \int_{t_0}^{t} \exp[-a(t-\tau)]\, bx(\tau)d\tau$$

If, for $t < t_0$, $x(t) = 0$, then the solution consists of only the homogeneous part:

$$y(t) = y_0\exp[-a(t-t_0)], \qquad t < t_0$$

Combining the solutions for $t \geq t_0$ and $t < t_0$, we have

$$y(t) = y_0\exp[-a(t-t_0)] + \left\{\int_{t_0}^{t} \exp[-a(t-\tau)]\, bx(\tau)d\tau\right\}u(t-t_0) \tag{2.5.6}$$

Since a linear system has the property that zero input produces zero output, the previous system is nonlinear if $y_0 \neq 0$. This can be easily seen by letting $x(t) = 0$ in Equation (2.5.6) to yield

$$y(t) = y_0\exp[-a(t-t_0)]$$

If $y_0 = 0$, the system is not only linear, but also time invariant. (Verify this.)

2.5.2 Basic System Components

Any finite-dimensional, linear, time-invariant, continuous-time system given by Equation (2.5.1) with $M \leq N$ can be realized or simulated using adders, subtractors, scalar multipliers, and integrators. These components can be implemented using resistors, capacitors, and operational amplifiers.

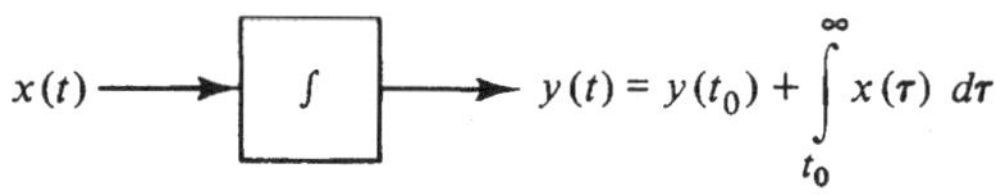

Figure 2.5.1 The integrator.

The Integrator. A basic element in the theory and practice of system engineering is the integrator. Mathematically, the input/output relation describing the integrator, shown in Figure 2.5.1, is

$$y(t) = y(t_0) + \int_{t_0}^{t} x(\tau)\,d\tau, \qquad t \geq t_0 \tag{2.5.7}$$

The input/output differential equation of the integrator is

$$\frac{dy(t)}{dt} = x(t) \tag{2.5.8}$$

If $y(t_0) = 0$, then the integrator is said to be at rest.

Adders, Subtractors, and Scalar Multipliers. These operators are illustrated in Figure 2.5.2.

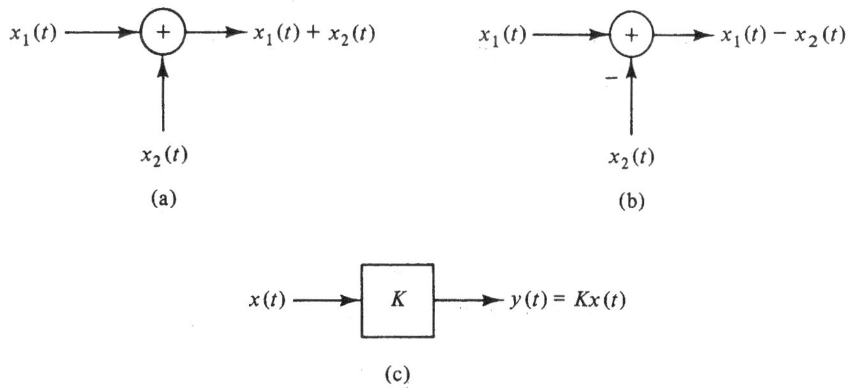

Figure 2.5.2 Basic components: (a) adder, (b) subtractor, and (c) scalar multiplier.

Example 2.5.2

We will find the differential equation describing the system of Figure 2.5.3. Let us denote the output of the first summer as $v_1(t)$, that of the second summer as $v_2(t)$ and that of the first integrator as $y_1(t)$. Then

$$v_2(t) = y'(t) = y_1(t) + 4y(t) + 4x(t) \tag{2.5.9}$$

Differentiate this equation and note that $y_1'(t) = v_1(t)$ to get

$$y''(t) = v_2'(t) = v_1(t) + 4y'(t) + 4x'(t) \tag{2.5.10}$$

which on substituting $v_1(t) = -y(t) + 2x(t)$ yields

$$y''(t) = 4y'(t) - y(t) + 4x'(t) + 2x(t) \tag{2.5.11}$$

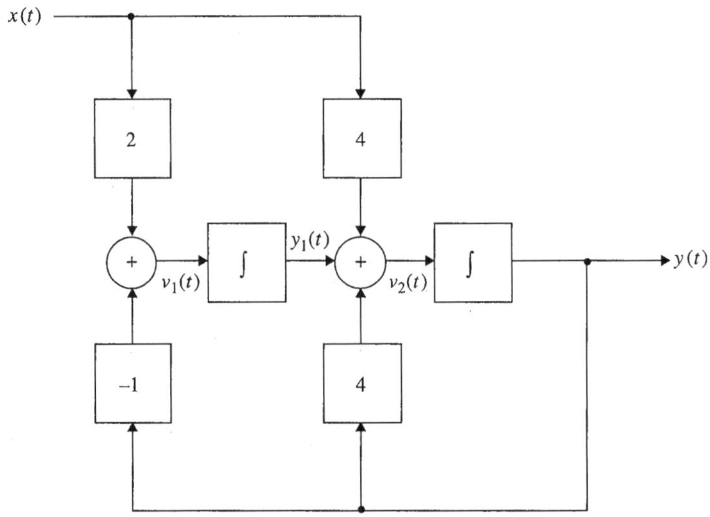

Figure 2.5.3 Realization of the system in Example 2.5.2.

2.5.3 Simulation Diagrams for Continuous-Time Systems

Consider the linear, time-invariant system described by Equation (2.5.2). This system can be realized in several different ways. Depending on the application, a particular one of these realizations may be preferable. In this section, we derive two different canonical realizations; each canonical form leads to a different realization, but the two are equivalent. To derive the first canonical form, we assume that $M = N$ and rewrite Equation (2.5.2) as

$$D^N(y - b_N x) + D^{N-1}(a_{N-1}y - b_{N-1}x) + \cdots + D(a_1 y - b_1 x) + a_0 y - b_0 x = 0 \tag{2.5.12}$$

Multiplying by D^{-N} and rearranging gives

$$y = b_N x + D^{-1}(b_{N-1}x - a_{N-1}y) + \cdots + D^{-(N-1)}(b_1 x - a_1 y) + D^{-N}(b_0 x - a_0 y) \tag{2.5.13}$$

from which the flow diagram in Figure 2.5.4 can be drawn, starting with output $y(t)$ at the right and working to the left. The operator D^{-k} stands for integrating k times.

Another useful simulation diagram can be obtained by converting the Nth-order differential equation into two equivalent equations. Let

$$\left(D^N + \sum_{j=0}^{N-1} a_j D^j\right)v(t) = x(t) \tag{2.5.14}$$

Then

$$y(t) = \left(\sum_{i=0}^{N} b_i D^i\right)v(t) \tag{2.5.15}$$

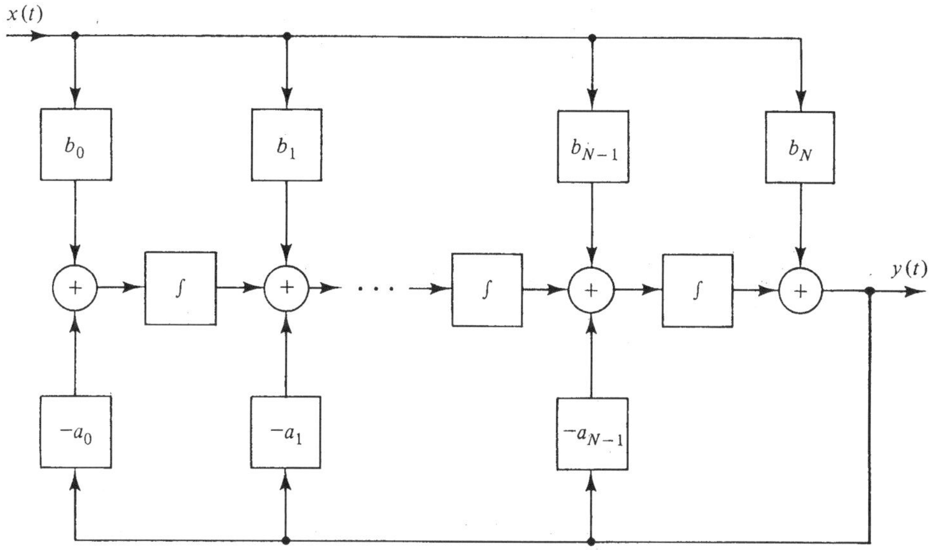

Figure 2.5.4 Simulation diagram using the first canonical form.

To verify that these two equations are equivalent to Equation (2.5.2), we substitute Equation (2.5.15) into the left side of Equation (2.5.2) to obtain

$$\left(D^N + \sum_{j=0}^{N-1} a_j D^j\right) y(t) = \left(\sum_{i=0}^{N} b_i D^{(i+N)} + \sum_{j=0}^{N-1} a_j \sum_{i=0}^{N} b_i D^{(i+j)}\right) v(t)$$

$$= \left(\sum_{i=0}^{N} b_i \left(D^{(i+N)} + \sum_{j=0}^{N-1} a_j D^{(i+j)}\right)\right) v(t)$$

$$= \left(\sum_{i=0}^{N} b_i D^i\right) x(t)$$

The variables $v^{(N-1)}(t), \ldots, v(t)$ that are used in constructing $y(t)$ and $x(t)$ in Equations (2.5.14) and (2.5.15), respectively, are produced by successively integrating $v^{(N)}(t)$. The simulation diagram corresponding to Equations (2.5.14) and (2.5.15) is given in Figure 2.5.5. We refer to this form of representation as the second canonical form.

Note that in the second canonical form, the input of any integrator is exactly the same as the output of the preceding integrator. For example, if the outputs of two successive integrators (counting from the right-hand side) are denoted by v_m and v_{m+1}, respectively, then

$$v'_m(t) = v_{m+1}(t)$$

This fact is used in Section 2.6.4 to develop state-variable representations that have useful properties.

Example 2.5.3

We obtain a simulation diagram for the LTI system described by the linear constant-coefficient differential equation:

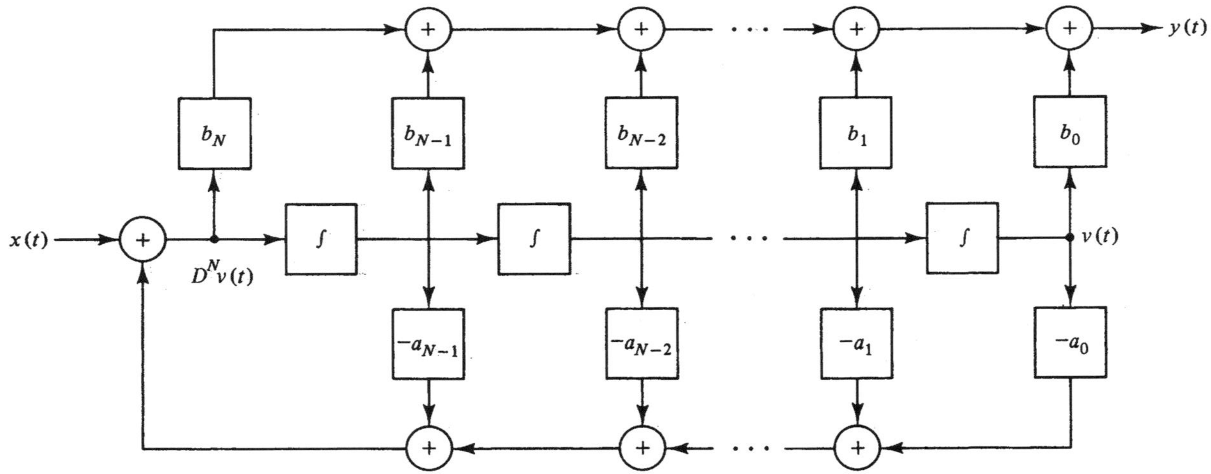

Figure 2.5.5 Simulation diagram using the second canonical form.

$$y''(t) + 3y'(t) + 4y(t) = 2x''(t) - 3x'(t) + x(t)$$

where $y(t)$ is the output, and $x(t)$ is the input to the system. To get the first canonic form, we rewrite this equation as

$$D^2y(t) = 2x(t) + D^{-1}[-3x(t) - 3y(t)] + D^{-2}[x(t) - 4y(t)]$$

Integrating both sides twice with respect to t yields

$$y(t) = 2x(t) + D^{-1}[-3x(t) - 3y(t)] + D^{-2}[x(t) - 4y(t)]$$

The simulation diagram for this representation is given in Figure 2.5.6(a). For the second canonic form, we set

$$v''(t) + 3v'(t) + 4v(t) = x(t)$$

and

$$y(t) = 2v''(t) - 3v'(t) + v(t)$$

which leads to the simulation diagram of Figure 2.5.6(b).

In Section 2.6, we demonstrate how to use the two canonical forms just described to derive two different state-variable representations.

2.5.4 Finding the Impulse Response

The system impulse response can be determined from the differential equation describing the system. In later chapters, we find the impulse response using transform techniques. We defined the impulse response $h(t)$ as the response $y(t)$ when $x(t) = \delta(t)$ and $y(t) = 0, -\infty < t < 0$. The following examples demonstrate the procedure for determining the impulse response from the system differential equation.

Example 2.5.4

Consider the system governed by

$$2y'(t) + 4y(t) = 3x(t)$$

Setting $x(t) = \delta(t)$ results in the response $y(t) = h(t)$. Therefore, $h(t)$ should satisfy the differential equation

$$2h'(t) + 4h(t) = 3\delta(t) \tag{2.5.16}$$

The homogeneous part of the solution to this first-order differential equation is

$$h(t) = C\exp[-2t]u(t) \tag{2.5.17}$$

We predict that the particular solution is zero, the motivation for this being that $h(t)$ cannot contain a delta function. Otherwise, $h'(t)$ would have a derivative of a delta function that is not a part of the right-hand side of Equation (2.5.16). To find the constant C, we substitute Equation (2.5.17) into Equation (2.5.16) to get

$$2\frac{d}{dt}(C\exp[-2t]u(t)) + 4C\exp[-2t]u(t) = 3\,\delta(t)$$

Simplifying this expression results in

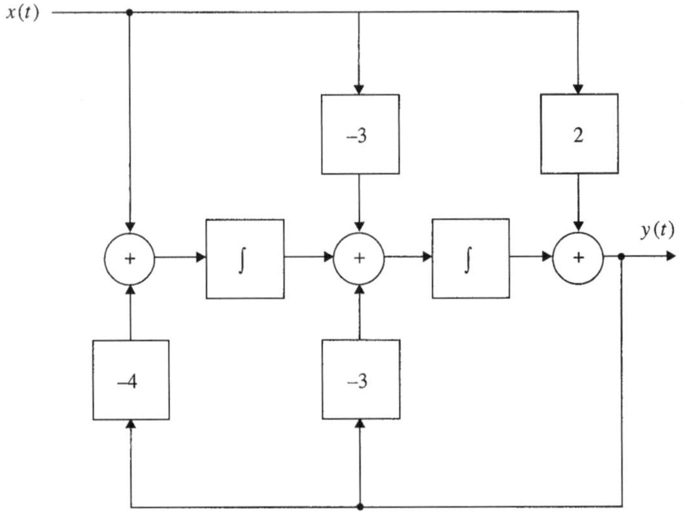

(a)

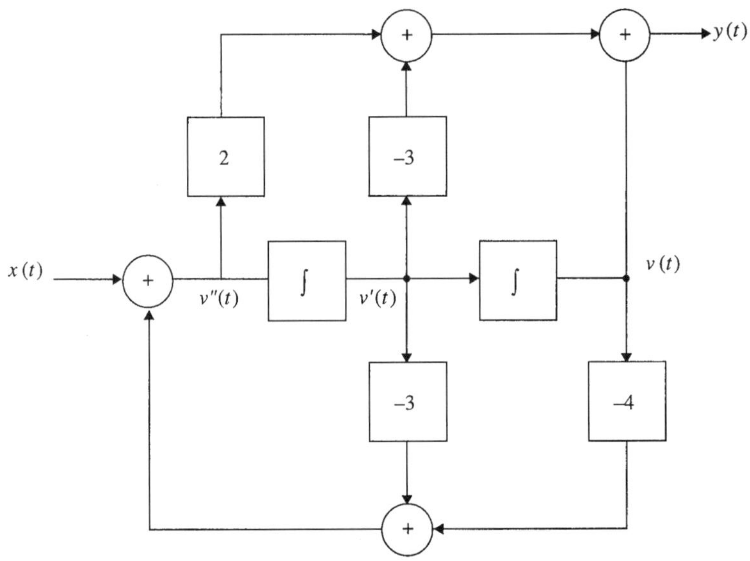

(b)

Figure 2.5.6 Simulation diagram of the second-order system in Example 2.5.3.

$$2C\exp[-2t]\,\delta(t) = 3\delta(t)$$

which, after applying the sampling property of the δ function, is equivalent to

$$2C\,\delta(t) = 3\delta(t)$$

so that $C = 1.5$. We therefore have

$$h(t) = 1.5 \exp[-2t]u(t)$$

In general, it can be shown that for $x(t) = \delta(t)$, the particular solution of Equation (2.5.2) is of the form

$$h_p(t) = \begin{cases} \sum_{i=0}^{M-N} C_i \delta^{(i)}(t), & M \geq N \\ 0, & M < N \end{cases}$$

where $\delta^{(i)}(t)$ is the ith derivative of the δ function. Since, in most cases of practical interest, $N \geq M$, it follows that the particular solution is at most a δ function.

Example 2.5.5

Consider the first-order system

$$y'(t) + 3y(t) = 2x(t)$$

The system impulse response should satisfy the following differential equation:

$$h'(t) + 3h(t) = 2\delta(t)$$

The homogeneous solution of this equation is of the form $C_1 \exp[-3t]u(t)$. Let us assume a particular solution of the form $h_p(t) = C_2\delta(t)$. The general solution is therefore

$$h(t) = C_1 \exp[-3t]u(t) + C_2\delta(t)$$

Substituting in the differential equation gives

$$C_1[-3\exp(-3t)u(t) + \exp(-3t)\delta(t)] + C_2\delta'(t) + 3[C_1 \exp(-3t)u(t) + C_2\delta(t)] = 2\delta(t)$$

Equate coefficients of $\delta(t)$ and $\delta'(t)$ on both sides and use the sifting property of the δ-function to get

$$C_1 = 2, \qquad C_2 = 0$$

This is in conformity with our previous discussion since $M < N$ in this example and so we should expect that $C_2 = 0$. We therefore have

$$h(t) = 2\exp[-3t]u(t) \tag{2.5.18}$$

Example 2.5.6

Consider the second-order system

$$y''(t) + 2y'(t) + 2y(t) = x''(t) + 3x'(t) + 3x(t)$$

The characteristic roots of this differential equation are equal to $-1 \pm j1$ so that the homogeneous solution is of the form $[C_1\exp(-t)\cos t + C_2\exp(-t)\sin t]u(t)$. Since $M = N$ in this case, we should expect that $h_p(t) = C_3\delta(t)$. Thus the impulse response is of the form

$$h(t) = [C_1\exp(-t)\cos t + C_2\exp(-t)\sin t]u(t) + C_3\delta(t)$$

so that

$$h'(t) = [C_2 - C_1]\exp(-t)\cos t\, u(t) - [C_2 + C_1]\exp(-t)\sin t\, u(t) + C_1\delta(t) + C_3\delta'(t)$$

and

$$h''(t) = -2C_2\exp(-t)\cos t\, u(t) + 2C_1\exp(-t)\sin t\, u(t) + (C_2 - C_1)\delta(t) + C_1\delta'(t) + C_3\delta''(t)$$

We now substitute these in the system differential equation. Collecting like terms in $\delta(t)$, $\delta'(t)$ and $\delta''(t)$ and solving for the coefficients C_i gives

$$C_1 = 1, \quad C_2 = 0, \quad \text{and} \quad C_3 = 1$$

so that the impulse response is

$$h(t) = \exp[-t]\cos t\, u(t) + \delta(t) \tag{2.5.19}$$

In Chapters 4 and 5, we use transform methods to find the impulse response in a much easier manner.

2.6 STATE-VARIABLE REPRESENTATION

In our previous discussions, we characterized linear time-invariant systems either by their impulse response functions or by differential equations relating their inputs and outputs. Frequently, the impulse response is the most convenient method of describing a system. By knowing the input of the system over the interval $-\infty < t < \infty$, we can obtain the output of the system by forming the convolution integral. In the case of the differential-equation representation, the output is determined in terms of a set of initial conditions. If we want to find the output over some interval $t_0 \le t < t_1$, we must know not only the input over this interval, but also a certain number of initial conditions that must be sufficient to describe how any past inputs (i.e., for $t \le t_0$) affect the output of the system during that interval.

In this section, we discuss the method of state-variable representation of systems. The representation of systems in this form has many advantages:

1. It provides an insight into the behavior of the system that neither the impulse response nor the differential-equation method does.
2. It can be easily adapted for solution by analog or digital computer techniques.
3. It can be extended to nonlinear and time-varying systems.
4. It allows us to handle systems with many inputs and outputs.

The computer solution feature by itself is the reason that state-variable methods are widely used in analyzing highly complex systems.

We define the state of the system as the minimal amount of information that is sufficient to determine the output of the system for all $t \ge t_0$, provided that the input to the system is also known for all times $t \ge t_0$. The variables that contain this information are called the state variables. Given the state of the system at t_0 and the input from t_0 to t_1, we can find both the output and the state at t_1. Note that this definition of the state of the system applies only to causal systems (systems in which future inputs cannot affect the output).

2.6.1 State Equations

Consider the single-input, single-output, second-order, continuous-time system described by Equation (2.5.11). Figure 2.5.3 depicts a realization of the system. Since integrators are elements with memory (i.e., they contain information about the past history of the system), it is natural to choose the outputs of integrators as the state of the system at any time t. Note that a continuous-time system of dimension N is realized by N integrators and is, therefore, completely represented by N state variables. It is often advantageous to think of the state variables as the components of an N-dimensional vector referred to as state vector $\mathbf{v}(t)$. Throughout the book, **boldface** lowercase letters are used to denote vectors, and **boldface** uppercase letters are used to denote matrices. In the example under consideration, we define the components of the state vector $\mathbf{v}(t)$ as

$$v_1(t) = y(t)$$

$$v_2(t) = v_1'(t)$$

$$v_2'(t) = -a_0v_1(t) - a_1v_2(t) + b_0x(t)$$

Expressing $\mathbf{v}'(t)$ in terms of $\mathbf{v}(t)$ yields

$$\begin{bmatrix} v_1'(t) \\ v_2'(t) \end{bmatrix} = \begin{bmatrix} 0 & 1 \\ -a_0 & -a_1 \end{bmatrix} \begin{bmatrix} v_1(t) \\ v_2(t) \end{bmatrix} + \begin{bmatrix} 0 \\ b_0 \end{bmatrix} x(t) \tag{2.6.1}$$

The output $y(t)$ can be expressed in terms of the state vector $\mathbf{v}(t)$ as

$$y(t) = [1 \quad 0] \begin{bmatrix} v_1(t) \\ v_2(t) \end{bmatrix} \tag{2.6.2}$$

In this representation, Equation (2.6.1) is called the state equation and Equation (2.6.2) is called the output equation.

In general, a state-variable description of an N-dimensional, single-input, single-output linear, time-invariant system is written in the form

$$\mathbf{v}'(t) = \mathbf{A}\,\mathbf{v}(t) + \mathbf{b}x(t) \tag{2.6.3}$$

$$y(t) = \mathbf{c}\,\mathbf{v}(t) + dx(t) \tag{2.6.4}$$

where $\mathbf{A}$ is an $N \times N$ square matrix that has constant elements, $\mathbf{b}$ is an $N \times 1$ column vector, and $\mathbf{c}$ is a $1 \times N$ row vector. In the most general case, $\mathbf{A}$, $\mathbf{b}$, $\mathbf{c}$, and d are functions of time, so that we have a time-varying system. The solution of the state equation for such systems generally requires the use of a computer. In this book, we restrict our attention to time-invariant systems in which all the coefficients are constant.

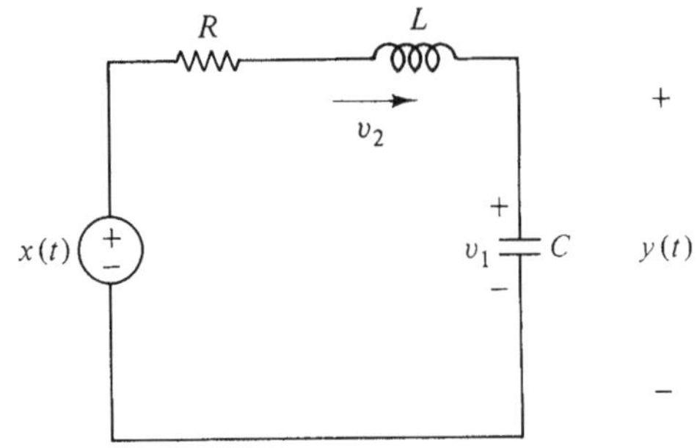

Figure 2.6.1 RLC circuit for Example 2.6.1.

Example 2.6.1

Consider the RLC series circuit shown in Figure 2.6.1. By choosing the voltage across the capacitor and the current through the inductor as the state variables, we obtain the following state equations:

$$C\frac{dv_1(t)}{dt} = v_2(t)$$

$$L\frac{dv_2(t)}{dt} = x(t) - Rv_2(t) - v_1(t)$$

$$y(t) = v_1(t)$$

In matrix form, these become

$$\mathbf{v}'(t) = \begin{bmatrix} 0 & \frac{1}{C} \\ \frac{-1}{L} & \frac{-R}{L} \end{bmatrix}\mathbf{v}(t) + \begin{bmatrix} 0 \\ \frac{1}{L} \end{bmatrix}x(t)$$

$$y(t) = [1 \quad 0]\mathbf{v}(t)$$

If we assume that $C = 1/2$ and $L = R = 1$, we have

$$\mathbf{v}'(t) = \begin{bmatrix} 0 & 2 \\ -1 & -1 \end{bmatrix}\mathbf{v}(t) + \begin{bmatrix} 0 \\ 1 \end{bmatrix}x(t)$$

$$y(t) = [1 \quad 0]\mathbf{v}(t)$$

2.6.2 Time-Domain Solution of the State Equations

Consider the single-input, single-output, linear, time-invariant, continuous-time system described by the following state equations:

$$\mathbf{v}'(t) = \mathbf{A}\mathbf{v}(t) + \mathbf{b}x(t) \tag{2.6.5}$$

$$y(t) = \mathbf{c}\mathbf{v}(t) + dx(t) \tag{2.6.6}$$

The state vector $\mathbf{v}(t)$is an explicit function of time, but it also depends implicitly on the initial state $\mathbf{v}(t_0) = \mathbf{v}_0$, the initial time t_0, and the input $x(t)$. Solving the state equations means finding that functional dependence. We can then compute the output $y(t)$ by using Equation (2.6.6).

As a natural generalization of the solution to the scalar first-order differential equation, we would expect the solution to the homogeneous matrix-differential equation to be of the form

$$\mathbf{v}(t) = \exp[\mathbf{A}t]\mathbf{v}_0$$

where $\exp[\mathbf{A}t]$ is an $N \times N$ matrix exponential of functions of time and is defined by the matrix power series

$$\exp[\mathbf{At}] = \mathbf{I} + \mathbf{A}t + \mathbf{A}^2\frac{t^2}{2!} + \mathbf{A}^3\frac{t^3}{3!} + \cdots + \mathbf{A}^k\frac{t^k}{k!} + \cdots \tag{2.6.7}$$

where $\mathbf{I}$ is the $N \times N$ identity matrix. Using this definition, we can establish the following properties:

$$\exp[\mathbf{A}(t_1 + t_2)] = \exp[\mathbf{A}t_1]\exp[\mathbf{A}t_2] \tag{2.6.8}$$

$$[\exp[\mathbf{A}t]]^{-1} = \exp[-\mathbf{A}t] \tag{2.6.9}$$

To prove Equation (2.6.8), we expand $\exp[\mathbf{A}t_1]$ and $\exp[\mathbf{A}t_2]$ in power series and multiply out terms to obtain

$$\begin{aligned}\exp[\mathbf{A}t_1]\exp[\mathbf{A}t_2] &= \left[\mathbf{I} + \mathbf{A}t_1 + \mathbf{A}^2\frac{t_1^2}{2!} + \cdots + \mathbf{A}^k\frac{t^k}{k!} + \cdots\right] \times \\ &\quad \left[\mathbf{I} + \mathbf{A}t_2 + \mathbf{A}^2\frac{t_2^2}{2!} + \mathbf{A}^3\frac{t_2^3}{3!} + \cdots + \mathbf{A}^k\frac{t_2^k}{k!} + \cdots\right] \\ &= \exp[\mathbf{A}(t_1 + t_2)]\end{aligned}$$

By setting $t_2 = -t_1 = t$, it follows that

$$\exp[-\mathbf{A}t]\exp[\mathbf{A}t] = \mathbf{I}$$

so that

$$\exp[-\mathbf{A}t] = [\exp[\mathbf{A}t]]^{-1}$$

It is well known that the scalar exponential $\exp[at]$ is the only function which possesses the property that its derivative and integral are also exponential functions with scaled amplitudes. This observation holds true for the matrix exponential as well. We require that $\mathbf{A}$ have an inverse for the integral to exist. To show that the derivative of $\exp[\mathbf{A}t]$ is also a matrix exponential, we differentiate Equation (2.6.7) with respect to t to get

$$\begin{aligned}\frac{d}{dt}\exp[\mathbf{A}t] &= \mathbf{0} + \mathbf{A} + \frac{2t}{2!}\mathbf{A}^2 + \frac{3t^2}{3!}\mathbf{A}^3 + \cdots + \frac{kt^{k-1}}{k!}\mathbf{A}^k + \cdots \\ &= \left[\mathbf{I} + \mathbf{A}t + \mathbf{A}^2\frac{t^2}{2!} + \mathbf{A}^3\frac{t^3}{3!} + \cdots + \mathbf{A}^k\frac{t^k}{k!} + \cdots\right]\mathbf{A} \\ &= \mathbf{A}\left[\mathbf{I} + \mathbf{A}t + \mathbf{A}^2\frac{t^2}{2!} + \mathbf{A}^3\frac{t^3}{3!} + \cdots + \mathbf{A}^k\frac{t^k}{k!} + \cdots\right]\end{aligned}$$

Thus,

$$\frac{d}{dt}\exp[\mathbf{A}t] = \exp[\mathbf{A}t]\mathbf{A} = \mathbf{A}\exp[\mathbf{A}t] \tag{2.6.10}$$

Now, multiplying Equation (2.6.5) on the left by $\exp[-\mathbf{A}t]$ and rearranging terms, we obtain

$$\exp[-\mathbf{A}t][\mathbf{v}'(t) - \mathbf{A}\,\mathbf{v}(t)] = \exp[-\mathbf{A}t]\mathbf{b}x(t)$$

Using Equation (2.6.10), we can write the last equation as

$$\frac{d}{dt}(\exp[-\mathbf{A}t]\mathbf{v}(t)) = \exp[-\mathbf{A}t]\mathbf{b}x(t) \tag{2.6.11}$$

Integrating both sides of Equation (2.6.11) between t_0 and t gives

$$\exp[-\mathbf{A}t]\mathbf{v}(t) + \exp[-\mathbf{A}t_0]\mathbf{v}_0 = \int_{t_0}^{t}\exp[-\mathbf{A}\tau]\mathbf{b}x(\tau)d\tau \tag{2.6.12}$$

Multiplying Equation (2.6.12) by $\exp[\mathbf{A}t]$ and rearranging terms, we obtain the complete solution of Equation (2.6.5) in the form

$$\mathbf{v}(t) = \exp[\mathbf{A}(t - t_0)]\mathbf{v}_0 + \int_{t_0}^{t}\exp[\mathbf{A}(t - \tau)]\mathbf{b}x(\tau)d\tau \tag{2.6.13}$$

The matrix exponential $\exp[\mathbf{A}t]$ is called the state transition matrix and is denoted by $\mathbf{\Phi}(t)$. The complete output response $y(t)$ is obtained by substituting Equation (2.6.13) into Equation (2.6.6). The result is

$$y(t) = \mathbf{c}\mathbf{\Phi}(t - t_0)\mathbf{v}_0 + \int_{t_0}^{t}\mathbf{c}\mathbf{\Phi}(t - \tau)\mathbf{b}x(\tau)d\tau + dx(t), \quad t \geq t_0 \tag{2.6.14}$$

Using the sifting property of the unit impulse $\delta(t)$, we can rewrite Equation (2.6.14) as

$$y(t) = \mathbf{c}\mathbf{\Phi}(t - t_0)\mathbf{v}_0 + \int_{t_0}^{t}\{\mathbf{c}\mathbf{\Phi}(t - \tau)\mathbf{b} + d\,\delta(t - \tau)\}\,x(\tau)d\tau, \qquad t \geq t_0 \tag{2.6.15}$$

Observe that the complete solution is the sum of two terms. The first term is the response when the input $x(t)$ is zero and is called the zero-input response. The second term is the response when the initial state $\mathbf{v}_0$ is zero and is called the zero-state response. Further inspection of the zero-state response reveals that this term is the convolution of input $x(t)$ with $\mathbf{c}\mathbf{\Phi}(t) + d\,\delta(t)$. Comparing this result with Equation (2.3.3), we conclude that the impulse response of the system is

$$h(t) = \begin{cases} \mathbf{c}\mathbf{\Phi}(t)\mathbf{b} + d\,\delta(t) & t \geq 0 \\ 0 & \text{otherwise} \end{cases} \tag{2.6.16}$$

That is, the impulse response is composed of two terms. The first term is due to the contribution of the state-transition matrix, and the second term is a straight-through path from input to output. Equation (2.6.16) can be used to compute the impulse response directly from the coefficient matrices of the state model of the system.

Example 2.6.2

Consider the linear, time-invariant, continuous-time system described by the differential equation

$$y''(t) + y'(t) - 2y(t) = x(t)$$

The state-space model for this system is

$$\mathbf{v}'(t) = \begin{bmatrix} 0 & 1 \\ 2 & -1 \end{bmatrix} \mathbf{v}(t) + \begin{bmatrix} 0 \\ 1 \end{bmatrix} x(t)$$

$$y(t) = [1 \quad 0] \quad \mathbf{v}(t)$$

so that

$$\mathbf{A} = \begin{bmatrix} 0 & 1 \\ 2 & -1 \end{bmatrix}, \quad \mathbf{b} = \begin{bmatrix} 0 \\ 1 \end{bmatrix}, \quad \text{and} \quad \mathbf{c} = [1 \quad 0]$$

To determine the zero-input response of the system to a specified initial-condition vector

$$\mathbf{v}_0 = \begin{bmatrix} 1 \\ 0 \end{bmatrix}$$

we have to calculate $\mathbf{\Phi}(t)$. The powers of the matrix $\mathbf{A}$ are

$$\mathbf{A}^2 = \begin{bmatrix} 2 & -1 \\ -2 & 3 \end{bmatrix} \quad \mathbf{A}^3 = \begin{bmatrix} -2 & 3 \\ 6 & -5 \end{bmatrix} \quad \mathbf{A}^4 = \begin{bmatrix} 6 & -5 \\ -10 & 11 \end{bmatrix} \quad \cdots$$

so that

$$\mathbf{\Phi}(t) = \begin{bmatrix} 1 & 0 \\ 0 & 1 \end{bmatrix} + \begin{bmatrix} 0 & t \\ 2t & -t \end{bmatrix} + \begin{bmatrix} t^2 & \dfrac{-t^2}{2} \\ -t^2 & \dfrac{3t^2}{2} \end{bmatrix} + \begin{bmatrix} \dfrac{-t^3}{3} & \dfrac{t^3}{2} \\ t^3 & \dfrac{-5}{6}t^3 \end{bmatrix} + \begin{bmatrix} \dfrac{t^4}{4} & \dfrac{-5}{24}t^4 \\ \dfrac{-5}{12}t^4 & \dfrac{11}{24}t^4 \end{bmatrix} + \cdots$$

$$= \begin{bmatrix} 1 + t^2 - \dfrac{t^3}{3} + \dfrac{t^4}{4} + \cdots & t - \dfrac{t^2}{2} + \dfrac{t^3}{2} - \dfrac{5}{24}t^4 + \cdots \\ 2t - t^2 + t^3 - \dfrac{5}{12}t^4 + \cdots & 1 - t + \dfrac{3}{2}t^2 - \dfrac{5}{6}t^3 + \dfrac{11}{24}t^4 + \cdots \end{bmatrix}$$

The zero-input response of the system is

$$y(t) = [1 \quad 0] \begin{bmatrix} 1 + t^2 - \dfrac{t^3}{3} - \dfrac{t^4}{4} + \cdots & t - \dfrac{t^2}{2} + \dfrac{t^3}{2} - \dfrac{5}{24}t^4 + \cdots \\ 2t - t^2 + t^3 - \dfrac{5}{12}t^4 + \cdots & 1 - t + \dfrac{3}{2}t^2 - \dfrac{5}{6}t^3 + \dfrac{11}{24}t^4 + \cdots \end{bmatrix} \begin{bmatrix} 1 \\ 0 \end{bmatrix}$$

$$= 1 + t^2 - \frac{t^3}{3} - \frac{t^4}{4} + \cdots$$

The impulse response of the system is

$$h(t) = [1 \quad 0] \begin{bmatrix} 1 + t^2 - \frac{t^3}{3} - \frac{t^4}{4} + \cdots & t - \frac{t^2}{2} + \frac{t^3}{2} - \frac{5}{24}t^4 + \cdots \\ 2t - t^2 + t^3 - \frac{5}{12}t^4 + \cdots & 1 - t + \frac{3}{2}t^2 - \frac{5}{6}t^3 + \frac{11}{24}t^4 + \cdots \end{bmatrix} \begin{bmatrix} 0 \\ 1 \end{bmatrix}$$

$$= t - \frac{t^2}{2} + \frac{t^3}{2} - \frac{5}{24}t^4 + \cdots$$

Note that for $x(t) = 0$, the state at time t is

$$\mathbf{v}(t) = \mathbf{\Phi}(t)\,\mathbf{v}_0$$

$$= \begin{bmatrix} 1 + t^2 - \frac{t^3}{3} + \frac{t^4}{4} + \cdots \\ 2t - t^2 + t^3 - \frac{5}{12}t^4 + \cdots \end{bmatrix}$$

Example 2.6.3

Given the continuous-time system

$$\mathbf{v}'(t) = \begin{bmatrix} -1 & 0 & 0 \\ 0 & -4 & 4 \\ 0 & -1 & 0 \end{bmatrix} \mathbf{v}(t) + \begin{bmatrix} 1 \\ 1 \\ 1 \end{bmatrix} x(t)$$

$$y(t) = [-1 \quad 2 \quad 0] \quad \mathbf{v}(t)$$

we compute the transition matrix and the impulse response of the system. By using Equation (2.6.7), we have

$$\mathbf{\Phi}(t) = \begin{bmatrix} 1 & 0 & 0 \\ 0 & 1 & 0 \\ 0 & 0 & 1 \end{bmatrix} + \begin{bmatrix} -t & 0 & 0 \\ 0 & -4t & 4t \\ 0 & -t & 0 \end{bmatrix} + \begin{bmatrix} \frac{t^2}{2} & 0 & 0 \\ 0 & 6t^2 & -8t^2 \\ 0 & 2t^2 & -2t^2 \end{bmatrix} + \cdots$$

$$= \begin{bmatrix} 1 - t + \frac{t^2}{2} + \cdots & 0 & 0 \\ 0 & 1 - 4t + 6t^2 + \cdots & 4t - 8t^2 + \cdots \\ 0 & -t + 2t^2 + \cdots & 1 - 2t^2 + \cdots \end{bmatrix}$$

Using Equation (2.6.16), we find the system impulse response to be

$$h(t) = [-1 \quad 2 \quad 0] \quad \mathbf{\Phi}(t) \begin{bmatrix} 1 \\ 1 \\ 1 \end{bmatrix} = 3 - 11t + \frac{33}{2}t^2 + \cdots$$

It is clear from Equations (2.6.13), (2.6.15), and (2.6.16) that in order to determine $\mathbf{v}(t)$, $y(t)$, or $h(t)$, we have to first obtain $\exp[\mathbf{A}t]$. The preceding two examples demonstrate how to use the power series method to find $\mathbf{\Phi}(t) = \exp[\mathbf{A}t]$. Although the method is straightforward and the form is acceptable, the major problem is that it is usually not possible to recognize a closed form corresponding to this solution. Another method that can be used comes from the Cayley-Hamilton theorem, which states that any arbitrary $N \times N$ matrix $\mathbf{A}$ satisfies its characteristic equation, that is,

$$\det(\mathbf{A} - \lambda\mathbf{I}) = 0$$

The Cayley-Hamilton theorem gives a means of expressing any power of a matrix $\mathbf{A}$ in terms of a linear combination of $\mathbf{A}^m$ for $m = 0, 1, \ldots, N - 1$.

Example 2.6.4

Given

$$\mathbf{A} = \begin{bmatrix} 5 & 4 \\ 1 & 2 \end{bmatrix}$$

it follows that

$$\det(\mathbf{A} - \lambda\mathbf{I}) = \lambda^2 - 7\lambda + 6$$

and the given matrix satisfies

$$\mathbf{A}^2 - 7\mathbf{A} + 6\mathbf{I} = \mathbf{0}$$

Therefore, $\mathbf{A}^2$ can be expressed in terms of $\mathbf{A}$ and $\mathbf{I}$ by

$$\mathbf{A}^2 = 7\mathbf{A} - 6\mathbf{I} \tag{2.6.17}$$

Also, $\mathbf{A}^3$ can be found by multiplying Equation (2.6.17) by $\mathbf{A}$ and then using Equation (2.6.17) again:

$$\begin{aligned} \mathbf{A}^3 &= 7\mathbf{A}^2 - 6\mathbf{A} = 7(7\mathbf{A} - 6\mathbf{I}) - 6\mathbf{A} \\ &= 43\mathbf{A} - 42\mathbf{I} \end{aligned}$$

Similarly, any power of $\mathbf{A}$ can be found as a linear combination of $\mathbf{A}$ and $\mathbf{I}$ by this method. Further, we can determine $\mathbf{A}^{-1}$, if it exists, by multiplying Equation (2.6.17) by $\mathbf{A}^{-1}$ and rearranging terms to obtain

$$\mathbf{A}^{-1} = \frac{1}{6}[7\mathbf{I} - \mathbf{A}]$$

It follows from our previous discussion that we can use the Cayley-Hamilton theorem to write $\exp[\mathbf{A}t]$ as a linear combination of the terms $(\mathbf{A}t)^i$, $i = 0, 1, 2, \ldots, N - 1$, so that

$$\exp[\mathbf{A}t] = \sum_{i=0}^{N-1} \gamma_i(t)\, \mathbf{A}^i \tag{2.6.18}$$

If $\mathbf{A}$ has distinct eigenvalues λ_i, we can obtain $\gamma_i(t)$ by solving the set of equations

$$\exp[\lambda_j t] = \sum_{i=0}^{N-1} \gamma_i(t)\, \lambda_j^i \qquad j = 1, \ldots, N \tag{2.6.19}$$

For the case of repeated eigenvalues, the procedure is a little more complex, as we will learn later. (See Appendix C for details.)

Example 2.6.5

Suppose that we want to find the transition matrix for the system with

$$\mathbf{A} = \begin{bmatrix} -3 & 1 & 0 \\ 1 & -3 & 0 \\ 0 & 0 & -3 \end{bmatrix}$$

using the Cayley-Hamilton method. First, we calculate the eigenvalues of $\mathbf{A}$ as $\lambda_1 = -2$, $\lambda_2 = -3$, and $\lambda_3 = -4$. It follows from the Cayley-Hamilton theorem that we can write $\exp[\mathbf{A}t]$ as

$$\exp[\mathbf{A}t] = \gamma_0(t)\mathbf{I} + \gamma_1(t)\mathbf{A} + \gamma_2(t)\mathbf{A}^2$$

where the coefficients $\gamma_0(t)$, $\gamma_1(t)$, and $\gamma_2(t)$ are the solution of the set of equations

$$\exp[-2t] = \gamma_0(t) - 2\gamma_1(t) + 4\gamma_2(t)$$
$$\exp[-3t] = \gamma_0(t) - 3\gamma_1(t) + 9\gamma_2(t)$$
$$\exp[-4t] = \gamma_0(t) - 4\gamma_1(t) + 16\gamma_2(t)$$

from which

$$\gamma_0(t) = 3\exp[-4t] - 8\exp[-3t] + 6\exp[-2t]$$

$$\gamma_1(t) = \frac{5}{2}\exp[-4t] - 6\exp[-3t] + \frac{7}{2}\exp[-2t]$$

$$\gamma_2(t) = \frac{1}{2}(\exp[-4t] - 2\exp[-3t] + \exp[-2t])$$

Thus, $\exp[\mathbf{A}t]$ is

$$\exp[\mathbf{A}t] = \gamma_0(t)\begin{bmatrix} 1 & 0 & 0 \\ 0 & 1 & 0 \\ 0 & 0 & 1 \end{bmatrix} + \gamma_1(t)\begin{bmatrix} -3 & 1 & 0 \\ 1 & -3 & 0 \\ 0 & 0 & -3 \end{bmatrix} + \gamma_2(t)\begin{bmatrix} 10 & -6 & 0 \\ -6 & 10 & 0 \\ 0 & 0 & 9 \end{bmatrix}$$

$$= \begin{bmatrix} \frac{1}{2}\exp[-4t] + \frac{1}{2}\exp[-2t] & -\frac{1}{2}\exp[-4t] + \frac{1}{2}\exp[-2t] & 0 \\ -\frac{1}{2}\exp[-4t] + \frac{1}{2}\exp[-2t] & \frac{1}{2}\exp[-4t] + \frac{1}{2}\exp[-2t] & 0 \\ 0 & 0 & \exp[-3t] \end{bmatrix}$$

Example 2.6.6

Let us repeat Example 2.6.5 for the system with

$$\mathbf{A} = \begin{bmatrix} -1 & 0 & 0 \\ 0 & -4 & 4 \\ 0 & -1 & 0 \end{bmatrix}$$

This matrix has $\lambda_1 = -1$, and $\lambda_2 = \lambda_3 = -2$. Thus,

$$\mathbf{\Phi}(t) = \exp[\mathbf{A}t] = \gamma_0(t)\,\mathbf{I} + \gamma_1(t)\,\mathbf{A} + \gamma_2(t)\,\mathbf{A}^2$$

The coefficients $\gamma_0(t)$, $\gamma_1(t)$, and $\gamma_2(t)$ are obtained by using

$$\exp[\lambda t] = \gamma_0(t) + \gamma_1(t)\lambda + \gamma_2(t)\lambda^2 \tag{2.6.20}$$

However, when we use $\lambda = -1, -2$, and -2 in this equation, we get

$$\exp[-t] = \gamma_0(t) - \gamma_1(t) + \gamma_2(t)$$
$$\exp[-2t] = \gamma_0(t) - 2\gamma_1(t) + 4\gamma_2(t)$$
$$\exp[-2t] = \gamma_0(t) - 2\gamma_1(t) + 4\gamma_2(t)$$

Since one eigenvalue is repeated, we have only two equations in three unknowns. To completely determine $\gamma_0(t)$, $\gamma_1(t)$, and $\gamma_2(t)$, we need another equation, which we can generate by differentiating Equation (2.6.20) with respect to λ to obtain

$$t\exp[\lambda t] = \gamma_1(t) + 2\gamma_2(t)\lambda$$

Thus, the coefficients $\gamma_0(t)$, $\gamma_1(t)$, and $\gamma_2(t)$ are obtained as the solution to the following three equations:

$$\exp[-t] = \gamma_0(t) - \gamma_1(t) + \gamma_2(t)$$
$$\exp[-2t] = \gamma_0(t) - 2\gamma_1(t) + 4\gamma_2(t)$$
$$t\exp[-2t] = \gamma_1(t) - 4\gamma_2(t)$$

Solving for $\gamma_i(t)$ yields

$$\gamma_0(t) = 4\exp[-t] - 3\exp[-2t] - 2t\exp[-2t]$$
$$\gamma_1(t) = 4\exp[-t] - 4\exp[-2t] - 3t\exp[-2t]$$
$$\gamma_2(t) = \exp[-t] - \exp[-2t] - t\exp[-2t]$$

so that

$$\mathbf{\Phi}(t) = \gamma_0(t)\begin{bmatrix} 1 & 0 & 0 \\ 0 & 1 & 0 \\ 0 & 0 & 0 \end{bmatrix} + \gamma_1(t)\begin{bmatrix} -1 & 0 & 0 \\ 0 & -4 & 4 \\ 0 & -1 & 0 \end{bmatrix} + \gamma_2(t)\begin{bmatrix} 1 & 0 & 0 \\ 0 & 12 & -16 \\ 0 & 4 & -4 \end{bmatrix}$$

$$= \begin{bmatrix} \exp[-t] & 0 & 0 \\ 0 & \exp[-2t] - 2t\exp[-2t] & 4t\exp[-2t] \\ 0 & -t\exp[-2t] & -4\exp[-t] + 4\exp[-2t] + 4t\exp[-2t] \end{bmatrix}$$

Other methods for calculating $\mathbf{\Phi}(t)$ also are available. The reader should keep in mind that no one method is easiest for all applications.

The state transition matrix possesses several properties, some of which are as follows:

1. Transition property

$$\mathbf{\Phi}(t_2 - t_0) = \mathbf{\Phi}(t_2 - t_1)\mathbf{\Phi}(t_1 - t_0) \tag{2.6.21}$$

2. Inversion property

$$\mathbf{\Phi}(t_0 - t) = \mathbf{\Phi}^{-1}(t - t_0) \tag{2.6.22}$$

3. Separation property

$$\mathbf{\Phi}(t - t_0) = \mathbf{\Phi}(t)\mathbf{\Phi}^{-1}(t_0) \tag{2.6.23}$$

These properties can be easily established by using the properties of the matrix exponential $\exp[\mathbf{A}t]$, namely, Equations (2.6.8) and (2.6.9). For instance, the transition property follows from

$$\begin{aligned}\mathbf{\Phi}(t_2 - t_0) &= \exp[\mathbf{A}(t_2 - t_0)] \\ &= \exp[\mathbf{A}(t_2 - t_1 + t_1 - t_0)] \\ &= \exp[\mathbf{A}(t_2 - t_1)]\exp[\mathbf{A}(t_1 - t_0)] \\ &= \mathbf{\Phi}(t_2 - t_1)\mathbf{\Phi}(t_1 - t_0)\end{aligned}$$

The inversion property follows directly from Equation (2.6.9). Finally, the separation property is obtained by substituting $t_2 = t$ and $t_1 = 0$ in Equation (2.6.21) and then using the inversion property.

2.6.3 State Equations in First Canonical Form

In Section 2.5, we discussed techniques for deriving two different canonical simulation diagrams for an LTI system. These diagrams can be used to develop two state-variable representations. The state equation in the first canonical form is obtained by choosing as a state variable the output of each integrator in Figure 2.5.4. In this case, the state equations have the form

$$\begin{aligned}y(t) &= v_1(t) + b_N x(t) \\ v_1'(t) &= -a_{N-1}\,y(t) + v_2(t) + b_{N-1}\,x(t) \\ v_2'(t) &= -a_{N-2}\,y(t) + v_3(t) + b_{N-2}\,x(t) \\ &\;\vdots \\ v_{N-1}'(t) &= -a_1\,y(t) + v_N(t) + b_1\,x(t) \\ v_N'(t) &= -a_0\,y(t) + b_0\,x(t)\end{aligned} \tag{2.6.24}$$

By using the first equation in Equation (2.6.24) to eliminate $y(t)$, the differential equations for the state variables can be written in the matrix form

$$\begin{bmatrix} v_1'(t) \\ v_2'(t) \\ \vdots \\ v_N'(t) \end{bmatrix} = \begin{bmatrix} -a_{N-1} & 1 & 0 & \cdots & 0 \\ -a_{N-2} & 0 & 1 & \cdots & 0 \\ \vdots & \vdots & \vdots & \vdots & \vdots \\ -a_1 & 0 & 0 & \cdots & 1 \\ -a_0 & 0 & 0 & \cdots & 0 \end{bmatrix} \begin{bmatrix} v_1(t) \\ v_2(t) \\ \vdots \\ v_{N-1}(t) \\ v_N(t) \end{bmatrix} + \begin{bmatrix} b_{N-1} - a_{N-1}b_N \\ b_{N-2} - a_{N-2}b_N \\ \vdots \\ b_1 - a_1 b_N \\ b_0 - a_0 b_N \end{bmatrix} x(t) \tag{2.6.25}$$

We call this the first canonical form for the state equations. Note that this form contains ones above the diagonal, and the first column of matrix $\mathbf{A}$ consists of the negatives of the coefficients a_i. Also, the output $y(t)$ can be written in terms of the state vector $\mathbf{v}(t)$ as

$$y(t) = [1 \quad 0 \quad \cdots \quad 0] \begin{bmatrix} v_1(t) \\ v_2(t) \\ \vdots \\ v_N(t) \end{bmatrix} + b_N x(t) \tag{2.6.26}$$

Observe that this form of state-variable representation can be written down directly from the original Equation (2.5.1)

Example 2.6.7

The first-canonical-form state-variable representation of the LTI system described by

$$2y''(t) + 4y'(t) + 3y(t) = 4x'(t) + 2x(t)$$

is

$$\begin{bmatrix} v_1'(t) \\ v_2'(t) \end{bmatrix} = \begin{bmatrix} -2 & 1 \\ -\dfrac{3}{2} & 0 \end{bmatrix} \begin{bmatrix} v_1(t) \\ v_2(t) \end{bmatrix} + \begin{bmatrix} 2 \\ 1 \end{bmatrix} x(t)$$

$$y(t) = [1 \quad 0] \begin{bmatrix} v_1(t) \\ v_2(t) \end{bmatrix}$$

Example 2.6.8

The LTI system described by

$$y'''(t) - 2y''(t) + y'(t) + 4y(t) = x'''(t) + 5x(t)$$

has the first canonical representation

$$\begin{bmatrix} v_1'(t) \\ v_2'(t) \\ v_3'(t) \end{bmatrix} = \begin{bmatrix} 2 & 1 & 0 \\ -1 & 0 & 1 \\ -4 & 0 & 0 \end{bmatrix} \begin{bmatrix} v_1(t) \\ v_2(t) \\ v_3(t) \end{bmatrix} + \begin{bmatrix} 2 \\ -1 \\ 1 \end{bmatrix} x(t)$$

$$y(t) = [1 \quad 0 \quad 0] \begin{bmatrix} v_1(t) \\ v_2(t) \\ v_3(t) \end{bmatrix} + x(t)$$

2.6.4 State Equations in Second Canonical Form

Another state-variable form can be obtained from the simulation diagram of Figure 2.5.5. Here, again, the state variables are chosen to be the output of each integrator. The equations for the state variables are now

$$
\begin{aligned}
v_1'(t) &= v_2(t)\\
v_2'(t) &= v_3(t)\\
&\vdots\\
v_{N-1}'(t) &= v_N(t)\\
v_N'(t) &= -a_{N-1}\, v_N(t) - a_{N-2}\, v_{N-1}(t) - \cdots - a_0\, v_1(t) + x(t)\\
y(t) &= b_0\, v_1(t) + b_1\, v_2(t) + \cdots + b_{N-1}\, v_N(t) +\\
&\qquad b_N(x(t) - a_0\, v_1(t) - a_1\, v_2(t) - \cdots - a_{N-1}\, v_N(t))
\end{aligned}
\tag{2.6.27}
$$

In matrix form, Equation (2.6.27) can be written as

$$
\frac{d}{dt}\begin{bmatrix} v_1(t)\\ v_2(t)\\ \vdots\\ v_N(t)\end{bmatrix} = \begin{bmatrix} 0 & 1 & 0 & \cdots & 0\\ 0 & 0 & 1 & \vdots & 0\\ \vdots & \vdots & \vdots & \vdots & \vdots\\ 0 & 0 & 0 & \cdots & 1\\ -a_0 & -a_1 & -a_2 & \cdots & -a_{N-1}\end{bmatrix}\begin{bmatrix} v_1(t)\\ v_2(t)\\ \vdots\\ v_N(t)\end{bmatrix} + \begin{bmatrix} 0\\ 0\\ \vdots\\ 1\end{bmatrix} x(t) \tag{2.6.28}
$$

$$
y(t) = [(b_0 - a_0 b_N)(b_1 - a_1 b_N)\cdots(b_{N-1} - a_{N-1} b_N)]\begin{bmatrix} v_1(t)\\ v_2(t)\\ \vdots\\ v_N(t)\end{bmatrix} + b_N x(t) \tag{2.6.29}
$$

This representation is called the second canonical form. Note that here the ones are above the diagonal, but the a's go across the bottom row of the $N \times N$ transition matrix. The second canonical state representation form can be written directly upon inspection of the original differential equation describing the system.

Example 2.6.9

The second canonical form of the state equation of the system described by

$$
y'''(t) - 2y''(t) + y'(t) + 4y(t) = x'''(t) + 5x(t)
$$

is

$$
\begin{bmatrix} v_1'(t)\\ v_2'(t)\\ v_3'(t)\end{bmatrix} = \begin{bmatrix} 0 & 1 & 0\\ 0 & 0 & 1\\ -4 & -1 & 2\end{bmatrix}\begin{bmatrix} v_1(t)\\ v_2(t)\\ v_3(t)\end{bmatrix} + \begin{bmatrix} 0\\ 0\\ 1\end{bmatrix} x(t)
$$

$$
y(t) = [-3 \quad -1 \quad 2]\begin{bmatrix} v_1(t)\\ v_2(t)\\ v_3(t)\end{bmatrix} + x(t)
$$

The first and second canonical forms are only two of many possible state-variable representations of a continuous-time system. In other words, the state-variable representation of a continuous-time system is not unique. For an N-dimensional system, there are an infinite number of state models that represent that system. However, all N-dimensional state models are equivalent in the sense that they have exactly the same input/output relationship. Mathematically, a set of state equations with state vector $\mathbf{v}(t)$ can be transformed to a new set with state vector $\mathbf{q}(t)$ by using a transformation $\mathbf{P}$ such that

$$\mathbf{q}(t) = \mathbf{P}\,\mathbf{v}(t) \tag{2.6.30}$$

where $\mathbf{P}$ is an invertible $N \times N$ matrix so that $\mathbf{v}(t)$ can be obtained from $\mathbf{q}(t)$. It can be shown (see Problem 2.34) that the new state and output equations are

$$\mathbf{q}'(t) = \mathbf{A}_1\,\mathbf{q}(t) + \mathbf{b}_1\,x(t) \tag{2.6.31}$$

$$y(t) = \mathbf{c}_1\,\mathbf{q}(t) + d_1\,x(t) \tag{2.6.32}$$

where

$$\mathbf{A}_1 = \mathbf{P}\,\mathbf{A}\,\mathbf{P}^{-1}, \quad \mathbf{b}_1 = \mathbf{P}\mathbf{b}, \quad \mathbf{c}_1 = \mathbf{c}\,\mathbf{P}^{-1}, \quad d_1 = d \tag{2.6.33}$$

The only restriction on $\mathbf{P}$ is that its inverse exist. Since there are an infinite number of such matrices, we conclude that we can generate an infinite number of equivalent N-dimensional state models.

If we envisage $\mathbf{v}(t)$ as a vector with N coordinates, the transformation in Equation (2.6.30) represents a coordinate transformation that takes the old state coordinates and maps them to the new state coordinates. The new state model can have one or more of the coefficients $\mathbf{A}_1$, $\mathbf{b}_1$, and $\mathbf{c}_1$ in a special form. Such forms result in a significant simplification in the solution of certain classes of problems: examples of these forms are the diagonal form and the two canonical forms discussed in this chapter.

Example 2.6.10

The state equations of a certain system are given by

$$\begin{bmatrix} v_1'(t) \\ v_2'(t) \end{bmatrix} = \begin{bmatrix} 4 & 2 \\ 2 & 4 \end{bmatrix} \begin{bmatrix} v_1(t) \\ v_2(t) \end{bmatrix} + \begin{bmatrix} 1 \\ 2 \end{bmatrix} x(t)$$

We need to find the state equations for this system in terms of the new state variables q_1 and q_2, where

$$\begin{bmatrix} q_1(t) \\ q_2(t) \end{bmatrix} = \begin{bmatrix} 1 & 1 \\ 1 & -1 \end{bmatrix} \begin{bmatrix} v_1(t) \\ v_2(t) \end{bmatrix}$$

The equation for the state variable $\mathbf{q}$ is given by Equation (2.6.31), where

$$\begin{aligned}
\mathbf{A}_1 = \mathbf{P}\mathbf{A}\mathbf{P}^{-1} &= \begin{bmatrix} 1 & 1 \\ 1 & -1 \end{bmatrix} \begin{bmatrix} 4 & 2 \\ 2 & 4 \end{bmatrix} \begin{bmatrix} 1 & 1 \\ 1 & -1 \end{bmatrix}^{-1} \\
&= \begin{bmatrix} 1 & 1 \\ 1 & -1 \end{bmatrix} \begin{bmatrix} 4 & 2 \\ 2 & 4 \end{bmatrix} \begin{bmatrix} \frac{1}{2} & \frac{1}{2} \\ \frac{1}{2} & -\frac{1}{2} \end{bmatrix} \\
&= \begin{bmatrix} 6 & 0 \\ 0 & 2 \end{bmatrix}
\end{aligned}$$

and

$$\mathbf{b}_1 = \mathbf{Pb} = \begin{bmatrix} 1 & 1 \\ 1 & -1 \end{bmatrix} \begin{bmatrix} 1 \\ 2 \end{bmatrix} = \begin{bmatrix} 3 \\ -1 \end{bmatrix}$$

Example 2.6.11

Let us find the matrix **P** that transforms the second-canonical-form state equations

$$\begin{bmatrix} v_1'(t) \\ v_2'(t) \end{bmatrix} = \begin{bmatrix} 0 & 1 \\ -2 & -3 \end{bmatrix} \begin{bmatrix} v_1(t) \\ v_2(t) \end{bmatrix} + \begin{bmatrix} 0 \\ 1 \end{bmatrix} x(t)$$

into the first-canonical-form state equations

$$\begin{bmatrix} q_1'(t) \\ q_2'(t) \end{bmatrix} = \begin{bmatrix} -3 & 1 \\ -2 & 0 \end{bmatrix} \begin{bmatrix} q_1(t) \\ q_2(t) \end{bmatrix} + \begin{bmatrix} 7 \\ 2 \end{bmatrix} x(t)$$

We desire the transformation such that $\mathbf{PAP}^{-1} = \mathbf{A}_1$ or

$$\mathbf{PA} = \mathbf{A}_1\mathbf{P}$$

Substituting for **A** and $\mathbf{A}_1$, we obtain

$$\begin{bmatrix} p_{11} & p_{12} \\ p_{21} & p_{22} \end{bmatrix} \begin{bmatrix} 0 & 1 \\ -2 & -3 \end{bmatrix} = \begin{bmatrix} -3 & 1 \\ -2 & 0 \end{bmatrix} \begin{bmatrix} p_{11} & p_{12} \\ p_{21} & p_{22} \end{bmatrix}$$

Equating the four elements on the two sides yields

$$\begin{aligned} -2p_{12} &= -3p_{11} + p_{21} \\ p_{11} - 3p_{12} &= -3p_{12} + p_{22} \\ -2p_{22} &= -2p_{11} \\ p_{21} - 3p_{22} &= -2p_{12} \end{aligned}$$

The reader will immediately recognize that the second and third equations are identical. Similarly, the first and fourth equations are identical. Hence, two equations may be discarded. This leaves us with only two equations and four unknowns. Note, however, that the constraint $\mathbf{Pb} = \mathbf{b}_1$ provides us with the following two additional equations:

$$p_{12} = 7$$

$$p_{22} = 2$$

Solving the four equations simultaneously yields

$$\mathbf{P} = \begin{bmatrix} 2 & 7 \\ -8 & 2 \end{bmatrix}$$

Example 2.6.12

If $\mathbf{A}_1$ is a diagonal matrix with entries λ_i, it can easily be verified that the transition matrix $\exp[\mathbf{A}_1 t]$ is also a diagonal matrix with entries $\exp[\lambda_i t]$ and is hence easily evaluated. We can use this result to find the transition matrix for any other representation with $\mathbf{A} = \mathbf{PA}_1\mathbf{P}^{-1}$, since

$$\begin{aligned}\exp[\mathbf{A}t] &= \mathbf{I} + \mathbf{A}t + \frac{1}{2!}\mathbf{A}^2t^2 + \cdots \\ &= \mathbf{I} + \mathbf{P}\mathbf{A}_1\mathbf{P}^{-1}t + \frac{1}{2!}\mathbf{P}\mathbf{A}_1^2\mathbf{P}^{-1}t^2 + \cdots \\ &= \mathbf{P}\left[\mathbf{I} + \mathbf{A}_1t + \frac{1}{2!}\mathbf{A}_1^2t^2 + \cdots\right]\mathbf{P}^{-1} = \mathbf{P}\exp[\mathbf{A}_1t]\mathbf{P}^{-1}\end{aligned}$$

For the matrices $\mathbf{A}$ and $\mathbf{A}_1$ of Example 2.6.10, it follows that

$$\exp[\mathbf{A}_1t] = \begin{bmatrix} \exp(6t) & 0 \\ 0 & \exp(2t) \end{bmatrix}$$

so that

$$\begin{aligned}\exp[\mathbf{A}t] &= \begin{bmatrix} \exp(6t) & 0 \\ 0 & \mathrm{cxp}(2t) \end{bmatrix} \\ &= \frac{1}{2}\begin{bmatrix} e^{6t} + e^{2t} & e^{6t} - e^{2t} \\ e^{6t} - e^{2t} & e^{6t} + e^{2t} \end{bmatrix}\end{aligned}$$

2.6.5 Stability Considerations

Earlier in this section, we found a general expression for the state vector $\mathbf{v}(t)$ of the system with state matrix $\mathbf{A}$ and initial state $\mathbf{v}_0$. The solution of this system consists of two components, the first (zero-input) due to the initial state $\mathbf{v}_0$ and the second (zero-state) due to input $x(t)$. For the continuous-time system to be stable, it is required that not only the output, but also all signals internal to the system, remain bounded when a bounded input is applied. If at least one of the state variables grows without bound, then the system is unstable.

Since the set of eigenvalues of the matrix $\mathbf{A}$ determines the behavior of $\exp[\mathbf{A}t]$, and since $\exp[\mathbf{A}t]$ is used in evaluating the two components in the expression for the state vector $\mathbf{v}(t)$, we expect the eigenvalues of $\mathbf{A}$ to play an important role in determining the stability of the system. Indeed, there exists a technique to test the stability of continuous-time systems without solving for the state vector. This technique follows from the Cayley-Hamilton theorem. We saw earlier that, using this theorem, we can write the elements of $\exp[\mathbf{A}t]$, and hence the components of the state vector, as functions of the exponentials $\exp[\lambda_1 t]$, $\exp[\lambda_2 t]$, $\ldots$, $\exp[\lambda_N t]$, where λ_i, $i = 1, 2, \ldots, N$, are the eigenvalues of the matrix $\mathbf{A}$. For these terms to be bounded, the real part of λ_i, $i = 1, 2, \ldots, N$, must be negative. Thus, the condition for stability of a continuous-time system is that all eigenvalues of the state-transition matrix should have negative real parts.

The foregoing conclusion also follows from the fact that the eigenvalues of $\mathbf{A}$ are identical with the roots of the characteristic equation associated with the differential equation describing the model.

Example 2.6.13

Consider the continuous-time system whose state matrix is

$$\mathbf{A} = \begin{bmatrix} 2 & -1 \\ 4 & -3 \end{bmatrix}$$

The eigenvalues of **A** are $\lambda_1 = -2$ and $\lambda_2 = 1$, and hence, the system is unstable.

Example 2.6.14

Consider the system described by the equations

$$\mathbf{v}'(t) = \begin{bmatrix} 1 & 0 \\ -3 & -2 \end{bmatrix}\mathbf{v}(t) + \begin{bmatrix} 1 \\ 0 \end{bmatrix}x(t)$$

$$y(t) = [1 \quad 1]\mathbf{v}(t)$$

A simulation diagram of this system is shown in Figure 2.6.2. The system can thus be considered as the cascade of the two systems shown inside the dashed lines.

The eigenvalues of **A** are $\lambda_1 = 1$ and $\lambda_2 = -2$. Hence, the system is unstable. The transition matrix of the system is

$$\exp[\mathbf{A}t] = \gamma_0(t)\mathbf{I} + \gamma_1(t)\mathbf{A} \tag{2.6.34}$$

where $\gamma_0(t)$ and $\gamma_1(t)$ are the solutions of

$$\exp[t] = \gamma_0(t) + \gamma_1(t)$$

$$\exp[-2t] = \gamma_0(t) - 2\gamma_1(t)$$

Solving these two equations simultaneously yields

$$\gamma_0(t) = \frac{2}{3}\exp[t] + \frac{1}{3}\exp[-2t]$$

$$\gamma_1(t) = \frac{1}{3}\exp[t] - \frac{1}{3}\exp[-2t]$$

Substituting into Equation (2.6.34), we obtain

$$\exp[\mathbf{A}t] = \begin{bmatrix} \exp[t] & 0 \\ -\exp[t] + \exp[-2t] & \exp[-2t] \end{bmatrix}$$

Let us now look at the response of the system to a unit-step input when the system is initially (at time $t_0 = 0$) relaxed, i.e., the initial state vector $\mathbf{v}_0$ is the zero vector. The output of the system is then

$$y(t) = \int_0^t \mathbf{c}\exp[\mathbf{A}(t-\tau)]\,\mathbf{b}\,x(\tau)\,d\tau$$

$$= \left(\frac{1}{2} - \frac{1}{2}\exp[-2t]\right)u(t)$$

The state vector at any time $t > 0$ is

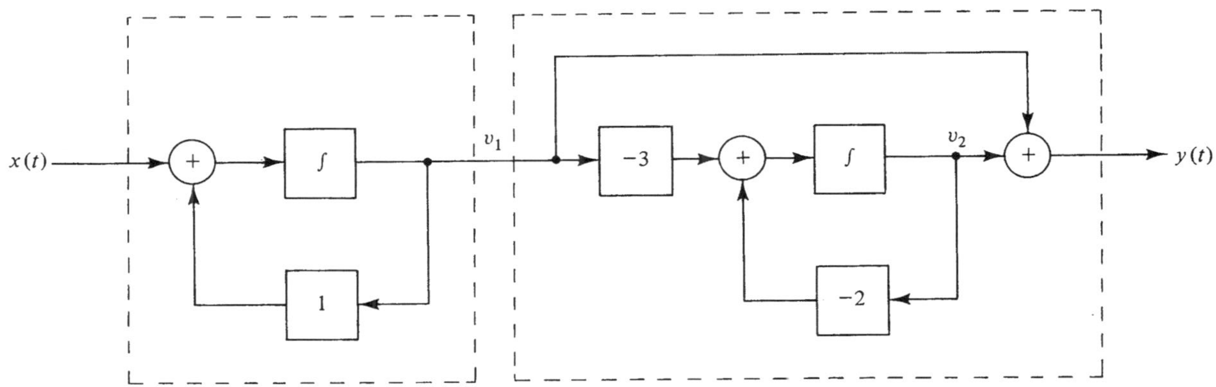

Figure 2.6.2 Simulation diagram for the system of Example 2.6.13.

$$\mathbf{v}(t) = \int_0^t \exp[\mathbf{A}(t-\tau)]\,\mathbf{b}\,x(\tau)d\tau$$

$$= \begin{bmatrix} (\exp[t] - 1)u(t) \\ (3/2 - \exp[t] - 1/2\exp[-2t])u(t) \end{bmatrix}$$

It is clear by looking at $y(t)$ that the output of the system is bounded, whereas inspection of the state variables reveals that the internal signals in the system are not bounded. To pave the way to explain what has happened, let us look at the input/output differential equation. From the output and state equations of the model, we have

$$\begin{aligned} y'(t) &= v_1'(t) + v_2'(t) \\ &= v_1(t) + x(t) - 3v_1(t) - 2v_2(t) \\ &= -2v_1(t) + x(t) - 2[y(t) - v_1(t)] \\ &= -2y(t) + x(t) \end{aligned}$$

The solution of the last first-order differential equation does not contain any terms that grow without bound. It is thus clear that the unstable term $\exp[t]$ that appears in the state variables $v_1(t)$ and $v_2(t)$ does not appear in the output $y(t)$. This term has, in some sense, been "cancelled out" at the output of the second system.

The preceding example demonstrates again the importance of the state-variable representation. State-variable models allow us to examine the internal nature of the system. Often, many important aspects of the system may go unnoticed in the computation or observation of only the output variable. In short, the state-variable techniques have the advantage that all internal components of the system can be made apparent.

2.7 SUMMARY

- A continuous-time system is a transformation that operates on a continuous-time input signal to produce a continuous-time output signal.
- A system is linear if it follows the principle of superposition.
- A system is time invariant if a time shift in the input signal causes an identical time shift in the output.
- A system is memoryless if the present value of the output $y(t)$ depends only on the present value of the input $x(t)$.
- A system is causal if the output $y(t_0)$ depends only on values of the input $x(t)$ for $t \le t_0$.
- A system is invertible if, by observing the output, we can determine the input.
- A system is BIBO stable if bounded inputs result in bounded outputs.
- A linear, time-invariant (LTI) system is completely characterized by its impulse response $h(t)$.
- The output $y(t)$ of an LTI system is the convolution of the input $x(t)$ with the impulse response of the system:

$$y(t) = x(t) * h(t) = \int_{-\infty}^{\infty} x(\tau)h(t-\tau)d\tau$$

- The convolution operation gives only the zero-state response of the system.
- The convolution operator is commutative, associative, and distributive.
- The step response of a linear system with impulse response $h(t)$ is

$$s(t) = \int_{-\infty}^{t} h(\tau)d\tau$$

- An LTI system is causal if $h(t) = 0$ for $t < 0$. The system is stable if and only if $\int_{-\infty}^{\infty} |h(\tau)|d\tau < \infty$.
- An LTI system is described by a linear, constant-coefficient, differential equation of the form

$$\left(D^N + \sum_{i=0}^{N-1} a_i D^i\right)y(t) = \left(\sum_{i=0}^{M} b_i D^i\right)x(t)$$

- A simulation diagram is a block-diagram representation of a system with components consisting of scalar multipliers (amplifiers), summers, or integrators.
- A system can be simulated or realized in several different ways. All these realizations are equivalent. Depending on the application, a particular one of these realizations may be preferable.
- The state equation of an LTI system in state-variable form is

$$\mathbf{v}'(t) = \mathbf{A}\,\mathbf{v}(t) + \mathbf{b}\,x(t)$$

- The output equation of an LTI system in state-variable form is

$$y(t) = \mathbf{c}\,\mathbf{v}(t) + dx(t)$$

- The matrix $\mathbf{\Phi}(t) = \exp[\mathbf{A}t]$ is called the state-transition matrix.
- The state-transition matrix has the following properties:

Transition property: $\mathbf{\Phi}(t_2 - t_0) = \mathbf{\Phi}(t_2 - t_1)\mathbf{\Phi}(t_1 - t_0)$

Inversion property: $\mathbf{\Phi}(t_0 - t) = \mathbf{\Phi}^{-1}(t - t_0)$

Separation property: $\mathbf{\Phi}(t - t_0) = \mathbf{\Phi}(t)\mathbf{\Phi}^{-1}(t_0)$

- The time-domain solution of the state equation is

$$y(t) = \mathbf{c}\mathbf{\Phi}(t - t_0)\mathbf{v}_0 + \int_{t_0}^{t} \mathbf{c}\mathbf{\Phi}(t-\tau)\mathbf{b}x(\tau)d\tau + dx(t), \quad t \geq t_0$$

- $\mathbf{\Phi}(t)$ can be evaluated using the Cayley-Hamilton theorem, which states that any matrix $\mathbf{A}$ satisfies its own characteristic equation.
- The matrix $\mathbf{A}$ in the first canonical form contains ones above the diagonal, and the first column consists of the negatives of the coefficients a_i in Equation (2.5.2).

- The matrix $\mathbf{A}$ in the second canonical form contains ones above the diagonal, and the a_i's go across the bottom row.
- A continuous-time system is stable if and only if all the eigenvalues of the transition matrix $\mathbf{A}$ have negative real parts.

2.8 CHECKLIST OF IMPORTANT TERMS

Causal system
Cayley-Hamilton theorem
Convolution integral
First canonical form
Impulse response of linear system
Impulse response of LTI system
Integrator
Inverse system
Linear system
Linear, time-invariant system
Memoryless system
Multiplier
Output equation
Scalar multiplier
Second canonical form
Simulation diagram
Stable system
State-transition matrix
State variable
Subtractor
Summer
Time-invariant system

2.9 PROBLEMS

2.1. Determine whether the systems described by the following input/output relationships are linear or nonlinear, causal or noncausal, time invariant or time variant, and memoryless or with memory.

(a) $y(t) = 2x(t) + 3$

(b) $y(t) = 2x^2(t) + 3x(t)$

(c) $y(t) = Ax(t)$

(d) $y(t) = Atx(t)$

(e) $y(t) = \begin{cases} x(t), & t \geq 0 \\ -x(t), & t < 0 \end{cases}$

(f) $y(t) = \int_{-\infty}^{t} x(\tau)\,d\tau$

(g) $y(t) = \int_{0}^{t} x(\tau)\,d\tau, \quad t \geq 0$

(h) $y(t) = x(t-5)$

(i) $y(t) = \exp[x(t)]$

(j) $y(t) = x(t)\,x(t-2)$

(k) $y(t) = \dfrac{1}{T}\int_{t-T/2}^{t+T/2} x(\tau)\,d\tau$

(l) $\dfrac{dy(t)}{dt} + 2y(t) = 2x^2(t)$

2.2. Use the model for $\delta(t)$ given by

$$\delta(t) = \lim_{\Delta \to 0} \frac{1}{\Delta} \operatorname{rect}(t/\Delta)$$

to prove Equation (2.3.1).

2.3. Evaluate the following convolutions:

(a) $\operatorname{rect}(t - a/a) * \delta(t - b)$
(b) $\operatorname{rect}(t/a) * \operatorname{rect}(t/a)$
(c) $\operatorname{rect}(t/a) * u(t)$
(d) $\operatorname{rect}(t/a) * \operatorname{sgn}(t)$
(e) $u(t) * u(t)$
(f) $t[u(t) - u(t-1)] * u(t)$
(g) $\operatorname{rect}(t/a) * r(t)$
(h) $r(t) * [\operatorname{sgn}(t) + u(-t-1)]$
(i) $[u(t+1) - u(t-1)]\operatorname{sgn}(t) * u(t)$
(j) $u(t) * \delta'(t)$

2.4. Graphically determine the convolution of the pairs of signals shown in Figure P2.4.

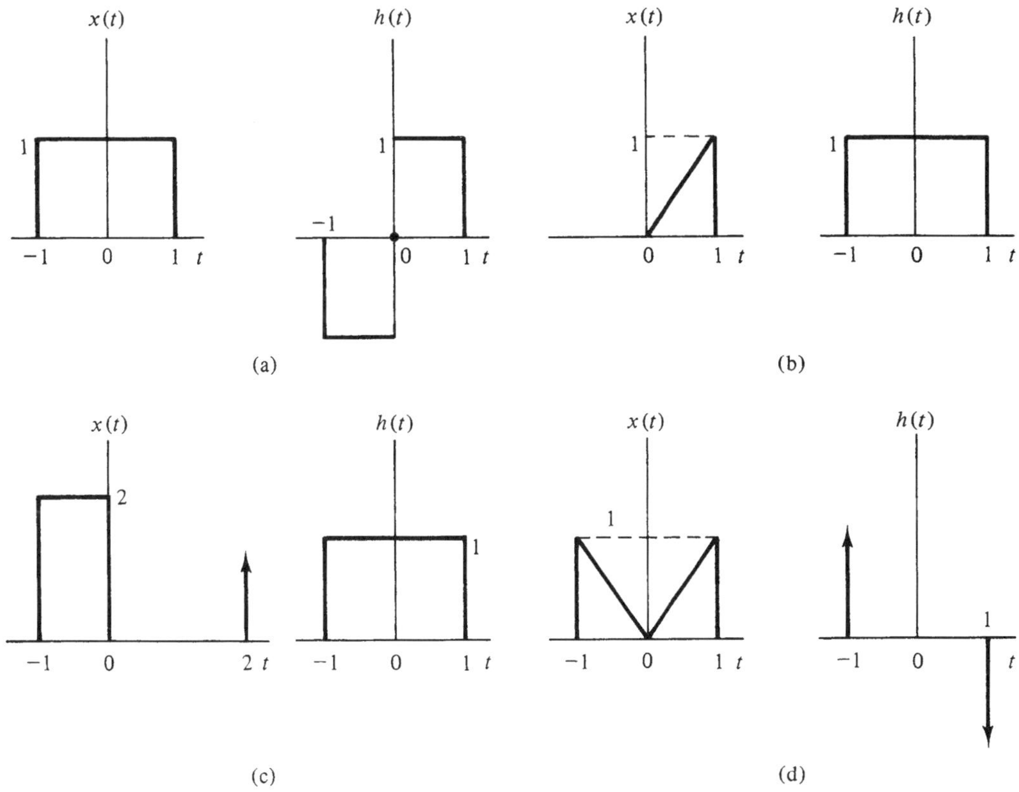

Figure P2.4

2.5. Use the convolution integral to find the response $y(t)$ of an LTI system with impulse response $h(t)$ to input $x(t)$:

(a) $x(t) = \exp[-t]u(t)$ $\qquad h(t) = \exp[-2t]u(t)$

(b) $x(t) = t\exp[-t]u(t)$ $\qquad h(t) = u(t)$

(c) $x(t) = \exp[-t]u(t) + u(t)$ $\qquad h(t) = u(t)$

(d) $x(t) = u(t)$ $\qquad h(t) = \exp[-2t]u(t) + \delta(t)$

(e) $x(t) = \exp[-at]u(t)$ $\qquad h(t) = u(t) - \exp[-at]u(t-b)$

(f) $x(t) = \delta(t-1) + \exp[-t]u(t)$ $\qquad h(t) = \exp[-2t]u(t)$

2.6. The cross correlation of two different signals is defined as

$$R_{xy}(t) = \int_{-\infty}^{\infty} x(\tau)\,y(\tau - t)\,d\tau = \int_{-\infty}^{\infty} x(\tau + t)\,y(\tau)\,d\tau$$

(a) Show that

$$R_{xy}(t) = x(t) * y(-t)$$

(b) Show that the cross correlation does not obey the commutative law.

(c) Show that $R_{xy}(t)$ is symmetric $(R_{xy}(t) = R_{yx}(-t))$.

2.7. Find the cross correlation between a signal $x(t)$ and the signal $y(t) = x(t-1) + n(t)$ for $B/A = 0$, 0.1, and 1, where $x(t)$ and $n(t)$ are as shown in Figure P2.7.

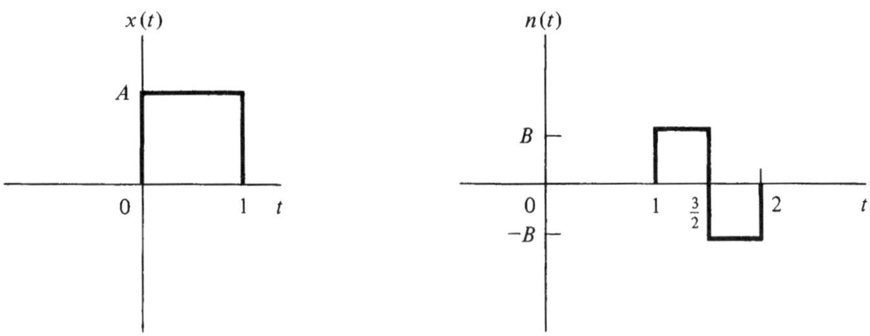

Figure P2.7

2.8. The autocorrelation is a special case of cross correlation with $y(t) = x(t)$. In this case,

$$R_x(t) = R_{xx}(t) = \int_{-\infty}^{\infty} x(\tau)x(\tau + t)\,d\tau$$

(a) Show that

$$R_x(0) = E, \qquad \text{the energy of } x(t)$$

(b) Show that

$$R_x(t) \le R_x(0) \qquad \text{(use the Schwarz inequality)}$$

(c) Show that the autocorrelation of $z(t) = x(t) + y(t)$ is

$$R_z(t) = R_x(t) + R_y(t) + R_{xy}(t) + R_{yx}(t)$$

2.9. Consider an LTI system whose impulse response is $h(t)$. Let $x(t)$ and $y(t)$ be the input and output of the system, respectively. Show that

$$R_y(t) = R_x(t) * h(t) * h(-t)$$

2.10. The input to an LTI system with impulse response $h(t)$ is the complex exponential $\exp[j\omega t]$. Show that the corresponding output is

$$y(t) = \exp[j\omega t]\, H(\omega)$$

where

$$H(\omega) = \int_{-\infty}^{\infty} h(t) \exp[-j\omega t]\,dt$$

2.11. Determine whether the continuous-time LTI systems characterized by the following impulse responses are causal or noncausal, and stable or unstable. Justify your answers.

(a) $h(t) = \exp[-3t] \sin(t) u(t)$

(b) $h(t) = \exp[4t] u(-t)$

(c) $h(t) = (-t) \exp[-t] u(-t)$

(d) $h(t) = \exp[-|2t|]$

(e) $h(t) = |(t-2)| \exp[-|2t|]$

(f) $h(t) = \text{rect}[t/2]$

(g) $h(t) = \delta(t) + \exp[-3t] u(t)$

(h) $h(t) = \delta'(t) + \exp[-2t]$

(i) $h(t) = \delta'(t) + \exp[-|2t|]$

(j) $h(t) = (1 - t)\,\text{rect}(t/3)$

2.12. For each of the following impulse responses, determine whether it is invertible. For those that are, find the inverse system.

(a) $h(t) = \delta(t + 2)$

(b) $h(t) = u(t)$

(c) $h(t) = \delta(t - 3)$

(d) $h(t) = \text{rect}(t/4)$

(e) $h(t) = \exp[-t] u(t)$

2.13. Consider the two systems shown in Figures P2.13(a) and P2.13(b). System I operates on $x(t)$ to give an output $y_1(t)$ that is optimum according to some desired criterion. System II first operates on $x(t)$ with an invertible operation (subsystem I) to obtain $z(t)$ and then operates on $z(t)$ to obtain an output $y_2(t)$ by an operation that is optimum according to the same criterion as in system I.

(a) Can system II perform better than system I? (Remember the assumption that system I is the optimum operation on $x(t)$.)

(b) Replace the optimum operation on $z(t)$ by two subsystems, as shown in Figure P2.13(c). Now the overall system works as well as system I. Can the new system be better than system II? (Remember that system II performs the optimum operation on $z(t)$.)

(c) What do you conclude from parts (a) and (b)?

(d) Does the system have to be linear for part (c) to be true?

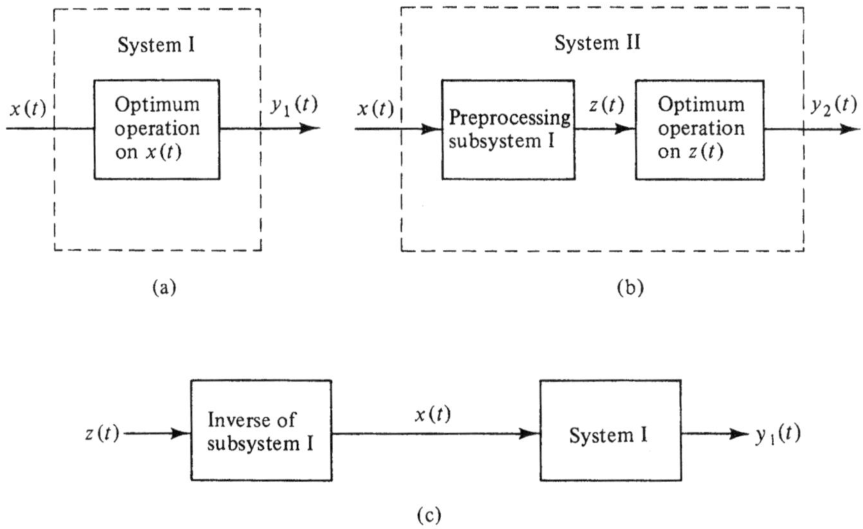

Figure P2.13

2.14. Determine whether the system in Figure P2.14 is BIBO stable.

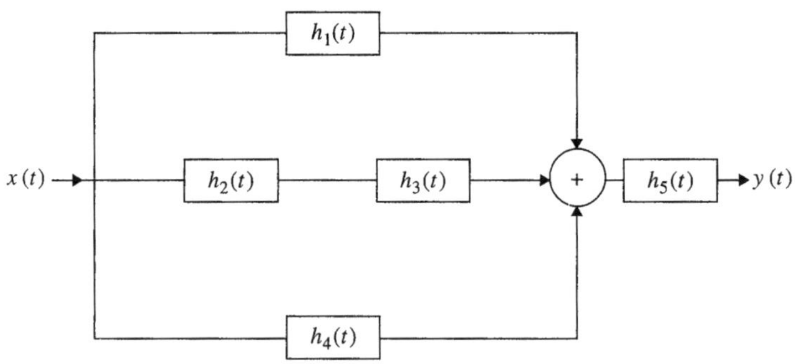

Figure P2.14

$$h_1(t) = \exp[-2t]u(t)$$
$$h_2(t) = \exp[-2t]u(t)$$
$$h_3(t) = \exp[-t]u(t)$$
$$h_4(t) = \delta(t)$$
$$h_5(t) = \exp[-3t]u(t)$$

2.15. The input $x(t)$ and output $y(t)$ of a linear, time-invariant system are as shown in Figure P2.15. Sketch the responses to the following inputs:

(a) $x(t + 2)$

(b) $2x(t) + 3x(-t)$

(c) $x(t - 1/2) - x(t + 1/2)$

(d) $\dfrac{dx(t)}{dt}$

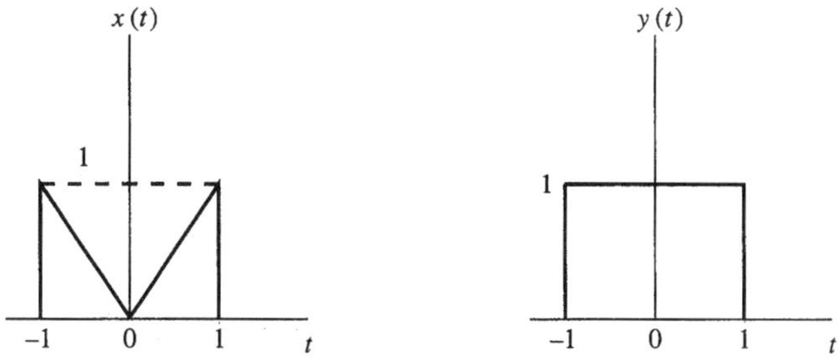

Figure P2.15

2.16. Find the impulse response of the initially relaxed system shown in Figure P2.16.

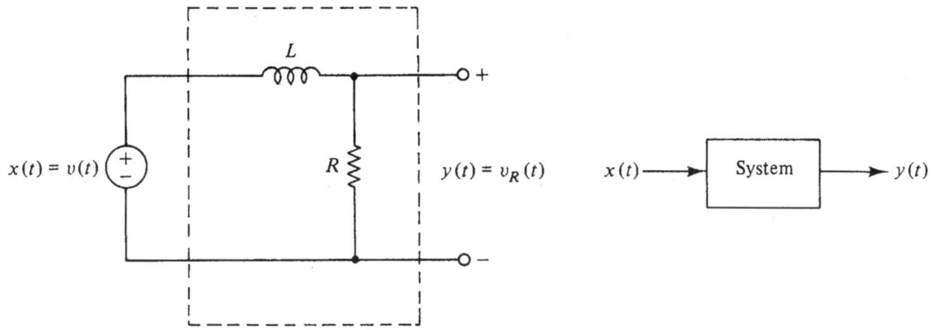

Figure P2.16

2.17. Find the impulse response of the initially relaxed system shown in Figure P2.17. Use this result to find the output of the system when the input is

(a) $u\left(t - \dfrac{\theta}{2}\right)$

(b) $u\left(t + \dfrac{\theta}{2}\right)$

(c) $\text{rect}\,(t/\theta)$, where $\theta = 1/RC$

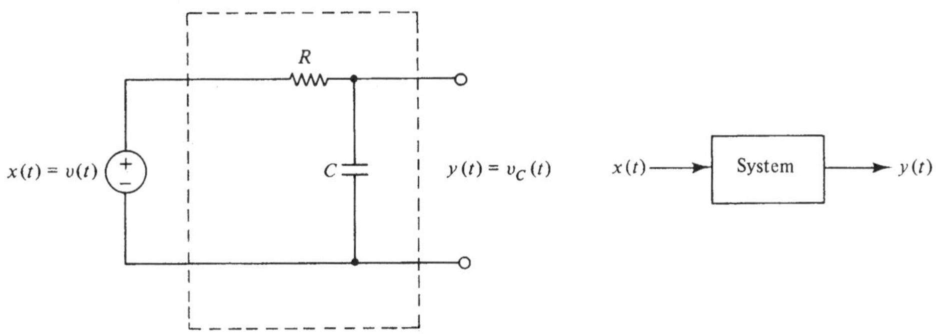

Figure P2.17

2.18. Repeat Problem 2.17 for the circuit shown in Figure P2.18.

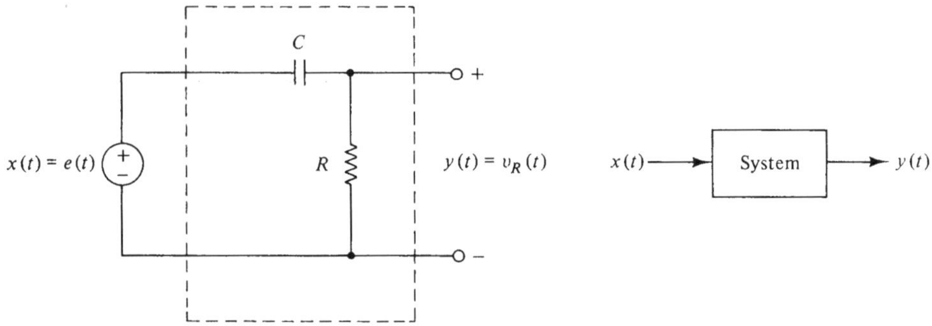

Figure P2.18

2.19. Show that any system that can be described by a differential equation of the form

$$\frac{d^N y(t)}{dt^N} + \sum_{k=0}^{N-1} a_k(t) \frac{d^k y(t)}{dt^k} = \sum_{k=0}^{M} b_k(t) \frac{d^k x(t)}{dt^k}$$

is linear. (Assume zero initial conditions.)

2.20. Show that any system that can be described by the differential equation in Problem 2.19 is time invariant. Assume that all the coefficients are constants.

2.21. A vehicle of mass M is traveling on a paved surface with coefficient of friction k. Assume that the position of the car at time t, relative to some reference, is $y(t)$ and the driving force applied to the vehicle is $x(t)$. Use Newton's second law of motion to write the differential equation describing the system. Show that the system is an LTI system. Can this system be time varying?

2.22. Consider a pendulum of length l and mass M as shown in Figure P2.22. The displacement from the equilibrium position is $l\theta$; hence, the acceleration is $l\theta''$. The input $x(t)$ is the force applied to the mass M tangential to the direction of motion of the mass. The restoring force is the tangential component $Mg \sin \theta$. Neglect the mass of the rod and the air resistance. Use Newton's second law of motion to write the differential equation describing the system. Is this system linear? As an approximation, assume that θ is small enough that $\sin \theta = \theta$. Is the system now linear?

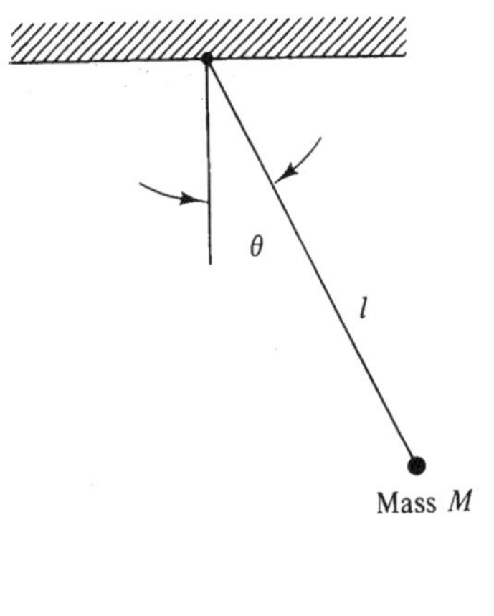

Figure P2.22

2.23. For the system realized by the interconnection shown in Figure P2.23, find the differential equation relating the input $x(t)$ to the output $y(t)$.

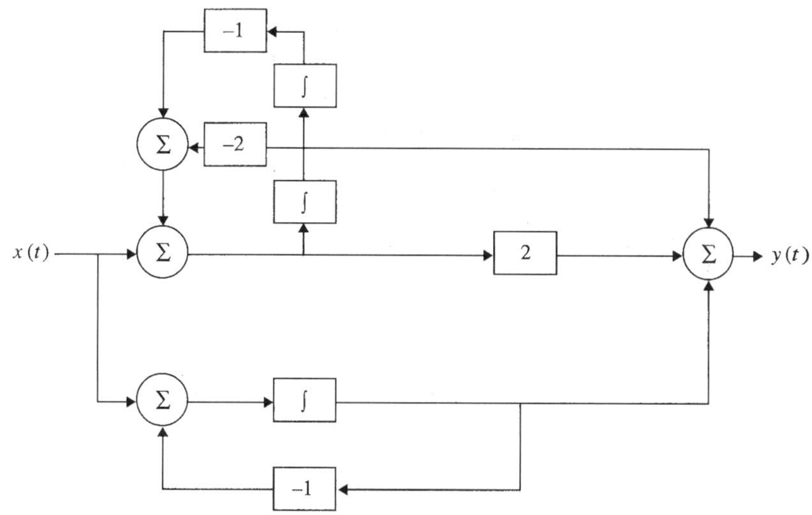

Figure P2.23

2.24. For the system simulated by the diagram shown in Figure P2.24, determine the differential equation describing the system.

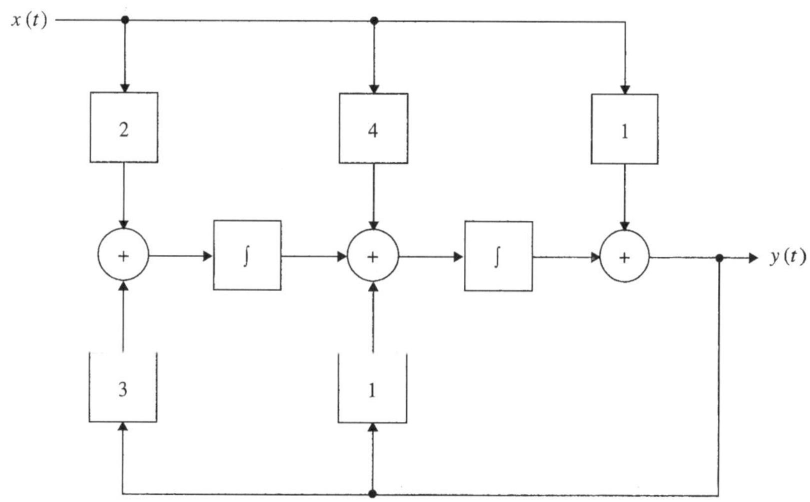

Figure P2.24

2.25. Consider the series RLC circuit shown in Figure P2.25.

(a) Derive the second-order differential equation that describes the system.

(b) Determine the first- and second-canonical-form simulation diagrams.

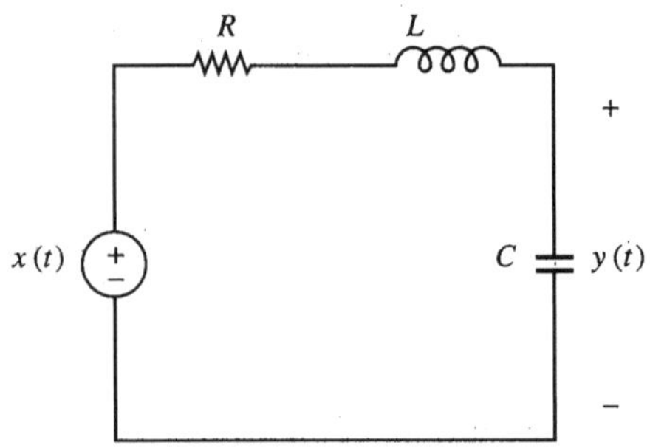

Figure P2.25

2.26. Given an LTI system described by

$$y'''(t) + 3y''(t) - y'(t) - 2y(t) = 3x''(t) - x(t)$$

Find the first- and second-canonical-form simulation diagrams.

2.27. Find the impulse response of the initially relaxed system shown in Figure P2.27.

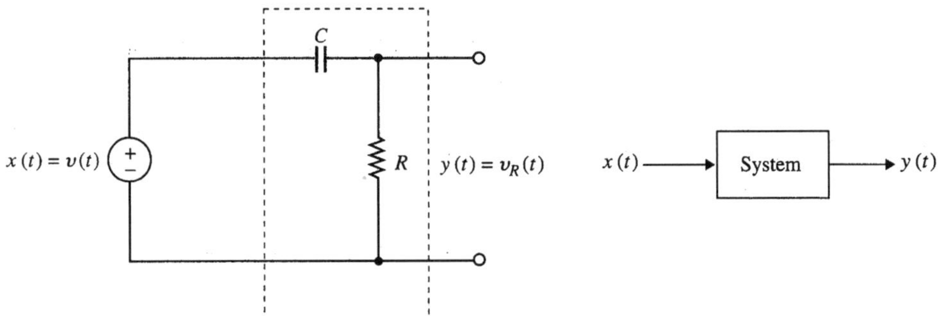

Figure P2.27

2.28. Find the state equations in the first and second canonical forms for the system described by the differential equation

$$y''(t) + 2.5y'(t) + y(t) = x'(t) + x(t)$$

2.29. For the circuit shown in Figure P2.29, choose the inductor current and the capacitor voltage as state variables, and write the state equations.

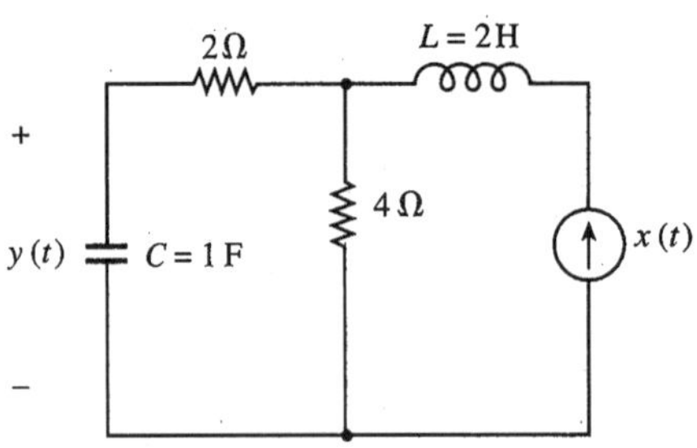

Figure P2.29

2.30. Repeat Problem 2.28 for the system described by the differential equation

$$y'''(t) + y''(t) - 2y(t) = x'(t) - 2x(t)$$

2.31. Calculate $\exp[\mathbf{A}t]$ for the following matrices. Use both the series-expansion and Cayley-Hamilton methods.

(a) $\mathbf{A} = \begin{bmatrix} -1 & 0 & 0 \\ 0 & -2 & 0 \\ 0 & 0 & -3 \end{bmatrix}$

(b) $\mathbf{A} = \begin{bmatrix} -1 & 2 & -1 \\ 0 & -1 & 0 \\ 0 & 0 & -1 \end{bmatrix}$

(c) $\mathbf{A} = \begin{bmatrix} -1 & 1 & -1 \\ 0 & 1 & -4 \\ 0 & 1 & -3 \end{bmatrix}$

2.32. Using state-variable techniques, find the impulse response for the system described by the differential equation

$$y''(t) + 6y'(t) + 8y(t) = x'(t) + x(t)$$

Assume that the system is initially relaxed, i.e., $y'(0) = 0$ and $y''(0) = 0$.

2.33. Use state-variable techniques to find the impulse response of the system described by

$$y''(t) + 7y'(t) + 12y(t) = x''(t) - 3x'(t) + 4x(t)$$

Assume that the system is initially relaxed, i.e., $y'(0) = 0$ and $y(0) = 0$.

2.34. Consider the system described by

$$\mathbf{v}'(t) = \mathbf{A}\,\mathbf{v}(t) + \mathbf{b}x(t)$$
$$y(t) = \mathbf{c}\,\mathbf{v}(t) + dx(t)$$

Select the change of variable given by

$$\mathbf{z}(t) = \mathbf{P}\,\mathbf{v}(t)$$

where $\mathbf{P}$ is a square matrix with inverse $\mathbf{P}^{-1}$. Show that the new state equations are

$$\mathbf{z}'(t) = \mathbf{A}_1\,\mathbf{z}(t) + \mathbf{b}_1\,x(t)$$
$$y(t) = \mathbf{c}_1\,\mathbf{z}(t) + d_1\,x(t)$$

where

$$\mathbf{A}_1 = \mathbf{PAP}^{-1}$$
$$\mathbf{b}_1 = \mathbf{Pb}$$
$$\mathbf{c}_1 = \mathbf{cP}^{-1}$$
$$d_1 = d$$

2.35. Consider the system described by the differential equation

$$y''(t) + 3y'(t) + 2y(t) = x'(t) - x(t)$$

(a) Write the state equations in the first canonical form.

(b) Write the state equations in the second canonical form.

(c) Use Problem 2.34 to find the matrix $\mathbf{P}$ which will transform the first canonical form into the second.

(d) Find the state equations if we transform the second canonical form using the matrix

$$\mathbf{P} = \begin{bmatrix} 2 & 1 \\ -1 & -1 \end{bmatrix}$$

Chapter 3

Fourier Series

3.1 INTRODUCTION

As we have seen in the previous chapter, we can obtain the response of a linear system to an arbitrary input by representing it in terms of basic signals. The specific signals used were the shifted δ-functions. Often, it is convenient to choose a set of orthogonal waveforms as the basic signals. There are several reasons for doing this. First, it is mathematically convenient to represent an arbitrary signal as a weighted sum of orthogonal waveforms, since many of the calculations involving signals are simplified by using such a representation. Second, it is possible to visualize the signal as a vector in an orthogonal coordinate system, with the orthogonal waveforms being coordinates. Finally, representation in terms of orthogonal basis functions provides a convenient means of solving for the response of linear systems to arbitrary inputs. In this chapter, we will consider the representation of an arbitrary signal over a finite interval in terms of some set of orthogonal basis functions.

For periodic signals, a convenient choice for an orthogonal basis is the set of harmonically related complex exponentials. The choice of these waveforms is appropriate, since such complex exponentials are periodic, are relatively easy to manipulate mathematically, and yield results that have a meaningful physical interpretation. The representation of a periodic signal in terms of complex exponentials, or equivalently, in terms of sine and cosine waveforms, leads to the Fourier series that are used extensively in all fields of science and engineering. The Fourier series is named after the French physicist Jean Baptiste Fourier (1768–1830), who was the first to suggest that periodic signals could be represented by a sum of sinusoids.

So far, we have only considered time-domain descriptions of continuous-time signals and systems. In this chapter, we introduce the concept of frequency-domain representations. We learn how to decompose periodic signals into their frequency components. The results can be extended to aperiodic signals, as will be shown in Chapter 4.

Periodic signals occur in a wide range of physical phenomena. A few examples of such signals are acoustic and electromagnetic waves of most types, the vertical displacement of a mechanical pendulum, the periodic vibrations of musical instruments, and the beautiful patterns of crystal structures.

In the present chapter, we discuss basic concepts, facts, and techniques in connection with Fourier series. Illustrative examples and some important engineering applications are included. We begin by considering orthogonal basis functions in Section 3.2. In Section 3.3, we consider periodic signals and develop procedures for resolving such signals into a linear combination of complex exponential functions. In Section 3.4, we discuss the sufficient conditions for a periodic signal to be represented in terms of a Fourier series. These conditions are known as the Dirichlet conditions. Fortunately, all the periodic signals that we deal with in practice obey these conditions. As with any other mathematical tool, Fourier series possess several useful properties. These properties are developed in Section 3.5. Understanding such properties helps us move easily from the time domain to the frequency domain and vice versa. In Section 3.6, we use the properties of the Fourier series to find the response of LTI systems to periodic signals. The effects of truncating the Fourier series and the Gibbs phenomenon are discussed in Section 3.7. We will see that whenever we attempt to reconstruct a discontinuous signal from its Fourier series, we encounter a strange behavior in the form of signal overshoot at the discontinuities. This overshoot effect does not go away even when we increase the number of terms used in reconstructing the signal.

3.2 ORTHOGONAL REPRESENTATIONS OF SIGNALS

Orthogonal representations of signals are of general importance in solving many engineering problems. Two of the reasons this is so are that it is mathematically convenient to represent arbitrary signals as a weighted sum of orthogonal waveforms, since many of the calculations involving signals are simplified by using such a representation and that it is possible to visualize the signal as a vector in an orthogonal coordinate system, with the orthogonal waveforms being the unit coordinates.

A set of signals ϕ_i, $i = 0, \pm 1, \pm 2, \ldots$, is said to be orthogonal over an interval (a, b) if

$$\int_a^b \phi_l(t)\,\phi_k^* dt = \begin{cases} E_k, & l = k \\ 0, & l \neq k \end{cases}$$
$$= E_k \delta(l - k) \tag{3.2.1}$$

where $\phi_k^*(t)$ stands for the complex conjugate of the signal and $\delta(l - k)$, called the Kronecker delta function, is defined as

$$\delta(l - k) = \begin{cases} 1, & l = k \\ 0, & l \neq k \end{cases} \tag{3.2.2}$$

If $\phi_i(t)$ corresponds to a voltage or a current waveform associated with a 1-ohm resistive load, then, from Equation (1.4.1), E_k is the energy dissipated in the load in $b - a$ seconds due to signal $\phi_k(t)$. If the constants E_k are all equal to 1, the $\phi_i(t)$ are said to be orthonormal signals. Normalizing any set of signals $\phi_i(t)$ is achieved by dividing each signal by $\sqrt{E_i}$.

Example 3.2.1

The signals $\phi_m(t) = \sin mt, m = 1, 2, 3, \ldots$, form an orthogonal set on the interval $-\pi < t < \pi$ because

$$\int_{-\pi}^{\pi} \phi_m(t)\phi_n^*(t)\,dt = \int_{-\pi}^{\pi} (\sin mt)(\sin nt)\,dt$$

$$= \frac{1}{2}\int_{-\pi}^{\pi} \cos(m-n)t\,dt - \frac{1}{2}\int_{-\pi}^{\pi} \cos(m+n)t\,dt$$

$$= \begin{cases} \pi, & m = n \\ 0, & m \neq n \end{cases}$$

Since the energy in each signal equals π, the following set of signals constitutes an orthonormal set over the interval $-\pi < t < \pi$:

$$\frac{\sin t}{\sqrt{\pi}}, \quad \frac{\sin 2t}{\sqrt{\pi}}, \quad \frac{\sin 3t}{\sqrt{\pi}}, \ldots$$

Example 3.2.2

The signals $\phi_k(t) = \exp[j(2\pi kt)/T]$, $k = 0, \pm 1, \pm 2, \ldots$, form an orthogonal set on the interval $(0, T)$ because

$$\int_0^T \phi_l(t)\phi_k^*(t)\,dt = \int_0^T \exp\left[\frac{j(2\pi lt)}{T}\right]\exp\left[\frac{-j(2\pi kt)}{T}\right]dt$$

$$= \begin{cases} T, & l = k \\ 0, & l \neq k \end{cases}$$

and hence, the signals $(1/\sqrt{T})\exp[(j2\pi kt)/T]$ constitute an orthonormal set over the interval $0 < t < T$.

Example 3.2.3

The three signals shown in Figure 3.2.1 are orthonormal, since they are mutually orthogonal and each has unit energy.

Orthonormal sets are useful in that they lead to a series representation of signals in a relatively simple fashion. Let $\phi_i(t)$ be an orthonormal set of signals on an interval $a < t < b$, and let $x(t)$ be a given signal with finite energy over the same interval. We can represent $x(t)$ in terms of $\{\phi_i\}$ by a convergent series as

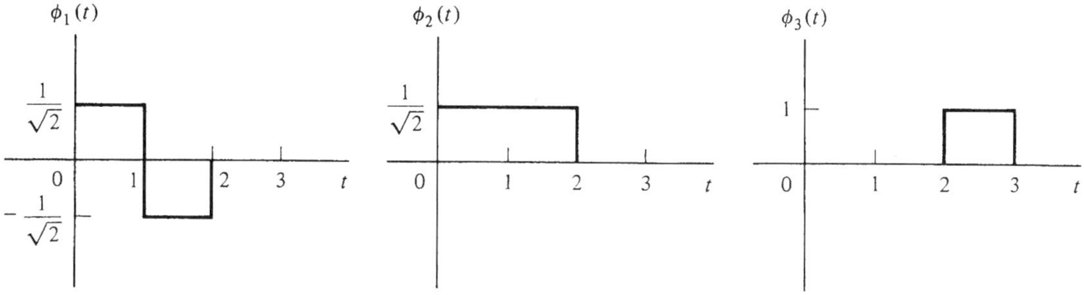

Figure 3.2.1 Three orthonormal signals.

$$x(t) = \sum_{i=-\infty}^{\infty} c_i \phi_i(t) \tag{3.2.3}$$

where

$$c_k = \int_a^b x(t)\phi_k^*(t)\,dt, \quad k = 0, \pm 1, \pm 2, \ldots \tag{3.2.4}$$

Equation (3.2.4) follows by multiplying Equation (3.2.3) by $\phi_k^*(t)$ and integrating the result over the range of definition of $x(t)$. Note that the coefficients can be computed independently of each other. If the set $\phi_i(t)$ is only an orthogonal set, then Equation (3.2.4) takes the form (see Problem 3.5)

$$c_i = \frac{1}{E_i}\int_a^b x(t)\phi_i^*(t)\,dt \tag{3.2.5}$$

The series representation of Equation (3.2.3) is called a generalized Fourier series of $x(t)$, and the constants c_i, $i = 0, \pm 1, \pm 2, \ldots$, are called the Fourier coefficients with respect to the orthogonal set $\{\phi_i(t)\}$.

In general, the representation of an arbitrary signal in a series expansion of the form of Equation (3.2.3) requires that the sum on the right side be an infinite sum. In practice, however, we can use only a finite number of terms on the right side. When we truncate the infinite sum on the right to a finite number of terms, we get an approximation $\hat{x}(t)$ to the original signal $x(t)$. When we use only M terms, the representation error is

$$e_M(t) = x(t) - \sum_{i=1}^{M} c_i \phi_i(t) \tag{3.2.6}$$

The energy in this error is

$$E_e(M) = \int_a^b |e_M(t)|^2 dt = \int_a^b \Big|x(t) - \sum_{i=1}^{M} c_i \phi_i(t)\Big|^2 dt \tag{3.2.7}$$

It can be shown that for any M, the choice of c_k according to Equation (3.2.4) minimizes the energy in the error e_M. (See Problem 3.4.)

Certain classes of signals—finite-length digital communication signals, for example—permit expansion in terms of a finite number of orthogonal functions $\{\phi_i(t)\}$. In this case, $i = 1, 2, \ldots, N$, where N is the dimension of the set of signals. The series representation is then reduced to

$$x(t) = \mathbf{x}^T \boldsymbol{\phi}(t) \tag{3.2.8}$$

where the vectors $\mathbf{x}$ and $\boldsymbol{\phi}(t)$ are defined as

$$\begin{aligned} \mathbf{x} &= [c_1, c_2, \ldots c_N]^T \\ \boldsymbol{\phi}(t) &= [\phi_1(t), \phi_2(t), \ldots \phi_N(t)]^T \end{aligned} \tag{3.2.9}$$

and the superscript T denotes vector transposition. The normalized energy of $x(t)$ over the interval $a < t < b$ is

$$\begin{aligned} E_x &= \int_a^b |x(t)|^2 dt = \int_a^b |\sum_{i=1}^{N} c_i \phi_i(t)|^2 dt \\ &= \sum_{k=1}^{N} \sum_{n=1}^{N} c_k c_i^* \int_a^b \phi_k(t) \phi_i^*(t)\, dt \\ &= \sum_{i=1}^{N} |c_i|^2 E_i \end{aligned} \tag{3.2.10}$$

This result relates the energy of the signal $x(t)$ to the sum of the squares of the orthogonal series coefficients, modified by the energy in each coordinate, E_i. If orthonormal signals are used, we have $E_i = 1$, and Equation (3.2.10) reduces to

$$E_x = \sum_{i=1}^{N} |c_i|^2$$

In terms of the coefficient vector $\mathbf{x}$, this can be written as

$$E_x = (\mathbf{x}^*)^T \mathbf{x} = \mathbf{x}^\dagger\, \mathbf{x} \tag{3.2.11}$$

where $\dagger$ denotes the complex conjugate transpose $[(\ )^*]^T$. This is a special case of what is known as Parseval's theorem, which we discuss in more detail in Section 3.5.6.

Example 3.2.4

In this example, we examine the representation of a finite-duration signal in terms of an orthogonal set of basis signals. Consider the four signals defined over the interval (0, 3), as shown in Figure 3.2.2(a). These signals are not orthogonal, but it is possible to represent them in terms of the three orthogonal signals shown in Figure 3.2.2(b), since combinations of these three basis signals can be used to represent any of the four signals in Figure 3.2.2(a).

The coefficients that represent the signal $x_1(t)$, obtained by using Equation (3.2.4), are

$$c_{11} = \int_0^3 x_1(t)\phi_1^*(t)\, dt = 2$$

$$c_{12} = \int_0^3 x_1(t)\phi_2^*(t)\, dt = 0$$

$$c_{13} = \int_0^3 x_1(t)\phi_3^*(t)\, dt = 1$$

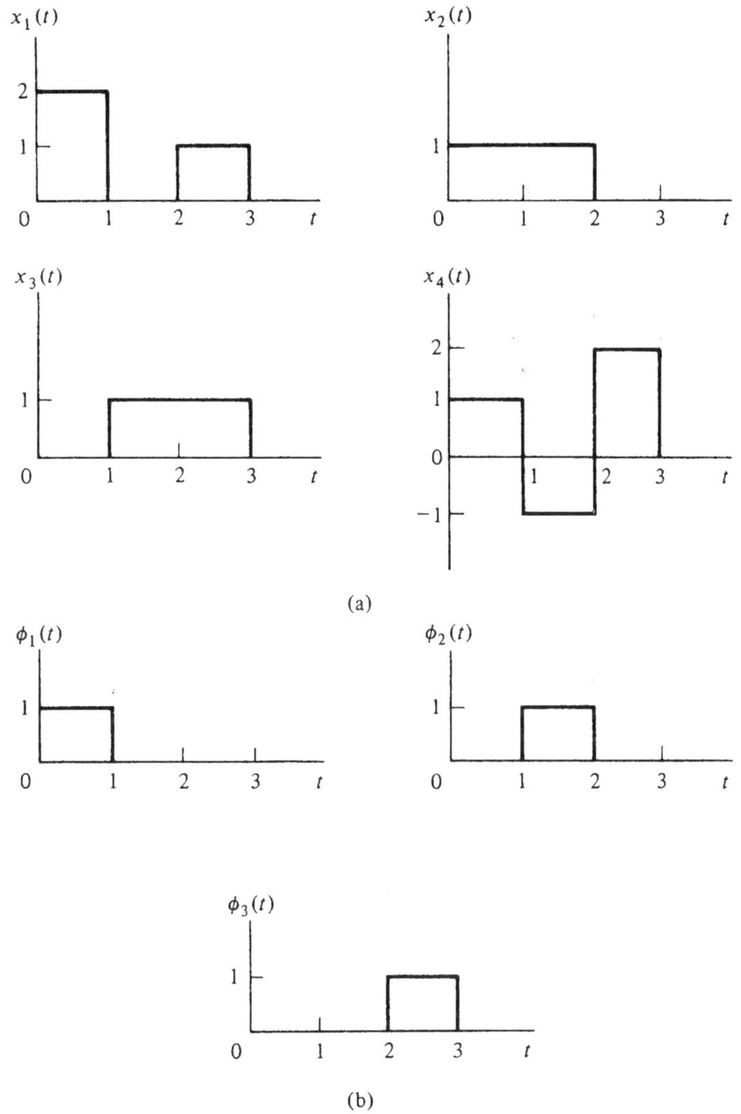

Figure 3.2.2 Orthogonal representations of digital signals.

In vector notation, $\mathbf{x}_1 = [2,\ 0,\ 1]^T$. Similarly, we can calculate the coefficients for $x_2(t)$, $x_3(t)$, and $x_4(t)$, and these become

$$x_{21} = 1, \quad x_{22} = 1, \quad x_{23} = 0, \quad \text{or} \quad \mathbf{x}_2 = [1, 1, 0]^T$$
$$x_{31} = 0, \quad x_{32} = 1, \quad x_{33} = 1, \quad \text{or} \quad \mathbf{x}_3 = [0, 1, 1]^T$$
$$x_{41} = 1, \quad x_{42} = -1, \quad x_{43} = 2, \quad \text{or} \quad \mathbf{x}_3 = [1, -1, 2]^T$$

Since there are only three basis signals required to completely represent $x_i(t)$, $i = 1, 2, 3, 4$, we now can think of these four signals as vectors in three-dimensional space. We would like to emphasize that the choice of the basis is not unique, and many other possibilities exist. For example, if we choose

$$\phi_1(t) = \frac{1}{\sqrt{2}} x_2(t), \quad \phi_2(t) = \frac{1}{\sqrt{3}} [2u(t-1) - u(t) - u(t-3)]$$

and

$$\phi_3(t) = \frac{1}{\sqrt{6}} x_4(t)$$

then

$$\mathbf{x}_1 = \left[\sqrt{2}, \ -\frac{\sqrt{3}}{3}, 2\frac{\sqrt{6}}{3}\right]^T, \quad \mathbf{x}_2 = [\sqrt{2}, 0, 0]^T$$

$$\mathbf{x}_3 = \left[\frac{\sqrt{2}}{2}, 2\frac{\sqrt{3}}{3}, \frac{\sqrt{6}}{6}\right]^T, \quad \mathbf{x}_4 = [0, 0, \sqrt{6}]^T$$

In closing this section, we should emphasize that the results presented are general, and the main purpose of the section is to introduce the reader to a way of representing signals in terms of other bases in a formal way. In Chapter 4, we will see that if the signal satisfies some restrictions, then we can write it in terms of an orthonormal basis (interpolating signals), with the series coefficients being samples of the signal obtained at appropriate time intervals.

3.3 THE EXPONENTIAL FOURIER SERIES

Recall from Chapter 1 that a signal is periodic if, for some positive nonzero value of T,

$$x(t) = x(t + nT), \quad n = 1, 2, \ldots \tag{3.3.1}$$

The quantity T for which Equation (3.3.1) is satisfied is referred to as the fundamental period, whereas $2\pi/T$ is referred to as the fundamental radian frequency and is denoted by ω_0. The graph of a periodic signal is obtained by periodic repetition of its graph in any interval of length T, as shown in Figure 3.3.1. From Equation (3.3.1), it follows that $2T, 3T, \ldots$, are also periods of $x(t)$. As was demonstrated in Chapter 1, if two signals, $x_1(t)$ and $x_2(t)$, are periodic with period T, then

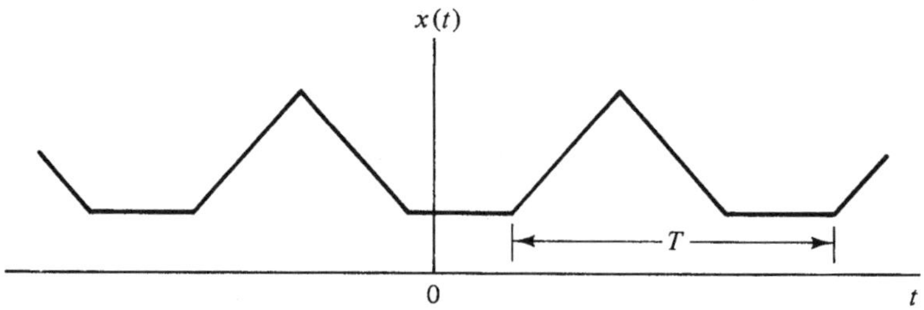

Figure 3.3.1 Periodic signal.

$$x_3(t) = ax_1(t) + bx_2(t) \tag{3.3.2}$$

is also periodic with period T.

Familiar examples of periodic signals are the sine, cosine, and complex exponential functions. Note that a constant signal $x(t) = c$ is also a periodic signal in the sense of the definition, because Equation (3.3.1) is satisfied for every positive T.

In this section, we consider the representation of periodic signals by an orthogonal set of basis functions. We saw in Section 3.2 that the set of complex exponentials $\phi_n(t) = \exp[j2n\pi t/T]$ forms an orthogonal set. If we select such a set as basis functions, then, according to Equation (3.2.3),

$$x(t) = \sum_{n=-\infty}^{\infty} c_n \exp\left[j\frac{2n\pi t}{T}\right] \tag{3.3.3}$$

where, from Equation (3.2.4), the c_n are complex constants and are given by

$$c_n = \frac{1}{T}\int_0^T x(t)\exp\left[-j\frac{2n\pi t}{T}\right]dt \tag{3.3.4}$$

Each term of the series has a period T and fundamental radian frequency $2\pi/T = \omega_0$. Hence, if the series converges, its sum is periodic with period T. Such a series is called the complex exponential Fourier series, and the c_n are called the Fourier coefficients. Note that because of the periodicity of the integrand, the interval of integration in Equation (3.3.4) can be replaced by any other interval of length T—for instance, by the interval $t_0 \le t \le t_0 + T$, where t_0 is arbitrary. We denote integration over an interval of length T by the symbol $\int_{\langle T\rangle}$. We observe that even though an infinite number of frequencies are used to synthesize the original signal in the Fourier-series expansion, they do not constitute a continuum; each frequency term is a multiple of $\omega_0/2\pi$. The frequency corresponding to $n = 1$ is called the fundamental, or first, harmonic; $n = 2$ corresponds to the second harmonic, and so on. The coefficients c_n define a complex-valued function of the discrete frequencies $n\omega_0$, where $n = 0, \pm 1, \pm 2, \ldots$. The dc component, or the full-cycle time average, of $x(t)$ is equal to c_0 and is obtained by setting $n = 0$ in Equation (3.3.4). Calculated values of c_0 can be checked by inspecting $x(t)$, a recommended practice to test the validity of the result obtained by integration. The plot of $|c_n|$ versus $n\omega_0$ displays the amplitudes of the various frequency components constituting $x(t)$. Such a plot is therefore called the amplitude, or magnitude spectrum, of the periodic signal $x(t)$. The locus of the tips of the magnitude lines is called the envelope of the magnitude spectrum. Similarly, the phase of the sinusoidal components making up $x(t)$ is equal to $\measuredangle\, c_n$ and the plot of $\measuredangle\, c_n$ versus $n\omega_0$ is called the phase spectrum of $x(t)$. In sum, the amplitude and phase spectra of any given periodic signal are defined in terms of the magnitude and phase of c_n. Since the spectra consist of a set of lines representing the magnitude and phase at $\omega = n\omega_0$, they are referred to as line spectra.

For real-valued (noncomplex) signals, the complex conjugate of c_n is

$$\begin{aligned} c_n^* &= \left[\frac{1}{T}\int_{\langle T\rangle} x(t)\exp\left[\frac{-j2n\pi t}{T}\right]dt\right]^* \\ &= \frac{1}{T}\int_{\langle T\rangle} x(t)\exp\left[\frac{-j2(-n)\pi t}{T}\right]dt \\ &= c_{-n} \end{aligned} \tag{3.3.5}$$

Hence,

$$|c_{-n}| = |c_n| \quad \text{and} \quad \measuredangle c_{-n} = -\measuredangle c_n \tag{3.3.6}$$

which means that the amplitude spectrum has even symmetry and the phase spectrum has odd symmetry. This property for real-valued signals allows us to regroup the exponential series into complex-conjugate pairs, except for c_0, as follows:

$$\begin{aligned} x(t) &= c_0 + \sum_{m=-\infty}^{-1} c_m \exp\left[\frac{j2m\pi t}{T}\right] + \sum_{m=1}^{\infty} c_m \exp\left[\frac{j2m\pi t}{T}\right] \\ &= c_0 + \sum_{n=1}^{\infty} c_{-n}\exp\left[\frac{-j2n\pi t}{T}\right] + \sum_{n=1}^{\infty} c_n \exp\left[\frac{j2n\pi t}{T}\right] \\ &= c_0 + \sum_{n=1}^{\infty}\left(c_{-n}\exp\left[\frac{-j2n\pi t}{T}\right] + c_n\exp\left[\frac{j2n\pi t}{T}\right]\right) \\ &= c_0 + \sum_{n=1}^{\infty} 2\,\mathrm{Re}\left\{c_n \exp\left[\frac{j2n\pi t}{T}\right]\right\} \\ &= c_0 + \sum_{n=1}^{\infty}\left(2\,\mathrm{Re}\{c_n\}\cos\frac{2n\pi t}{T} - 2\,\mathrm{Im}\{c_n\}\sin\frac{2n\pi t}{T}\right) \end{aligned} \tag{3.3.7}$$

Here, $\mathrm{Re}\{\cdot\}$ and $\mathrm{Im}\{\cdot\}$ denote the real and imaginary parts of the arguments, respectively. Equation (3.3.7) can be written as

$$x(t) = a_0 + \sum_{n=1}^{\infty}\left[a_n\cos\frac{2n\pi t}{T} + b_n\sin\frac{2n\pi t}{T}\right] \tag{3.3.8}$$

The expression for $x(t)$ in Equation (3.3.8) is called the trigonometric Fourier series for the periodic signal $x(t)$. The coefficients a_0, a_n, and b_n are given by

$$a_0 = c_0 = \frac{1}{T}\int_{\langle T\rangle} x(t)\,dt \tag{3.3.9a}$$

$$a_n = 2\,\mathrm{Re}\{c_n\} = \frac{2}{T}\int_{\langle T\rangle} x(t)\cos\frac{2n\pi t}{T}\,dt \tag{3.3.9b}$$

$$b_n = -2\,\mathrm{Im}\{c_n\} = \frac{2}{T}\int_{\langle T\rangle} x(t)\sin\frac{2n\pi t}{T}\,dt \tag{3.3.9c}$$

In terms of the magnitude and phase of c_n, the real-valued signal $x(t)$ can be expressed as

$$x(t) = c_0 + \sum_{n=1}^{\infty} 2|c_n| \cos\left(\frac{2n\pi t}{T} + \measuredangle c_n\right)$$
$$= c_0 + \sum_{n=1}^{\infty} A_n \cos\left(\frac{2n\pi t}{T} + \phi_n\right) \tag{3.3.10}$$

where

$$A_n = 2|c_n| \tag{3.3.11}$$

and

$$\phi_n = \measuredangle c_n \tag{3.3.12}$$

Equation (3.3.10) represents an alternative form of the Fourier series that is more compact and meaningful than Equation (3.3.8). Each term in the series represents an oscillator needed to generate the periodic signal $x(t)$.

A display of $|c_n|$ and $\measuredangle c_n$ versus n or $n\omega_0$ for both positive and negative values of n is called a two-sided amplitude spectrum. A display of A_n and ϕ_n versus positive n or $n\omega_0$ is called a one-sided spectrum. Two-sided spectra are encountered most often in theoretical treatments because of the convenient nature of the complex Fourier series. It must be emphasized that the existence of a line at a negative frequency does not imply that the signal is made of negative frequency components, since, for every component $c_n \exp[j2n\pi t/T]$, there is an associated one of the form $c_{-n} \exp[-j2n\pi t/T]$. These complex signals combine to create the real component $a_n \cos(2n\pi t/T) + b_n \sin(2n\pi t/T)$. Note that, from the definition of a definite integral, it follows that if $x(t)$ is continuous or even merely piecewise continuous (continuous except for finitely many jumps in the interval of integration), the Fourier coefficients exist, and we can compute them by the indicated integrals.

Let us illustrate the practical use of the previous equations by the following examples. We will see numerous other examples in subsequent sections.

Example 3.3.1

Suppose we want to find the line spectra for the periodic signal shown in Figure 3.3.2. The signal $x(t)$ has the analytic representation

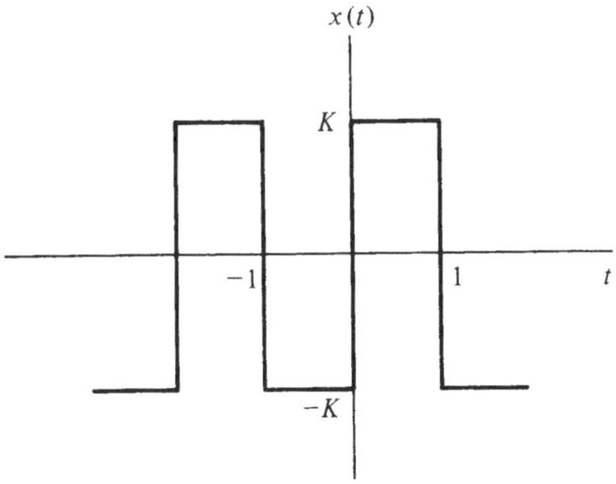

Figure 3.3.2 Signal $x(t)$ for Example 3.3.1.

$$x(t) = \begin{cases} -K, & -1 < t < 0 \\ K, & 0 < t < 1 \end{cases}$$

and $x(t+2) = x(t)$. Therefore, $\omega_0 = 2\pi/2 = \pi$. Signals of this type can occur as external forces acting on mechanical systems, as electromotive forces in electric circuits, etc. The Fourier coefficients are

$$\begin{aligned} c_n &= \frac{1}{2}\int_{-1}^{1} x(t)\exp[-jn\pi t]\,dt \\ &= \frac{1}{2}\left[\int_{-1}^{0} -K\exp[-jn\pi t]\,dt + \int_{0}^{1} K\exp[-jn\pi t]\,dt\right] \\ &= \frac{K}{2}\left(\frac{1-\exp[jn\pi]}{jn\pi} + \frac{\exp[-jn\pi]-1}{-jn\pi}\right) \\ &= \frac{K}{jn\pi}\left\{1 - \frac{1}{2}(\exp[jn\pi] + \exp[-jn\pi])\right\} \end{aligned} \tag{3.3.13}$$

$$= \begin{cases} \dfrac{2K}{jn\pi}, & n \text{ odd} \\ 0, & n \text{ even} \end{cases} \tag{3.3.14}$$

The amplitude spectrum is

$$|c_n| = \begin{cases} \dfrac{2K}{|n|\pi}, & n \text{ odd} \\ 0, & n \text{ even} \end{cases}$$

The dc component, or the average value of the periodic signal $x(t)$, is obtained by setting $n = 0$. When we substitute $n = 0$ into Equation (3.3.13), we obtain an undefined result. This can be circumvented by using l'Hôpital's rule, yielding $c_0 = 0$. This can be checked by noticing that $x(t)$ is an odd function and that the area under the curve represented by $x(t)$ over one period of the signal is zero.

The phase spectrum of $x(t)$ is given by

$$\measuredangle c_n = \begin{cases} -\dfrac{\pi}{2}, & n = (2m-1), m = 1, 2, \ldots \\ 0, & n = 2m, m = 0, 1, 2 \ldots \\ \dfrac{\pi}{2}, & n = -(2m-1), m = 1, 2, \ldots \end{cases}$$

The line spectra of $x(t)$ are displayed in Figure 3.3.3. Note that the amplitude spectrum has even symmetry, the phase spectrum odd symmetry.

Example 3.3.2

A sinusoidal voltage $E \sin\omega_0 t$ is passed through a half-wave rectifier that clips the negative portion of the waveform, as shown in Figure 3.3.4. Such signals may be encountered

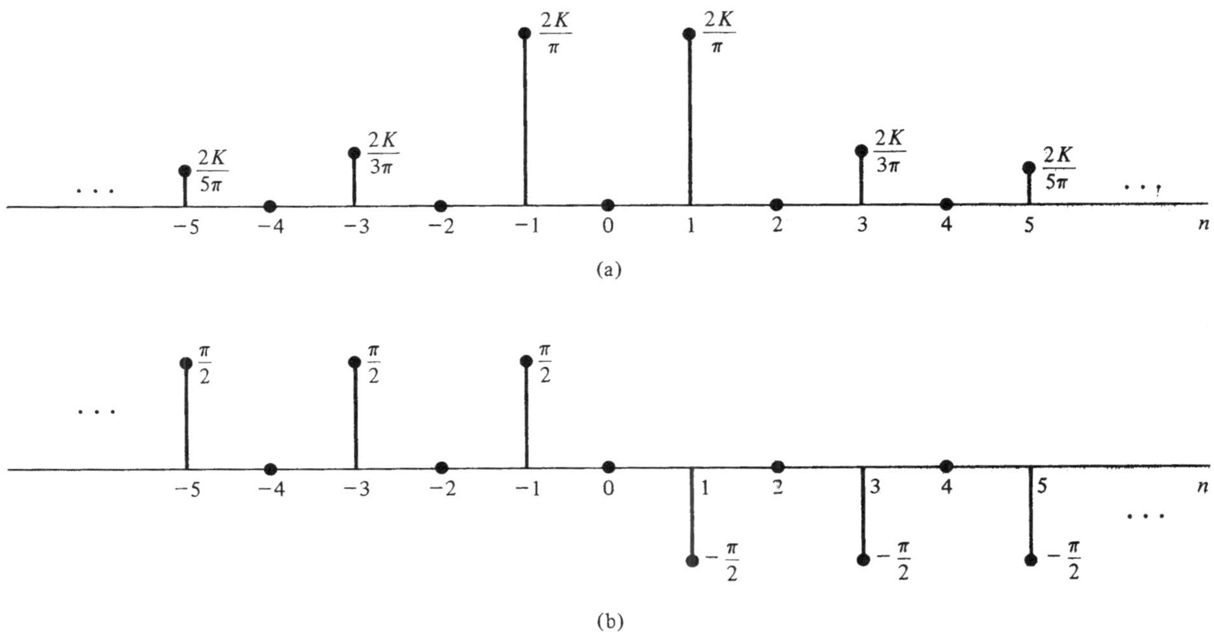

Figure 3.3.3 Line spectra for the periodic signal $x(t)$ of Example 3.3.1. (a) Magnitude spectrum and (b) phase spectrum.

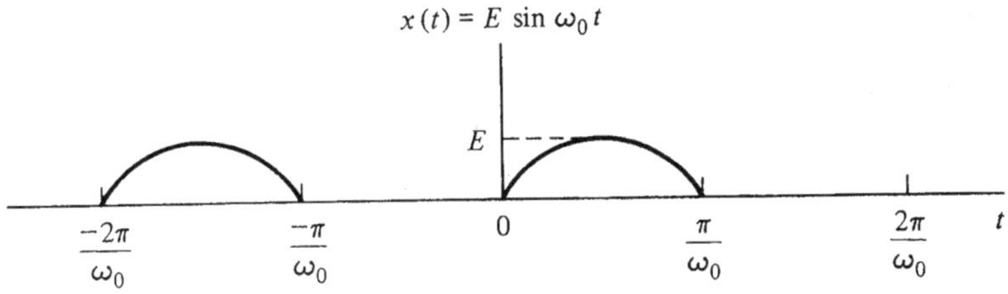

Figure 3.3.4 Signal $x(t)$ for Example 3.3.2.

in rectifier design problems. Rectifiers are circuits that produce direct current (dc) from alternating current (ac).

The analytic representation of $x(t)$ is

$$x(t) = \begin{cases} 0, & \text{when } \dfrac{-\pi}{\omega_0} < t < 0 \\ E \sin \omega_0 t, & \text{when } 0 < t < \dfrac{\pi}{\omega_0} \end{cases}$$

and $x(t + 2\pi/\omega_0) = x(t)$. Since $x(t) = 0$ when $-\pi/\omega_0 < t < 0$, we obtain, from Equation (3.3.4),

$$\begin{aligned} c_n &= \frac{1}{T}\int_0^T E \sin \omega_0 t \exp\left[-j\frac{2n\pi t}{T}\right] dt \\ &= \frac{E\omega_0}{2\pi}\int_0^{\pi/\omega_0} \frac{1}{2j}\left[\exp[j\omega_0 t] - \exp[-j\omega_0 t]\right] \exp[-jn\omega_0 t]\, dt \\ &= \frac{E\omega_0}{4\pi j}\int_0^{\pi/\omega_0} \left[\exp[-j\omega_0(n-1)t] - \exp[-j\omega_0(n+1)t]\right] dt \\ &= \frac{E\exp[-jn\pi/2]}{2\pi(1-n^2)}\left(\exp\left[\frac{-jn\pi}{2}\right] + \exp\left[\frac{jn\pi}{2}\right]\right) \\ &= \frac{E}{\pi(1-n^2)}\cos(n\pi/2)\exp[-jn\pi/2], \quad n \neq \pm 1 \end{aligned} \tag{3.3.15}$$

$$= \begin{cases} \dfrac{E}{\pi(1-n^2)}, & n \text{ even} \\ 0, & n \text{ odd}, \quad n \neq \pm 1 \end{cases} \tag{3.3.16}$$

Setting $n = 0$, we obtain the dc component, or the average value of the periodic signal, as $c_0 = E/\pi$. This result can be verified by calculating the area under one half cycle of a sine wave and dividing by T. To determine the coefficients c_1 and c_{-1} which correspond to the first harmonic, we note that we cannot substitute $n = \pm 1$ in Equation (3.3.15), since this yields an indeterminate quantity. So we use Equation (3.3.4) instead with $n = \pm 1$, which results in

$$c_1 = \frac{E}{4j} \quad \text{and} \quad c_{-1} = \frac{-E}{4j}$$

The line spectra of $x(t)$ are displayed in Figure 3.3.5.

In general, a rectifier is used to convert an ac signal to a dc signal. Ideally, rectified output $x(t)$ should consist only of a dc component. Any ac component contributes to the rip-

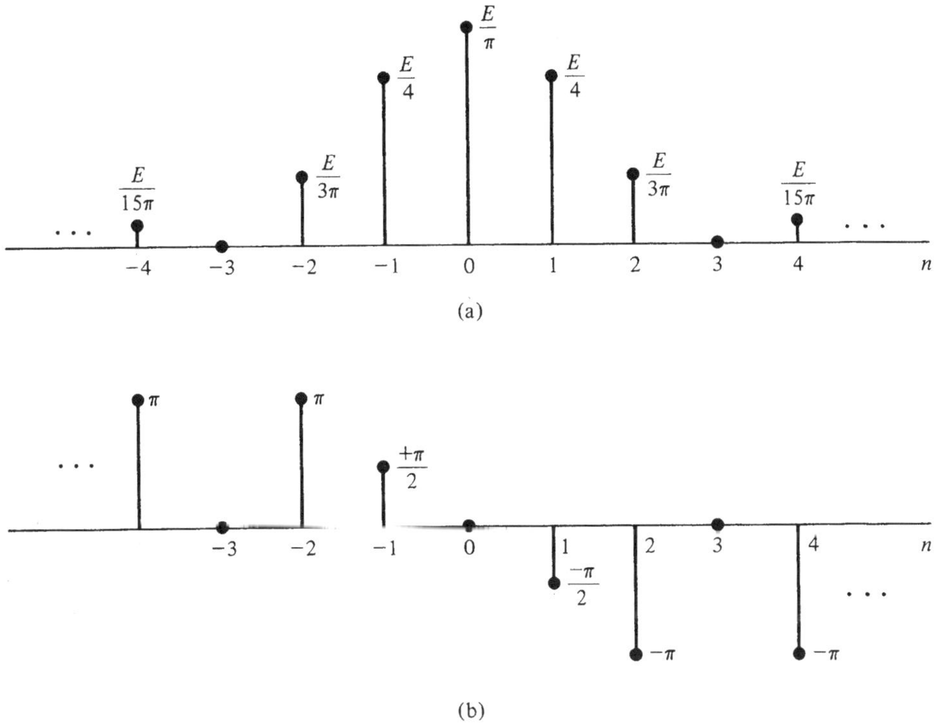

Figure 3.3.5 Line spectra for $x(t)$ of Example 3.3.2. (a) Magnitude spectrum and (b) phase spectrum.

ple (deviation from pure dc) in the signal. As can be seen from Figure 3.3.5, the amplitudes of the harmonics decrease rapidly as n increases, so that the main contribution to the ripple comes from the first harmonic. The ratio of the amplitudes of the first harmonic to the dc component can be used as a measure of the amount of ripple in the rectified signal. In this example, the ratio is equal to $\pi/4$. More complex circuits can be used that produce less ripple. (See Example 3.6.4).

Example 3.3.3

Consider the square-wave signal shown in Figure 3.3.6. The analytic representation of $x(t)$ is

$$x(t) = \begin{cases} 0, & \text{when } \dfrac{-T}{2} < t < \dfrac{-\tau}{2} \\ K, & \text{when } \dfrac{-\tau}{2} < t < \dfrac{\tau}{2} \\ 0, & \text{when } \dfrac{\tau}{2} < t < \dfrac{T}{2} \end{cases}$$

and $x(t + T) = x(t)$. Signals of this type can be produced by pulse generators and are used extensively in radar and sonar systems. From Equation (3.3.4), we obtain

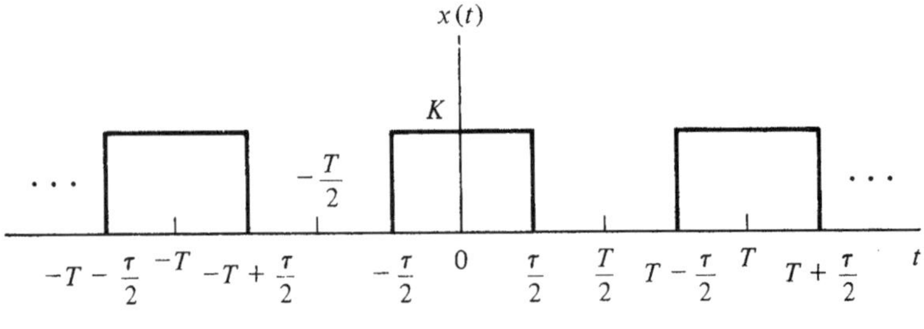

Figure 3.3.6 Signal $x(t)$ for Example 3.3.3.

$$
\begin{aligned}
c_n &= \frac{1}{T}\int_{-T/2}^{T/2} x(t)\exp\left[\frac{-j2n\pi t}{T}\right]dt = \frac{1}{T}\int_{-\tau/2}^{\tau/2} K\exp\left[\frac{-j2n\pi t}{T}\right]dt \\
&= \frac{K}{j2n\pi}\left[\exp\left[\frac{jn\pi\tau}{T}\right] - \exp\left[\frac{-jn\pi\tau}{T}\right]\right] \\
&= \frac{K}{n\pi}\sin\frac{n\pi\tau}{T} \\
&= \frac{K\tau}{T}\operatorname{sinc}\frac{n\tau}{T}
\end{aligned}
\tag{3.3.17}
$$

where $\operatorname{sinc}(\lambda) = \sin(\pi\lambda)/\pi\lambda$. The sinc function plays an important role in Fourier analysis and in the study of LTI systems. It has a maximum value at $\lambda = 0$ and approaches zero as λ approaches infinity, oscillating through positive and negative values. It goes through zero at $\lambda = \pm 1, \pm 2, \ldots$.

Let us investigate the effect of changing T on the frequency spectrum of $x(t)$. For fixed τ, increasing T reduces the amplitude of each harmonic as well as the fundamental frequency and, hence, the spacing between harmonics. However, the shape of the spectrum is dependent only on the shape of the pulse and does not change as T increases, except for the amplitude factor. A convenient measure of the frequency spread (known as the bandwidth) is the distance from the origin to the first zero-crossing of the sinc function. This distance is equal to $2\pi/\tau$ and is independent of T. Other measures of the frequency width of the spectrum are discussed in detail in Section 4.5.

We conclude that as the period increases, the amplitude becomes smaller and the spectrum becomes denser, whereas the shape of the spectrum remains the same and does not depend on the repetition period T. The amplitude spectra of $x(t)$ with $\tau = 1$ and $T = 5$, 10, and 15 are displayed in Figure 3.3.7.

Example 3.3.4

In this example, we show that $x(t) = t^2, -\pi < t < \pi$, with $x(t + 2\pi) = x(t)$ has the Fourier series representation

$$
x(t) = \frac{\pi^2}{3} - 4\left(\cos t - \frac{1}{4}\cos 2t + \frac{1}{9}\cos 3t - \ldots\right)
\tag{3.3.18}
$$

Note that $x(t)$ is periodic with period 2π and fundamental frequency $\omega_0 = 1$. The complex Fourier series coefficients are

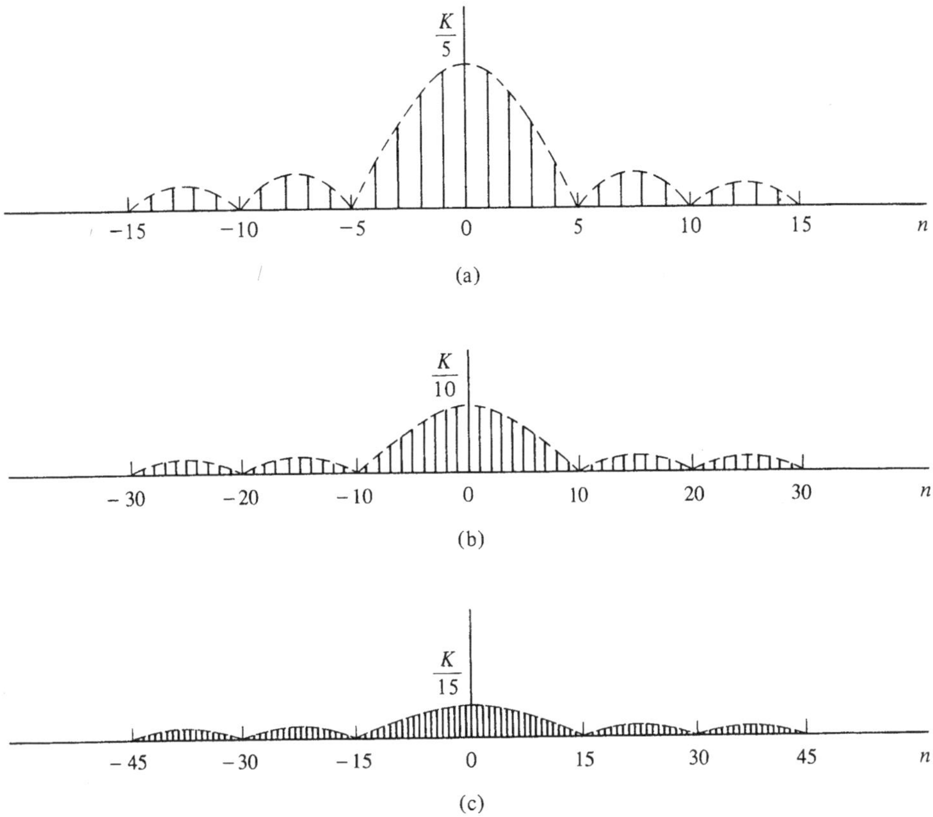

Figure 3.3.7 Line spectra for the $x(t)$ in Example 3.3.3. (a) Magnitude spectrum for $\tau = 1$ and $T = 5$. (b) Magnitude spectrum for $\tau = 1$ and $T = 10$. (c) Magnitude spectrum for $\tau = 1$ and $T = 15$.

$$c_n = \frac{1}{2\pi}\int_{-\pi}^{\pi} t^2 \exp[-jnt]\,dt$$

Integrating by parts twice yields

$$c_n = \frac{2\cos n\pi}{n^2}, \quad n \neq 0$$

The term c_0 is obtained from

$$c_0 = \frac{1}{2\pi}\int_{-\pi}^{\pi} t^2\,dt$$

$$= \frac{\pi^2}{3}$$

From Equations (3.3.9b) and (3.3.9c),

$$a_n = 2\,\mathrm{Re}\{c_n\} = \frac{4}{n^2}\cos n\pi$$

$$b_n = -2\,\mathrm{Im}\{c_n\} = 0$$

because c_n is real. Substituting into Equation (3.3.8), we obtain Equation (3.3.18).

Example 3.3.5

Consider

$$x(t) = 1 + 2\sin\left(\frac{7\pi}{3}t\right) - 3\cos\left(\frac{7\pi}{3}t\right) + \sin(7\pi t) + \cos\left(\frac{28\pi}{3}t\right)$$

It can easily be verified that this signal is periodic with period $T = 6/7$ s so that $\omega_0 = 7\pi/3$. Thus $7\pi = 3\omega_0$ and $28\pi/3 = 4\omega_0$. While we can find the Fourier series coefficients for this example by using Equation (3.3.4), it is much easier in this case to represent the sine and cosine signals in terms of exponentials and write $x(t)$ in the form

$$x(t) = 1 + \frac{1}{j}\exp[j\omega_0 t] - \frac{1}{j}\exp[-j\omega_0 t] - \frac{3}{2}\exp[j\omega_0 t] - \frac{3}{2}\exp[-j\omega_0 t]$$

$$+\frac{1}{2j}\exp[j3\omega_0 t] - \frac{1}{2j}\exp[-j3\omega_0 t] + \frac{1}{2}\exp[j4\omega_0 t] + \frac{1}{2}\exp[-j4\omega_0 t]$$

Comparison with Equation (3.3.3) yields

$$c_0 = 1$$

$$c_1 = c_{-1}^* = -\left(\frac{3}{2} + j1\right)$$

$$c_2 = c_{-2}^* = -\frac{j}{2}$$

$$c_4 = c_{-4}^* = \frac{1}{2}$$

with all other c_n being zero.

Since the amplitude spectrum of $x(t)$ contains only a finite number of components, it is called a band-limited signal.

3.4 DIRICHLET CONDITIONS

The work of Fourier in representing a periodic signal as a trigonometric series is a remarkable accomplishment. His results indicate that a periodic signal, such as the signal with discontinuities in Figure 3.3.2, can be expressed as a sum of sinusoids. Since sinusoids are infinitely smooth signals (i.e., they have ordinary derivatives of arbitrary high order), it is difficult to believe that these discontinuous signals can be expressed in that manner. Of course, the key here is that the sum is an infinite sum, and the signal has to satisfy some general conditions. Fourier believed that any periodic signal could be expressed as a sum of sinusoids. However, this turned out not to be the case. Fortunately, the class of functions which can be represented by a Fourier series is large and sufficiently general that most conceivable periodic signals arising in engineering applications do have a Fourier-series representation.

For the Fourier series to converge, the signal $x(t)$ must possess the following properties, which are known as the Dirichlet conditions, over any period:

1. $x(t)$ is absolutely integrable: that is,

$$\int_{h}^{h+T} |x(t)|\, dt < \infty$$

2. $x(t)$ has only a finite number of maxima and minima.
3. The number of discontinuities in $x(t)$ must be finite.

These conditions are sufficient, but not necessary. Thus if a signal $x(t)$ satisfies the Dirichlet conditions, then the corresponding Fourier series is convergent and its sum is $x(t)$, except at any point t_0 at which $x(t)$ is discontinuous. At the points of discontinuity, the sum of the series is the average of the left- and right-hand limits of $x(t)$ at t_0; that is,

$$x(t_0) = \frac{1}{2}[x(t_0^+) + x(t_0^-)] \tag{3.4.1}$$

Example 3.4.1

Consider the periodic signal in Example 3.3.1. The trigonometric Fourier series coefficients are given by

$$a_n = 2\,\mathrm{Re}\{c_n\} = 0$$

$$b_n = -2\,\mathrm{Im}\{c_n\} = \frac{2K}{n\pi}(1 - \cos n\pi)$$

$$= \begin{cases} \dfrac{4K}{n\pi}, & n \text{ odd} \\ 0, & n \text{ even} \end{cases}$$

so that $x(t)$ can be written as

$$x(t) = \frac{4K}{\pi}\left(\sin \pi t + \frac{1}{3}\sin 3\pi t + \cdots + \frac{1}{n}\sin n\pi t + \cdots\right) \tag{3.4.2}$$

We notice that at $t = 0$ and $t = 1$, two points of discontinuity of $x(t)$, the sum in Equation (3.4.2) has a value of zero, which is equal to the arithmetic mean of the values $-K$ and K of $x(t)$. Furthermore, since the signal satisfies the Dirichlet conditions, the series converges, and $x(t)$ is equal to the sum of the infinite series. Setting $t = 1/2$ in Equation (3.4.2), we obtain

$$x\left(\frac{1}{2}\right) = K = \frac{4K}{\pi}\left(1 - \frac{1}{3} + \frac{1}{5} + \cdots (-1)^{n-1}\frac{1}{2n-1} + \cdots\right)$$

or

$$\sum_{n-1}^{\infty} (-1)^{n-1} \frac{1}{2n-1} = \frac{\pi}{4}$$

Example 3.4.2

Consider the periodic signal in Example 3.3.3 with $\tau = 1$ and $T = 2$. The trigonometric Fourier-series coefficients are given by

$$a_n = 2\,\mathrm{Re}\{c_n\} = K \operatorname{sinc} \frac{n}{2}$$

$$b_n = -2\,\mathrm{Im}\{c_n\} = 0$$

Thus, $a_0 = K/2$, $a_{n=0}$ when n is even, $a_n = 2K/n\pi$ when $n = 1, 5, 9, \ldots$, $a_n = -2K/n\pi$ when $n = 3, 7, 11, \ldots$, and $b_n = 0$ for $n = 1, 2, \ldots$. Hence, $x(t)$ can be written as

$$x(t) = \frac{K}{2} + \frac{2K}{\pi}\left[\cos \pi t - \frac{1}{3}\cos 3\pi t + \frac{1}{5}\cos 5\pi t + \cdots\right] \tag{3.4.3}$$

Since $x(t)$ satisfies the Dirichlet conditions over the interval $[-1, 1]$, the sum in Equation (3.4.3) converges at all points in that interval, except at $t = \pm 1/2$, the points of discontinuity of the signal. At the points of discontinuity, the right-hand side in Equation (3.4.3) has the value $K/2$, which is the arithmetic average of the values K and zero of $x(t)$.

Example 3.4.3

Consider the periodic signal $x(t)$ in Example 3.3.4. The trigonometric Fourier-series coefficients are

$$a_0 = \frac{\pi^2}{3}$$

$$a_n = \frac{4}{n^2}\cos n\pi, \quad n \neq 0$$

$$b_n = 0$$

Hence, $x(t)$ can be written as

$$x(t) = \frac{\pi^2}{3} + 4\sum_{n=1}^{\infty} \frac{(-1)^n}{n^2}\cos nt \tag{3.4.4}$$

For this example, the Dirichlet conditions are satisfied. Further, $x(t)$ is continuous at all t. Thus the sum in Equation (3.4.4) converges to $x(t)$ at all points. Evaluating $x(t)$ at $t = \pi$ gives

$$x(\pi) = \pi^2 = \frac{\pi^2}{3} + 4\sum_{n=1}^{\infty} \frac{1}{n^2}$$

or

$$\sum_{n=1}^{\infty} \frac{1}{n^2} = \frac{\pi^2}{6}$$

It is important to realize that the Fourier-series representation of a periodic signal $x(t)$ cannot be differentiated term by term to give the representation of $dx(t)/dt$. To demonstrate this fact, consider the signal in Example 3.3.3. The derivative of this signal in one period is a pair of oppositely directed δ functions. These are not obtainable by direct differentiation of the Fourier-series representation of $x(t)$. However, a Fourier series can be integrated term by term to yield a valid representation of $\int x(t)\,dt$.

3.5 PROPERTIES OF FOURIER SERIES

In this section, we consider a number of properties of the Fourier series. These properties provide us with a better understanding of the notion of the frequency spectrum of a continuous-time signal. In addition, many of the properties are often useful in reducing the complexity involved in computing the Fourier-series coefficients.

3.5.1 Least Squares Approximation Property

If we were to construct a periodic signal $x(t)$ from a set of exponentials, how many terms must we use to obtain a reasonable approximation? If $x(t)$ is a band-limited signal, we can use a finite number of exponentials. Otherwise, using only a finite number of terms—say, M—results in an approximation of $x(t)$. The difference between $x(t)$ and its approximation is the error in the approximation. We want the approximation to be "close" to $x(t)$ in some sense. For the best approximation, we must minimize some measure of the error. A useful and mathematically tractable criterion we use is the average of the total squared error over one period, also known as the mean-squared value of the error. This criterion of approximation is also known as the approximation of $x(t)$ by least squares. The least-squares approximation property of the Fourier series relates quantitatively the energy of the difference signal to the error difference between the specified signal $x(t)$ and its truncated Fourier-series approximation. Specifically, the property shows that the Fourier-series coefficients are the best choice (in the mean-square sense) for the coefficients of the truncated series.

Now suppose that $x(t)$ can be approximated by a truncated series of exponentials in the form

$$x_N(t) = \sum_{n=-N}^{N} d_n \exp[jn\omega_0 t] \tag{3.5.1}$$

We want to select coefficients d_n such that the error, $x(t) - x_N(t)$, has a minimum mean-square value. If we use the Fourier series representation for $x(t)$, we can write the error signal as

$$\begin{aligned} \varepsilon(t) &= x(t) - x_N(t) \\ &= \sum_{-\infty}^{\infty} c_n \exp[jn\omega_0 t] - \sum_{-N}^{N} d_n \exp[jn\omega_0 t] \end{aligned} \tag{3.5.2}$$

Let us define coefficients

$$g_n = \begin{cases} c_n, & |n| > N \\ c_n - d_n, & -N < n < N \end{cases} \tag{3.5.3}$$

so that Equation (3.5.2) can be written as

$$\varepsilon(t) = \sum_{-\infty}^{\infty} g_n \exp[jn\omega_0 t] \tag{3.5.4}$$

Now, $\varepsilon(t)$ is a periodic signal with period $T = 2\pi/\omega_0$, since each term in the summation is periodic with the same period. It therefore follows that Equation (3.5.4) represents the Fourier-series expansion of $\varepsilon(t)$. As a measure of how well $x_N(t)$ approximates $x(t)$, we use the mean-square error, defined as

$$\text{MSE} = \frac{1}{T}\int_{\langle T\rangle} |\varepsilon(t)|^2 dt$$

Substituting for $\varepsilon(t)$ from Equation (3.5.4), we can write

$$\begin{aligned} \text{MSE} &= \frac{1}{T}\int_{\langle T\rangle} \left(\sum_{n=-\infty}^{\infty} g_n \exp[jn\omega_0 t]\right)\left(\sum_{m=-\infty}^{\infty} g_m^* \exp[-jm\omega_0 t]\right) dt \\ &= \sum_{n=-\infty}^{\infty}\sum_{m=-\infty}^{\infty} g_n g_m^* \left\{\frac{1}{T}\int_{\langle T\rangle} \exp[j(n-m)\omega_0 t]\,dt\right\} \end{aligned} \tag{3.5.5}$$

Since the term in braces on the right-hand side is zero for $n \neq m$ and is 1 for $m = n$, Equation (3.5.5) reduces to

$$\begin{aligned} \text{MSE} &= \sum_{n=-\infty}^{\infty} |g_n|^2 \\ &= \sum_{n=-N}^{N} |c_n - d_n|^2 + \sum_{|n|>N} |c_n|^2 \end{aligned} \tag{3.5.6}$$

Each term in Equation (3.5.6) is positive; so, to minimize the MSE, we must select

$$d_n = c_n \tag{3.5.7}$$

This makes the first summation vanish, and the resulting error is

$$(\text{MSE})_{\min} = \sum_{|n|>N} |c_n|^2 \tag{3.5.8}$$

Equation (3.5.8) demonstrates the fact that the mean-square error is minimized by selecting the coefficients d_n in the finite exponential series of Equation (3.5.1) to be identical with the Fourier-series coefficients c_n. That is, if the Fourier-series expansion of the signal $x(t)$ is truncated at any given value of N, it approximates $x(t)$ with smaller mean-square error than any other exponential series with the same number of terms. Furthermore, since the error is the sum of positive terms, the error decreases monotonically as the number of terms used in the approximation increases.

Example 3.5.1

Consider the approximation of the periodic signal $x(t)$ shown in Figure 3.4.2 by a set of $2N + 1$ exponentials. In order to see how the approximation error varies with the number of terms, we consider the approximation of $x(t)$ based on three terms, then seven terms, then nine terms, and so on. (Note that $x(t)$ contains only odd harmonics.) For $N = 1$ (three terms), the minimum mean-square error is

$$\begin{aligned}
(\text{MSE})_{\min} &= \sum_{|n|>1} |c_n|^2 \\
&= \sum_{|n|>1} \frac{4K^2}{n^2\pi^2} \sin^4 \frac{n\pi}{2} \\
&= \frac{4K^2}{\pi^2} \sum_{\substack{|n|>1,\\ n \text{ odd}}} \frac{1}{n^2} \\
&= \frac{8K^2}{\pi^2}\left(\frac{3\pi^2}{24} - 1\right) \\
&\simeq 0.189K^2
\end{aligned}$$

Similarly, for $N = 3$, it can be shown that

$$\begin{aligned}
(\text{MSE})_{\min} &= \frac{8K^2}{\pi^2}\left[\frac{3\pi^2}{24} - \frac{10}{9}\right] \\
&\simeq 0.01K^2
\end{aligned}$$

3.5.2 Effects of Symmetry

Unnecessary work (and corresponding sources of errors) in determining Fourier coefficients of periodic signals can be avoided if the signals possess any type of symmetry. The important types of symmetry are:

1. even symmetry, $x(t) = x(-t)$,
2. odd symmetry, $x(t) = -x(-t)$,
3. half-wave odd symmetry, $x(t) = -x\left(t + \dfrac{T}{2}\right)$

Each of these is illustrated in Figure 3.5.1.

Recognizing the existence of one or more of these symmetries simplifies the computation of the Fourier-series coefficients. For example, the Fourier series of an even signal $x(t)$ having period T is a "Fourier cosine series,"

$$x(t) = a_0 + \sum_{n=1}^{\infty} a_n \cos \frac{2n\pi t}{T}$$

with coefficients

$$a_0 = \frac{2}{T}\int_0^{T/2} x(t)\,dt, \quad \text{and} \quad a_n = \frac{4}{T}\int_0^{T/2} x(t) \cos \frac{2n\pi t}{T}\,dt$$

whereas the Fourier series of an odd signal $x(t)$ having period T is a "Fourier sine series,"

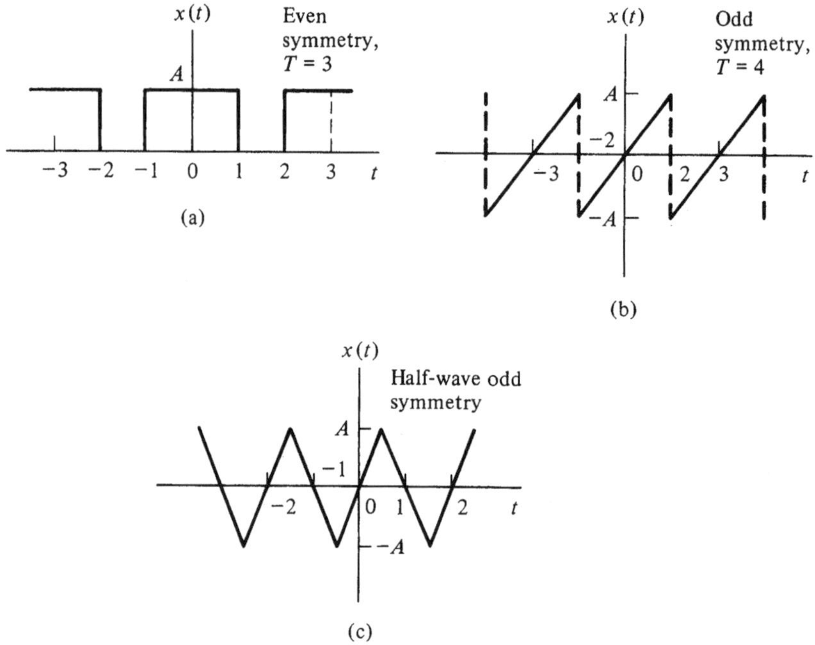

Figure 3.5.1 Types of symmetry.

$$x(t) = \sum_{n=1}^{\infty} b_n \sin \frac{2n\pi t}{T}$$

with coefficients

$$b_n = \frac{4}{T} \int_0^{T/2} x(t) \sin \frac{2n\pi t}{T} dt$$

The effects of these symmetries are summarized in Table 3.1, in which entries such as $a_0 \neq 0$ and $b_{2n+1} \neq 0$ are to be interpreted to mean that these coefficients are not necessarily zero, but may be so in specific examples.

In Example 3.3.1 $x(t)$ is an odd signal, and therefore, the c_n are imaginary ($a_n = 0$), whereas in Example 3.3.3 the c_n are real ($b_n = 0$) because $x(t)$ is an even signal.

TABLE 3.1
Effects of Symmetry

Symmetry	a_0	a_n	b_n	**Remarks**
Even	$a_0 \neq 0$	$a_n \neq 0$	$b_n = 0$	Integrate over $T/2$ only, and multiply the coefficients by 2.
Odd	$a_0 = 0$	$a_n = 0$	$b_n \neq 0$	Integrate over $T/2$ only, and multiply the coefficients by 2.
Half-wave odd	$a_0 = 0$	$a_{2n} = 0$ $a_{2n+1} \neq 0$	$b_{2n} = 0$ $b_{2n+1} \neq 0$	Integrate over $T/2$ only, and multiply the coefficients by 2.

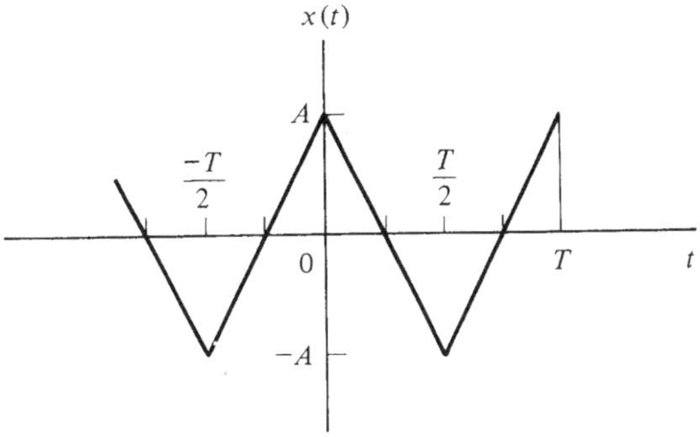

Figure 3.5.2 Signal $x(t)$ for Example 3.5.2.

Example 3.5.2

Consider the signal

$$x(t) = \begin{cases} A - \dfrac{4A}{T}t, & 0 < t < T/2 \\ \dfrac{4A}{T}t - 3A, & T/2 < t < T \end{cases}$$

which is shown in Figure 3.5.2.

Notice that $x(t)$ is both an even and a half-wave odd signal. Therefore, $a_0 = 0$, $b_n = 0$, and we expect to have no even harmonics. Computing a_n, we obtain

$$\begin{aligned} a_n &= \frac{4}{T}\int_0^{T/2}\left(A - \frac{4At}{T}\right)\cos\frac{2n\pi t}{T}dt \\ &= \frac{4A}{(n\pi)^2}(1 - \cos(n\pi)), \quad n \neq 0 \\ &= \begin{cases} 0, & n \text{ even} \\ \dfrac{8A}{(n\pi)^2} & n \text{ odd} \end{cases} \end{aligned}$$

Observe that a_0, which corresponds to the dc term (the zero harmonic), is zero because the area under one period of $x(t)$ evaluates to zero.

3.5.3 Linearity

Suppose that $x(t)$ and $y(t)$ are periodic with the same period. Let their Fourier-series expansions be given by

$$x(t) = \sum_{n=-\infty}^{\infty} \beta_n \exp[jn\omega_0 t] \tag{3.5.9a}$$

$$y(t) = \sum_{n=-\infty}^{\infty} \gamma_n \exp[jn\omega_0 t] \tag{3.5.9b}$$

and let

$$z(t) = k_1 x(t) + k_2 y(t)$$

where k_1 and k_2 are arbitrary constants. Then we can write

$$\begin{aligned} z(t) &= \sum_{n=-\infty}^{\infty} (k_1 \beta_n + k_2 \gamma_n) \exp[jn\omega_0 t] \\ &= \sum_{n=-\infty}^{\infty} \alpha_n \exp[jn\omega_0 t] \end{aligned}$$

The last equation implies that the Fourier coefficients of $z(t)$ are

$$\alpha_n = k_1 \beta_n + k_2 \gamma_n \tag{3.5.10}$$

3.5.4 Product of Two Signals

If $x(t)$ and $y(t)$ are periodic signals with the same period as in Equation (3.5.9), their product is

$$\begin{aligned} z(t) &= x(t)y(t) \\ &= \sum_{n=-\infty}^{\infty} \beta_n \exp[jn\omega_0 t] \sum_{m=-\infty}^{\infty} \gamma_m \exp[jm\omega_0 t] \\ &= \sum_{n=-\infty}^{\infty} \sum_{m=-\infty}^{\infty} \beta_n \gamma_m \exp[j(n+m)\omega_0 t] \\ &= \sum_{l=-\infty}^{\infty} \left(\sum_{m=-\infty}^{\infty} \beta_{l-m} \gamma_m \right) \exp[jl\omega_0 t] \end{aligned} \tag{3.5.11}$$

The sum in parentheses is known as the convolution sum of the two sequences β_m and γ_m. (More on the convolution sum is presented in Chapter 6.) Equation (3.5.11) indicates that the Fourier coefficients of the product signal $z(t)$ are equal to the convolution sum of the two sequences generated by the Fourier coefficients of $x(t)$ and $y(t)$. That is,

$$\sum_{m=-\infty}^{\infty} \beta_{l-m} \gamma_m = \frac{1}{T} \int_{\langle T \rangle} x(t) y(t) \exp[-jl\omega_0 t]\, dt$$

If $y(t)$ is replaced by $y^*(t)$, we obtain

$$z(t) = \sum_{l=-\infty}^{\infty} \left(\sum_{m=-\infty}^{\infty} \beta_{l+m} \gamma_m^* \right) \exp[jl\omega_0 t]$$

and

$$\sum_{m=-\infty}^{\infty} \beta_{l+m} \gamma_m^* = \frac{1}{T} \int_{\langle T \rangle} x(t) y^*(t) \exp[-jl\omega_0 t]\, dt \tag{3.5.12}$$

3.5.5 Convolution of Two Signals

For periodic signals with the same period, a special form of convolution, known as periodic or circular convolution, is defined by the integral

$$z(t) = \frac{1}{T}\int_{\langle T\rangle} x(\tau)y(t-\tau)\,d\tau \tag{3.5.13}$$

where the integral is taken over one period T. It is easy to show that $z(t)$ is periodic with period T and the periodic convolution is commutative and associative. (See Problem 3.22). Thus, we can write $z(t)$ in a Fourier-series representation with coefficients

$$\begin{aligned}\alpha_n &= \frac{1}{T}\int_0^T z(t)\exp[-jn\omega_0 t]\,dt \\ &= \frac{1}{T^2}\iint_0^T x(\tau)y(t-\tau)\exp[-jn\omega_0 t]\,dt\,d\tau \\ &= \frac{1}{T}\int_0^T x(\tau)\exp[-jn\omega_0\tau]\left\{\frac{1}{T}\int_0^T y(t-\tau)\exp[-jn\omega_0(t-\tau)]\,dt\right\}d\tau \end{aligned} \tag{3.5.14}$$

Using the change of variables $\sigma = t - \tau$ in the second integral, we obtain

$$\alpha_n = \frac{1}{T}\int_0^T x(\tau)\exp[-jn\omega_0\tau]\left[\frac{1}{T}\int_{-\tau}^{T-\tau} y(\sigma)\exp[-jn\omega_0\sigma]\,d\sigma\right]d\tau$$

Since $y(t)$ is periodic, the inner integral is independent of a shift τ and is equal to the Fourier-series coefficients of $y(t)$, γ_n. It follows that

$$\alpha_n = \beta_n\gamma_n \tag{3.5.15}$$

where β_n are the Fourier-series coefficients of $x(t)$.

Example 3.5.3

In this example, we compute the Fourier-series coefficients of the product and of the periodic convolution of the two signals shown in Figure 3.5.3.

The analytic representation of $x(t)$ is

$$x(t) = t, \quad 0 < t < 4, \quad x(t) = x(t+4)$$

Figure 3.5.3 Signals $x(t)$ and $y(t)$ for Example 3.5.3.

The Fourier-series coefficients of $x(t)$ are

$$\beta_n = \frac{1}{4}\int_0^4 t \exp\left[\frac{-jn\pi t}{2}\right] dt$$
$$= \frac{2j}{n\pi}$$

For the signal $y(t)$, the analytic representation is given in Example 3.3.3 with $\tau = 2$ and $T = 4$. The Fourier-series coefficients are

$$\gamma_n = \frac{1}{4}\int_{-1}^{1} K \exp\left[\frac{-jn\pi t}{2}\right] dt$$
$$= \frac{K}{n\pi}\sin\frac{n\pi}{2} = \frac{K}{2}\operatorname{sinc}\frac{n}{2}$$

From Equation (3.5.15), the Fourier coefficients of the convolution signal are

$$\alpha_n = \frac{2jK}{(n\pi)^2}\sin\frac{n\pi}{2}$$

and from Equation (3.5.11), the coefficients of the product signal are

$$\alpha_n = \sum_{m=-\infty}^{\infty} \beta_{n-m}\gamma_m$$
$$= \frac{2jK}{\pi^2}\sum_{m=-\infty}^{\infty} \frac{1}{m(n-m)}\sin\frac{m\pi}{2}$$

3.5.6 Parseval's Theorem

In Chapter 1, it was shown that the average power of a periodic signal $x(t)$ is

$$P = \frac{1}{T}\int_{\langle T\rangle} |x(t)|^2 dt$$

The square root of the average power, called the root-mean-square (or rms) value of $x(t)$, is a useful measure of the amplitude of a complicated waveform. For example, the complex exponential signal $x(t) = c_n \exp[jn\omega_0 t]$ with frequency $n\omega_0$ has $|c_n|^2$ as its average power. The relationship between the average power of a periodic signal and the power in its harmonics is one form (the conventional one) of Parseval's theorem.

We have seen that if $x(t)$ and $y(t)$ are periodic signals with the same period T and Fourier-series coefficients β_n and γ_n, respectively, then the product of $x(t)$ and $y(t)$ has Fourier-series coefficients (see Equation (3.5.12))

$$\alpha_n = \sum_{m=-\infty}^{\infty} \beta_{n+m}\gamma_m^*$$

The dc component, or the full-cycle average of the product over time, is

$$\alpha_0 = \frac{1}{T}\int_0^T x(t)y^*(t)\,dt$$

$$= \sum_{m=-\infty}^{\infty} \beta_m \gamma_m^* \qquad (3.5.16)$$

If we let $y(t) = x(t)$ in this expression, then $\beta_n = \gamma_n$, and Equation (3.5.16) becomes

$$\frac{1}{T}\int_{\langle T\rangle} |x(t)|^2\,dt = \sum_{m=-\infty}^{\infty} |\beta_m|^2 \qquad (3.5.17)$$

The left-hand side is the average power of the periodic signal $x(t)$. The result indicates that the total average power of $x(t)$ is the sum of the average power in each harmonic component. Even though power is a nonlinear quantity, we can use superposition of average powers in this particular situation, provided that all the individual components are harmonically related.

We now have two different ways of finding the average power of any periodic signal $x(t)$: in the time domain, using the left-hand side of Equation (3.5.17), and in the frequency domain, using the right-hand side of the same equation.

3.5.7 Shift in Time

If $x(t)$ has the Fourier-series coefficients c_n, then the signal $x(t - \tau)$ has coefficients d_n, where

$$d_n = \frac{1}{T}\int_{\langle T\rangle} x(t-\tau)\exp[-jn\omega_0 t]\,dt$$

$$= \exp[-j\omega_0\tau]\frac{1}{T}\int_{\langle T\rangle} x(\sigma)\exp[-jn\omega_0\sigma]\,d\sigma$$

$$= c_n \exp[-jn\omega_0\tau] \qquad (3.5.18)$$

Thus, if the Fourier-series representation of a periodic signal $x(t)$ is known relative to one origin, the representation relative to another origin shifted by τ is obtained by adding the phase shift $n\omega_0\tau$ to the phase of the Fourier coefficients of $x(t)$.

Example 3.5.4

Consider the periodic signal $x(t)$ shown in Figure 3.5.4. The signal can be written as the sum of the two periodic signals $x_1(t)$ and $x_2(t)$, each with period $2\pi/\omega_0$, where $x_1(t)$ is the

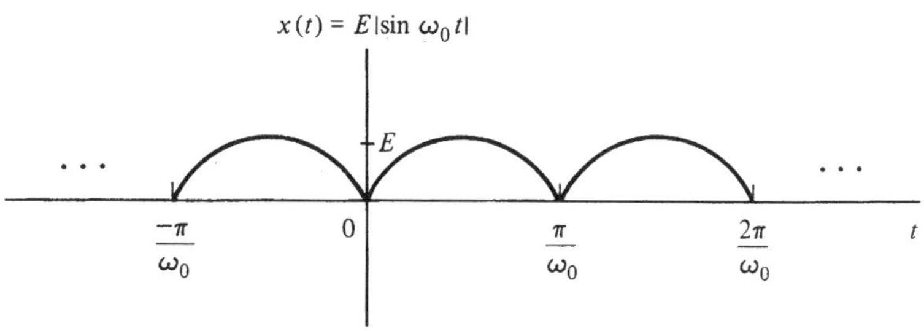

Figure 3.5.4 Signal $x(t)$ for Example 3.5.4.

half-wave rectified signal of Example 3.3.2 and $x_2(t) = x_1(t - \pi/\omega_0)$. Therefore, if β_n and γ_n are the Fourier coefficients of $x_1(t)$ and $x_2(t)$, respectively, then, according to Equation (3.5.18),

$$\gamma_n = \beta_n \exp\left[-jn\omega_0 \frac{\pi}{\omega_0}\right]$$
$$= \beta_n \exp[-jn\pi] = (-1)^n \beta_n$$

From Equation (3.5.10), the Fourier-series coefficients of $x(t)$ are

$$\alpha_n = \beta_n + (-1)^n \beta_n$$
$$= \begin{cases} 2\beta_n, & n \text{ even} \\ 0, & n \text{ odd} \end{cases}$$

where the Fourier-series coefficients of the periodic signal $x_1(t)$ can be determined as in Equation (3.3.16) as

$$\beta_n = \frac{\omega_0}{2\pi} \int_0^{\pi/\omega_0} E \sin \omega_0 t \exp[jn\omega_0 t]\, dt$$
$$= \begin{cases} \dfrac{E}{\pi(1 - n^2)}, & n \text{ even} \\ \dfrac{-jnE}{4}, & n = \pm 1 \\ 0, & \text{otherwise} \end{cases}$$

Thus,

$$\alpha_n = \begin{cases} \dfrac{2E}{\pi(1 - n^2)}, & n \text{ even} \\ 0, & n \text{ odd} \end{cases}$$

This result can be verified by directly computing the Fourier-series coefficients of $x(t)$.

3.5.8 Integration of Periodic Signals

If a periodic signal contains a nonzero average value ($c_0 \neq 0$), then the integration of this signal produces a component that increases linearly with time, and therefore, the resultant signal is aperiodic. However, if $c_0 = 0$, then the integrated signal is periodic, but might contain a dc component. Integrating both sides of Equation (3.3.3) yields

$$\int_{-\infty}^{t} x(\tau)\, d\tau = \sum_{n=-\infty}^{\infty} \frac{c_n}{jn\omega_0} \exp[jn\omega_0 t], \quad n \neq 0 \tag{3.5.19}$$

The relative amplitudes of the harmonics of the integrated signal compared with its fundamental are less than those for the original, unintegrated signal. In other words, integration attenuates (deemphasizes) the magnitude of the high-frequency components of the signal. High-frequency components of the signal are the main contributors to its sharp details, such as those occurring at the points of discontinuity or at discontinuous derivatives of the signal. Hence, integration smooths the signal, and this is one of the reasons it is sometimes called a smoothing operation.

3.6 SYSTEMS WITH PERIODIC INPUTS

Consider a linear, time-invariant, continuous-time system with impulse response $h(t)$. From Chapter 2, we know that the response resulting from an input $x(t)$ is

$$y(t) = \int_{-\infty}^{\infty} h(\tau)x(t-\tau)\,d\tau$$

For complex exponential inputs of the form

$$x(t) = \exp[j\omega t]$$

the output of the system is

$$y(t) = \int_{-\infty}^{\infty} h(\tau)\exp[j\omega(t-\tau)]\,d\tau$$

$$= \exp[j\omega t]\int_{-\infty}^{\infty} h(\tau)\exp[-j\omega\tau]\,d\tau$$

By defining

$$H(\omega) = \int_{-\infty}^{\infty} h(\tau)\exp[-j\omega\tau]\,d\tau \tag{3.6.1}$$

we can write

$$y(t) = H(\omega)\exp[j\omega t] \tag{3.6.2}$$

$H(\omega)$ is called the system (transfer) function and is a constant for fixed ω. Equation (3.6.2) is of fundamental importance because it tells us that the system response to a complex exponential is also a complex exponential, with the same frequency ω, scaled by the quantity $H(\omega)$. The magnitude $|H(\omega)|$ is called the magnitude function of the system, and $\measuredangle H(\omega)$ is known as the phase function of the system. Knowing $H(\omega)$, we can determine whether the system amplifies or attenuates a given sinusoidal component of the input and how much of a phase shift the system adds to that particular component.

To determine the response $y(t)$ of an LTI system to a periodic input $x(t)$ with the Fourier-series representation of Equation (3.3.3), we use the linearity property and Equation (3.6.2) to obtain

$$y(t) = \sum_{n=-\infty}^{\infty} H(n\omega_0)c_n \exp[jn\omega_0 t] \tag{3.6.3}$$

Equation (3.6.3) tells us that the output signal is the summation of exponentials with coefficients

$$d_n = H(n\omega_0)c_n \tag{3.6.4}$$

These coefficients are the outputs of the system in response to $c_n \exp[jn\omega_0 t]$. Note that since $H(n\omega_0)$ is a complex constant for each n, it follows that the output is also periodic with Fourier-series coefficients d_n. In addition, since the fundamental frequency of $y(t)$ is ω_0, which is the fundamental frequency of the input $x(t)$, the period of $y(t)$ is equal to the period of $x(t)$. Hence, the response of an LTI system to a periodic input with period T is periodic with the same period.

Example 3.6.1

Consider the system described by the input/output differential equation

$$y^{(n)}(t) + \sum_{i=0}^{n-1} p_i y^{(i)}(t) = \sum_{i=0}^{m} q_i x^{(i)}(t)$$

For input $x(t) = \exp[j\omega t]$, the corresponding output is $y(t) = H(\omega)\exp[j\omega t]$. Since every input and output should satisfy the system differential equation, substituting into the latter yields

$$\left[(j\omega)^n + \sum_{i=0}^{n-1} p_i(j\omega)^i\right] H(\omega)\exp[j\omega t] = \sum_{i=0}^{m} q_i(j\omega)^i \exp[j\omega t]$$

Solving for $H(\omega)$, we obtain

$$H(\omega) = \frac{\sum_{i=0}^{m} q_i(j\omega)^i}{(j\omega)^n + \sum_{i=0}^{n-1} p_i(j\omega)^i}$$

Example 3.6.2

Let us find the output voltage $y(t)$ of the system shown in Figure 3.6.1 if the input voltage is the periodic signal

$$x(t) = 4\cos t - 2\cos 2t$$

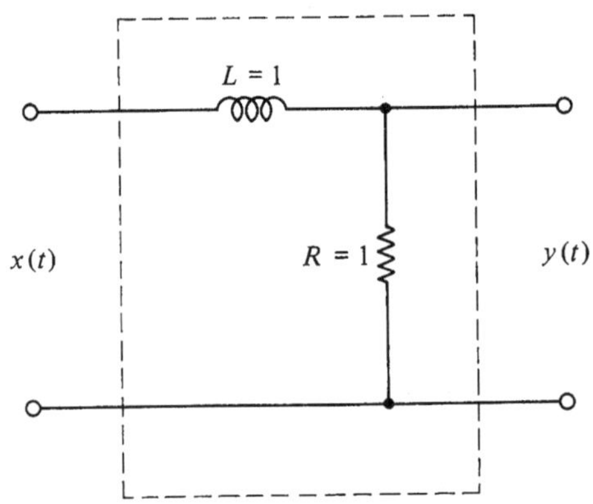

Figure 3.6.1 System for Example 3.6.2.

Applying Kirchhoff's voltage law to the circuit yields

$$\frac{dy(t)}{dt} + \frac{R}{L}y(t) = \frac{R}{L}x(t)$$

If we set $x(t) = \exp[j\omega t]$ in this equation, the output voltage is $y(t) = H(\omega)\exp[j\omega t]$. Using the system differential equation, we obtain

$$j\omega H(\omega)\exp[j\omega t] + \frac{R}{L}H(\omega)\exp[j\omega t] = \frac{R}{L}\exp[j\omega t]$$

Solving for $H(\omega)$ yields

$$H(\omega) = \frac{R/L}{R/L + j\omega}.$$

At any frequency $\omega = n\omega_0$, the system function is

$$H(n\omega_0) = \frac{R/L}{R/L + jn\omega_0}$$

For this example, $\omega_0 = 1$ and $R/L = 1$, so that the output is

$$\begin{aligned} y(t) &= \frac{2}{1+j1}\exp[jt] + \frac{2}{1-j1}\exp[-jt] - \frac{1}{1+j2}\exp[j2t] - \frac{1}{1-j2}\exp[-j2t] \\ &= 2\sqrt{2}\cos(t - 45°) - \frac{2}{\sqrt{5}}\cos(2t - 63°) \end{aligned}$$

Example 3.6.3

Consider the circuit shown in Figure 3.6.2. The differential equation governing the system is

$$i(t) = C\frac{dv(t)}{dt} + \frac{v(t)}{R}$$

For an input of the form $i(t) = \exp[j\omega t]$, we expect the output $v(t)$ to be $v(t) = H(\omega)\exp[j\omega t]$. Substituting into the differential equation yields

$$\exp[j\omega t] = Cj\omega H(\omega)\exp[j\omega t] + \frac{1}{R}H(\omega)\exp[j\omega t]$$

Canceling the $\exp[j\omega t]$ term and solving for $H(\omega)$, we have

$$H(\omega) = \frac{1}{1/R + j\omega C}$$

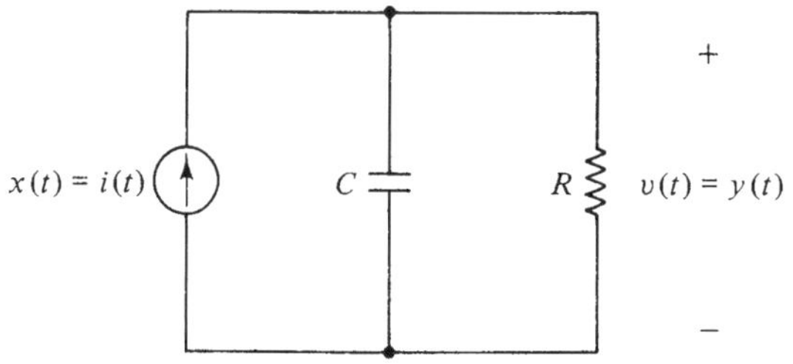

Figure 3.6.2 Circuit for Example 3.6.3.

Let us investigate the response of the system to a more complex input. Consider an input that is given by the periodic signal $x(t)$ in Example 3.3.1. The input signal is periodic with period 2 and $\omega_0 = \pi$, and we have found that

$$c_n = \begin{cases} \dfrac{2K}{jn\pi}, & n \text{ odd} \\ 0, & n \text{ even} \end{cases}$$

From Equation (3.5.3), the output of the system in response to this periodic input is

$$y(t) = \sum_{\substack{n=-\infty, \\ n \text{ odd}}}^{\infty} \frac{2K}{jn\pi} \frac{1}{1/R + jn\pi C} \exp[jn\pi t]$$

Example 3.6.4

Consider the system shown in Figure 3.6.3. Applying Kirchhoff's voltage law, we find that the differential equation describing the system is

$$x(t) = L\frac{d}{dt}\left(\frac{y(t)}{R} + C\frac{dy(t)}{dt}\right) + y(t)$$

which can be written as

$$LC\,y''(t) + \frac{L}{R}y'(t) + y(t) = x(t)$$

For an input in the form $x(t) = \exp[j\omega t]$, the output voltage is $y(t) = H(\omega)\exp[j\omega t]$. Using the system differential equation, we obtain

$$(j\omega)^2 LC\,H(\omega) + j\omega\frac{L}{R}H(\omega) + H(\omega) = 1$$

Solving for $H(\omega)$ yields

$$H(\omega) = \frac{1}{1 + j\omega L/R - \omega^2 LC}$$

with

$$|H(\omega)| = \frac{1}{\sqrt{(1 - \omega^2 LC)^2 + (\omega L/R)^2}}$$

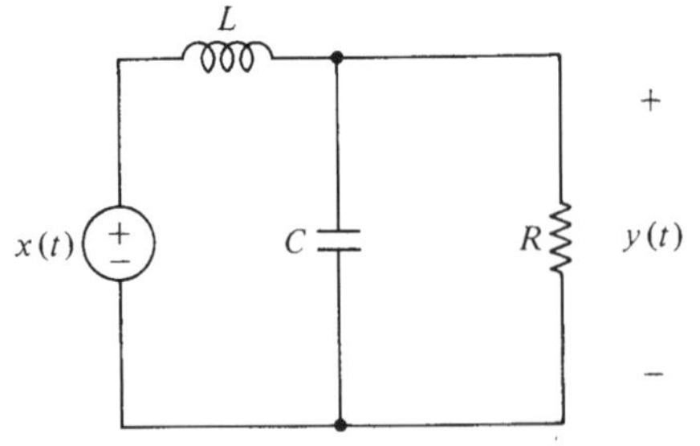

Figure 3.6.3 Circuit for Example 3.6.4.

and

$$\measuredangle H(\omega) = -\tan^{-1}\frac{\omega L/R}{1 - \omega^2 LC}$$

Now, suppose that the input to the system is the half-wave rectified signal in Example 3.3.2. Then the output of the system is periodic, with the Fourier-series representation given by Equation (3.6.3). Let us investigate the effect of the system on the harmonics of the input signal $x(t)$. Suppose that $\omega_0 = 120\pi$, $LC = 1.0 \times 10^{-4}$, and $L/R = 1.0 \times 10^{-4}$. For these values, the amplitude and phase of $H(n\omega_0)$ can be approximated respectively by

$$|H(n\omega_0)| = \frac{1}{(n\omega_0)^2 LC}$$

and

$$\measuredangle H(n\omega_0) = \frac{1}{n\omega_0 RC}$$

Note that the amplitude of $H(n\omega_0)$ decreases as rapidly as $1/n^2$. The amplitudes of the first few components d_n, $n = 0, 1, 2, 3, 4$, in the Fourier-series representation of $y(t)$ are as follows:

$$|d_0| = \frac{E}{\pi}, \quad |d_1| = 7.6 \times 10^{-2}\frac{E}{4}, \quad |d_2| = 1.8 \times 10^{-2}\frac{E}{3\pi}$$

$$|d_3| = 0, \quad |d_4| = 4.4 \times 10^{-3}\frac{E}{15\pi}$$

The dc component of the input $x(t)$ has been passed without any attenuation, whereas the first- and higher order harmonics have had in their amplitudes reduced. The amount of reduction increases as the order of the harmonic increases. As a matter of fact, the function of this circuit is to attenuate all the ac components of the half-wave rectified signal. Such an operation is an example of smoothing, or filtering. The ratio of the amplitudes of the first harmonic and the dc component is $7.6 \times 10^{-2}\pi/4$, in comparison with a value of $\pi/4$ for the unfiltered half-wave rectified waveform. As we mentioned before, complex circuits can be designed to produce better rectified signals. The designer is always faced with a trade-off between complexity and performance.

We have seen so far that when signal $x(t)$ is transmitted through an LTI system (a communication system, an amplifier, etc.) with transfer function $H(\omega)$, the output $y(t)$ is, in general, different from $x(t)$ and is said to be distorted. In contrast, an LTI system is said to be distortionless if the shapes of the input and the output are identical, to within a multiplicative constant. A delayed output that retains the shape of the input signal is also considered distortionless. Thus, the input/output relationship for a distortionless LTI should satisfy the equation

$$y(t) = Kx(t - t_d) \tag{3.6.5}$$

The corresponding transfer function $H(\omega)$ of the distortionless system will be of the form

$$H(\omega) = K\exp[-j\omega t_d] \tag{3.6.6}$$

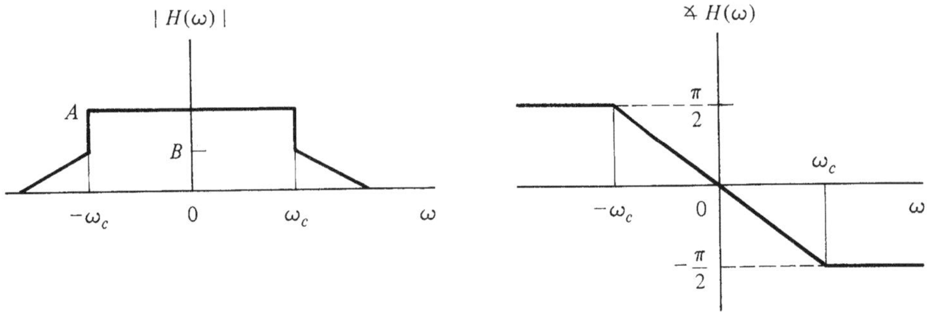

Figure 3.6.4 Magnitude and phase characteristics of a distortionless system.

Thus, the magnitude $|H(\omega)|$ is constant for all ω, while the phase shift is a linear function of frequency of the form $-t_d\omega$.

Let the input to a distortionless system be a periodic signal with Fourier series coefficients c_n. It follows from Equation (3.6.4) that the corresponding Fourier series coefficients for the output are given by

$$d_n = K\exp[-jn\omega_0 t_d]c_n \tag{3.6.7}$$

Thus, for a distortionless system, the quantities $|d_n|/|c_n|$ and $[\underline{/c_n} - \underline{/d_n}]/n$ must be constant for all n.

In practice, we cannot have a system that is distortionless over the entire range $-\infty < \omega < \infty$. Figure 3.6.4 shows the magnitude and phase characteristics of an LTI system that is distortionless in the frequency range $-\omega_c < \omega < \omega_c$.

Example 3.6.5

The input and output of an LTI system are

$$x(t) = 8\exp[j(\omega_0 t + 30°)] + 6\exp[j(3\omega_0 t - 15°)] - 2\exp[j(5\omega_0 t + 45°)]$$

$$y(t) = 4\exp[j(\omega_0 t - 15°)] - 3\exp[j(3\omega_0 t - 30°)] + \exp[j(5\omega_0 t)]$$

We want to determine whether these two signals have the same shape. Note that the ratio of the magnitudes of corresponding harmonics, $|d_n|/|c_n|$, has a value of 1/2 for all the harmonics. To compare the phases, we note that the quantity $[\measuredangle c_n - \measuredangle d_n]/n$ evaluates to $30° - (-15°) = 45°$ for the fundamental, $(-15° - 30° + 180°)/3 = 45°$ for the third harmonic, and $(45° + 180°)/5 = 45°$ for the fifth harmonic. It therefore follows that the two signals $x(t)$ and $y(t)$ have the same shape, except for a scale factor of 1/2 and a phase shift of $\pi/4$. This phase shift corresponds to a time shift of $\tau = \pi/4\omega_0$. Hence, $y(t)$ can be written as

$$y(t) = \frac{1}{2}x\left(t - \frac{\pi}{4\omega_0}\right)$$

The system is therefore distortionless for this choice of $x(t)$.

Example 3.6.6

Let $x(t)$ and $y(t)$ be the input and the output, respectively, of the simple RC circuit shown in Figure 3.6.5. Applying Kirchhoff's voltage law, we obtain

$$\frac{dy(t)}{dt} + \frac{1}{RC}y(t) = \frac{1}{RC}x(t)$$

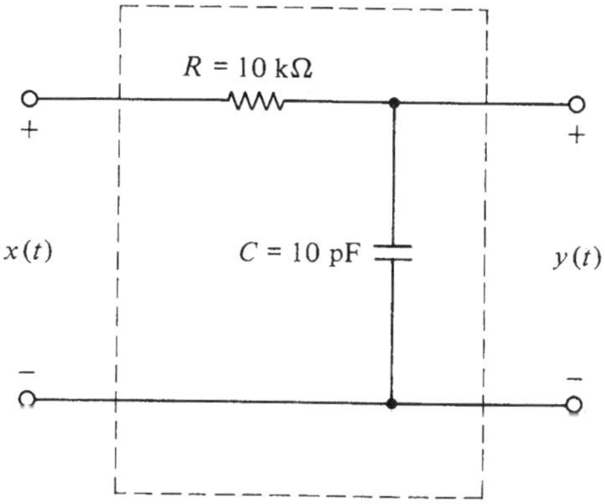

Figure 3.6.5 Circuit for Example 3.6.6.

Setting $x(t) = \exp[j\omega t]$ and recognizing that $y(t) = H(\omega)\exp[j\omega t]$, we have

$$j\omega H(\omega)\exp[j\omega t] + \frac{1}{RC}H(\omega)\exp[j\omega t] = \frac{1}{RC}\exp[j\omega t]$$

Solving for $H(\omega)$ yields

$$H(\omega) = \frac{1/RC}{1/RC + j\omega} = \frac{\eta}{\eta + j\omega}$$

where

$$\eta = \frac{1}{10^4 \times 10^{-11}} = 10^7\,s^{-1}$$

Hence,

$$|H(\omega)| = \frac{\eta}{\sqrt{\eta^2 + \omega^2}}$$

$$\measuredangle H(\omega) = -\tan^{-1}\frac{\omega}{\eta}$$

The amplitude and phase spectra of $H(\omega)$ are shown in Figure 3.6.6. Note that for $\omega \ll \eta$,

$$H(\omega) \simeq 1$$

and

$$\measuredangle H(\omega) \simeq -\frac{\omega}{\eta}$$

That is, the magnitude and phase characteristics are practically ideal. For example, for input

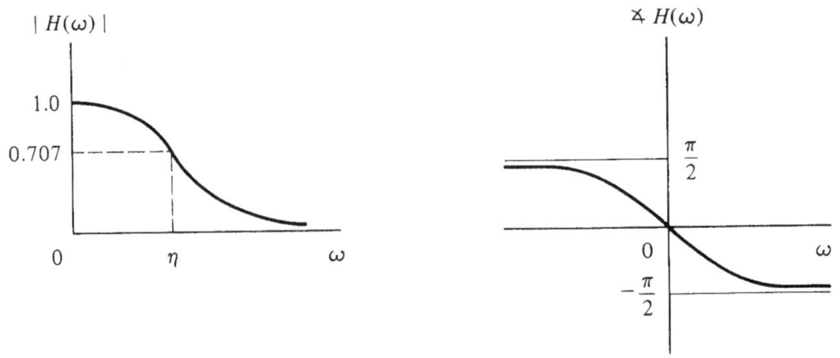

Figure 3.6.6 Magnitude and phase spectra of $H(\omega)$.

$$x(t) = A \exp[j10^3 t]$$

the system is practically distortionless with output

$$\begin{aligned} y(t) &= H(10^3)A \exp[j10^3 t] \\ &= \frac{10^7}{10^7 + j10^3} A \exp[j10^3 t] \\ &\simeq A \exp[j10^3(t - 10^{-7})] \end{aligned}$$

Hence, the time delay is $10^{-7}s$.

3.7 THE GIBBS PHENOMENON

Consider the signal in Example 3.3.1, where we have shown that $x(t)$ could be expressed as

$$x(t) = \frac{2K}{j\pi} \sum_{\substack{n=-\infty \\ n \text{ odd}}}^{\infty} \frac{1}{n} \exp[jn\pi t]$$

We wish to investigate the effect of truncating the infinite series. For this purpose, consider the truncated series

$$x_N(t) = \frac{2K}{j\pi} \sum_{\substack{n=-N, \\ n \text{ odd}}}^{N} \frac{1}{n} \exp[jn\pi t]$$

The truncated series is shown in Figure 3.7.1 for $N = 3$ and 5. Note that even with $N = 3$, $x_3(t)$ resembles the pulse train in Figure 3.3.2. Increasing N to 39, we obtain the approximation shown in Figure 3.7.2. It is clear that, except for the overshoot at the points of discontinuity, the latter figure is a much closer approximation to the pulse train $x(t)$ than is $x_5(t)$. In general, as N increases, the mean-square error between the approximation and the given signal decreases, and the approximation to the given sig-

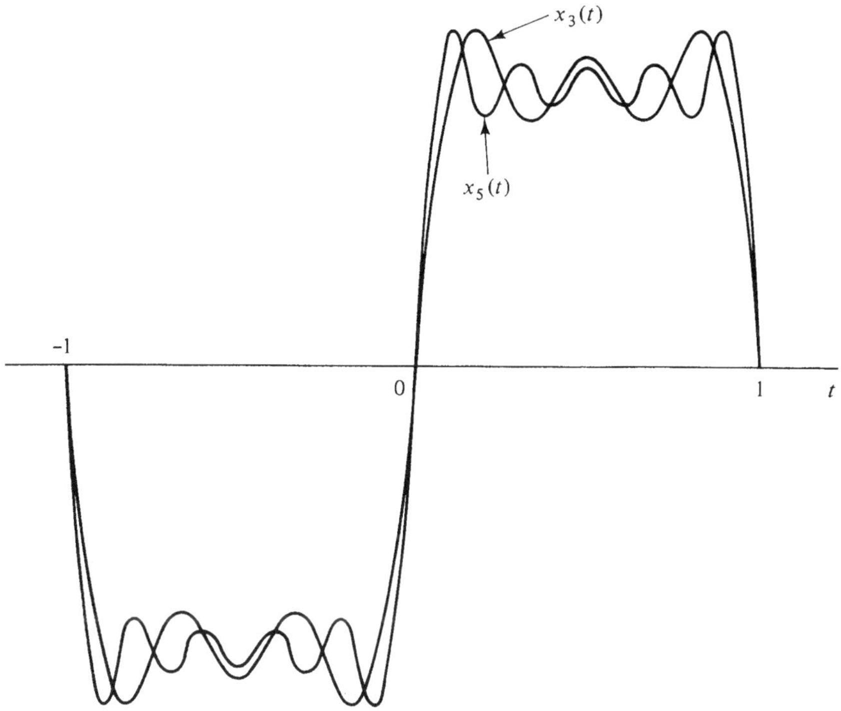

Figure 3.7.1 Signals $x_3(t)$ and $x_5(t)$.

nal improves everywhere except in the immediate vicinity of a finite discontinuity. In the neighborhood of points of discontinuity in $x(t)$, the Fourier-series representation fails to converge, even though the mean-square error in the representation approaches zero. A careful examination of the plot in Figure 3.7.2 reveals that the magnitude of the overshoot is approximately 9% higher than the signal $x(t)$. In fact, the 9% overshoot is always present and is independent of the number of terms used to approximate signal $x(t)$. This observation was first made by the mathematical physicist Josiah Willard Gibbs.

To obtain an explanation of this phenomenon, let us consider the general form of a truncated Fourier series:

$$\begin{aligned} x_N(t) &= \sum_{n=-N}^{N} c_n \exp[jn\omega_0 t] \\ &= \sum_{n=-N}^{N} \frac{1}{T}\int_{\langle T\rangle} x(\tau)\exp[-jn\omega_0\tau]\,d\tau \exp[jn\omega_0 t] \\ &= \frac{1}{T}\int_{\langle T\rangle} x(\tau)\left\{\sum_{n=-N}^{N} \exp[jn\omega_0(t-\tau)]\right\}d\tau \end{aligned} \tag{3.7.1}$$

It can be shown (see Problem 3.39) that the sum in braces is equal to

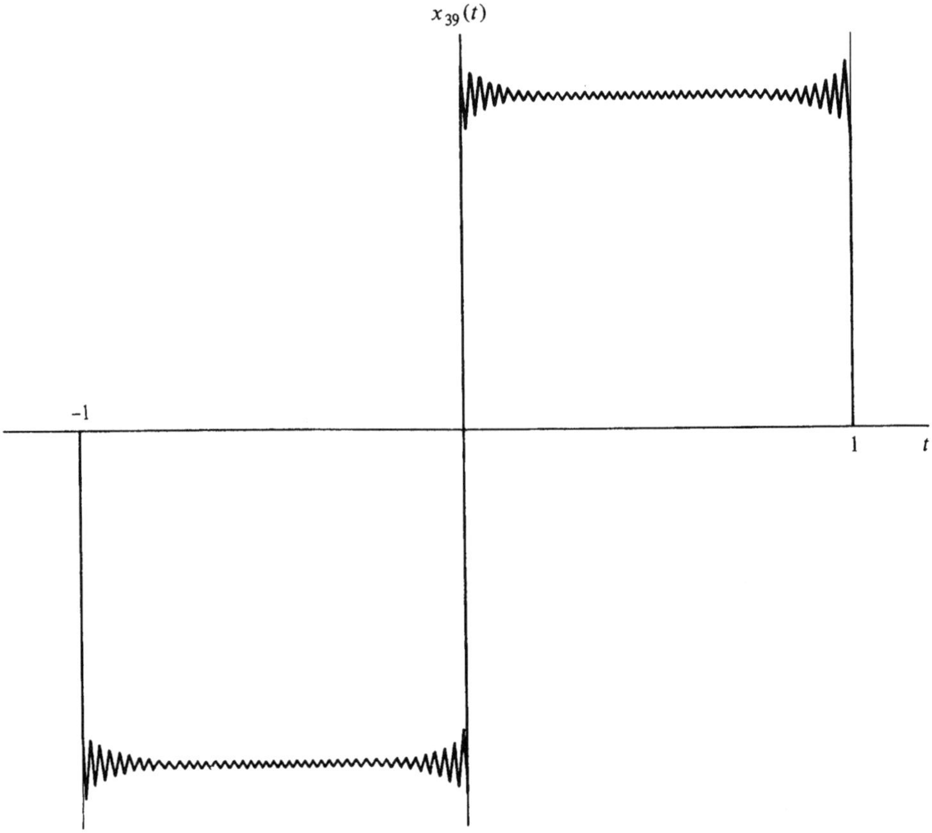

Figure 3.7.2 Signal $x_{39}(t)$.

$$g(t-\tau) \underline{\Delta} \sum_{n=-N}^{N} \exp[jn\omega_0(t-\tau)] = \frac{\sin\left[\left(N+\frac{1}{2}\right)\omega_0(t-\tau)\right]}{\sin\left(\omega_0 \frac{t-\tau}{2}\right)} \tag{3.7.2}$$

The signal $g(\sigma)$ with $N = 6$ is plotted in Figure 3.7.3. Notice the oscillatory behavior of the signal and the peaks at points $\sigma = m\pi, m = 1, 2, \ldots$.

Substituting into Equation (3.7.1) yields

$$\begin{aligned} x_N(t) &= \frac{1}{T}\int_{\langle T \rangle} x(\tau) \frac{\sin\left[\left(N+\frac{1}{2}\right)\omega_0(t-\tau)\right]}{\sin\left(\omega_0 \frac{t-\tau}{2}\right)} d\tau \\ &= \frac{1}{T}\int_{\langle T \rangle} x(t-\tau) \frac{\sin\left[\left(N+\frac{1}{2}\right)\omega_0\tau\right]}{\sin\left(\omega_0 \frac{\tau}{2}\right)} d\tau \end{aligned} \tag{3.7.3}$$

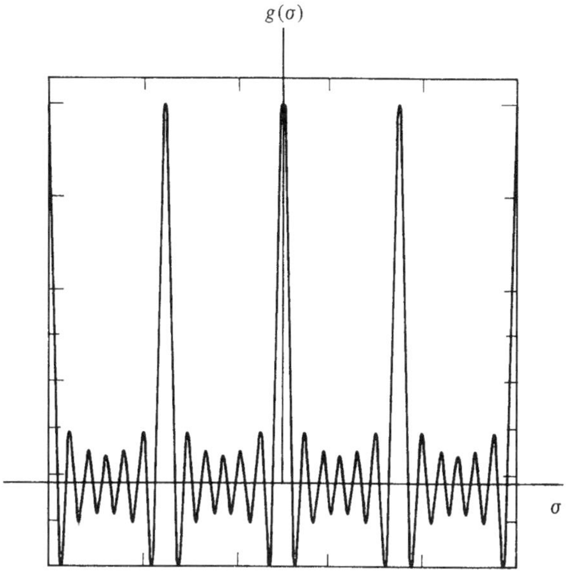

Figure 3.7.3 Signal $g(\sigma)$ for $N = 6$.

In Section 3.5.1, we showed that $x_N(t)$ converges to $x(t)$ (in the mean-square sense) as $N \to \infty$. In particular, for suitably large values of N, $x_N(t)$ should be a close approximation to $x(t)$. Equation (3.7.3) demonstrates the Gibbs phenomenon mathematically, by showing that truncating a Fourier series is the same as convolving the given $x(t)$ with the signal $g(t)$ defined in Equation (3.7.2). The oscillating nature of the signal $g(t)$ causes the ripples at the points of discontinuity.

Notice that, for any signal, the high-frequency components (high-order harmonics) of its Fourier series are the main contributors to the sharp details, such as those occurring at the points of discontinuity or at discontinuous derivatives of the signal.

3.8 SUMMARY

- Two functions $\phi_i(t)$ and $\phi_j(t)$ are orthogonal over an interval (a, b) if

$$\int_a^b \phi_i(t)\phi_j^*(t)\,dt = \begin{cases} E_i, & i = j \\ 0, & i \neq j \end{cases}$$

  and are orthonormal over an interval (a, b) if $E_i = 1$ for all i.
- Any arbitrary signal $x(t)$ can be expanded over an interval (a, b) in terms of the orthogonal basis functions $\{\phi_i(t)\}$ as

$$x(t) = \sum_{i=-\infty}^{\infty} c_i \phi_i(t)$$

  where

$$c_i = \frac{1}{E_i}\int_a^b x(t)\phi_i^*(t)\,dt$$

- The complex exponentials

$$\phi_n(t) = \exp\left[\frac{2\pi}{T} nt\right]$$

are orthogonal over the interval $[0, T]$.

- A periodic signal $x(t)$, of period T, can be expanded in an exponential Fourier series as

$$x(t) = \sum_{n=-\infty}^{\infty} c_n \exp\left[\frac{j2n\pi t}{T}\right]$$

- The fundamental radian frequency of a periodic signal is related to the fundamental period by

$$\omega_0 = \frac{2\pi}{T}$$

- The coefficients c_n are called Fourier-series coefficients and are given by

$$c_n = \frac{1}{T}\int_{\langle T\rangle} x(t) \exp\left[-\frac{j2n\pi t}{T}\right] dt$$

- The fundamental frequency ω_0 is called the first harmonic frequency, the frequency $2\omega_0$ is the second harmonic frequency, and so on.
- The plot of $|c_n|$ versus $n\omega_0$ is called the magnitude spectrum. The locus of the tips of the magnitude lines is called the envelope of the magnitude spectrum.
- The plot of $\measuredangle c_n$ versus $n\omega_0$ is called the phase spectrum.
- For periodic signals, both the magnitude and phase spectra are line spectra. For real-valued signals, the magnitude spectrum has even symmetry, and the phase spectrum has odd symmetry.
- If signal $x(t)$ is a real-valued signal, then it can be expanded in a trigonometric series of the form

$$x(t) = a_0 + \sum_{n=1}^{\infty}\left(a_n \cos\frac{2n\pi t}{T} + b_n \sin\frac{2n\pi t}{T}\right)$$

- The relation between the trigonometric-series coefficients and the exponential-series coefficients is given by

$$\begin{aligned} a_0 &= c_0 \\ a_n &= 2\,\mathrm{Re}\{c_n\} \\ b_n &= -2\,\mathrm{Im}\{c_n\} \\ c_n &= \frac{1}{2}(a_n - jb_n) \end{aligned}$$

- An alternative form of the Fourier series is

$$x(t) = A_0 + \sum_{n=1}^{\infty} A_n \cos\left(\frac{2n\pi t}{T} + \phi_n\right)$$

with

$$A_0 = c_0$$

and

$$A_n = 2|c_n|, \qquad \phi_n = \measuredangle c_n$$

- For the Fourier series to converge, the signal $x(t)$ must be absolutely integrable, have only a finite number of maxima and minima, and have a finite number of discontinuities over any period. This set of conditions is known as the Dirichlet conditions.
- If the signal $x(t)$ has even symmetry, then

$$b_n = 0, \quad n = 1, 2, \ldots$$

$$a_0 = \frac{2}{T}\int_{\langle T/2\rangle} x(t)\,dt$$

$$a_n = \frac{4}{T}\int_{\langle T/2\rangle} x(t)\cos\frac{2n\pi t}{T}\,dt$$

- If the signal $x(t)$ has odd symmetry, then

$$a_n = 0, \quad n = 0, 1, 2 \ldots$$

$$b_n = \frac{4}{T}\int_{\langle T/2\rangle} x(t)\sin\frac{2n\pi t}{T}\,dt$$

- If the signal $x(t)$ has half-wave odd symmetry, then

$$a_{2n} = 0, \quad n = 0, 1, \ldots$$

$$a_{2n+1} = \frac{4}{T}\int_{\langle T/2\rangle} x(t)\cos\frac{2(2n+1)\pi t}{T}\,dt$$

$$b_{2n} = 0, \quad n = 1, 2, \ldots$$

$$b_{2n+1} = \frac{4}{T}\int_{\langle T/2\rangle} x(t)\sin\frac{2(2n+1)\pi t}{T}\,dt$$

- If β_n and γ_n are, respectively, the exponential Fourier-series coefficients for two periodic signals $x(t)$ and $y(t)$ with the same period, then the Fourier-series coefficients for $z(t) = k_1 x(t) + k_2 y(t)$ are

$$\alpha_n = k_1\beta_n + k_2\gamma_n$$

whereas the Fourier-series coefficients for $z(t) = x(t)y(t)$ are

$$\alpha_n = \sum_{m=-\infty}^{\infty} \beta_{n-m}\gamma_m$$

- For periodic signals $x(t)$ and $y(t)$ with the same period T, the periodic convolution is defined as

$$z(t) = \frac{1}{T}\int_{\langle T\rangle} x(\tau)y(t-\tau)\,d\tau$$

- The Fourier-series coefficients of the periodic convolution of $x(t)$ and $y(t)$ are

$$\alpha_n = \beta_n \gamma_n$$

- One form of Parseval's theorem states that the average power in the signal $x(t)$ is related to the Fourier-series coefficients c_n as

$$P = \sum_{n=-\infty}^{\infty} |c_n|^2$$

- The system (transfer) function of an LTI system is defined as

$$H(\omega) = \int_{-\infty}^{\infty} h(\tau) \exp[-j\omega\tau]\, d\tau$$

- The magnitude of $H(\omega)$ is called the magnitude function (magnitude characteristic) of the system, and $\measuredangle H(\omega)$ is known as the phase function (phase characteristic) of the system.
- The response $y(t)$ of an LTI system to the periodic input $x(t)$ is

$$y(t) = \sum_{n=-\infty}^{\infty} H(n\omega_0) c_n \exp[jn\omega_0 t]$$

where ω_0 is the fundamental frequency and c_n are the Fourier series coefficients of the input $x(t)$.
- Representing $x(t)$ by a finite series results in an overshoot behavior at the points of discontinuity. The magnitude of the overshoot is approximately 9%. This phenomenon is known as the Gibbs phenomenon.

3.9 CHECKLIST OF IMPORTANT TERMS

Absolutely integrable signal
Dirichlet conditions
Distortionless system
Even harmonic
Exponential Fourier series
Fourier coefficients
Gibbs phenomenon
Half-wave odd symmetry
Laboratory form of Fourier series
Least squares approximation
Magnitude spectrum
Mean-square error
Minimum mean-square error
Odd harmonic
Orthogonal functions
Orthonormal functions
Parseval's theorem
Periodic convolution
Periodic signals
Phase spectrum
Transfer function
Trigonometric Fourier series

3.10 PROBLEMS

3.1. Express the set of signals shown in Figure P3.1 in terms of the orthonormal basis signals $\phi_1(t)$ and $\phi_2(t)$.

3.2. Given an arbitrary set of functions $x_1(t)$, $i = 1, 2, \ldots$, defined over an interval $[t_0, t_f]$, we can generate a set of orthogonal functions $\psi_i(t)$ by following the *Gram-Schmidt orthogonalization procedure*. Let us choose as the first basis function

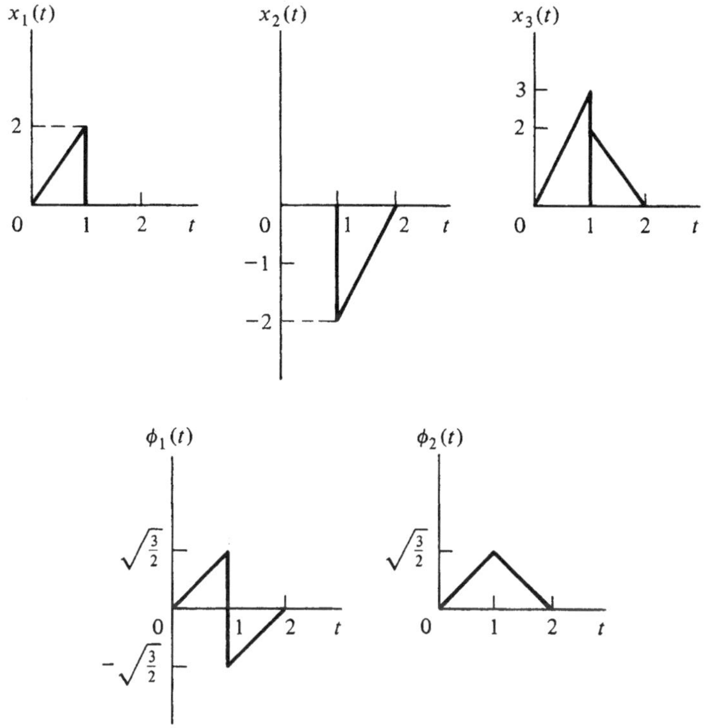

Figure P3.1

$$\psi_1(t) = x_1(t)$$

We then choose as our second basis function

$$\psi_2(t) = x_2(t) + a_1\psi_1(t)$$

where a_1 is determined so as to make $\psi_2(t)$ orthogonal to $\psi_1(t)$. We can continue this procedure by choosing

$$\psi_3(t) = x_3(t) + b_1\psi_1(t) + b_2\psi_2(t)$$

with b_1 and b_2 determined from the requirement that $\psi_3(t)$ must be orthogonal to both $\psi_1(t)$ and $\psi_2(t)$. Subsequent functions can be generated in a similar manner.

For any two signals $x(t)$, $y(t)$, let

$$\langle x(t), y(t)\rangle = \int_{t_0}^{t_f} x(t)y(t)\,dt$$

and let $E_i = \langle x_i(t), x_i(t)\rangle$.

(a) Verify that the coefficients a_1, b_1, and b_2 are given by

$$a_1 = -\frac{\langle x_1(t), x_2(t)\rangle}{E_1}$$

$$b_1 = -\frac{\langle x_1(t), x_3(t)\rangle}{E_1}$$

$$b_2 = -\frac{\langle x_1(t), x_2(t)\rangle\langle x_1(t), x_3(t)\rangle - E_1\langle x_2(t), x_3(t)\rangle}{\langle x_1(t), x_2(t)\rangle^2 - E_1 E_2}$$

(b) Use your results from Part (a) to generate a set of orthogonal functions from the set of signals shown in Figure P3.2.

(c) Obtain a set of orthonormal functions $\phi_i(t)$, $i = 1, 2, 3$, from the set $\psi_i(t)$ that you determined in Part (b).

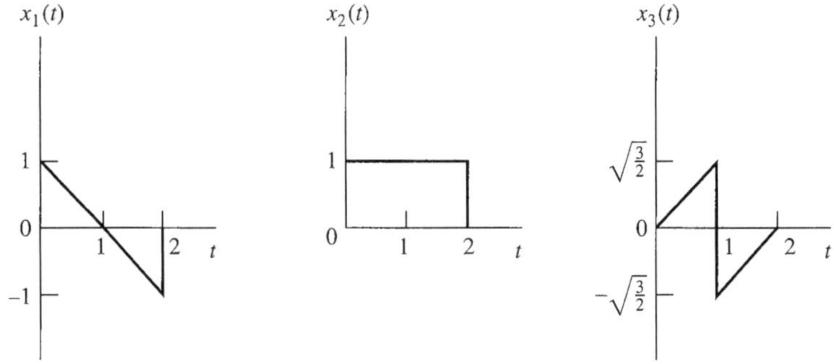

Figure P3.2

3.3. Consider the set of functions

$$x_1(t) = e^{-t}u(t), x_2(t) = e^{-2t}u(t), x_3(t) = e^{-3t}u(t)$$

(a) Use the method of Problem 3.2 to generate a set of orthonormal functions $\phi_i(t)$ from $x_i(t)$, $i = 1, 2, 3$.

(b) Let $\hat{x}(t) = \sum_{i=1}^{3} c_i \phi_i(t)$ be the approximation of $x(t) = 3e^{-4t}u(t)$ in terms of $\phi_i(t)$, and let $\varepsilon(t)$ denote the approximation error. Determine the accuracy of $\hat{x}(t)$ by computing the ratio of the energies in $\varepsilon(t)$ and $x(t)$.

3.4. **(a)** Assuming that all ϕ_i are real-valued functions, prove that Equation (3.2.4) minimizes the energy in the error given by Equation (3.2.7). (*Hint:* Differentiate Equation (3.2.7) with respect to some particular c_j, set the result equal to zero, and solve.)

(b) Can you extend the result in Part (a) to complex functions?

3.5. Show that if the set $\phi_k(t)$, $k = 0, \pm 1, \pm 2, \ldots$, is an orthogonal set over the interval $(0, T)$ and

$$x(t) = \sum_{k=-\infty}^{\infty} c_k \phi_k(t) \tag{P3.5}$$

then

$$c_i = \frac{1}{E_i}\int_0^T x(t)\phi_i^*(t)\,dt$$

where

$$E_i = \int_0^T |\phi_i(t)|^2\,dt$$

3.6. Walsh functions are a set of orthonormal functions defined over the interval $[0, 1)$ that take on values of ± 1 over this interval. Walsh functions are characterized by their *sequency*, which is defined as one-half the number of zero-crossings of the function over the interval [0, 1). Figure P3.6 shows the first seven Walsh-ordered Walsh functions $\text{wal}_w(k, t)$, arranged in order of increasing sequency.

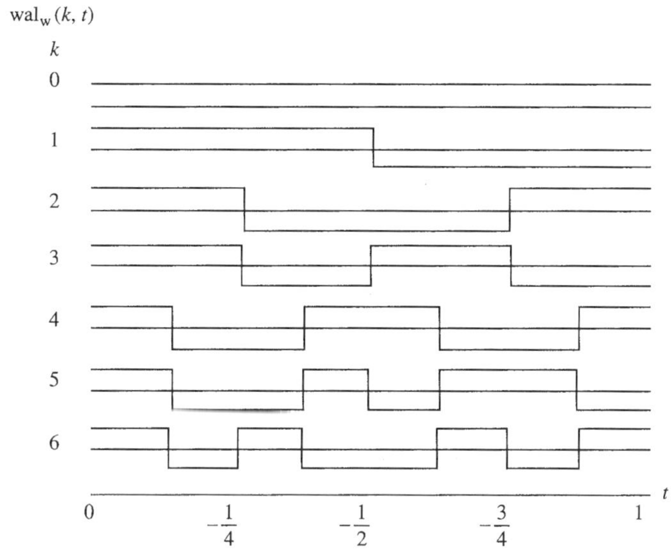

Figure P3.6

(a) Verify that the Walsh functions shown are orthonormal over [0, 1).

(b) Suppose we want to represent the signal $x(t) = t[u(t) - u(t-1)]$ in terms of the Walsh functions as

$$x_N(t) = \sum_{k=0}^{N} c_k \,\text{wal}_w(k, t)$$

Find the coefficients c_k for $N = 6$.

(c) Sketch $x_N(t)$ for $N = 3$ and 6.

3.7. For the periodic signal

$$x(t) = 2 + \frac{1}{2}\cos(t + 45°) + 2\cos(3t) - 2\sin(4t + 30°)$$

(a) Find the exponential Fourier series.

(b) Sketch the magnitude and phase spectra as a function of ω.

3.8. The signal shown in Figure P3.8 is created when a cosine voltage or current waveform is rectified by a single diode, a process known as half-wave rectification. Deduce the exponential Fourier-series expansion for the half-wave rectified signal.

3.9. Find the trigonometric Fourier-series expansion for the signal in Problem 3.8.

3.10. The signal shown in Figure P3.10 is created when a sine voltage or current waveform is rectified by a a circuit with two diodes, a process known as full-wave rectification. Deduce the exponential Fourier-series expansion for the full-wave rectified signal.

3.11. Find the trigonometric Fourier-series expansion for the signal in Problem 3.10.

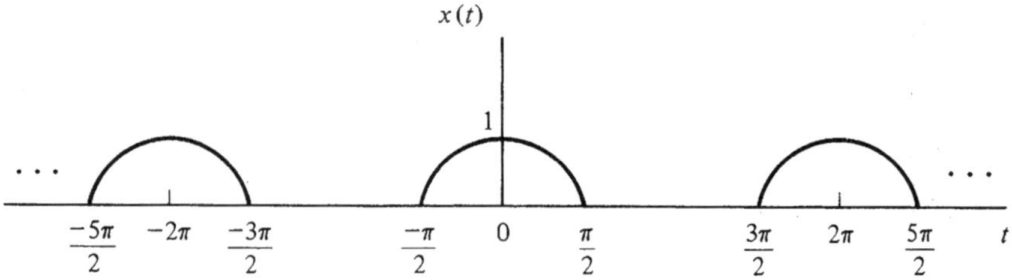

Figure P3.8

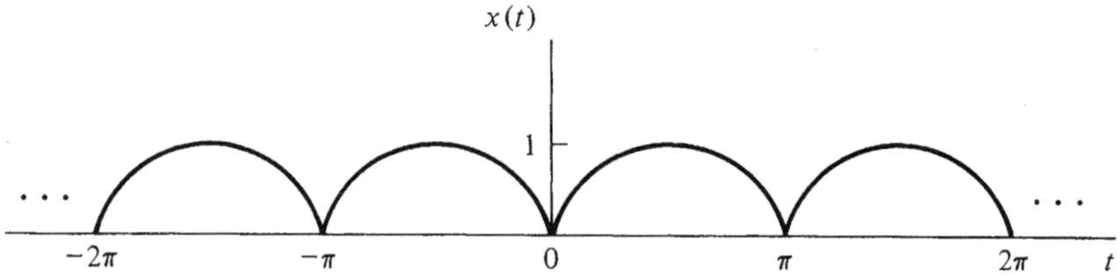

Figure P3.10

3.12. Find the exponential Fourier-series representations of the signals shown in Figure P3.12. Plot the magnitude and phase spectrum for each case.

3.13. Find the trigonometric Fourier-series representations of the signals shown in Figure P3.12.

3.14. **(a)** Show that if a periodic signal is absolutely integrable, then $|c_n| < \infty$.

(b) Does the periodic signal $x(t) = \sin \dfrac{2\pi}{t}$ have a Fourier-series representation? Why?

(c) Does the periodic signal $x(t) = \tan 2\pi t$ have a Fourier-series representation? Why?

3.15. **(a)** Show that $x(t) = t^2, -\pi < t \leq \pi, x(t + 2\pi) = x(t)$ has the Fourier series

$$x(t) = \frac{\pi^2}{3} - 4\left(\cos t - \frac{1}{4}\cos 2t + \frac{1}{9}\cos 3t - + \cdots\right)$$

(b) Set $t = 0$ to obtain

$$\sum_{n=1}^{\infty} \frac{(-1)^{n+1}}{n^2} = \frac{\pi^2}{12}$$

3.16. The Fourier coefficients of a periodic signal with period T are

$$c_n = \begin{cases} 0, & n = 0 \\ (1 + \exp[-\frac{jn\pi}{3}] - 2\exp[-j\pi n]), & n \neq 0 \end{cases}$$

Does this represent a real signal? Why or why not? From the form of c_n, deduce the time signal $x(t)$. *Hint:* Use

$$\int \exp[-jn\omega t]\delta(t - t_1)\,dt = \exp[-jn\omega t_1]$$

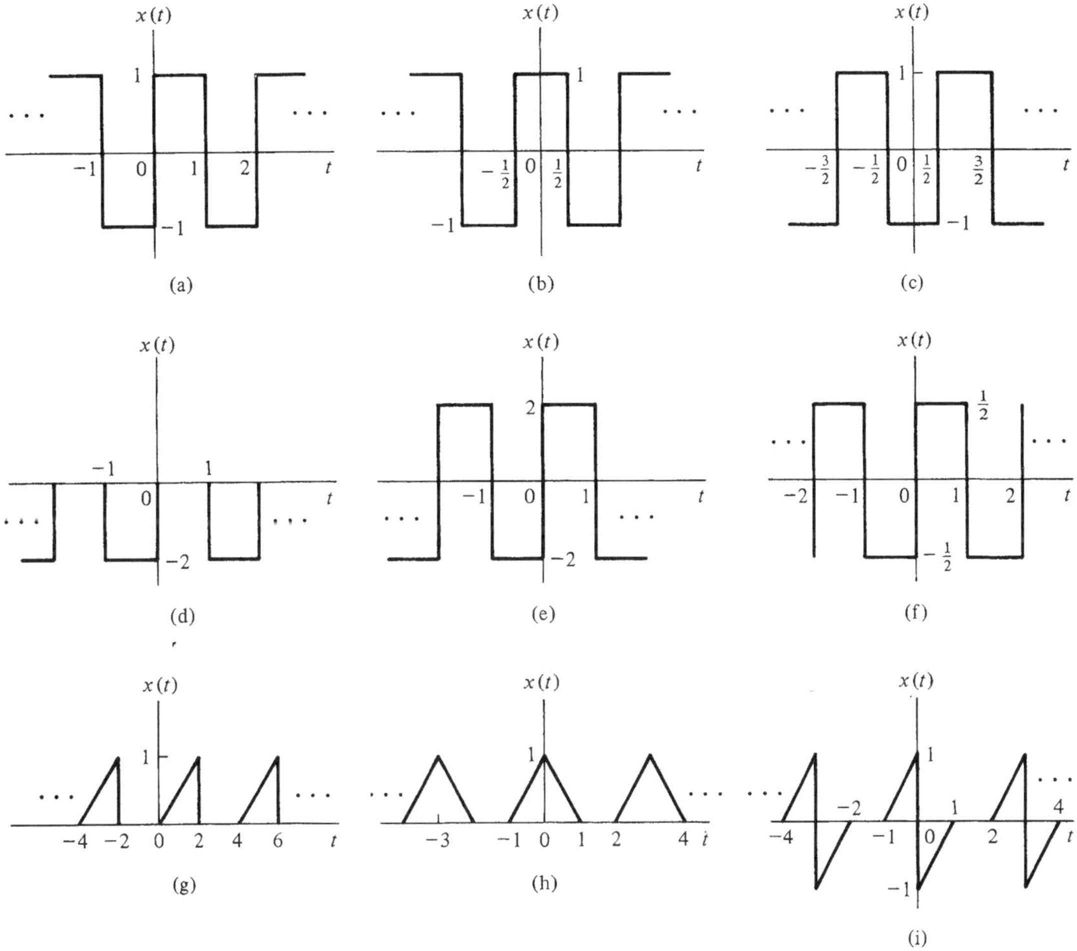

Figure P3.12

3.17. **(a)** Plot the signal

$$x(t) = \frac{1}{4} + \sum_{n=1}^{M} \frac{2}{n\pi} \sin \frac{n\pi}{4} \cos 2n\pi t$$

for $M = 1, 3$, and 5.

(b) Predict the form of $x(t)$ as $M \to \infty$.

3.18. Find the exponential Fourier series for the impulse trains shown in Figure P3.18.

3.19. The waveforms in Problem 3.18 can be considered to be periodic with period N for N any integer. Find the exponential Fourier series coefficients for the case $N = 3$.

3.20. The Fourier series coefficients for a periodic signal $x(t)$ with period T are

$$c_n = \left[\frac{\sin(n\pi/T)}{n\pi}\right]^2$$

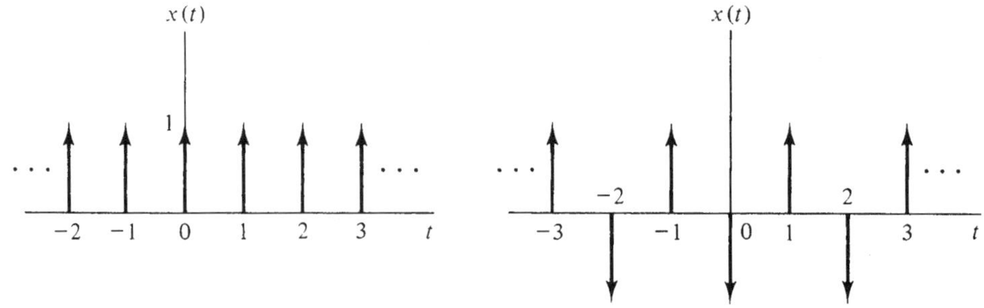

Figure P3.18

(a) Find T such that $c_5 = 1/150$ if T is large, so that $\sin(n\pi/T) \simeq n\pi/T$.
(b) Determine the energy in $x(t)$ and in

$$\hat{x}(t) = \sum_{n=-2}^{2} c_n e^{jn\omega_0 t}$$

3.21. Specify the types of symmetry for the signals shown in Figure P3.21. Specify also which terms in the trigonometric Fourier series are zero.

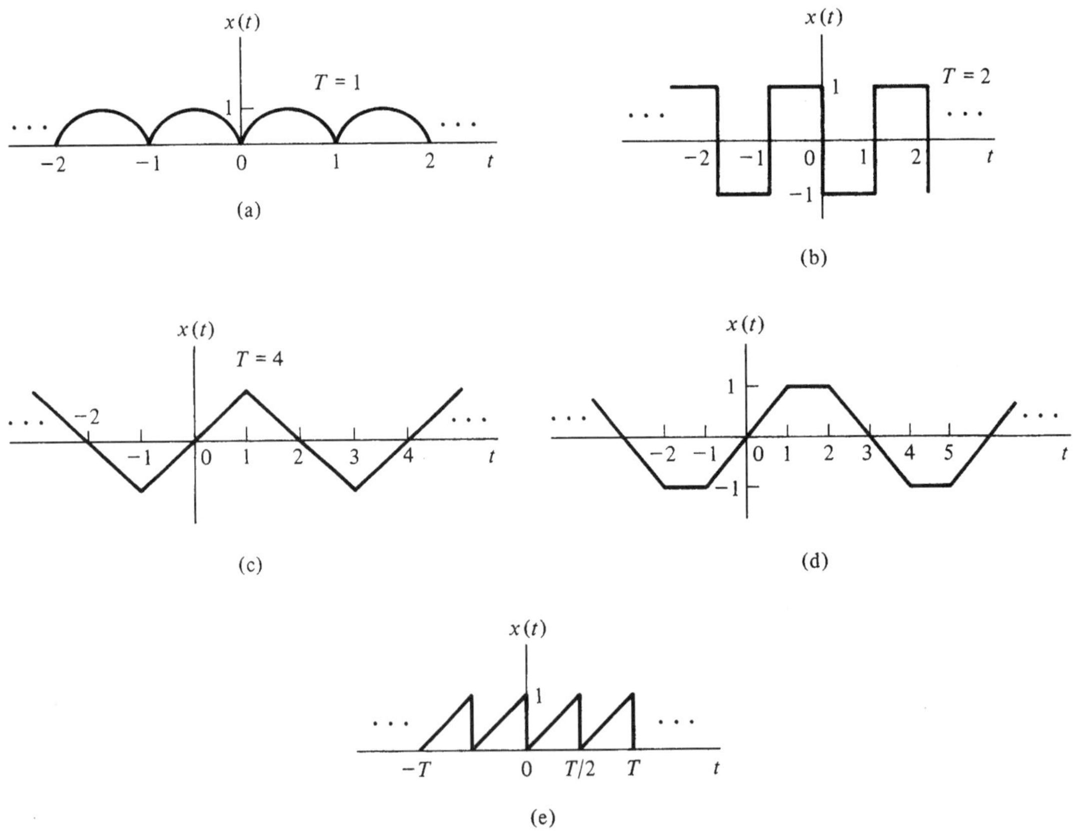

Figure P3.21

3.22. Periodic or circular convolution is a special case of general convolution. For periodic signals with the same period T, periodic convolution is defined by the integral

$$z(t) = \frac{1}{T}\int_{\langle T\rangle} x(\tau)y(t-\tau)\,d\tau$$

(a) Show that $z(t)$ is periodic. Find its period.

(b) Show that periodic convolution is commutative and associative.

3.23. Find the periodic convolution $z(t) = x(t)^*y(t)$ of the two signals shown in Figure P3.23. Verify Equation (3.5.15) for these signals.

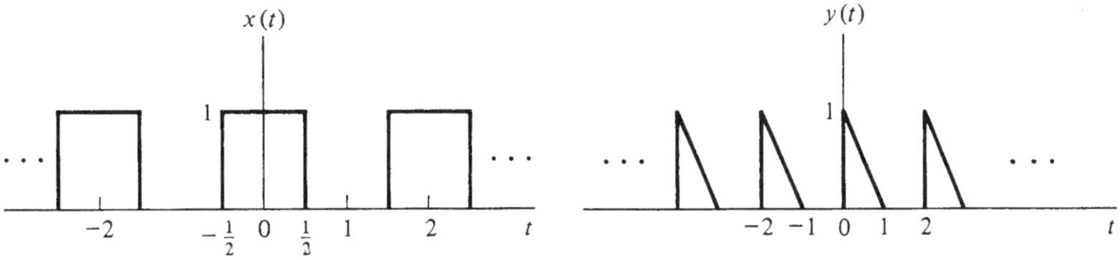

Figure P3.23

3.24. Consider the periodic signal $x(t)$ that has the exponential Fourier-series expansion

$$x(t) = \sum_{n=-\infty}^{\infty} c_n \exp[jn\omega_0 t], \quad c_0 = 0$$

(a) Integrate term by term to obtain the Fourier-series expansion of $y(t) = \int x(t)\,dt$, and show that $y(t)$ is periodic, too.

(b) How do the amplitudes of the harmonics of $y(t)$ compare to the amplitudes of the harmonics of $x(t)$?

(c) Does integration deemphasize or accentuate the high-frequency components?

(d) From Part (c), is the integrated waveform smoother than the original waveform?

3.25. The Fourier-series representation of the triangular signal in Figure P3.25(a) is

$$x(t) = \frac{8}{\pi^2}\left(\sin t - \frac{1}{9}\sin 3t + \frac{1}{25}\sin 5t - \frac{1}{49}\sin 7t + \cdots\right)$$

Use this result to obtain the Fourier series for the signal in Figure P3.25(b).

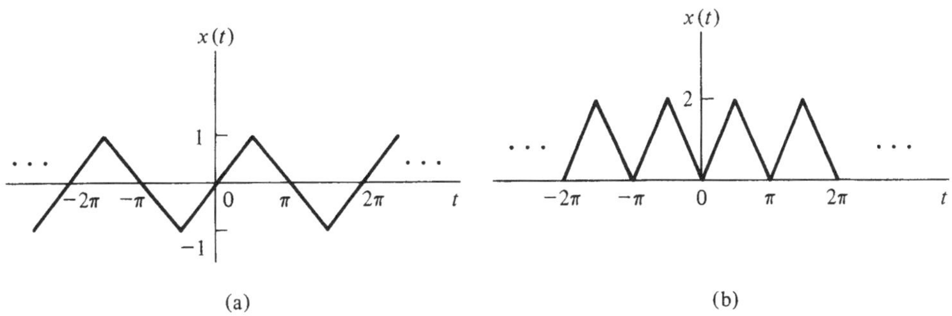

Figure P3.25

3.26. A voltage $x(t)$ is applied to the circuit shown in Figure P3.26. If the Fourier coefficients of $x(t)$ are given by

$$c_n = \frac{1}{n^2 + 1} \exp\left[jn \frac{\pi}{3} \right]$$

(a) Prove that $x(t)$ must be a real signal of time.

(b) What is the average value of the signal?

(c) Find the first three nonzero harmonics of $y(t)$.

(d) What does the circuit do to the high-frequency terms of the input?

(e) Repeat Parts (c) and (d) for the case where $y(t)$ is the voltage across the resistor instead.

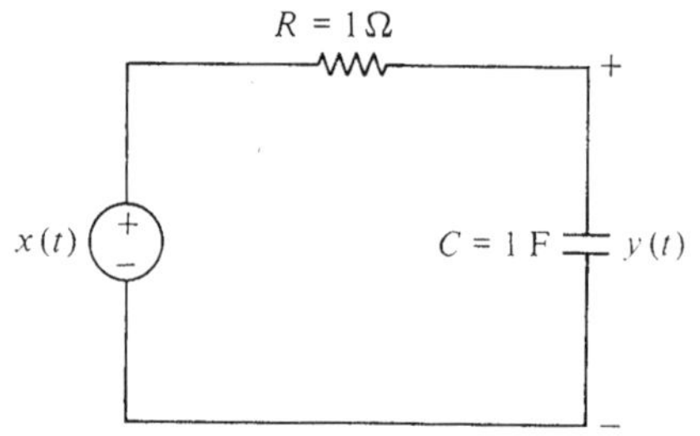

Figure P3.26

3.27. Find the voltage $y(t)$ across the capacitor in Figure P3.26 if the input is

$$x(t) = 1 + 3\cos(t + 30°) + \cos(2t)$$

3.28. The input

$$x(t) = \sum_{n=-\infty}^{\infty} c_n \exp[j\omega_0 t]$$

is applied to four different systems. If the resulting outputs are

$$y_1(t) = \sum_{n=-10}^{10} |c_n| \exp[j(n\omega_0 t + \phi - 3n\omega_0)]$$

$$y_2(t) = \sum_{n=-\infty}^{\infty} c_n \exp[j(n\omega_0 (t - t_0) - 3n\omega_0)]$$

$$y_3(t) = \sum_{n=-\infty}^{\infty} \exp[-\omega_0 |n|] c_n \exp[j(n\omega_0 t - 3n\omega_0)]$$

$$y_4(t) = \sum_{n=-\infty}^{\infty} \exp[-j\omega_0 |n|] c_n \exp[j(n\omega_0 t)]$$

Determine what type of distortion each system has.

3.29 For the circuit shown in Figure P3.29,

(a) Determine the transfer function $H(\omega)$.

(b) Sketch both $|H(\omega)|$ and $\measuredangle H(\omega)$.

(c) Consider the input $x(t) = 10 \exp[j\omega t]$. What is the highest frequency ω you can use such that

$$\frac{|y(t) - x(t)|}{|x(t)|} < 0.01$$

(d) What is the highest frequency ω you can use such that $\measuredangle H(\omega)$ deviates from the ideal linear characteristics by less than 0.02?

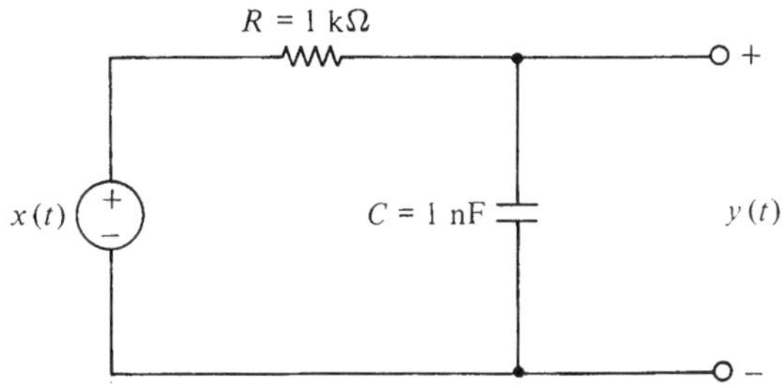

Figure P3.29

3.30. Nonlinear devices can be used to generate harmonics of the input frequency. Consider the nonlinear system described by

$$y(t) = Ax(t) + Bx^2(t)$$

Find the response of the system to $x(t) = a_1 \cos \omega_0 t + a_2 \cos 2\omega_0 t$. List all new harmonics generated by the system, along with their amplitudes.

3.31. The square-wave signal of Example 3.3.3 with $K = 1$, $T = 500\,\mu s$, and $\tau = 100\,\mu s$ is passed through an ideal-low pass filter with cutoff $f_c = 4.2$kHz and applied to the LTI system whose frequency response $H(\omega)$ is shown in Figure P3.31. Find the response of the system.

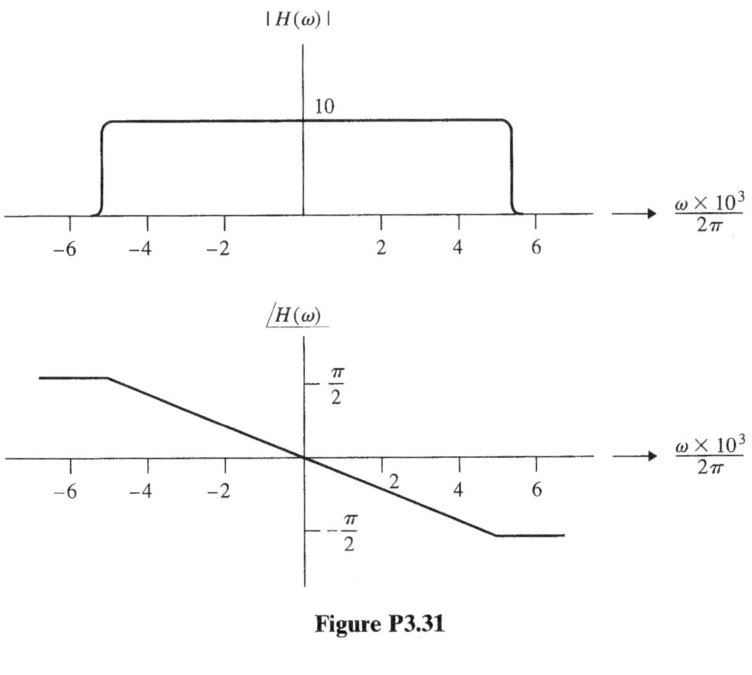

Figure P3.31

3.32. The triangular waveform of Example 3.5.2 with period $T = 4$ and peak amplitude $A = 10$ is applied to a series combination of a resistor $R = 100\ \Omega$ and an inductor $L = 0.1$ H. Determine the power dissipated in the resistor.

3.33. A first-order system is modeled by the differential equation

$$\frac{dy(t)}{dt} + 3y(t) = 2x(t)$$

If the input is the waveform of Example 3.3.2, find the amplitudes of the first three harmonics in the output.

3.34. Repeat Problem 3.33 for the system

$$y''(t) + 3y'(t) + 2y(t) = x'(t) + x(t)$$

3.35. For the system shown in Figure P3.35, the input $x(t)$ is periodic with period T. Show that at any time $t > T_1$ after the input is switched on, $y_c(t)$ and $y_s(t)$ approximate Re$\{c_n\}$ and Im$\{c_n\}$, respectively. Indeed, if T_1 is an integer multiple of the period T of the input signal $x(t)$, then the outputs are precisely equal to the desired values. Discuss the outputs for the following cases:

(a) $T_1 = T$

(b) $T_1 = lT$

(c) $T_1 \gg T$, but $T_1 \neq T$

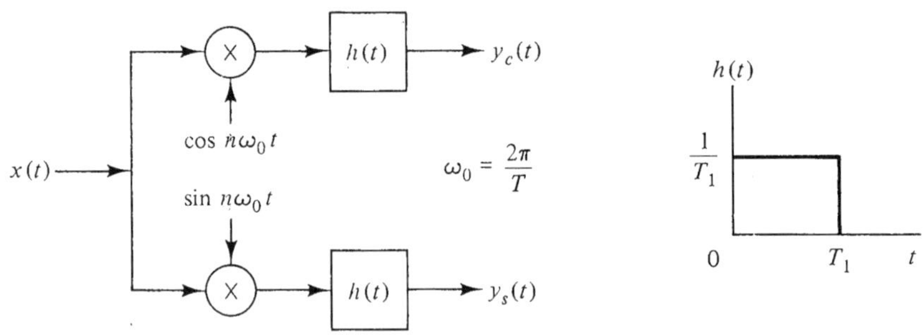

Figure P3.35

3.36. Consider the circuit shown in Figure P3.36. The input is the half-wave rectified signal of Problem 3.8. Find the amplitude of the second and fourth harmonics of the output $y(t)$.

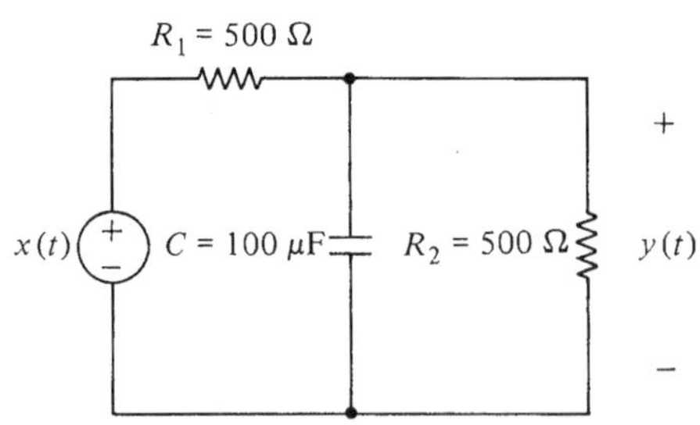

Figure P3.36

3.37. Consider the circuit shown in Figure P3.37. The input is the half-wave rectified signal of Problem 3.8. Find the amplitude of the second and fourth harmonics of output $y(t)$.

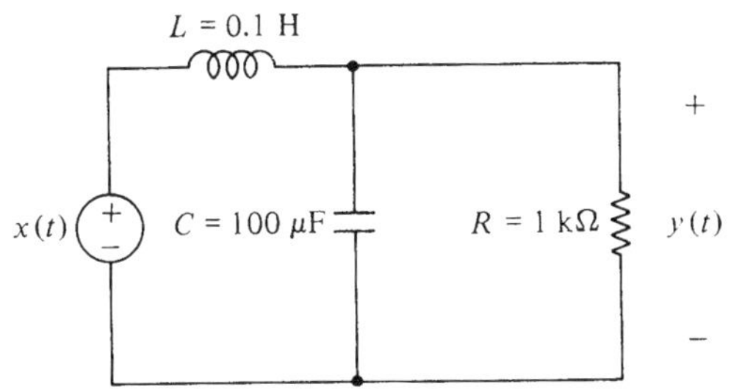

Figure P3.37

3.38. (a) Determine the dc component and the amplitude of the second harmonic of the output signal $y(t)$ in the circuits in Figures P3.36 and P3.37 if the input is the full-wave rectified signal of Problem 3.10.

(b) Find the first harmonic of the output signal $y(t)$ in the circuits in Figures P3.36 and P3.37 if the input is the triangular waveform of Problem 3.32.

3.39. Show that the following are identities:

(a) $$\sum_{n=-N}^{N} \exp[jn\omega_0 t] = \frac{\sin\left[\left(N + \frac{1}{2}\right)\omega_0 t\right]}{\sin(\omega_0 t/2)}$$

(b) $$\frac{1}{T}\int_{-T/2}^{T/2} \frac{\sin\left(N + \frac{1}{2}\right)\omega_0 t}{\sin(\omega_0 t/2)}\, dt = 1$$

3.40. For the signal $x(t)$ depicted in Example 3.3.3, keep T fixed and discuss the effect of varying τ (with the restriction $\tau < T$) on the Fourier coefficients.

3.41. Consider the signal $x(t)$ shown in Figure 3.3.6. Determine the effect on the amplitude of the second harmonic of $x(t)$ when there is a very small error in measuring τ. To do this, let $\tau = \tau_0 - \varepsilon$, where $\varepsilon \ll \tau_0$, and find the second harmonic dependence on ε. Find the percentage change in $|c_2|$ when $T = 10$, $\tau = 1$, and $\varepsilon = 0.1$.

3.42. A truncated sinusoidal waveform is shown in Figure P3.42.

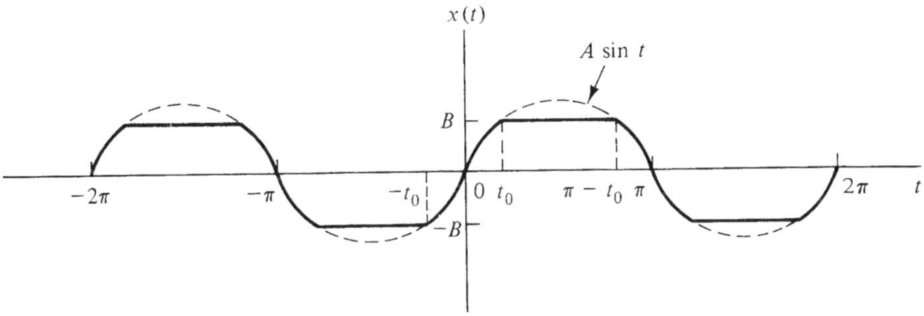

Figure P3.42

(a) Determine the Fourier-series coefficients.

(b) Calculate the amplitude of the third harmonic for $B = A/2$.

(c) Solve for t_0 such that $|c_3|$ is maximum. This method is used to generate harmonic content from a sinusoidal waveform.

3.43. For the signal $x(t)$ shown in Figure P3.43, find the following:

(a) Determine the Fourier-series coefficients.

(b) Solve for the optimum value of t_0 for which $|c_3|$ is maximum.

(c) Compare the result with part (c) of Problem 3.42.

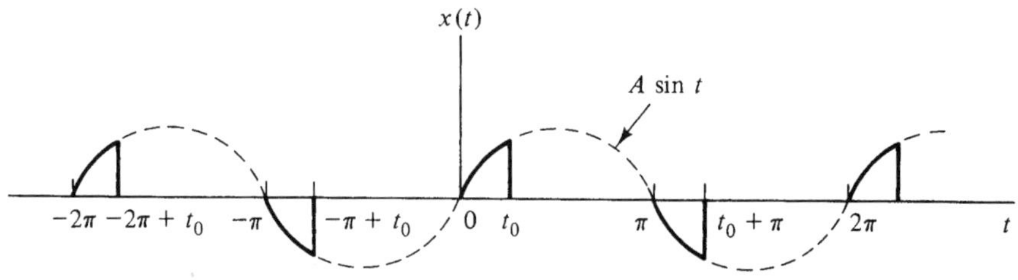

Figure P3.43

3.44. The signal $x(t)$ shown in Figure P3.44 is the output of a smoothed half-wave rectified signal. The constants t_1, t_2, and A satisfy the following relations:

$$\omega t_1 = \pi - \tan^{-1}(\omega RC)$$

$$A = \sin \omega t_1 \exp\left[\frac{t_1}{RC}\right]$$

$$A \exp\left[-\frac{t_2}{RC}\right] = \sin \omega t_2$$

$$RC = 0.1s$$

$$\omega = 2\pi \times 60 = 377 \text{ rad/s}$$

(a) Verify that $\omega t_1 = 1.5973$ rad, $A = 1.0429$, and $\omega t_2 = 7.316$ rad.

(b) Determine the exponential Fourier-series coefficients.

(c) Find the ratio of the amplitudes of the first harmonic and the dc component.

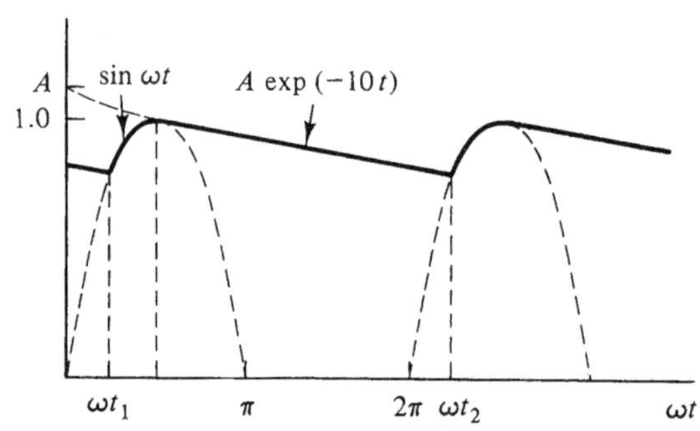

Figure P3.44

3.11 COMPUTER PROBLEMS

3.45. The Fourier-series coefficients can be computed numerically. This becomes advantageous when an analytical expression for $x(t)$ is not known and $x(t)$ is available as numerical data or when the integration is difficult to perform. Show that

$$a_0 \simeq \frac{1}{M} \sum_{m=1}^{M} x(m\,\Delta t)$$

$$a_n \simeq \frac{2}{M} \sum_{m=1}^{M} x(m\,\Delta t) \cos \frac{2\pi mn}{M}$$

$$b_n \simeq \frac{2}{M} \sum_{m=1}^{M} x(m\,\Delta t) \sin \frac{2\pi mn}{M}$$

where $x(m\,\Delta t)$ are M equally spaced data points representing $x(t)$ over $(0, T)$ and Δt is the interval between data points such that

$$\Delta t = T/M$$

(Hint: Approximate the integral with a summation of rectangular strips, each of width Δt.)

3.46. Consider the triangular signal of Example 3.5.2 with $A = \pi/2$ and $T = 2\pi$.

(a) Use the method of Problem 3.45 to compute numerically the first five harmonics from N equally spaced points per period of this waveform. Assume that $N = 100$.

(b) Compare the numerical values obtained in (a) with the actual values.

(c) Repeat for values of $N = 20, 40, 60,$ and 80. Comment on your results.

3.47. The signal of Figure P3.47 can be represented as

$$x(t) = \frac{4}{\pi} \sum_{\substack{n=1,\\ n=\text{odd}}}^{\infty} \frac{1}{n} \sin n\pi t$$

Using the approximation

$$\hat{x}_N(t) = \frac{4}{\pi} \sum_{\substack{n=1,\\ n=\text{odd}}}^{N} \frac{1}{n} \sin n\pi t$$

write a computer program to calculate and sketch the error function

$$e_N(t) = x(t) - \hat{x}_N(t)$$

from $t = 0$ to $t = 2$ for $N = 1, 3, 5,$ and 7.

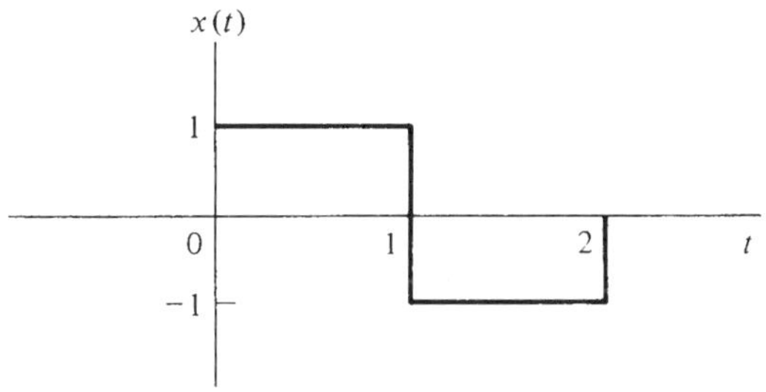

Figure P3.47

3.48. The integral-squared error (error energy) remaining in the approximation of Problem 3.47 after N terms is

$$\int_0^2 |e_N(t)|^2\,dt = \int_0^2 |x(t)|^2\,dt - \sum_{\substack{n=1,\\ n=\text{odd}}}^{N} \left|\frac{4}{n\pi}\right|^2$$

Calculate the integral-squared error for $N = 11, 21, 32, 41, 51, 101,$ and 201.

3.49. Write a program to compute numerically the coefficients of the series expansion in terms of $\text{wal}_w(k, t)$, $0 \le k \le 6$, of the signal $x(t) = t[u(t) - u(t-1)]$. Compare your results with those of Problem 3.6.

Chapter 4

The Fourier Transform

4.1 INTRODUCTION

We saw in Chapter 3 that the Fourier series is a powerful tool in treating various problems involving periodic signals. We first illustrated this fact in Section 3.6, where we demonstrated how an LTI system processes a periodic input to produce the output response. More precisely, at any frequency $n\omega_0$, we showed that the amplitude of the output is equal to the product of the amplitude of the periodic input signal, $|c_n|$, and the magnitude of the system function $|H(\omega)|$ evaluated at $\omega = n\omega_0$, and the phase of the output is equal to the sum of the phase of the periodic input signal, $\measuredangle c_n$, and the system phase $\measuredangle H(\omega)$ evaluated at $\omega = n\omega_0$.

In Chapter 3, we were able to decompose any periodic signal with period T in terms of infinitely many harmonically related complex exponentials of the form $\exp[jn\omega_0 t]$. All such harmonics have the common period $T = 2\pi/\omega_0$. In this chapter, we consider another powerful mathematical technique, called the Fourier transform, for describing both periodic and nonperiodic signals for which no Fourier series exists. Like the Fourier-series coefficients, the Fourier transform specifies the spectral content of a signal, thus providing a frequency-domain description of the signal. Besides being useful in analytically representing aperiodic signals, the Fourier transform is a valuable tool in the analysis of LTI systems.

It is perhaps difficult to see how some typical aperiodic signals, such as

$$u(t), \quad \exp[-t]u(t), \quad \text{rect}(t/T)$$

could be made up of complex exponentials. The problem is that complex exponentials exist for all time and have constant amplitudes, whereas typical aperiodic signals do not possess these properties. In spite of this, we will see that such aperiodic signals do have harmonic content; that is, they can be expressed as the superposition of harmonically related exponentials.

In Section 4.2, we use the Fourier series as a stepping-stone to develop the Fourier transform and show that the latter can be considered an extension of the Fourier series. In Section 4.3, we consider the properties of the Fourier transform that make it useful in LTI system analysis and provide examples of the calculation of some elementary transform pairs. In Section 4.4, we discuss some applications related to the use of Fourier transform theory in communication systems, signal processing, and control systems. In Section 4.5, we introduce the concepts of bandwidth and duration of a signal and discuss several measures for these quantities. Finally, in that same section, the uncertainty principle is developed and its significance is discussed.

4.2 THE CONTINUOUS-TIME FOURIER TRANSFORM

In Chapter 3, we presented the Fourier series as a method for analyzing periodic signals. We saw that the representation of a periodic signal in terms of a weighted sum of complex exponentials was useful in obtaining the steady state response of stable, linear, time-invariant systems to periodic inputs. Fourier series analysis has somewhat limited application in that it is restricted to inputs which are periodic, while many signals of interest are aperiodic. We can develop a method, known as the Fourier transform, for representing aperiodic signals by decomposing such signals into a set of weighted exponentials, in a manner analogous to the Fourier series representation of periodic signals. We will use a heuristic development invoking physical arguments where necessary, to circumvent rigorous mathematics. As we see in the next subsection, in the case of aperiodic signals, the sum in the Fourier series becomes an integral and each exponential has essentially zero amplitude, but the totality of all these infinitesimal exponentials produces the aperiodic signal.

4.2.1 Development of the Fourier Transform

The generalization of the Fourier series to aperiodic signals was suggested by Fourier himself and can be deduced from an examination of the structure of the Fourier series for periodic signals as the period T approaches infinity. In making the transition from the Fourier series to the Fourier transform, where necessary, we use a heuristic development invoking physical arguments to circumvent some very subtle mathematical concepts. After taking the limit, we will find that the magnitude spectrum of an aperiodic signal is not a line spectrum (as with a periodic signal), but instead occupies a continuum of frequencies. The same is true of the corresponding phase spectrum.

To clarify how the change from discrete to continuous spectra takes place, consider the periodic signal $\tilde{x}(t)$ shown in Figure 4.2.1. Now think of keeping the waveform of one period of $\tilde{x}(t)$ unchanged, but carefully and intentionally increase T. In the limit as $T \to \infty$, only a single pulse remains because the nearest neighbors have been moved to infinity. We saw in Chapter 3 that increasing T has two effects on the spectrum of

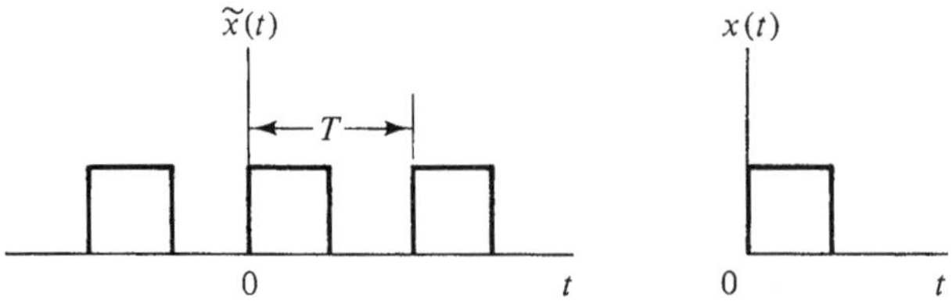

Figure 4.2.1 Allowing the period T to increase to obtain the aperiodic signal.

$\tilde{x}(t)$: The amplitude of the spectrum decreases as $1/T$, and the spacing between lines decreases as $2\pi/T$. As T approaches infinity, the spacing between lines approaches zero. This means that the spectral lines move closer, eventually becoming a continuum. The overall shapes of the magnitude and phase spectra are determined by the shape of the single pulse that remains in the new signal $x(t)$, which is aperiodic.

To investigate what happens mathematically, we use the exponential form of the Fourier series representation for $\tilde{x}(t)$; i.e.,

$$\tilde{x}(t) = \sum_{n=-\infty}^{\infty} c_n \exp[jn\omega_0 t] \tag{4.2.1}$$

where

$$c_n = \frac{1}{T}\int_{-T/2}^{T/2} \tilde{x}(t)\exp[-jn\omega_0 t]\,dt \tag{4.2.2}$$

In the limit as $T \to \infty$, we see that $\omega_0 = 2\pi/T$ becomes an infinitesimally small quantity, $d\omega$, so that

$$\frac{1}{T} \to \frac{d\omega}{2\pi}$$

We argue that in the limit, $n\omega_0$ should be a continuous variable. Then, from Equation (4.2.2), the Fourier coefficients per unit frequency interval are

$$\frac{c_n}{d\omega} = \frac{1}{2\pi}\int_{t=-\infty}^{\infty} \tilde{x}(t)\exp[-j\omega t]\,dt \tag{4.2.3}$$

Substituting Equation (4.2.3) into Equation (4.2.1), and recognizing that in the limit the sum becomes an integral and $\tilde{x}(t)$ approaches $x(t)$, we obtain

$$x(t) = \int_{\omega=-\infty}^{\infty}\left[\int_{t=-\infty}^{\infty} x(t)\exp[-j\omega t]\,dt\right]\exp[j\omega t]\frac{d\omega}{2\pi} \tag{4.2.4}$$

The inner integral, in brackets, is a function of ω only, not t. Denoting this integral by $X(\omega)$, we can write Equation (4.2.4) as

$$x(t) = \frac{1}{2\pi}\int_{-\infty}^{\infty} X(\omega)\exp[j\omega t]\,d\omega \tag{4.2.5}$$

where

$$X(\omega) = \int_{-\infty}^{\infty} x(t) \exp[-j\omega t]\, dt \tag{4.2.6}$$

Equations (4.2.5) and (4.2.6) constitute the Fourier-transform pair for aperiodic signals that most electrical engineers use. (Some communications engineers prefer to write the frequency variable in hertz rather than rad/s; this can be done by an obvious change of variables.) $X(\omega)$ is called the Fourier transform of $x(t)$ and plays the same role for aperiodic signals that c_n plays for periodic signals. Thus, $X(\omega)$ is the spectrum of $x(t)$ and is a continuous function defined for all values of ω, whereas c_n is defined only for discrete frequencies. Therefore, as mentioned earlier, an aperiodic signal has a continuous spectrum rather than a line spectrum. $X(\omega)$ specifies the weight of the complex exponentials used to represent the waveform in Equation (4.2.5) and, in general, is a complex function of the variable ω. Thus, it can be written as

$$X(\omega) = |X(\omega)| \exp[j\phi(\omega)] \tag{4.2.7}$$

The magnitude of $X(\omega)$ plotted against ω is called the magnitude spectrum of $x(t)$, and $|X(\omega)|^2$ is called the energy spectrum. The angle of $X(\omega)$ plotted versus ω is called the phase spectrum.

In Chapter 3, we saw that for any periodic signal $x(t)$, there is a one-to-one correspondence between $x(t)$ and the set of Fourier coefficients c_n. Here, too, it can be shown that there is a one-to-one correspondence between $x(t)$ and $X(\omega)$, denoted by

$$x(t) \leftrightarrow X(\omega)$$

which is meant to imply that for every $x(t)$ having a Fourier transform, there is a unique $X(\omega)$ and vice versa. Some sufficient conditions for the signals to have a Fourier transform are discussed later. We emphasize that while we have used a real-valued signal $x(t)$ as an artifice in the development of the transform pair, the Fourier-transform relations hold for complex signals as well. With few exceptions, however, we will be concerned primarily with real-valued signals of time.

As a notational convenience, $X(\omega)$ is often denoted by $\mathscr{F}\{x(t)\}$ and is read "the Fourier transform of $x(t)$." In addition, we adhere to the convention that the Fourier transform is represented by a capital letter that is the same as the lowercase letter denoting the time signal. For example,

$$\mathscr{F}\{h(t)\} = H(\omega) = \int_{-\infty}^{\infty} h(t) \exp[-j\omega t]\, dt$$

Before we examine further the general properties of the Fourier transform and its physical meaning, let us introduce a set of sufficient conditions for the existence of the Fourier transform.

4.2.2 Existence of the Fourier Transform

The signal $x(t)$ is said to have a Fourier transform in the ordinary sense if the integral in Equation (4.2.6) converges (i.e., exists). Since

$$\left|\int y(t)\, dt\right| \le \int |y(t)|\, dt$$

and $|\exp[-j\omega t]| = 1$, it follows that the integral in Equation (4.2.6) exists if

1. $x(t)$ is absolutely integrable and
2. $x(t)$ is "well behaved."

The first condition means that

$$\int_{-\infty}^{\infty} |x(t)|\, dt < \infty \tag{4.2.8}$$

A class of signals that satisfy Equation (4.2.8) is energy signals. Such signals, in general, are either time limited or asymptotically time limited in the sense that $x(t) \to 0$ as $t \to \pm\infty$. The Fourier transform of power signals (a class of signals defined in Chapter 1 to have infinite energy content, but finite average power) can also be shown to exist, but to contain impulses. Therefore, any signal that is either a power or an energy signal has a Fourier transform.

"Well behaved" means that the signal is not too "wiggly" or, more correctly, that it is of bounded variation. This, simply stated, means that $x(t)$ can be represented by a curve of finite length in any finite interval of time, or alternatively, that the signal has a finite number of discontinuities, minima, and maxima within any finite interval of time. At a point of discontinuity, t_0, the inversion integral in Equation (4.2.5) converges to $\frac{1}{2}[x(t_0^+) + x(t_0^-)]$; otherwise it converges to $x(t)$. Except for impulses, most signals of interest are well behaved and satisfy Equation (4.2.8).

The conditions just given for the existence of the Fourier transform of $x(t)$ are sufficient conditions. This means that there are signals that violate either one or both conditions and yet possess a Fourier transform. Examples are power signals (unit-step signal, periodic signals, etc.) that are not absolutely integrable over an infinite interval and impulse trains that are not "well behaved" and are neither power nor energy signals, but still have Fourier transforms. We can include signals that do not have Fourier transforms in the ordinary sense by generalization to transforms in the limit. For example, to obtain the Fourier transform of a constant, we consider $x(t) = \text{rect}(t/\tau)$ and let $\tau \to \infty$ after obtaining the Fourier transform.

4.2.3 Examples of the Continuous-Time Fourier Transform

In this section, we compute the transform of some commonly encountered time signals.

Example 4.2.1

The Fourier transform of the rectangular pulse $x(t) = \text{rect}(t/\tau)$ is

$$\begin{aligned} X(\omega) &= \int_{-\infty}^{\infty} x(t) \exp[-j\omega t]\, dt \\ &= \int_{-\tau/2}^{\tau/2} \exp[-j\omega t]\, dt \\ &= \frac{1}{-j\omega}\left(\exp\left[\frac{-j\omega\tau}{2}\right] - \exp\left[\frac{j\omega\tau}{2}\right]\right) \end{aligned}$$

This can be simplified to

$$X(\omega) = \frac{2}{\omega}\sin\frac{\omega\tau}{2} = \tau\ \mathrm{sinc}\frac{\omega\tau}{2\pi} = \tau\ \mathrm{Sa}\frac{\omega\tau}{2}$$

Since $X(\omega)$ is a real-valued function of ω, its phase is zero for all ω. $X(\omega)$ is plotted in Figure 4.2.2 as a function of ω.

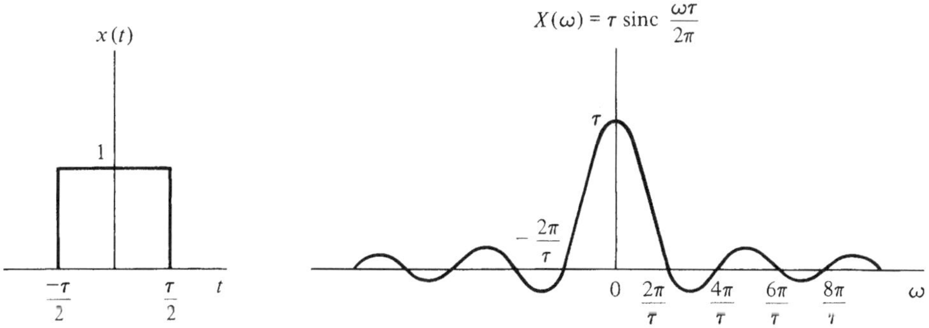

Figure 4.2.2 Fourier transform of a rectangular pulse.

Clearly, the spectrum of the rectangular pulse extends over the range $-\infty < \omega < \infty$. However, from Figure 4.2.2, we see that most of the spectral content of the pulse is contained in the interval $-2\pi/\tau < \omega < 2\pi/\tau$. This interval is labeled the main lobe of the sinc signal. The other portion of the spectrum represents what are called the side lobes of the spectrum. Increasing τ results in a narrower main lobe, whereas a smaller τ produces a Fourier transform with a wider main lobe.

Example 4.2.2

Consider the triangular pulse defined as

$$\Delta(t/\tau) = \begin{cases} 1 - \dfrac{|t|}{\tau}, & |t| \le \tau \\ 0, & |t| > \tau \end{cases}$$

This pulse is of unit height, centered about $t = 0$, and of width 2τ. Its Fourier transform is

$$\begin{aligned} X(\omega) &= \int_{-\infty}^{\infty} \Delta(t/\tau)\exp[-j\omega t]\,dt \\ &= \int_{-\tau}^{0}\left(1 + \frac{t}{\tau}\right)\exp[-j\omega t]\,dt + \int_{0}^{\tau}\left(1 - \frac{t}{\tau}\right)\exp[-j\omega t]\,dt \\ &= \int_{0}^{\tau}\left(1 - \frac{t}{\tau}\right)\exp[j\omega t]\,dt + \int_{0}^{\tau}\left(1 - \frac{t}{\tau}\right)\exp[-j\omega t]\,dt \\ &= 2\int_{0}^{\tau}\left(1 - \frac{t}{\tau}\right)\cos\omega t\,dt \end{aligned}$$

After performing the integration and simplifying the expression, we obtain

$$\Delta(t/\tau) \leftrightarrow \tau \operatorname{sinc}^2 \frac{\omega\tau}{2\pi} = \tau \operatorname{Sa}^2 \frac{\omega\tau}{2}$$

Example 4.2.3

The Fourier transform of the one-sided exponential signal

$$x(t) = \exp[-\alpha t]u(t), \quad \alpha > 0$$

is obtained from Equation (4.2.6) as

$$\begin{aligned} X(\omega) &= \int_{-\infty}^{\infty} (\exp[-\alpha t]u(t)\exp[-j\omega t])\,dt \\ &= \int_{0}^{\infty} \exp[-(\alpha + j\omega)t]\,dt \\ &= \frac{1}{\alpha + j\omega} \end{aligned} \tag{4.2.9}$$

Example 4.2.4

In this example, we evaluate the Fourier transform of the two-sided exponential signal

$$x(t) = \exp[-\alpha|t|], \quad \alpha > 0$$

From Equation (4.2.6), the transform is

$$\begin{aligned} X(\omega) &= \int_{-\infty}^{0} \exp[\alpha t]\exp[-j\omega t]\,dt + \int_{0}^{\infty} \exp[-\alpha t]\exp[-j\omega t]\,dt \\ &= \frac{1}{\alpha - j\omega} + \frac{1}{\alpha + j\omega} \\ &= \frac{2\alpha}{\alpha^2 + \omega^2} \end{aligned}$$

Example 4.2.5

The Fourier transform of the impulse function is readily obtained from Equation (4.2.6) by making use of Equation (1.6.7):

$$\mathscr{F}\{\delta(t)\} = \int_{-\infty}^{\infty} \delta(t)\exp[-j\omega t]\,dt = 1$$

We thus have the pair

$$\delta(t) \leftrightarrow 1 \tag{4.2.10}$$

Using the inversion formula, we must clearly have

$$\delta(t) = \frac{1}{2\pi}\int_{-\infty}^{\infty} \exp[j\omega t]\,d\omega \tag{4.2.11}$$

Equation (4.2.11) states that the impulse signal theoretically consists of equal-amplitude sinusoids of all frequencies. This integral is obviously meaningless, unless we interpret $\delta(t)$ as a function specified by its properties rather than an ordinary function having definite values for every t, as we demonstrated in Chapter 1. Equation (4.2.11) can also be written in the limit form

$$\delta(t) = \lim_{\alpha\to\infty} \frac{\sin \alpha t}{\pi t} \tag{4.2.12}$$

This result can be established by writing Equation (4.2.11) as

$$\begin{aligned}\delta(t) &= \frac{1}{2\pi}\lim_{\alpha\to\infty}\int_{-\alpha}^{\alpha} \exp[j\omega t]\,d\omega \\ &= \frac{1}{2\pi}\lim_{\alpha\to\infty}\frac{2\sin\alpha t}{t} \\ &= \lim_{\alpha\to\infty}\frac{\sin\alpha t}{\pi t}\end{aligned}$$

Example 4.2.6

We can easily show that $\int_{-\infty}^{\infty} \exp[j\omega t]\,d\omega/2\pi$ "behaves" like the unit-impulse function by putting it inside an integral; i.e., we evaluate an integral of the form

$$\int_{-\infty}^{\infty}\left[\frac{1}{2\pi}\int_{-\infty}^{\infty} \exp[j\omega t]\,d\omega\right] g(t)\,dt$$

where $g(t)$ is any arbitrary well-behaved signal that is continuous at $t = 0$ and possesses a Fourier transform $G(\omega)$. Interchanging the order of integration, we have

$$\frac{1}{2\pi}\int_{-\infty}^{\infty}\left[\int_{-\infty}^{\infty} g(t)\exp[j\omega t]\,dt\right]d\omega = \frac{1}{2\pi}\int_{-\infty}^{\infty} G(-\omega)\,d\omega$$

From the inversion formula it follows that

$$\frac{1}{2\pi}\int_{-\infty}^{\infty} G(-\omega)\,d\omega = \frac{1}{2\pi}\int_{-\infty}^{\infty} G(\omega)\,d\omega = g(0)$$

That is, $(1/2\pi)\int_{-\infty}^{\infty} \exp[j\omega t]\,d\omega$ "behaves" like an impulse at $t = 0$.

Another transform pair follows from interchanging the roles of t and ω in Equation (4.2.11). The result is

$$\delta(\omega) = \frac{1}{2\pi}\int_{-\infty}^{\infty} \exp[j\omega t]\,dt$$

or

$$1 \leftrightarrow 2\pi\,\delta(\omega) \tag{4.2.13}$$

In words, the Fourier transform of a constant is an impulse in the frequency domain. The factor 2π arises because we are using radian frequency. If we were to write the transform in terms of frequency in hertz, the factor 2π would disappear ($\delta(\omega) = \delta(f)/2\pi$).

Example 4.2.7

In this example, we use Equation (4.2.12) and Example 4.2.1 to prove Equation (4.2.13). By letting τ go to ∞ in Example 4.2.1, we find that the signal $x(t)$ approaches 1 for all values of t. On the other hand, from Equation (4.2.12), the limit of the transform of rect (t/τ) becomes

$$\lim_{\tau \to \infty} \frac{2}{\omega} \frac{\sin \omega\tau}{2} = 2\pi\,\delta(\omega)$$

Example 4.2.8

Consider the exponential signal $x(t) = \exp[j\omega_0 t]$. The Fourier transform of this signal is

$$X(\omega) = \int_{-\infty}^{\infty} \exp[j\omega_0 t] \exp[-j\omega t]\, dt$$

$$= \int_{-\infty}^{\infty} \exp[-j(\omega - \omega_0)t]\, dt$$

Using the result leading to Equation (4.2.13), we obtain

$$\exp[j\omega_0 t] \leftrightarrow 2\pi\,\delta(\omega - \omega_0) \tag{4.2.14}$$

This is expected, since $\exp[j\omega_0 t]$ has energy concentrated at ω_0.

Periodic signals are power signals, and we anticipate, according to the discussion in Section 4.2.2, that their Fourier transforms contain impulses (delta functions). In Chapter 3, we examined the spectrum of periodic signals by computing the Fourier-series coefficients. We found that the spectrum consists of a set of lines located at $\pm n\omega_0$, where ω_0 is the fundamental frequency of the periodic signal. In the following example, we find the Fourier transform of periodic signals and show that the spectra of periodic signals consist of trains of impulses.

Example 4.2.9

Consider the periodic signal $x(t)$ with period T; thus, $\omega_0 = 2\pi/T$. Assume that $x(t)$ has the Fourier-series representation

$$x(t) = \sum_{n=-\infty}^{\infty} c_n \exp[jn\omega_0 t]$$

Hence, taking the Fourier transform of both sides yields

$$X(\omega) = \sum_{n=-\infty}^{\infty} c_n \mathcal{F}\{\exp[jn\omega_0 t]\}$$

Using Equation (4.2.14), we obtain

$$X(\omega) = \sum_{n=-\infty}^{\infty} 2\pi c_n \delta(\omega - n\omega_0) \tag{4.2.15}$$

Thus, the Fourier transform of a periodic signal is simply an impulse train with impulses located at $\omega = n\omega_0$, each of which has a strength $2\pi c_n$, and all impulses are separated from each other by ω_0. Note that because the signal $x(t)$ is periodic, the magnitude spectrum $|X(\omega)|$ is a train of impulses of strength $2\pi|c_n|$, whereas the spectrum obtained through the use of the Fourier series is a line spectrum with lines of finite amplitude $|c_n|$. Note that the Fourier transform is not a periodic function: Even though the impulses are separated by the same amount, their weights are all different.

Example 4.2.10

Consider the periodic signal

$$x(t) = \sum_{n=-\infty}^{\infty} \delta(t - nT)$$

which has period T. To find the Fourier transform, we first have to compute the Fourier-series coefficients. From Equation (3.3.4), the Fourier-series coefficients are

$$c_n = \frac{1}{T}\int_{\langle T\rangle} x(t)\exp\left[-\frac{j2\pi nt}{T}\right]dt = \frac{1}{T}$$

since $x(t) = \delta(t)$ in any interval of length T. Thus, the impulse train has the Fourier series representation

$$x(t) = \sum_{n=-\infty}^{\infty} \frac{1}{T}\exp\left[\frac{j2\pi nt}{T}\right]$$

By using Equation (4.2.14), we find that the Fourier transform of the impulse train is

$$X(\omega) = \frac{2\pi}{T}\sum_{n=-\infty}^{\infty} \delta\left(\omega - \frac{2\pi n}{T}\right) \tag{4.2.16}$$

That is, the Fourier transformation of a sequence of impulses in the time domain yields a sequence of impulses in the frequency domain.

A brief listing of some other Fourier pairs is given in Table 4.1.

4.3 PROPERTIES OF THE FOURIER TRANSFORM

A number of useful properties of the Fourier transform allow some problems to be solved almost by inspection. In this section, we shall summarize many of these properties, some of which may be more or less obvious to the reader.

4.3.1 Linearity

$$x_1(t) \leftrightarrow X_1(\omega)$$

$$x_2(t) \leftrightarrow X_2(\omega)$$

then

$$ax_1(t) + bx_2(t) \leftrightarrow aX_1(\omega) + bX_2(\omega) \tag{4.3.1}$$

TABLE 4.1
Some Selected Fourier Transform Pairs

$x(t)$	$X(\omega)$
1. 1	$2\pi\,\delta(\omega)$
2. $u(t)$	$\pi\delta(\omega) + \dfrac{1}{j\omega}$
3. $\delta(t)$	1
4. $\delta(t - t_0)$	$\exp[-j\omega t_0]$
5. $\mathrm{rect}(t/\tau)$	$\tau\,\mathrm{sinc}\,\dfrac{\omega\tau}{2\pi} = \dfrac{2\sin\omega\tau/2}{\omega}$
6. $\dfrac{\omega_B}{\pi}\,\mathrm{sinc}\,\dfrac{\omega_B t}{\pi} = \dfrac{\sin\omega_B t}{\pi t}$	$\mathrm{rect}(\omega/2\omega_B)$
7. $\mathrm{sgn}\,t$	$\dfrac{2}{j\omega}$
8. $\exp[j\omega_0 t]$	$2\pi\,\delta(\omega - \omega_0)$
9. $\sum_{n=-\infty}^{\infty} a_n \exp[jn\omega_0 t]$	$2\pi \sum_{n=-\infty}^{\infty} a_n \delta(\omega - n\omega_0)$
10. $\cos\omega_0 t$	$\pi[\delta(\omega - \omega_0) + \delta(\omega + \omega_0)]$
11. $\sin\omega_0 t$	$\dfrac{\pi}{j}[\delta(\omega - \omega_0) - \delta(\omega + \omega_0)]$
12. $(\cos\omega_0 t)u(t)$	$\dfrac{\pi}{2}[\delta(\omega - \omega_0) + \delta(\omega + \omega_0)] + \dfrac{j\omega}{\omega_0^2 - \omega^2}$
13. $(\sin\omega_0 t)u(t)$	$\dfrac{\pi}{2_j}[\delta(\omega - \omega_0) - \delta(\omega + \omega_0)] + \dfrac{\omega_0}{\omega_0^2 - \omega^2}$
14. $\cos\omega_0 t\,\mathrm{rect}(t/\tau)$	$\tau\,\mathrm{sinc}\,\dfrac{(\omega - \omega_0)\tau}{2\pi}$
15. $\exp[-at]u(t), \quad \mathrm{Re}\{a\} > 0$	$\dfrac{1}{a + j\omega}$
16. $t\exp[-at]u(t), \quad \mathrm{Re}\{a\} > 0$	$\left(\dfrac{1}{a + j\omega}\right)^2$
17. $\dfrac{t^{n-1}}{(n-1)!}\exp[-at]u(t), \quad \mathrm{Re}\{a\} > 0$	$\dfrac{1}{(a + j\omega)^n}$
18. $\exp[-a\lvert t\rvert], \quad a > 0$	$\dfrac{2a}{a^2 + \omega^2}$
19. $\lvert t\rvert\exp[-a\lvert t\rvert], \quad \mathrm{Re}\{a\} > 0$	$\dfrac{4aj\omega}{a^2 + \omega^2}$

TABLE 4.1 *(continued)*

	$x(t)$	$X(\omega)$
20.	$\dfrac{1}{a^2+t^2}$, $\text{Re}\{a\}>0$	$\dfrac{\pi}{a}\exp[-a\|\omega\|]$
21.	$\dfrac{t}{a^2+t^2}$, $\text{Re}\{a\}>0$	$\dfrac{-j\pi\omega\exp[-a\|\omega\|]}{2a}$
22.	$\exp[-at^2]$, $a>0$	$\sqrt{\dfrac{\pi}{a}}\exp\left[\dfrac{-\omega^2}{4a}\right]$
23.	$\Delta(t/\tau)$	$\tau\,\text{sinc}^2\dfrac{\omega\tau}{2\pi}$
24.	$\sum_{n=-\infty}^{\infty}\delta(t-nT)$	$\dfrac{2\pi}{T}\sum_{n=-\infty}^{\infty}\delta\left(\omega-\dfrac{2n\pi}{T}\right)$

where a and b are arbitrary constants. This property is the direct result of the linearity of the operation of integration. The linearity property can be easily extended to a linear combination of an arbitrary number of components and simply means that the Fourier transform of a linear combination of an arbitrary number of signals is the same linear combination of the transform of the individual components.

Example 4.3.1

Suppose we want to find the Fourier transform of $\cos\omega_0 t$. The cosine signal can be written as a sum of two exponentials as follows:

$$\cos\omega_0 t = \frac{1}{2}[\exp[j\omega_0 t] + \exp[-j\omega_0 t]]$$

From Equation (4.2.14) and the linearity property of the Fourier transform,

$$\mathcal{F}\{\cos\omega_0 t\} = \pi[\delta(\omega-\omega_0) + \delta(\omega+\omega_0)]$$

Similarly, the Fourier transform of $\sin\omega_0 t$ is

$$\mathcal{F}\{\sin\omega_0 t\} = \frac{\pi}{j}[\delta(\omega-\omega_0) - \delta(\omega+\omega_0)]$$

4.3.2 Symmetry

If $x(t)$ is a real-valued time signal, then

$$X(-\omega) = X^*(\omega) \tag{4.3.2}$$

where * denotes the complex conjugate. This property, referred to as conjugate symmetry, follows from taking the conjugate of both sides of Equation (4.2.6) and using the fact that $x(t)$ is real.

Now, if we express $X(\omega)$ in the polar form, we have

$$X(\omega) = |X(\omega)| \exp[j\phi(\omega)] \tag{4.3.3}$$

Taking the complex conjugate of both sides of Equation (4.3.3) yields

$$X^*(\omega) = |X(\omega)| \exp[-j\phi(\omega)]$$

Replacing each ω by $-\omega$ in Equation (4.3.3) results in

$$X(-\omega) = |X(-\omega)| \exp[j\phi(-\omega)]$$

By Equation (4.3.2), the left-hand sides of the last two equations are equal. It then follows that

$$|X(\omega)| = |X(-\omega)| \tag{4.3.4}$$

$$\phi(\omega) = -\phi(-\omega) \tag{4.3.5}$$

i.e., the magnitude spectrum is an even function of frequency, and the phase spectrum is an odd function of frequency.

Example 4.3.2

From Equations (4.3.4) and (4.3.5), the inversion formula, Equation (4.2.5), which is written in terms of complex exponentials, can be changed to an expression involving real cosinusoidal signals. Specifically, for real $x(t)$,

$$\begin{aligned} x(t) &= \frac{1}{2\pi}\int_{-\infty}^{\infty} X(\omega)\exp[j\omega t]\,d\omega \\ &= \frac{1}{2\pi}\int_{-\infty}^{0} X(\omega)\exp[j\omega t]\,d\omega + \frac{1}{2\pi}\int_{0}^{\infty} X(\omega)\exp[j\omega t]\,d\omega \\ &= \frac{1}{2\pi}\int_{0}^{\infty} |X(\omega)|\,(\exp[j(\omega t + \phi(\omega))] + \exp[-j(\omega t + \phi(\omega))])\,d\omega \\ &= \frac{1}{2\pi}\int_{0}^{\infty} 2|X(\omega)|\cos[\omega t + \phi(\omega)]\,d\omega \end{aligned}$$

Equations (4.3.4) and (4.3.5) ensure that the exponentials of the form $\exp[j\omega t]$ combine properly with those of the form $\exp[-j\omega t]$ to produce real sinusoids of frequency ω for use in the expansion of real-valued time signals. Thus, a real-valued signal $x(t)$ can be written in terms of the amplitudes and phases of real sinusoids that constitute the signal.

Example 4.3.3

Consider an even and real-valued signal $x(t)$. Its transform $X(\omega)$ is

$$\begin{aligned} X(\omega) &= \int_{-\infty}^{\infty} x(t)\exp[-j\omega t]\,dt \\ &= \int_{-\infty}^{\infty} x(t)(\cos\omega t - j\sin\omega t)\,dt \end{aligned}$$

Since $x(t)\cos\omega t$ is an even function of t and $x(t)\sin\omega t$ is an odd function of t, we have

$$X(\omega) = 2\int_0^\infty x(t)\cos\omega t\,dt$$

which is a real and even function of ω. Therefore, the Fourier transform of an even and real-valued signal in the time domain is an even and real-valued signal in the frequency domain.

4.3.3 Time Shifting

If

$$x(t) \leftrightarrow X(\omega)$$

then

$$x(t - t_0) \leftrightarrow X(\omega)\exp[-j\omega t_0] \tag{4.3.6a}$$

Similarly,

$$x(t)\,e^{j\omega_0 t} \leftrightarrow X(\omega - \omega_0) \tag{4.3.6b}$$

The proofs of these properties follow from Equation (4.2.6) after suitable substitution of variables. Using the polar form, Equation (4.3.3), in Equation (4.3.6a) yields

$$\mathcal{F}\{x(t - t_0)\} = |X(\omega)|\exp[j(\phi(\omega) - \omega t_0)]$$

The last equation indicates that shifting in time does not alter the amplitude spectrum of the signal. The only effect of such shifting is to introduce a phase shift in the transform that is a linear function of ω. The result is reasonable because we have already seen that, to delay or advance a sinusoid, we have only to adjust the phase. In addition, the energy content of a waveform does not depend on its position in time.

4.3.4 Time Scaling

If

$$x(t) \leftrightarrow X(\omega)$$

then

$$x(\alpha t) \leftrightarrow \frac{1}{|\alpha|}X\left(\frac{\omega}{\alpha}\right) \tag{4.3.7}$$

where α is a real constant. The proof of this follows directly from the definition of the Fourier transform and the appropriate substitution of variables.

Aside from the amplitude factor of $1/|\alpha|$, linear scaling in time by a factor of α corresponds to linear scaling in frequency by a factor of $1/\alpha$. The result can be interpreted physically by considering a typical signal $x(t)$ and its Fourier transform $X(\omega)$, as shown in Figure 4.3.1. If $|\alpha| < 1$, $x(\alpha t)$ is expanded in time, and the signal varies more slowly (becomes smoother) than the original. These slower variations deemphasize the high-frequency components and manifest themselves in more appreciable low-frequency sinusoidal components. That is, expansion in the time domain implies compression in

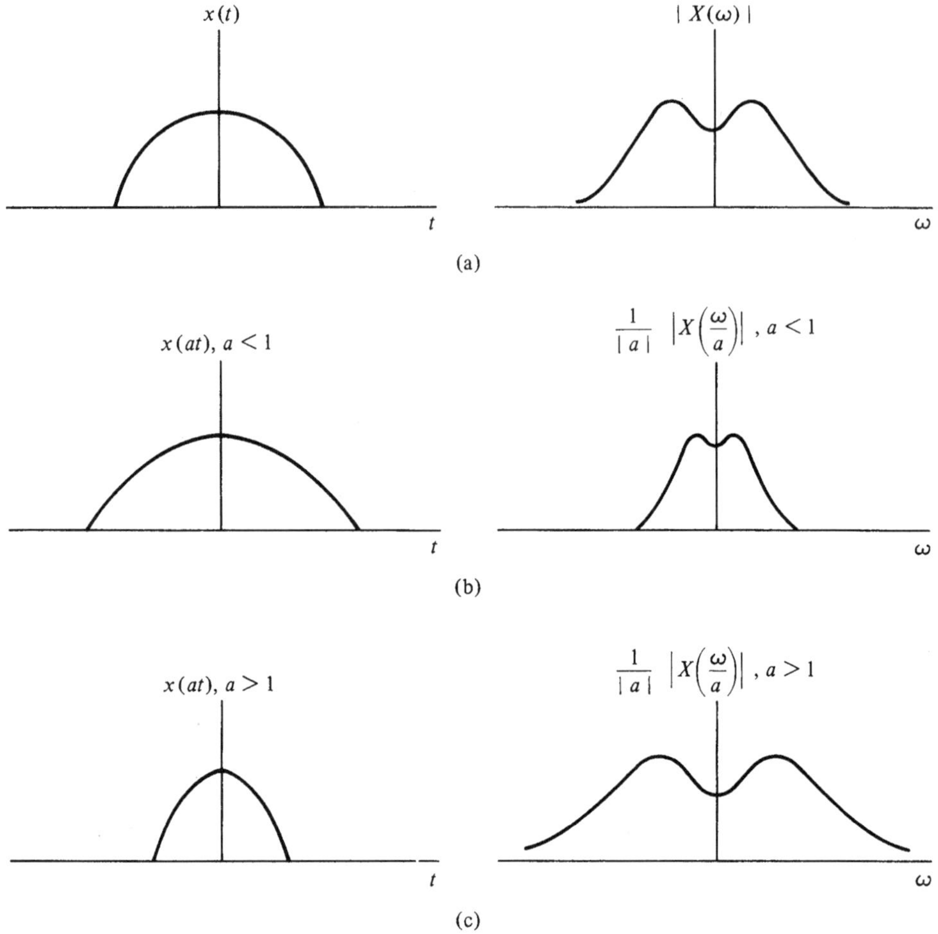

Figure 4.3.1 Examples of the time-scaling property: (a) The original signal and its magnitude spectrum, (b) the time-expanded signal and its magnitude spectrum, and (c) the time-compressed signal and the resulting magnitude spectrum.

the frequency domain and vice versa. If $|\alpha| > 1$, $x(\alpha t)$ is compressed in time and must vary rapidly. Faster variations in time are manifested by the presence of higher frequency components.

The notion of time expansion and frequency compression has found application in areas such as data transmission from space probes to receiving stations on Earth. To reduce the amount of noise superimposed on the required signal, it is necessary to keep the bandwidth of the receiver as small as possible. One means of accomplishing this is to reduce the bandwidth of the signal, store the data collected by the probe, and then play the data back at a slower rate. Because the time-scaling factor is known, the signal can be reproduced at the receiver.

Example 4.3.4

Suppose we want to determine the Fourier transform of the pulse $x(t) = \alpha\,\text{rect}(\alpha t/\tau)$, $\alpha > 0$. The Fourier transform of $\text{rect}(t/\tau)$ is, by Example 4.2.1,

$$\mathcal{F}\{\text{rect}(t/\tau)\} = \tau\,\text{sinc}\frac{\omega\tau}{2\pi}$$

By Equation (4.3.7), the Fourier transform of $\alpha\,\text{rect}(\alpha t/\tau)$ is

$$\mathcal{F}\{\alpha\,\text{rect}(\alpha t/\tau)\} = \tau\,\text{sinc}\frac{\omega\tau}{2\alpha\pi}$$

Note that as we increase the value of the parameter α, the rectangular pulse becomes narrower and higher and approaches an impulse as $\alpha \to \infty$. Correspondingly, the main lobe of the Fourier transform becomes wider, and in the limit $X(\omega)$ approaches a constant value for all ω. On the other hand, as α approaches zero, the rectangular signal approaches 1 for all t, and the transform approaches a delta signal. (See Example 4.2.7.)

The inverse relationship between time and frequency is encountered in a wide variety of science and engineering applications. In Section 4.5, we will cover one application of this relationship, namely, the uncertainty principle.

4.3.5 Differentiation

If

$$x(t) \leftrightarrow X(\omega)$$

then

$$\frac{dx(t)}{dt} \leftrightarrow j\omega X(\omega) \tag{4.3.8}$$

The proof of this property is obtained by direct differentiation of both sides of Equation (4.2.5) with respect to t. The differentiation property can be extended to yield

$$\frac{d^n x(t)}{dt^n} \leftrightarrow (j\omega)^n X(\omega) \tag{4.3.9}$$

We must be careful when using the differentiation property. First of all, the property does not ensure the existence of $\mathcal{F}\{dx(t)/dt\}$. However, if $\mathcal{F}$ exists, it is given by $j\omega X(\omega)$. Second, one cannot always infer that $X(\omega) = \mathcal{F}\{dx(t)/dt\}/j\omega$.

Since differentiation in the time domain corresponds to multiplication by $j\omega$ in the frequency domain, one might conclude that integration in the time domain should involve division by $j\omega$ in the frequency domain. However, this is true only for a certain class of signals. To demonstrate it, consider the signal $y(t) = \int_{-\infty}^{t} x(\tau)\,d\tau$. With $Y(\omega)$

as its transform, we conclude from $dy(t)/dt = x(t)$ and Equation (4.3.8) that $j\omega Y(\omega) = X(\omega)$. For $Y(\omega)$ to exist, $y(t)$ should satisfy the conditions listed in Section 4.2.2. This is equivalent to $y(\infty) = 0$, i.e., $\int_{-\infty}^{\infty} x(\tau)\,d\tau = X(0) = 0$. In this case,

$$\int_{-\infty}^{t} x(\tau)\,d\tau \leftrightarrow \frac{1}{j\omega} X(\omega) \tag{4.3.10}$$

This equation implies that integration in the time domain attenuates (deemphasizes) the magnitude of the high-frequency components of the signal. Hence, an integrated signal is smoother than the original signal. This is why integration is sometimes called a smoothing operation.

If $X(0) \neq 0$, then signal $x(t)$ has a dc component, so that according to Equation (4.2.13), the transform will contain an impulse. As we will show later (see Example 4.3.10), in this case

$$\int_{-\infty}^{t} x(\tau)\,d\tau \leftrightarrow \pi X(0)\,\delta(\omega) + \frac{1}{j\omega} X(\omega) \tag{4.3.11}$$

Example 4.3.5

Consider the unit-step function. As we saw in Section 1.6, this function can be written as

$$\begin{aligned} u(t) &= \frac{1}{2} + \left[u(t) - \frac{1}{2}\right] \\ &= \frac{1}{2} + \frac{1}{2}\operatorname{sgn} t \end{aligned}$$

The first term has $\pi\delta(\omega)$ as its transform. Although $\operatorname{sgn} t$ does not have a derivative in the regular sense, in Section 1.6 we defined the derivatives of discontinuous signals in terms of the delta function. As a consequence,

$$\frac{d}{dt}\left\{\frac{1}{2}\operatorname{sgn} t\right\} = \delta(t)$$

Since $\operatorname{sgn} t$ has a zero dc component (it is an odd signal), applying Equation (4.3.10) yields

$$j\omega \mathscr{F}\left\{\frac{1}{2}\operatorname{sgn} t\right\} = 1$$

or

$$\mathscr{F}\left\{\frac{1}{2}\operatorname{sgn} t\right\} = \frac{1}{j\omega} \tag{4.3.12}$$

By the linearity of the Fourier transform, we obtain

$$u(t) \leftrightarrow \pi\delta(\omega) + \frac{1}{j\omega} \tag{4.3.13}$$

Therefore, the Fourier transform of the unit-step function contains an impulse at $\omega = 0$ corresponding to the average value of 1/2. It also has all the high-frequency components of the signum function, reduced by one-half.

4.3.6 Energy of Aperiodic Signals

In Section 3.5.6, we related the total average power of a periodic signal to the average power of each frequency component in the Fourier series of the signal. We did this through Parseval's theorem. Now we would like to find the analogous relationship for aperiodic signals, which are energy signals. Thus, in this section, we show that the energy of aperiodic signals can be computed from their transform $X(\omega)$. The energy is defined as

$$E = \int_{-\infty}^{\infty} |x(t)|^2 dt = \int_{-\infty}^{\infty} x(t)x^*(t)\, dt$$

Using Equation (4.2.5) in this equation results in

$$E = \int_{-\infty}^{\infty} x(t)\left[\frac{1}{2\pi}\int_{-\infty}^{\infty} X^*(\omega)\exp[-j\omega t]\,d\omega\right]dt$$

Interchanging the order of integration gives

$$E = \frac{1}{2\pi}\int_{-\infty}^{\infty} X^*(\omega)\left[\int_{-\infty}^{\infty} x(t)\exp[-j\omega t]\,dt\right]d\omega$$

$$= \frac{1}{2\pi}\int_{-\infty}^{\infty} |X(\omega)|^2 d\omega$$

We can therefore write

$$\int_{-\infty}^{\infty} |x(t)|^2 dt = \frac{1}{2\pi}\int_{-\infty}^{\infty} |X(\omega)|^2 d\omega \tag{4.3.14}$$

This relation is Parseval's relation for aperiodic signals. It says that the energy of an aperiodic signal can be computed in the frequency domain by computing the energy per unit frequency, $\mathscr{E}(\omega) = |X(\omega)|^2/2\pi$, and integrating over all frequencies. For this reason, $\mathscr{E}(\omega)$ is often referred to as the energy-density spectrum, or, simply, the energy spectrum of the signal, since it measures the frequency distribution of the total energy of $x(t)$. We note that the energy spectrum of a signal depends on the magnitude of the spectrum and not on the phase. This fact implies that there are many signals that may have the same energy spectrum. However, for a given signal, there is only one energy spectrum. The energy in an infinitesimal band of frequencies $d\omega$ is, then, $\mathscr{E}(\omega)\,d\omega$, and the energy contained within a band $\omega_1 \leq \omega \leq \omega_2$ is

$$\Delta E = \int_{\omega_1}^{\omega_2} \frac{1}{2\pi}|X(\omega)|^2 d\omega \tag{4.3.15}$$

That is, $|X(\omega)|^2$ not only allows us to calculate the total energy of $x(t)$ using Parseval's relation, but also permits us to calculate the energy in any given frequency band. For real-valued signals, $|X(\omega)|^2$ is an even function, and Equation (4.3.14) can be reduced to

$$E = \frac{1}{\pi}\int_{0}^{\infty} |X(\omega)|^2 d\omega \tag{4.3.16}$$

Periodic signals, as defined in Chapter 1, have infinite energy, but finite average power. A function that describes the distribution of the average power of the signal as a function of frequency is called the power-density spectrum, or, simply, the power spectrum. In the following, we develop an expression for the power spectral density of power signals, and in Section 4.3.9 we give an example to demonstrate how to compute the power spectral density of a periodic signal. Let $x(t)$ be a power signal, and define $x_\tau(t)$ as

$$x_\tau(t) = \begin{cases} x(t), & -\tau < t < \tau \\ 0, & \text{otherwise} \end{cases}$$
$$= x(t)\,\mathrm{rect}(t/2\tau)$$

We also assume that

$$x_\tau(t) \leftrightarrow X_\tau(\omega)$$

The average power in the signal $x(t)$ is

$$P = \lim_{\tau\to\infty}\left[\frac{1}{2\tau}\int_{-\tau}^{\tau}|x(t)|^2\,dt\right] = \lim_{\tau\to\infty}\left[\frac{1}{2\tau}\int_{-\infty}^{\infty}|x_\tau(t)|^2\,dt\right] \tag{4.3.17}$$

where the last equality follows from the definition of $x_\tau(t)$. Using Parseval's relation, we can write Equation (4.3.17) as

$$P = \frac{1}{2\pi}\lim_{\tau\to\infty}\left[\frac{1}{2\tau}\int_{-\infty}^{\infty}|X_\tau(\omega)|^2\,d\omega\right] = \frac{1}{2\pi}\int_{-\infty}^{\infty}\lim_{\tau\to\infty}\left[\frac{|X_\tau(\omega)|^2}{2\tau}\right]d\omega \tag{4.3.18}$$
$$= \frac{1}{2\pi}\int_{-\infty}^{\infty}S(\omega)\,d\omega$$

where

$$S(\omega) = \lim_{\tau\to\infty}\left[\frac{|X_\tau(\omega)|^2}{2\tau}\right] \tag{4.3.19}$$

$S(\omega)$ is referred to as the power-density spectrum, or, simply, power spectrum, of the signal $x(t)$ and represents the distribution, or density, of the power of the signal with frequency ω. As in the case of the energy spectrum, the power spectrum of a signal depends only on the magnitude of the spectrum and not on the phase.

Example 4.3.6

Consider the one-sided exponential signal

$$x(t) = \exp[-t]u(t)$$

From Equation (4.2.9),

$$|X(\omega)|^2 = \frac{1}{1+\omega^2}$$

The total energy in this signal is equal to 1/2 and can be obtained by using either Equation (1.4.2) or Equation (4.3.14). The energy in the frequency band $-4 < \omega < 4$ is

$$\Delta E = \frac{1}{\pi}\int_0^4 \frac{1}{1+\omega^2}\,d\omega$$

$$= \frac{1}{\pi}\tan^{-1}\omega\Big|_0^4 \simeq 0.422$$

Thus, approximately 84% of the total energy content of the signal lies in the frequency band $-4 < \omega < 4$. Note that the previous result could not be obtained with a knowledge of $x(t)$ alone.

4.3.7 Convolution

Convolution plays an important role in the study of LTI systems and their applications. The property is expressed as follows: If

$$x(t) \leftrightarrow X(\omega)$$

and

$$h(t) \leftrightarrow H(\omega)$$

then

$$x(t) * h(t) \leftrightarrow X(\omega)H(\omega) \tag{4.3.20}$$

The proof of this statement follows from the definition of the convolution integral, namely,

$$\mathscr{F}\{x(t) * h(t)\} = \int_{-\infty}^{\infty}\left[\int_{-\infty}^{\infty} x(\tau)h(t-\tau)\,d\tau\right]\exp[-j\omega t]\,dt$$

Interchanging the order of integration and noting that $x(\tau)$ does not depend on t, we have

$$\mathscr{F}\{x(t) * h(t)\} = \int_{-\infty}^{\infty} x(\tau)\left[\int_{-\infty}^{\infty} h(t-\tau)\exp[-j\omega t]\,dt\right]d\tau$$

By the shifting property, Equation (4.3.6a), the bracketed term is simply $H(\omega)$ $\exp[-j\omega\tau]$. Thus,

$$\mathscr{F}\{x(t) * h(t)\} = \int_{-\infty}^{\infty} x(\tau)\exp[-j\omega\tau]H(\omega)\,d\tau$$

$$= H(\omega)\int_{-\infty}^{\infty} x(\tau)\exp[-j\omega\tau]\,d\tau$$

$$= H(\omega)X(\omega)$$

Hence, convolution in the time domain is equivalent to multiplication in the frequency domain, which, in many cases, is convenient and can be done by inspection. The use of the convolution property for LTI systems is demonstrated in Figure 4.3.2. The amplitude and phase spectrum of the output $y(t)$ are related to those of the input $x(t)$ and the impulse response $h(t)$ in the following manner:

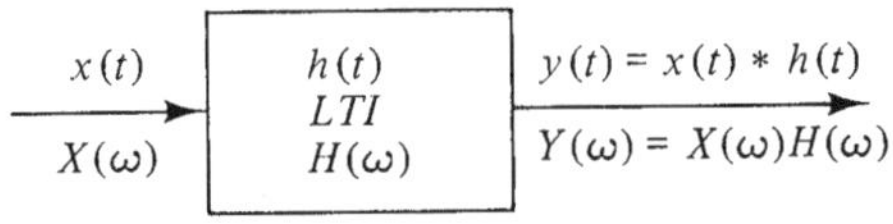

Figure 4.3.2. Convolution property of LTI system response.

$$|Y(\omega)| = |X(\omega)|\,|H(\omega)|$$
$$\measuredangle Y(\omega) = \measuredangle X(\omega) + \measuredangle H(\omega)$$

Thus, the amplitude spectrum of the input is modified by $|H(\omega)|$ to produce the amplitude spectrum of the output, and the phase spectrum of the input is changed by $\measuredangle H(\omega)$ to produce the phase spectrum of the output.

The quantity $H(\omega)$, the Fourier transform of the system impulse response, is generally referred to as the frequency response of the system.

As we have seen in Section 4.2.2, for $H(\omega)$ to exist, $h(t)$ has to satisfy two conditions. The first condition requires that the impulse response be absolutely integrable. This, in turn, implies that the LTI system is stable. Thus, assuming that $h(t)$ is "well behaved," as are essentially all signals of practical significance, we conclude that the frequency response of a stable LTI system exists. If, however, an LTI system is unstable, that is, if

$$\int_{-\infty}^{\infty} |h(t)|\, dt = \infty$$

then the response of the system to complex exponential inputs may be infinite, and the Fourier transform may not exist. Therefore, Fourier analysis is used to study LTI systems with impulse responses that possess Fourier transforms. Other, more general transform techniques are used to examine those unstable systems that do not have finite-value frequency responses. In Chapter 5, we discuss the Laplace transform, which is a generalization of the continuous-time Fourier transform.

Example 4.3.7

In this example, we demonstrate how to use the convolution property of the Fourier transform. Consider an LTI system with impulse response

$$h(t) = \exp[-at]u(t)$$

whose input is the unit step function $u(t)$. The Fourier transform of the output is

$$\begin{aligned}
Y(\omega) &= \mathcal{F}\{u(t)\}\mathcal{F}\{\exp[-at]u(t)\} \\
&= \left[\pi\delta(\omega) + \frac{1}{j\omega}\right]\left(\frac{1}{a + j\omega}\right) \\
&= \frac{\pi}{a}\delta(\omega) + \frac{1}{j\omega(a + j\omega)} \\
&= \frac{1}{a}\left[\pi\delta(\omega) + \frac{1}{j\omega}\right] - \frac{1}{a}\frac{1}{a + j\omega}
\end{aligned}$$

Taking the inverse Fourier transform of both sides results in

$$y(t) = \frac{1}{a}u(t) - \frac{1}{a}\exp[-at]u(t)$$

$$= \frac{1}{a}[1 - \exp[-at]]u(t)$$

Example 4.3.8

The Fourier transform of the triangle signal $\Delta(t/\tau)$ can be obtained by observing that the signal is the convolution of the rectangular pulse $(1/\sqrt{\tau})\,\mathrm{rect}(t/\tau)$ with itself; that is,

$$\Delta(t/\tau) = \frac{1}{\sqrt{\tau}}\mathrm{rect}(t/\tau) * \frac{1}{\sqrt{\tau}}\mathrm{rect}(t/\tau)$$

From Example 4.2.1 and Equation (4.3.20), it follows that

$$\mathcal{F}\{\Delta(t/\tau)\} = \left(\mathcal{F}\left\{\frac{1}{\sqrt{\tau}}\mathrm{rect}(t/\tau)\right\}\right)^2 = \tau\left(\mathrm{sinc}\,\frac{\omega\tau}{2}\right)^2$$

Example 4.3.9

An LTI system has an impulse response

$$h(t) = \exp[-at]u(t)$$

and output

$$y(t) = [\exp[-bt] - \exp[-ct]]u(t)$$

Using the convolution property, we find that the transform of the input is

$$X(\omega) = \frac{Y(\omega)}{H(\omega)}$$

$$= \frac{(c-b)(j\omega + a)}{(j\omega + b)(j\omega + c)}$$

$$= \frac{D}{j\omega + b} + \frac{E}{j\omega + c}$$

where

$$D = a - b \qquad \text{and} \qquad E = c - a$$

Therefore,

$$x(t) = [(a-b)\exp[-bt] + (c-a)\exp[-ct]]u(t)$$

Example 4.3.10

In this example, we use the relation

$$\int_{-\infty}^{t} x(\tau)\,d\tau = x(t) * u(t)$$

and the transform of $u(t)$ to prove the integration property, Equation (4.3.11). From Equation (4.3.13) and the convolution property, we have

$$\mathcal{F}\left\{\int_{-\infty}^{t} x(\tau)\,d\tau\right\} = \mathcal{F}\{x(t) * u(t)\} = X(\omega)\left[\pi\delta(\omega) + \frac{1}{j\omega}\right]$$

$$= \pi X(0)\delta(\omega) + \frac{X(\omega)}{j\omega}$$

The last equality follows from the sampling property of the delta function.

Another important relation follows as a consequence of using the convolution property to represent the spectrum of the output of an LTI system; that is,

$$Y(\omega) = X(\omega)H(\omega)$$

We then have

$$|Y(\omega)|^2 = |X(\omega)H(\omega)|^2 = |X(\omega)|^2|H(\omega)|^2 \tag{4.3.21}$$

This equation shows that the energy-spectrum density of the response of an LTI system is the product of the energy-spectrum density of the input signal and the square of the magnitude of the system function. The phase characteristic of the system does not affect the energy-spectrum density of the output, in spite of the fact that, in general, $H(\omega)$ is a complex quantity.

4.3.8 Duality

We sometimes have to find the Fourier transform of a time signal that has a form similar to an entry in the transform column in the table of Fourier transforms. We can find the desired transform by using the table backwards. To accomplish that, we write the inversion formula in the form

$$\int_{-\infty}^{\infty} X(\omega)\exp[+j\omega t]\,d\omega = 2\pi x(t)$$

Notice that there is a symmetry between this equation and Equation (4.2.6): The two equations are identical except for a sign change in the exponential, a factor of 2π, and an interchange of the variables involved. This type of symmetry leads to the duality property of the Fourier transform. This property states that if $x(t)$ has a transform $X(\omega)$, then

$$X(t) \leftrightarrow 2\pi x(-\omega) \tag{4.3.22}$$

We prove Equation (4.3.22) by replacing t with $-t$ in Equation (4.2.5) to get

$$2\pi x(-t) = \int_{\omega=-\infty}^{\infty} X(\omega)\exp[-j\omega t]\,d\omega$$

$$= \int_{\tau=-\infty}^{\infty} X(\tau)\exp[-j\tau t]\,d\tau$$

since ω is just a dummy variable for integration. Now replacing t by ω and τ by t gives Equation (4.3.22).

Example 4.3.11

Consider the signal

$$x(t) = \text{Sa}\,\frac{\omega_B t}{2} = \text{sinc}\,\frac{\omega_B t}{2\pi}$$

From Equation (4.2.6),

$$\mathcal{F}\left\{\text{Sa}\,\frac{\omega_B t}{2}\right\} = \int_{-\infty}^{\infty} \text{Sa}\,\frac{\omega_B t}{2}\exp[-j\omega t]\,dt$$

This is a very difficult integral to evaluate directly. However, we found in Example 4.2.1 that

$$\text{rect}(t/\tau) \leftrightarrow \tau\,\text{Sa}\,\frac{\omega\tau}{2}$$

Then according to Equation (4.3.22),

$$\mathcal{F}\left\{\text{Sa}\,\frac{\omega_B t}{2}\right\} = \frac{2\pi}{\omega_B}\,\text{rect}(-\omega/\omega_B) = \frac{2\pi}{\omega_B}\,\text{rect}(\omega/\omega_B)$$

because the rectangular pulse is an even signal. Note that the transform $X(\omega)$ is zero outside the range $-\omega_B/2 \leq \omega \leq \omega_B/2$, but that the signal $x(t)$ is not time limited. Signals with Fourier transforms that vanish outside a given frequency band are called band-limited signals (signals with no spectral content above a certain maximum frequency, in this case, $\omega_B/2$.). It can be shown that time limiting and frequency limiting are mutually exclusive phenomena; i.e., a time-limited signal $x(t)$ always has a Fourier transform that is not band limited. On the other hand, if $X(\omega)$ is band limited, then the corresponding time signal is never time limited.

Example 4.3.12

Differentiating Equation (4.2.6) n times with respect to ω, we readily obtain

$$(-jt)^n x(t) \leftrightarrow \frac{d^n X(\omega)}{d\omega^n} \tag{4.3.23}$$

that is, multiplying a time signal by t is equivalent to differentiating the frequency spectrum, which is the dual of differentiation in the time domain.

The previous two examples demonstrate that, in addition to its consequences in reducing the complexity of the calculation involved in determining some Fourier transforms, duality also implies that every property of the Fourier transform has a dual.

4.3.9 Modulation

If

$$x(t) \leftrightarrow X(\omega)$$
$$m(t) \leftrightarrow M(\omega)$$

then

$$x(t)m(t) \leftrightarrow \frac{1}{2\pi}[X(\omega) * M(\omega)] \tag{4.3.24}$$

Convolution in the frequency domain is carried out exactly like convolution in the time domain. That is,

$$X(\omega) * H(\omega) = \int_{\sigma=-\infty}^{\infty} X(\sigma)H(\omega - \sigma)\,d\sigma = \int_{\sigma=-\infty}^{\infty} H(\sigma)X(\omega - \sigma)\,d\sigma$$

This property is a direct result of combining two properties, the duality and the convolution properties, and it states that multiplication in the time domain corresponds to convolution in the frequency domain. Multiplication of the desired signal $x(t)$ by $m(t)$ is equivalent to altering or modulating the amplitude of $x(t)$ according to the variations in $m(t)$. This is the reason that the multiplication of two signals is often referred to as modulation. The symmetrical nature of the Fourier transform is clearly reflected in Equations (4.3.20) and (4.3.24): Convolution in the time domain is equivalent to multiplication in the frequency domain, and multiplication in the time domain is equivalent to convolution in the frequency domain. The importance of this property is that the spectrum of a signal such as $x(t)\cos\omega_0 t$ can be easily computed. These types of signals arise in many communications systems, as we shall see later. Since

$$\cos\omega_0 t = \frac{1}{2}[\exp[j\omega_0 t] + \exp[-j\omega_0 t]]$$

it follow that

$$\mathcal{F}\{x(t)\cos\omega_0 t\} = \frac{1}{2}[X(\omega - \omega_0) + X(\omega + \omega_0)]$$

This result constitutes the fundamental property of modulation and is useful in the spectral analysis of signals obtained from multipliers and modulators.

Example 4.3.13

Consider the signal

$$x_s(t) = x(t)p(t)$$

where $p(t)$ is the periodic impulse train with equal-strength impulses, as shown in Figure 4.3.3. Analytically, $p(t)$ can be written as

$$p(t) = \sum_{n=-\infty}^{\infty} \delta(t - nT)$$

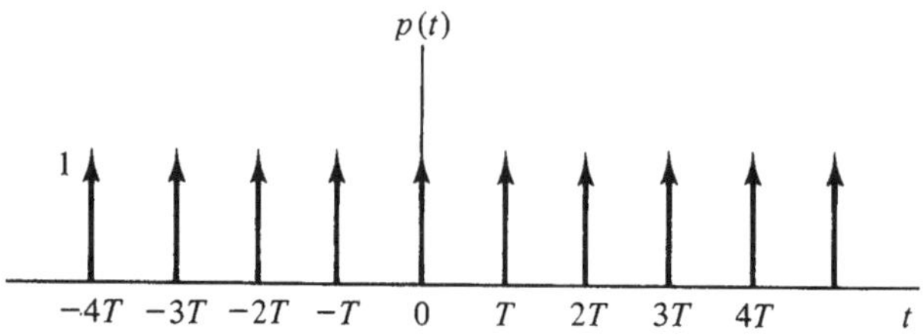

Figure 4.3.3 Periodic pulse train used in Example 4.3.13.

Using the sampling property of the delta function, we obtain

$$x_s(t) = \sum_{n=-\infty}^{\infty} x(nT)\delta(t - nT)$$

That is, $x_s(t)$ is a train of impulses spaced T seconds apart, the strength of the impulses being equal to the sample values of $x(t)$. Recall from Example 4.2.10 that the Fourier transform of the periodic impulse train $p(t)$ is itself a periodic impulse train; specifically,

$$P(\omega) = \frac{2\pi}{T} \sum_{n=-\infty}^{\infty} \delta\left(\omega - \frac{2\pi n}{T}\right)$$

Consequently, from the modulation property,

$$X_s(\omega) = \frac{1}{2\pi}[X(\omega) * P(\omega)]$$

$$= \frac{1}{T}\sum_{n=-\infty}^{\infty} X(\omega) * \delta\left(\omega - \frac{2\pi n}{T}\right) = \frac{1}{T}\sum_{n=-\infty}^{\infty} X\left(\omega - \frac{2\pi n}{T}\right)$$

That is, $X_s(\omega)$ consists of a periodically repeated replica of $X(\omega)$.

Example 4.3.14

Consider the system depicted in Figure 4.3.4, where

$$x(t) = \frac{\sin(\omega_B t/2)}{\pi t}$$

$$p(t) = \sum_{n=-\infty}^{\infty} \delta\left(t - \frac{n\pi}{\omega_B}\right)$$

$$h(t) = \frac{\sin(3\omega_B t/2)}{\pi t}$$

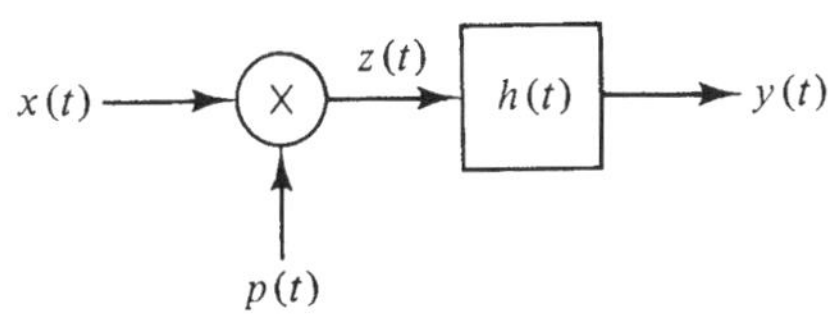

Figure 4.3.4 System for Example 4.3.14.

The Fourier transform of $x(t)$ is the rectangular pulse with width ω_B, and the Fourier transform of the product $x(t)p(t)$ consists of the periodically repeated replica of $X(\omega)$, as shown in Figure 4.3.5. Similarly, the Fourier transform of $h(t)$ is a rectangular pulse with width $3\omega_B$. According to the convolution property, the transform of the output of the system is

$$Y(\omega) = X_s(\omega)H(\omega)$$
$$= X(\omega)$$

or

$$y(t) = x(t)$$

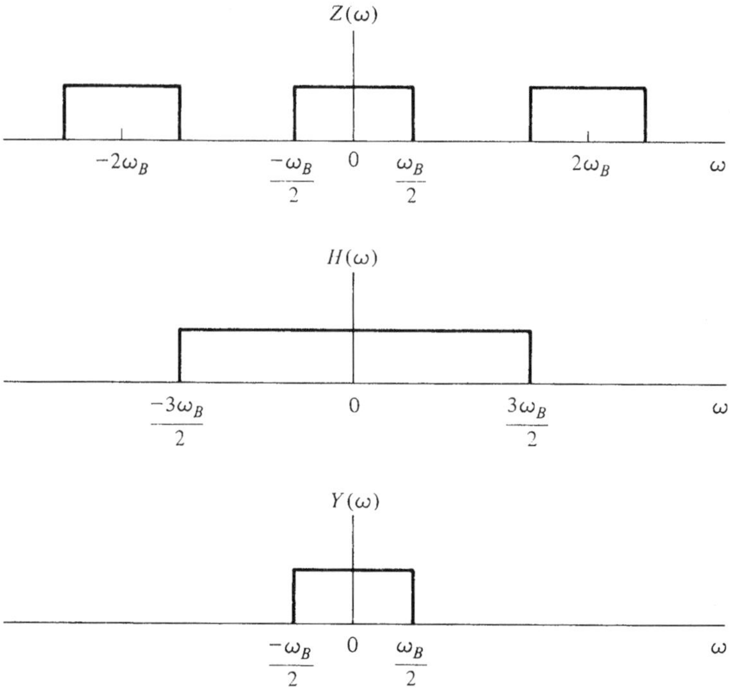

Figure 4.3.5 Spectra associated with signals for Example 4.3.14.

Note that since the system $h(t)$ blocked (i.e., filtered out) all the undesired components of $x_s(t)$ in order to obtain a scaled version of $x(t)$, we refer to such a system as a filter. Filters are important components of any communication or control system. In Chapter 10, we study the design of both analog and digital filters.

Example 4.3.15

In this example, we use the modulation property to show that the power spectrum of the periodic signal $x(t)$ with period T is

$$S(\omega) = 2\pi \sum_{n=-\infty}^{\infty} |c_n|^2 \delta(\omega - n\omega_0)$$

where c_n are the Fourier coefficients of $x(t)$ and

$$\omega_0 = \frac{2\pi}{T}$$

We begin by defining the truncated signal $x_\tau(t)$ as the product $x(t)\,\mathrm{rect}(t/2\tau)$. Using the modulation property, we find that

$$X_\tau(\omega) = \frac{1}{2\pi}[2\tau \,\mathrm{Sa}\,\omega\tau * X(\omega)]$$

$$= \frac{1}{2\pi}\int_{-\infty}^{\infty} 2\tau \,\mathrm{Sa}\,\mu\tau\, X(\omega - \mu)\,d\mu$$

Substituting Equation (4.2.15) for $X(\omega)$ and forming the function $|X_\tau(\omega)|^2$, we have

$$\frac{|X_\tau(\omega)|^2}{2\tau} = \sum_{n=-\infty}^{\infty} \sum_{m=-\infty}^{\infty} 2\tau c_n c_m^* \,\mathrm{Sa}[(\omega - n\omega_0)\tau]\,\mathrm{Sa}[(\omega - m\omega_0)\tau]$$

The power-density spectrum of the periodic signal $x(t)$ is obtained by taking the limit of the last expression as $\tau \to \infty$. It has been observed earlier that as $\tau \to \infty$, the transform of the rectangular signal approaches $\delta(\omega)$; therefore, we anticipate that the two sampling functions in the previous expression approach $\delta(\omega - k\omega_0)$, where $k = m$ and n. Also, observing that

$$\delta(\omega - n\omega_0)\delta(\omega - m\omega_0) = \begin{cases} \delta(\omega - n\omega_0), & m = n \\ 0, & \text{otherwise} \end{cases}$$

we calculate that the power-density spectrum of the periodic signal is

$$S(\omega) = \lim_{\tau \to \infty} \frac{|X_\tau(\omega)|^2}{2\tau}$$
$$= 2\pi \sum_{n=-\infty}^{\infty} |c_n|^2 \delta(\omega - n\omega_0)$$

For convenience, a summary of the foregoing properties of the Fourier transform is given in Table 4.2. These properties are used repeatedly in this chapter, and they should be thoroughly understood.

TABLE 4.2
Some Selected Properties of the Fourier Transform

1. Linearity	$\sum_{n=1}^{N} \alpha_n x_n(t)$	$\sum_{n=1}^{N} \alpha_n X_n(\omega)$	(4.3.1)
2. Complex conjugation	$x^*(t)$	$X^*(-\omega)$	(4.2.6)
3. Time shift	$x(t - t_0)$	$X(\omega)\exp[-j\omega t_0]$	(4.3.64)
4. Frequency shift	$x(t)\exp[j\omega_0 t]$	$X(\omega - \omega_0)$	(4.3.6b)
5. Time scaling	$x(at)$	$1/\lvert a\rvert X(\omega/a)$	(4.3.7)
6. Differentiation	$d^n x(t)/dt^n$	$(j\omega)^n X(\omega)$	(4.3.9)
7. Integration	$\int_{-\infty}^{t} x(\tau)\,d\tau$	$\frac{X(\omega)}{j(\omega)} + \pi X(0)\delta(\omega)$	(4.3.11)
8. Parseval's relation	$\int_{-\infty}^{\infty} \lvert x(t)\rvert^2 dt$	$\frac{1}{2\pi}\int_{-\infty}^{\infty} \lvert X(\omega)\rvert^2 d\omega$	(4.3.14)
9. Convolution	$x(t)*h(t)$	$X(\omega)H(\omega)$	(4.3.20)
10. Duality	$X(t)$	$2\pi\, x(-\omega)$	(4.3.22)
11. Multiplication by t	$(-jt)^n x(t)$	$\frac{d^n X(\omega)}{d\omega^n}$	(4.3.23)
12. Modulation	$x(t)m(t)$	$\frac{1}{2\pi} X(\omega) * M(\omega)$	(4.3.24)

4.4 APPLICATIONS OF THE FOURIER TRANSFORM

The continuous-time Fourier transform and its discrete counterpart, the discrete-time Fourier transform, which we study in detail in Chapter 7, are tools that find extensive applications in communication systems, signal processing, control systems, and many other varieties of physical and engineering disciplines. The important processes of amplitude modulation and frequency multiplexing provide examples of the use of Fourier-transform theory in the analysis and design of communication systems. The sampling theorem is considered to have the most profound effect on information transmission and signal processing, especially in the digital area. The design of filters and compensators that are employed in control systems cannot be done without the help of the Fourier transform. In this section, we discuss some of these applications in more detail.

4.4.1 Amplitude Modulation

The goal of all communication systems is to convey information from one point to another. Prior to sending the information signal through the transmission channel, the signal is converted to a useful form through what is known as modulation. Among the many reasons for employing this type of conversion are the following:

1. to transmit information efficiently
2. to overcome hardware limitations
3. to reduce noise and interference
4. to utilize the electromagnetic spectrum efficiently.

Consider the signal multiplier shown in Figure 4.4.1. The output is the product of the information-carrying signal $x(t)$ and the signal $m(t)$, which is referred to as the carrier signal. This scheme is known as amplitude modulation, which has many forms, depending on $m(t)$. We concentrate only on the case when $m(t) = \cos\omega_0 t$, which represents a practical form of modulation and is referred to as double-sideband (DSB) amplitude modulation. We will now examine the spectrum of the output (the modulated signal) in terms of the spectrum of both $x(t)$ and $m(t)$.

The output of the multiplier is

$$y(t) = x(t)\cos\omega_0 t$$

Since $y(t)$ is the product of two time signals, convolution in the frequency domain can be used to obtain its spectrum. The result is

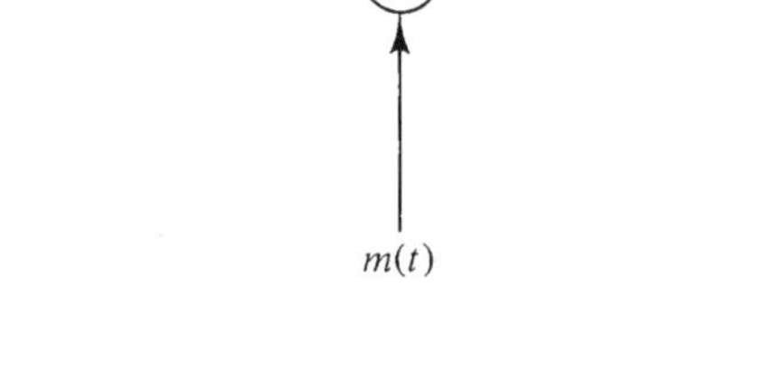

Figure 4.4.1 Signal multiplier.

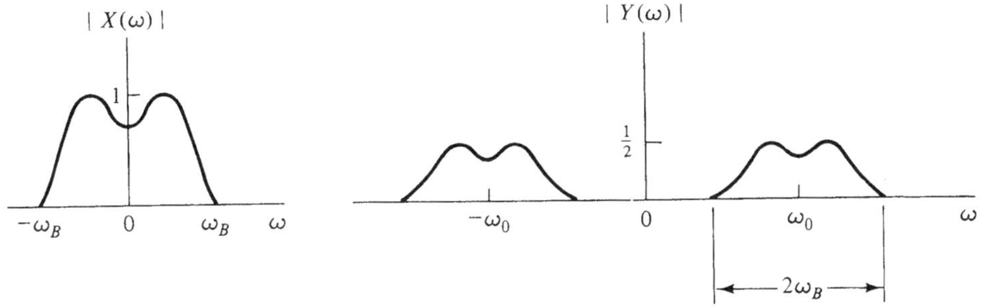

Figure 4.4.2 Magnitude spectra of information signal and modulated signal.

$$Y(\omega) = \frac{1}{2\pi} X(\omega) * \pi[\delta(\omega - \omega_0) + \delta(\omega + \omega_0)]$$

$$= \frac{1}{2}[X(\omega - \omega_0) + X(\omega + \omega_0)].$$

The magnitude spectra of $x(t)$ and $y(t)$ are illustrated in Figure 4.4.2. The part of the spectrum of $Y(\omega)$ centered at $+\omega_0$ is the result of convolving $X(\omega)$ with $\delta(\omega - \omega_0)$, and the part centered at $-\omega_0$ is the result of convolving $X(\omega)$ with $\delta(\omega + \omega_0)$. This process of shifting the spectrum of the signal by ω_0 is necessary because low-frequency (baseband) information signals cannot be propagated easily by radio waves.

The process of extracting the information signal from the modulated signal is referred to as demodulation. In effect, demodulation shifts back the message spectrum to its original low-frequency location. Synchronous demodulation is one of several techniques used to perform amplitude demodulation. A synchronous demodulator consists of a signal multiplier, with the multiplier inputs being the modulated signal and $\cos \omega_0 t$. The output of the multiplier is

$$z(t) = y(t) \cos \omega_0 t$$

Hence,

$$Z(\omega) = \frac{1}{2}[Y(\omega - \omega_0) + Y(\omega + \omega_0)]$$

$$= \frac{1}{2} X(\omega) + \frac{1}{4} X(\omega - 2\omega_0) + \frac{1}{4} X(\omega + 2\omega_0)$$

The result is shown in Figure 4.4.3(a). To extract the original information signal $x(t)$, the signal $z(t)$ is passed through the system with frequency response $H(\omega)$ shown in Figure 4.4.3(b). Such a system is referred to as a low-pass filter, since it passes only low-frequency components of the input signal and filters out all frequencies higher than ω_B, the cutoff frequency of the filter. The output of the low-pass filter is illustrated in Figure 4.4.3(c). Note that if $|H(\omega)| = 1$, $|\omega| < \omega_B$, and there were no transmission losses

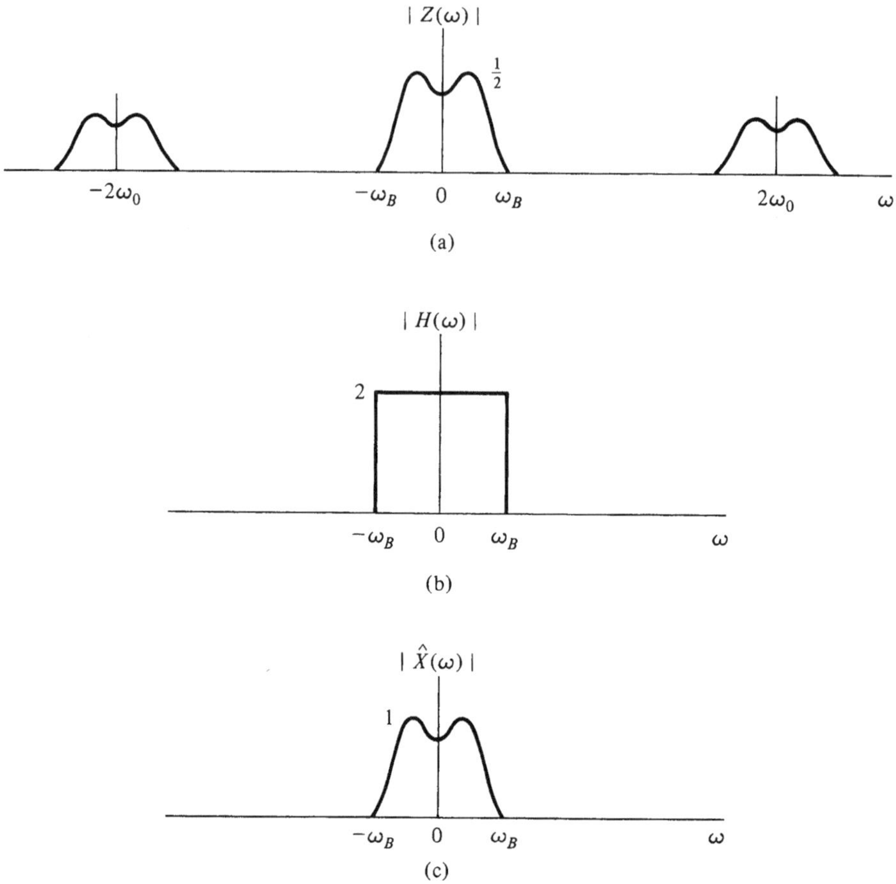

Figure 4.4.3 Demodulation process: (a) Magnitude spectrum of $z(t)$; (b) the low-pass-filter frequency response; and (c) the extracted information spectrum.

involved, then the energy of final signal is one-fourth that of the original signal because the total demodulated signal contains energy located at $\omega = 2\omega_0$ that is eventually discarded by the receiver.

4.4.2 Multiplexing

A very useful technique for simultaneously transmitting several information signals involves the assignment of a portion of the final frequency to each signal. This technique is known as frequency-division multiplexing (FDM), and we encounter it almost daily, often without giving it much thought. Larger cities usually have several AM radio and television stations, fire engines, police cruisers, taxicabs, mobile telephones, citizen band radios, and many other sources of radio waves. All these sources are frequency multiplexed into the radio spectrum by means of assigning distinct frequency bands to each signal. FDM is very similar to amplitude modulation. Consider three

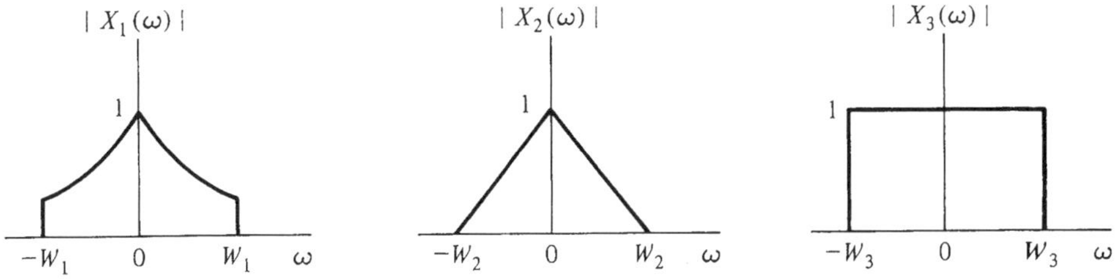

Figure 4.4.4 Magnitude spectra for $x_1(t)$, $x_2(t)$, and $x_3(t)$ for the FDM system.

band-limited signals with Fourier transforms, as shown in Figure 4.4.4. (Extension to n signals follows in a straightforward manner.)

If we modulate $x_1(t)$ with $\cos\omega_1 t$, $x_2(t)$ with $\cos\omega_2 t$, and $x_3(t)$ with $\cos\omega_3 t$, then, summing the three modulated signals, we obtain

$$y(t) = x_1(t)\cos\omega_1 t + x_2(t)\cos\omega_2 t + x_3(t)\cos\omega_3 t$$

The frequency spectrum of $y(t)$ is

$$Y(\omega) = \frac{1}{2}[X_1(\omega - \omega_1) + X_1(\omega + \omega_1)]$$

$$+ \frac{1}{2}[X_2(\omega - \omega_2) + X_2(\omega + \omega_2)]$$

$$+ \frac{1}{2}[X_3(\omega - \omega_3) + X_3(\omega + \omega_3)]$$

which has a spectrum similar to that in Figure 4.4.5. It is important here to make sure that the spectra do not overlap—that is, that $\omega_1 + W_1 < \omega_2 - W_2$ and $\omega_2 + W_2 < \omega_3 - W_3$. At the receiving end, some operations must be performed to recover the individual spectra.

Because of the form of $|Y(\omega)|$, in order to capture the spectrum of $x_1(t)$, we would need a system whose frequency response is equal to 1 for $\omega_1 - W_1 \leq \omega \leq \omega_1 + W_1$ and zero otherwise. Such a system is called a band-pass filter, since it passes only frequencies in the band $\omega_1 - W_1 \leq \omega \leq \omega_1 + W_1$ and suppresses all other frequencies.

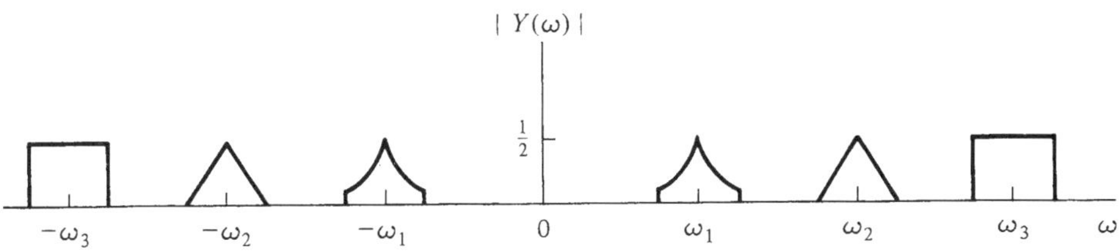

Figure 4.4.5 Magnitude spectrum of $y(t)$ for the FDM system.

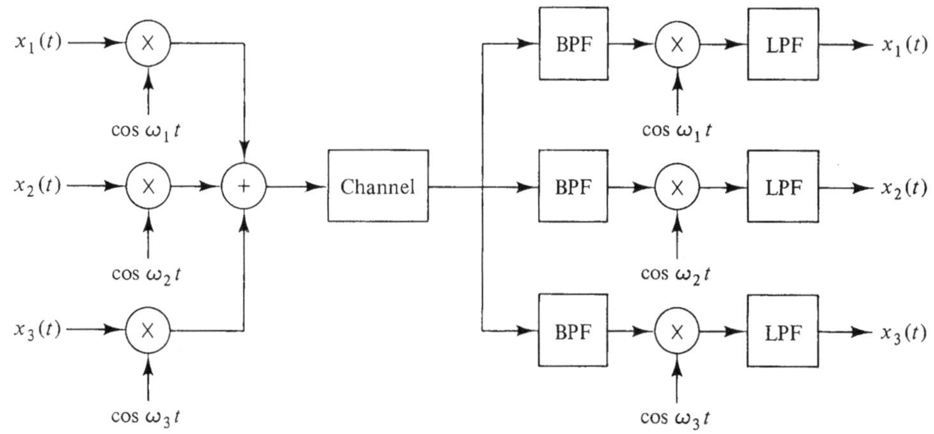

Figure 4.4.6 Frequency-division multiplexing (FDM) system. BPF = band-pass filter, LPF = low-pass filter.

The output of this filter is then processed as in the case of synchronous amplitude demodulation. A similar procedure can be used to extract $x_2(t)$ or $x_3(t)$. The overall system of modulation, multiplexing, transmission, demultiplexing, and demodulation is illustrated in Figure 4.4.6.

4.4.3 The Sampling Theorem

Of all the theorems and techniques pertaining to the Fourier transform, the one that has had the most impact on information transmission and processing is the sampling theorem. For a low-pass signal $x(t)$ which is band limited such that it has no frequency components above ω_B rad/s, the sampling theorem says that $x(t)$ is uniquely determined by its values at equally spaced points in time T seconds apart, provided that $T < \pi/\omega_B$. The sampling theorem allows us to completely reconstruct a band-limited signal from instantaneous samples taken at a rate $\omega_s = 2\pi/T$, provided that ω_s is at least as large as $2\omega_B$, which is twice the highest frequency present in the band-limited signal $x(t)$. The minimum sampling rate $2\omega_B$ is known as the Nyquist rate.

The process of obtaining a set of samples from a continuous function of time $x(t)$ is referred to as sampling. The samples can be considered to be obtained by passing $x(t)$ through a sampler, which is a switch that closes and opens instantaneously at sampling instants nT. When the switch is closed, we obtain a sample $x(nT)$. At all other times, the output of the sampler is zero. This ideal sampler is a fictitious device, since, in practice, it is impossible to obtain a switch that closes and opens instantaneously. We denote the output of the sampler by $x_s(t)$.

In order to arrive at the sampling theorem, we model the sampler output as

$$x_s(t) = x(t)p(t) \tag{4.4.1}$$

where

$$p(t) = \sum_{n=-\infty}^{\infty} \delta(t - nT) \tag{4.4.2}$$

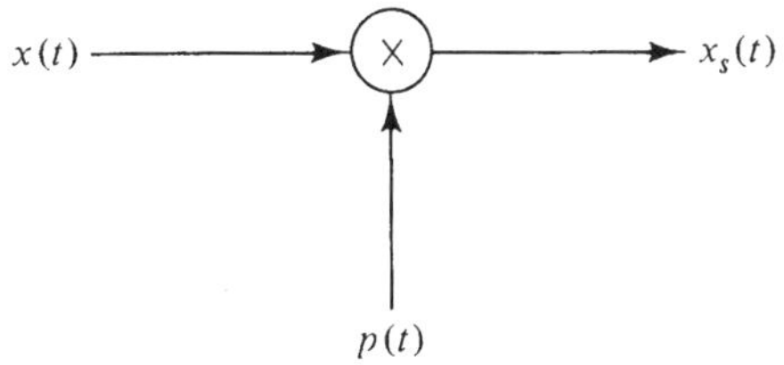

Figure 4.4.7 The ideal sampling process.

is the periodic impulse train. We provide a justification of this model later, in Chapter 8, where we discuss the sampling of continuous-time signals in greater detail. As can be seen from the equation, the sampled signal is considered to be the product (modulation) of the continuous-time signal $x(t)$ and the impulse train $p(t)$ and, hence, is usually referred to as the impulse modulation model for the sampling operation. This is illustrated in Figure 4.4.7.

From Example 4.2.10, it follows that

$$P(\omega) = \frac{2\pi}{T}\sum_{n=-\infty}^{\infty}\delta\left(\omega - \frac{2\pi n}{T}\right) = \frac{2\pi}{T}\sum_{n=-\infty}^{\infty}\delta(\omega - n\omega_s) \tag{4.4.3}$$

and hence,

$$\begin{aligned}X_s(\omega) &= \frac{1}{2\pi}X(\omega) * P(\omega)\\ &= \frac{1}{2\pi}\int_{-\infty}^{\infty}X(\sigma)P(\omega - \sigma)\,d\sigma\\ &= \frac{1}{T}\sum_{n=-\infty}^{\infty}X(\omega - n\omega_s) \end{aligned}\tag{4.4.4}$$

The signals $x(t)$, $p(t)$, and $x_s(t)$, are depicted together with their magnitude spectra in Figure 4.4.8, with $x(t)$ being a band-limited signal—that is, $X(\omega)$ is zero for $|\omega| > \omega_B$. As can be seen, $x_s(t)$, which is the sampled version of the continuous-time signal $x(t)$, consists of impulses spaced T seconds apart, each having an area equal to the sampled value of $x(t)$ at the respective sampling instant. The spectrum $X_s(\omega)$ of the sampled signal is obtained as the convolution of the spectrum of $X(\omega)$ with the impulse train $P(\omega)$ and, hence, consists of the periodic repetition at intervals ω_s of $X(\omega)$, as shown in the figure. For the case shown, ω_s is large enough so that the different components of $X_s(\omega)$ do not overlap. It is clear that if we pass the sampled signal $x_s(t)$ through an ideal low-pass filter which passes only those frequencies contained in $x(t)$, the spectrum of the filter output will be identical to $X(\omega)$, except for an amplitude scale factor of $1/T$ introduced by the sampling operation. Thus, to recover $x(t)$, we pass $x_s(t)$ through a filter with frequency response

$$\begin{aligned}H(\omega) &= \begin{cases} T, & |\omega| < \omega_B \\ 0, & \text{otherwise}\end{cases}\\ &= T\,\mathrm{rect}\left(\frac{\omega}{2\omega_B}\right)\end{aligned}\tag{4.4.5}$$

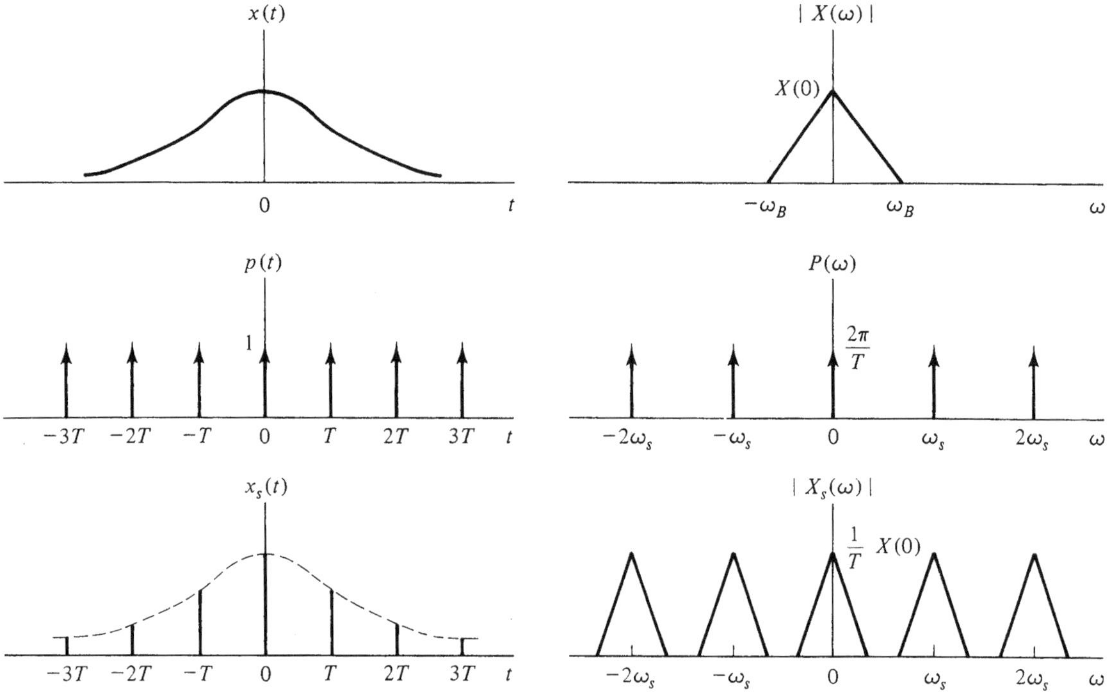

Figure 4.4.8 Time-domain signals and their respective magnitude spectra.

This filter is called an ideal reconstruction filter.

As the sampling frequency is reduced, the different components in the spectrum of $X_s(\omega)$ start coming closer together and eventually will overlap. As shown in Figure 4.4.9(a) if $\omega_s - \omega_B > \omega_s$, the components do not overlap, and the signal $x(t)$ can be recovered from $x_s(t)$ as described previously. If $\omega_s - \omega_B = 0$, the components just touch each other, as shown in Figure 4.4.9(b). If $\omega_s - \omega_B < \omega_B$, the components will overlap as shown in Figure 4.4.9(c). Then the resulting spectrum obtained by adding the overlapping components together no longer resembles $X(\omega)$ (Figure 4.9.(d)), and $x(t)$ can no longer be recovered from the sampled signal. Thus, to recover $x(t)$ from the sampled signal, it is clear that the sampling rate should be such that

$$\omega_s - \omega_B > \omega_B$$

Hence, signal $x(t)$ can be recovered from its samples only if

$$\omega_s > 2\omega_B \tag{4.4.6}$$

This is the sampling theorem (usually called the Nyquist theorem) that we referred to earlier. The minimum permissible value of ω_s is called the Nyquist rate.

The maximum time spacing between samples that can be used is

$$T = \frac{\pi}{\omega_B} \tag{4.4.7}$$

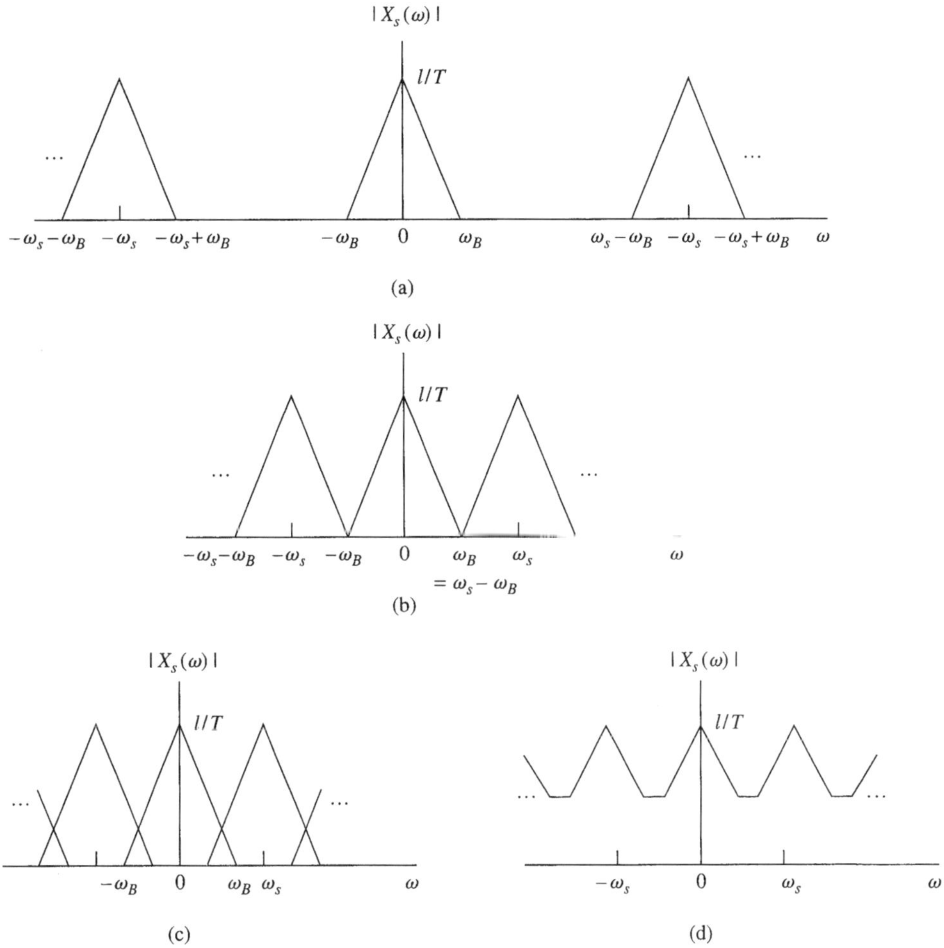

Figure 4.4.9 Effect of reducing the sampling frequency on $X_s(\omega)$.

If T does not satisfy Equation (4.4.7), the different components of $X_s(\omega)$ overlap, and we will not be able to recover $x(t)$ exactly. This is referred to as aliasing. If $x(t)$ is not band limited, there will always be aliasing, irrespective of the chosen sampling rate.

Example 4.4.1

The spectrum of a signal (for example, a speech signal) is essentially zero for all frequencies above 5 kHz. The Nyquist sampling rate for such a signal is

$$\begin{aligned}\omega_s &= 2\omega_B = 2(2\pi \times 5 \times 10^3) \\ &= 2\pi \times 10^4 \text{ rad/s}\end{aligned}$$

The sample spacing T is equal to $2\pi/\omega_s = 0.1$ ms.

Example 4.4.2

Instead of sampling the previous signal at the Nyquist rate of 10 kHz, let us sample it at a rate of 8 kHz. That is,

$$\omega_s = 2\pi \times 8 \times 10^3 \text{ rad/s}$$

The sampling interval T is equal to $2\pi/\omega_s = 0.125$ ms. If we filter the sampled signal $x_s(t)$ using a low-pass filter with a cutoff frequency of 4 kHz, the output spectrum contains high-frequency components of $x(t)$ superimposed on the low-frequency components, i.e., we have aliasing and $x(t)$ cannot be recovered.

In theory, if a signal $x(t)$ is not band limited, we can eliminate aliasing by low-pass filtering the signal before sampling it. Clearly, we will need to use a sampling frequency which is twice the bandwidth of the filter, ω_B. In practice, however, aliasing cannot be completely eliminated because, first, we cannot build a low-pass filter that cuts off all frequency components above a certain frequency and second, in many applications, $x(t)$ cannot be low-pass filtered without removing information from it. In such cases, we can reduce aliasing effects by sampling the signal at a high enough frequency that aliased components do not seriously distort the reconstructed signal. In some cases, the sampling frequency can be as large as 8 or 10 times the signal bandwidth.

Example 4.4.3

An analog bandpass signal, $x_a(t)$, which is bandlimited to the range $800 \leq f \leq 1200$Hz is input to the system in Figure 4.4.10(a) where $H(\omega)$ is an ideal low-pass filter with cutoff frequency of 200 Hz. Assume that the spectrum of $x_a(t)$ has a triangle shape symmetric about the center frequency as shown in Figure 4.4.10(b).

Figure 4.4.10(c) shows $X_m(\omega)$, the spectrum of the modulated signal, $x_m(t)$, while $X_b(\omega)$, that of the output of the low-pass filter (baseband signal), $x_b(t)$ is shown in Figure 4.4.10(d). If we now sample $x_b(t)$ at intervals T with $T < 1/400$ secs, as discussed earlier, the resulting spectrum $X_s(\omega)$ will be the aliased version of $X_a(\omega)$ and will thus consist of a set of triangular shaped pulses centered at frequencies $\omega = 2\pi k/T$, $k = 0, \pm 1, \pm 2$, etc. If one of these pulses is centered at $2\pi \times 1000$ rad/s, we can clearly recover $X_a(\omega)$ and hence $x_a(t)$ by passing the sampled signal through an ideal bandpass filter with center frequency 2000π rad/s and bandwidth of 800π rad/s. Figure 4.4.10(e) shows the spectrum of the sampled signal for $T = 1$ msec.

In general, we can recover $x_a(t)$ from the sampled signal by using a band-pass filter if $2\pi k/T = \omega_s$, that is if $1/T$ is an integer submultiple of the center frequency in Hz.

The fact that a band-limited signal that has been sampled at the Nyquist rate can be recovered from its samples can also be illustrated in the time domain using the concept of interpolation. From our previous discussion, we have seen that, since $x(t)$ can be obtained by passing $x_s(t)$ through the ideal reconstruction filter of Equation (4.4.5), we can write

$$X(\omega) = H(\omega)X_s(\omega) \tag{4.4.8}$$

The impulse response corresponding to $H(\omega)$ is

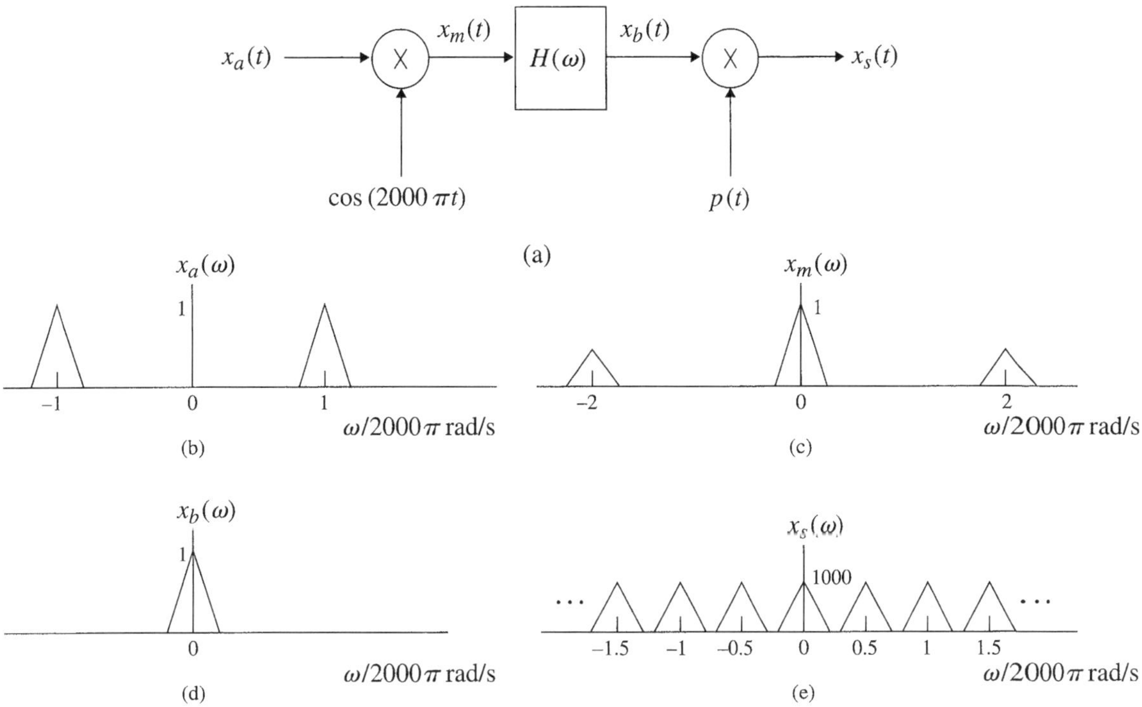

Figure 4.4.10 Spectra of the signals of Example 4.4.3.

$$h(t) = T\frac{\sin \omega_B t}{\pi t}$$

Taking the inverse Fourier transform of both sides of Equation (4.4.8), we obtain

$$\begin{aligned}
x(t) &= x_s(t) * h(t) \\
&= \left[x(t) \sum_{n=-\infty}^{\infty} \delta(t - nT)\right] * T\frac{\sin \omega_B t}{\pi t} \\
&= \sum_{n=-\infty}^{\infty} Tx(nT)\frac{\sin \omega_B (t - nT)}{\pi(t - nT)} \\
&= \sum_{n=-\infty}^{\infty} \frac{2\omega_B}{\omega_s} x(nT)\frac{\sin \omega_B (t - nT)}{\omega_B(t - nT)} \\
&= \sum_{n=-\infty}^{\infty} \frac{2\omega_B}{\omega_s} x(nT)\,\mathrm{Sa}(\omega_B(t - nT))
\end{aligned} \tag{4.4.9}$$

Equation (4.4.9) can be interpreted as using interpolation to reconstruct $x(t)$ from its samples $x(nT)$. The functions $\mathrm{Sa}[\omega_B(t - kT)]$ are called interpolating, or sampling, functions. Interpolation using sampling functions, as in Equation (4.4.9), is commonly referred to as band-limited interpolation.

4.4.4 Signal Filtering

Filtering is the process by which the essential and useful part of a signal is separated from extraneous and undesirable components that are generally referred to as noise. The term "noise" used here refers to either the undesired part of the signal, as in the case of amplitude modulation, or interference signals generated by the electronic devices themselves.

The idea of filtering using LTI systems is based on the convolution property of the Fourier transform discussed in Section 4.3, namely, that for LTI systems, the Fourier transform of the output is the product of the Fourier transform of the input and the frequency response of the system. An ideal frequency-selective filter is a filter that passes certain frequencies without any change and stops the rest. The range of frequencies that pass through is called the passband of the filter, whereas the range of frequencies that do not pass is referred to as the stop band. In the ideal case, $|H(\omega)| = 1$ in a passband, while $|H(\omega)| = 0$ in a stop band. Frequency-selective filters are classified according to the functions they perform. The most common types of filters are the following:

1. Low-pass filters are those characterized by a passband that extends from $\omega = 0$ to $\omega = \omega_c$, where ω_c is called the cutoff frequency of the filter. (See Figure 4.4.11(a).)
2. High-pass filters are characterized by a stop band that extends from $\omega = 0$ to $\omega = \omega_c$ and a passband that extends from $\omega = \omega_c$ to infinity. (See Figure 4.4.11(b).)
3. Band-pass filters are characterized by a passband that extends from $\omega = \omega_1$ to $\omega = \omega_2$, and all other frequencies are stopped. (See Figure 4.4.11(c).)
4. Band-stop filters stop frequencies extending from ω_1 to ω_2 and pass all other frequencies. (See Figure 4.4.11(d).)

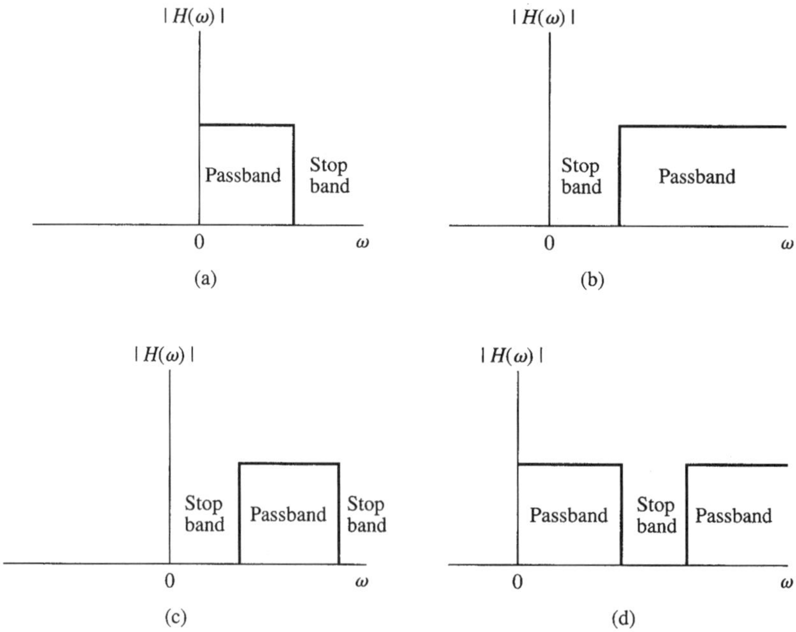

Figure 4.4.11 Most common classes of filters.

As is usual with spectra of real-valued signals, in Figure 4.4.11 we have shown $H(\omega)$ only for values of $\omega \geq 0$, since $H(\omega) = H(-\omega)$ for such signals.

Example 4.4.4

Consider the ideal low-pass filter with frequency response

$$H_{\text{lp}}(\omega) = \begin{cases} 1, & |\omega| < \omega_c \\ 0, & \text{elsewhere} \end{cases}$$

The impulse response of this filter corresponds to the inverse Fourier transform of the frequency response $H_{\text{lp}}(\omega)$ and is given by

$$h_{\text{lp}} = \frac{\omega_c}{\pi} \operatorname{sinc} \frac{\omega_c t}{\pi}$$

Clearly, this filter is noncausal and, hence, is not physically realizable.

The filters described so far are referred to as ideal filters because they pass one set of frequencies without any change and completely stop others. Since it is impossible to realize filters with characteristics like those shown in Figure 4.4.11, with abrupt changes from passband to stop band and vice versa, most of the filters we deal with in practice have some transition band, as shown in Figure 4.4.12.

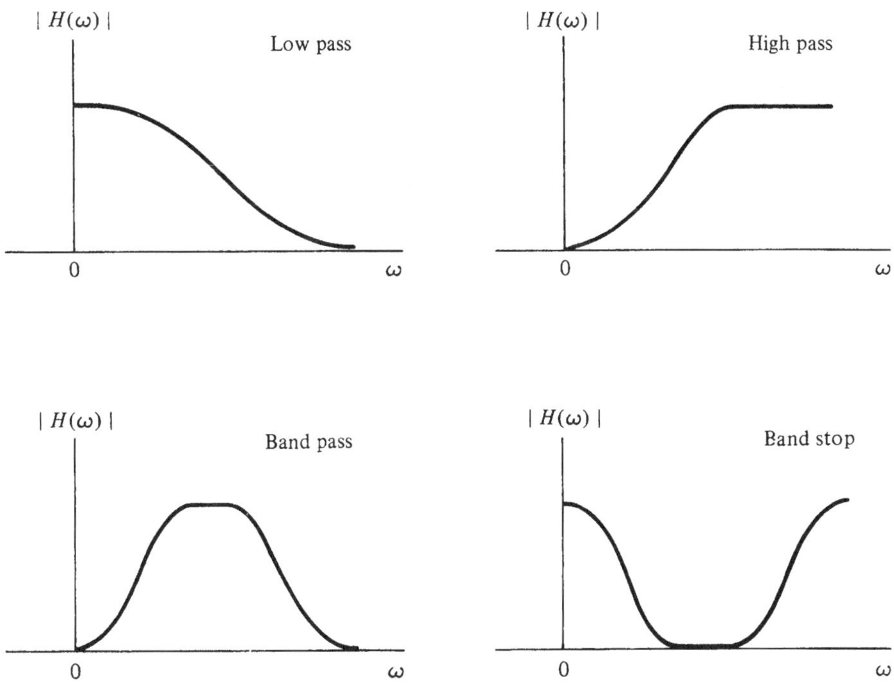

Figure 4.4.12 Practical filters.

Example 4.4.5

Consider the following RC circuit:

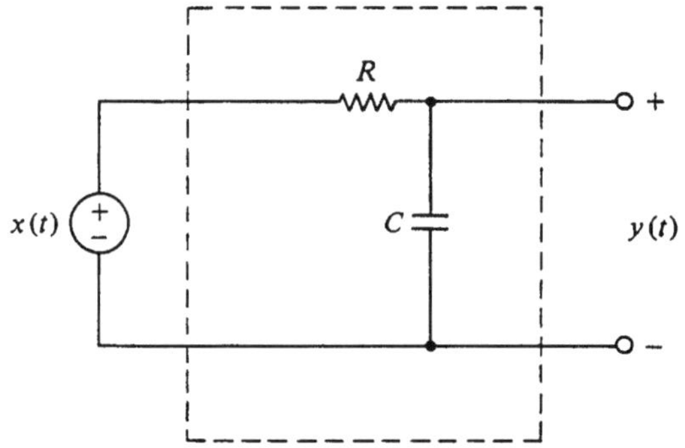

The impulse response of this circuit is (see Problem 2.17)

$$h(t) = \frac{1}{RC} \exp\left[\frac{-t}{RC}\right] u(t)$$

and the frequency response is

$$H(\omega) = \frac{1}{1 + j\omega RC}$$

The amplitude spectrum is given by

$$|H(\omega)|^2 = \frac{1}{1 + (\omega RC)^2}$$

and is shown in Figure 4.4.13. It is clear that the RC circuit with the output taken as the voltage across the capacitor performs as a low-pass filter. The frequency ω_c at which the magnitude spectrum $|H(\omega)| = H(0)/\sqrt{2}$ (3 dB below $H(0)$) is called the band edge, or the 3-dB cutoff frequency of the filter. (The transition between the passband and the stop band occurs near ω_c.) Setting $|H(\omega)| = 1/\sqrt{2}$, we obtain

$$\omega_c = \frac{1}{RC}$$

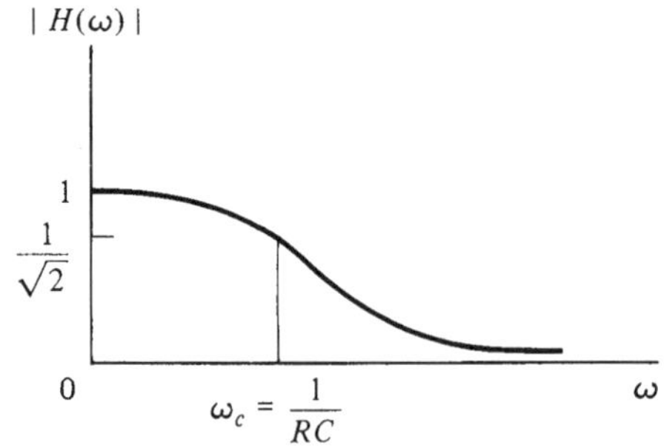

Figure 4.4.13 Magnitude spectrum of a low-pass RC circuit.

If we interchange the positions of the capacitor and the resistor, we obtain a system with impulse response (see Problem 2.18)

$$h(t) = \delta(t) - \frac{1}{RC}\exp\left[\frac{-t}{RC}\right]u(t)$$

and frequency response

$$H(\omega) = \frac{j\omega RC}{1 + j\omega RC}$$

The amplitude spectrum is given by

$$|H(\omega)|^2 = \frac{(\omega RC)^2}{1 + (\omega RC)^2}$$

and is shown in Figure 4.4.14. It is clear that the RC circuit with output taken as the voltage across the resistor performs as a high-pass filter. Again, by setting $|H(\omega)| = 1/\sqrt{2}$, the cutoff frequency of this high-pass filter can be determined as

$$\omega_c = \frac{1}{RC}$$

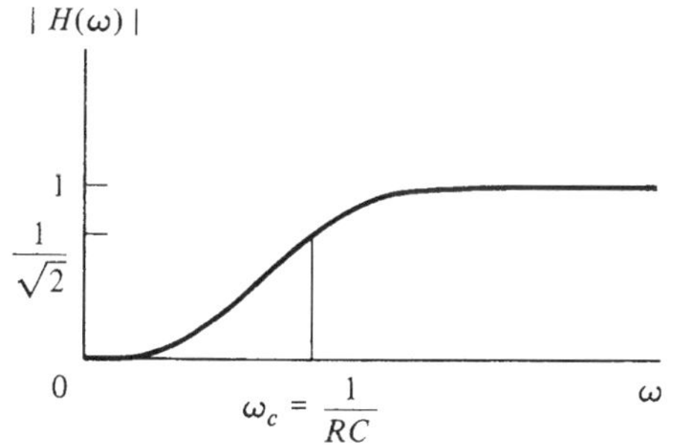

Figure 4.4.14 Magnitude spectrum of a high-pass RC circuit.

Filters can be classified as passive or active. Passive filters are made of passive elements (resistors, capacitors, and inductors), and active filters use operational amplifiers together with capacitors and resistors. The decision to use a passive filter in preference to an active filter in a certain application depends on several factors, such as the following:

1. *The range of frequency of operation of the filter.* Passive filters can operate at higher frequencies, whereas active filters are usually used at lower frequencies.
2. *The weight and size of the filter realization.* Active filters can be realized as an integrated circuit on a chip. Thus, they are superior when considerations of weight and size are important. This is a factor in the design of filters for low-frequency applications where passive filters require large inductors.

3. *The sensitivity of the filter* to parameter changes and *stability.* Components used in circuits deviate from their nominal values due to tolerances related to their manufacture or due to chemical changes because of thermal and aging effects. Passive filters are always superior to active filters when it comes to sensitivity.
4. *The availability of voltage sources for operational amplifiers.* Operational amplifiers require voltage sources ranging from 1 to about 12 volts for their proper operation. Whether such voltages are available without maintenance is an important consideration.

We consider the design of analog and discrete-time filters in more detail in Chapter 10.

4.5 DURATION-BANDWIDTH RELATIONSHIPS

In Section 4.3, we discussed the time-scaling property of the Fourier transform. We noticed that expansion in the time domain implies compression in the frequency domain, and conversely. In the current section, we give a quantitative measure to this observation. The width of the signal, in time or in frequency, can be formally defined in many different ways. No one way or set of ways is best for all purposes. As long as we use the same definition when working with several signals, we can compare their durations and spectral widths. If we change definitions, "conversion factors" are needed to compare the durations and spectral widths involved. The principal purpose of this section is to show that the width of a time signal in seconds (duration) is inversely related to the width of the Fourier transform of the signal in hertz (bandwidth). The spectral width of signals is a very important concept in communication systems and signal processing, for two main reasons. First, more and more users are being assigned to increasingly crowded radio frequency (RF) bands, so that the spectral width required for each band has to be considered carefully. Second, the spectral width of signals is important from the equipment design viewpoint, since the circuits have to have sufficient bandwidth to accommodate the signal, but reject the noise. The remarkable observation is that, independent of shape, there is a lower bound on the duration-bandwidth product of a given signal; this relationship is known as the uncertainty principle.

4.5.1 Definitions of Duration and Bandwidth

As we mentioned earlier, spectral representation is an efficient and convenient method of representing physical signals. Not only does it simplify some operations, but it also reveals the frequency content of the signal. One characterization of the signal is its spread in the frequency domain, or, simply, its bandwidth.

We will give some engineering definitions for the bandwidth of an arbitrary real-valued time signal. Some of these definitions are fairly generally applicable, and others are restricted to a particular application. The reader should keep in mind that there are also other definitions that might be useful, depending on the application.

The signal $x(t)$ is called a baseband (low-pass) signal if $|X(\omega)| = 0$ for $|\omega| \geq \omega_B$ and is called a band-pass signal centered at ω_0 if $|X(\omega)| = 0$ for $|\omega - \omega_0| \geq \omega_B$.(See

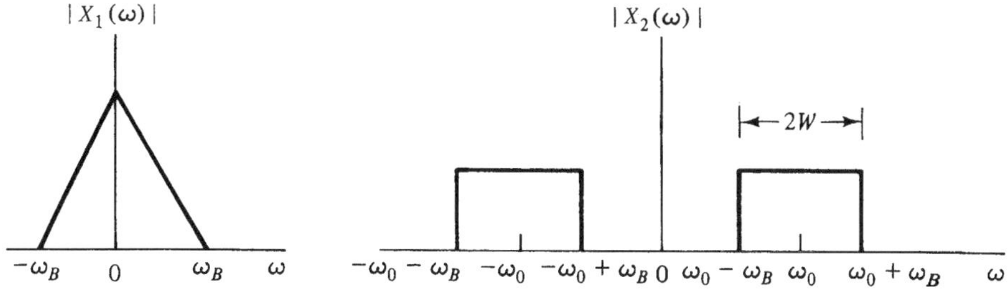

Figure 4.5.1 Amplitude spectra for baseband and band-pass signals.

Figure 4.5.1.) For baseband signals, we measure the bandwidth in terms of the positive frequency portion only.

Absolute Bandwidth. This notion is used in conjunction with band-limited signals and is defined as the region outside of which the spectrum is zero. That is, if $x(t)$ is a baseband signal and $|X(\omega)|$ is zero outside the interval $|\omega| < \omega_B$, then

$$B = \omega_B \tag{4.5.1}$$

But if $x(t)$ is a band-pass signal and $|X(\omega)|$ is zero outside the interval $\omega_1 < \omega < \omega_2$, then

$$B = \omega_2 - \omega_1 \tag{4.5.2}$$

Example 4.5.1

The signal $x(t) = \sin \omega_B t/\pi t$ is a baseband signal and has the Fourier transform $\text{rect}(\omega/2\omega_B)$. The bandwidth of this signal is then ω_B.

3-dB (Half-Power) Bandwidth. This idea is used with baseband signals that have only one maximum, located at the origin. The 3-dB bandwidth is defined as the frequency ω_1 such that

$$\frac{|X(\omega_1)|}{|X(0)|} = \frac{1}{\sqrt{2}} \tag{4.5.3}$$

Note that inside the band $0 < \omega < \omega_1$, the magnitude $|X(\omega)|$ falls no lower than $1/\sqrt{2}$ of its value at $\omega = 0$. The 3-dB bandwidth is also known as the half-power bandwidth because a voltage or current attenuation of 3 dB is equivalent to a power attenuation by a factor 2.

Example 4.5.2

The signal $x(t) = \exp[-t/T]u(t)$ is a baseband signal and has the Fourier transform

$$X(\omega) = \frac{1}{1/T + j\omega}$$

The magnitude spectrum of this signal is shown in Figure 4.5.2. Clearly, $X(0) = T$, and the 3-dB bandwidth is

$$B = \frac{1}{T}$$

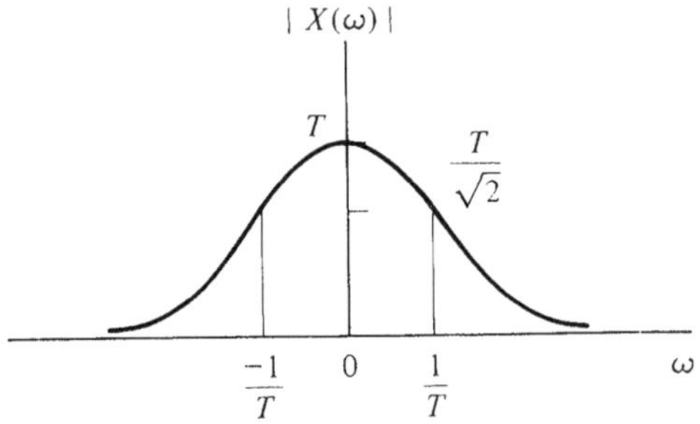

Figure 4.5.2 Magnitude spectrum for the signal in Example 4.5.2.

Equivalent Bandwidth. This definition is used in association with band-pass signals with unimodal spectra whose maxima are at the center of the frequency band. The equivalent bandwidth is the width of a fictitious rectangular spectrum such that the energy in that rectangular band is equal to the energy associated with the actual spectrum. In Section 4.3.6, we saw that the energy density is proportional to the square of the magnitude of the signal spectrum. If ω_m is the frequency at which the magnitude spectrum has its maximum, we let the energy in the equivalent rectangular band be

$$\text{Equivalent energy} = \frac{2B_{\text{eq}}|X(\omega_m)|^2}{2\pi} \tag{4.5.4}$$

The actual energy in the signal is

$$\text{Actual energy} = \frac{1}{2\pi}\int_{-\infty}^{\infty}|X(\omega)|^2 d\omega = \frac{1}{\pi}\int_{0}^{\infty}|X(\omega)|^2 d\omega \tag{4.5.5}$$

Setting Equation (4.5.4) equal to Equation (4.5.5), we have the formula that gives the equivalent bandwidth in hertz:

$$B_{\text{eq}} = \frac{1}{|X(\omega_m)|^2}\int_{0}^{\infty}|X(\omega)|^2 d\omega. \tag{4.5.6}$$

Example 4.5.3

The equivalent bandwidth of the signal in Example 4.5.2 is

$$B_{\text{eq}} = \frac{1}{T^2}\int_{0}^{\infty}\frac{1}{(1/T)^2 + \omega^2}d\omega = \frac{\pi}{2T}$$

Null-to-Null (Zero-Crossing) Bandwidth. This concept applies to non-band-limited signals and is defined as the distance between the first null in the envelope of the magnitude spectrum above ω_m and the first null in the envelope below ω_m, where ω_m is the radian frequency at which the magnitude spectrum is maximum. For baseband signals, the spectrum maximum is at $\omega = 0$, and the bandwidth is the distance between the first null and the origin.

Example 4.5.4

In Example 4.2.1, we showed that signal $x(t) = \text{rect}(t/T)$ has the Fourier transform

$$X(\omega) = T \,\text{sinc}\, \frac{\omega T}{2\pi}$$

The magnitude spectrum of this signal is shown in Figure 4.5.3. From the figure, the null-to-null bandwidth is

$$B = \frac{2\pi}{T}$$

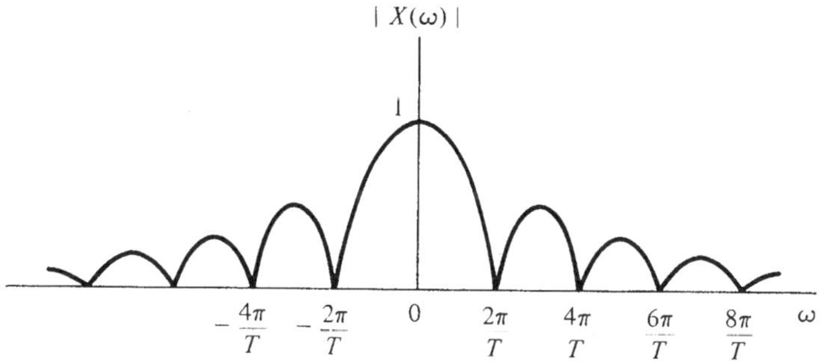

Figure 4.5.3 Magnitude spectrum for the signal in Example 4.5.4.

z% Bandwidth. This is defined such that

$$\int_{-B_z}^{B_z} |X(\omega)|^2 d\omega = \frac{z}{100} \int_{-\infty}^{\infty} |X(\omega)|^2 d\omega \tag{4.5.7}$$

For example, $z = 99$ defines the frequency band in which 99% of the total energy resides. This is similar to the Federal Communications Commission (FCC) definition of the occupied bandwidth, which states that the energy above the upper band edge ω_2 is $\frac{1}{2}$% and the energy below the lower band edge ω_1 is $\frac{1}{2}$%, leaving 99% of the total energy within the occupied band. The z% bandwidth is implicitly defined.

RMS (Gabor) Bandwidth. Probably the most analytically useful definitions of bandwidth are given by various moments of $X(\omega)$, or even better, of $|X(\omega)|^2$. The rms bandwidth of the signal is defined as

$$B_{\text{rms}}^2 = \frac{\displaystyle\int_{-\infty}^{\infty} \omega^2 |X(\omega)|^2 d\omega}{\displaystyle\int_{-\infty}^{\infty} |X(\omega)|^2 d\omega} \tag{4.5.8}$$

A dual characterization of a signal $x(t)$ can be given in terms of its duration T, which is a measure of the extent of $x(t)$ in the time domain. As with bandwidth, duration can be defined in several ways. The particular definition to be used depends on the application. Three of the more common definitions are as follows:

1. *Distance between successive zeros.* As an example, the signal

$$x(t) = \frac{\sin 2\pi Wt}{\pi t}$$

has duration $T = 1/W$.

2. *Time at which $x(t)$ drops to a given value.* For example, the exponential signal

$$x(t) = \exp[-t/\Delta]u(t)$$

has duration $T = \Delta$, measured as the time at which $x(t)$ drops to $1/e$ of its value at $t = 0$.

3. *Radius of gyration.* This measure is used with signals that are concentrated around $t = 0$ and is defined as

$$\begin{aligned} T &= 2 \times \text{radius of gyration} \\ &= 2\sqrt{\frac{\int_{-\infty}^{\infty} t^2 |x(t)|^2 dt}{\int_{-\infty}^{\infty} |x(t)|^2 dt}} \end{aligned} \tag{4.5.9}$$

For example, the signal

$$x(t) = \frac{1}{\sqrt{2\pi\Delta^2}} \exp\left[\frac{-t^2}{2\Delta^2}\right]$$

has a duration of

$$\begin{aligned} T &= 2\sqrt{\left[\frac{\Delta^2}{4\sqrt{2\pi\Delta^2}}\right] \Big/ \left[\frac{1}{2\sqrt{2\pi\Delta^2}}\right]} \\ &= \sqrt{2}\,\Delta \end{aligned}$$

4.5.2 The Uncertainty Principle

The uncertainty principle states that for any real signal $x(t)$ that vanishes at infinity faster than $1/\sqrt{t}$, that is,

$$\lim_{t \to \pm\infty} \sqrt{t}\, x(t) = 0 \tag{4.5.10}$$

and for which the duration is defined as in Equation (4.5.9) and the bandwidth is defined as in Equation (4.5.8), the product TB should satisfy the inequality

$$TB \geq 1 \tag{4.5.11}$$

In words, T and B cannot simultaneously be arbitrarily small: A short duration implies a large bandwidth, and a small-bandwidth signal must last a long time. This constraint has a wide domain of applications in communication systems, radar, and signal and speech processing.

The proof of Equation (4.5.11) follows from Parseval's formula, Equation (4.3.14), and Schwarz's inequality,

$$\left|\int_a^b y_1(t)y_2(t)\,dt\right|^2 \leq \int_a^b |y_1(t)|^2\,dt \int_a^b |y_2(t)|^2\,dt \tag{4.5.12}$$

where the equality holds if and only if $y_1(t)$ is proportional to $y_2(t)$—that is,

$$y_2(t) = ky_1(t) \tag{4.5.13}$$

Schwarz's inequality can be easily derived from

$$0 \leq \int_a^b |\theta y_1(t) - y_2(t)|^2\,dt = \theta^2 \int_a^b |y_1(t)|^2\,dt - 2\theta \int_a^b y_1(t)y_2^*(t)\,dt + \int_a^b |y_2(t)|^2\,dt$$

This equation is a nonnegative quadratic form in the variable θ. For the quadratic to be nonnegative for all values of θ, its discriminant must be nonpositive. Setting this condition establishes Equation (4.5.12). If the discriminant equals zero, then for some value $\theta = k$, the quadratic equals zero. This is possible only if $ky_1(t) - y_2(t) = 0$, and Equation (4.5.13) follows.

By using Parseval's formula, we can write the bandwidth of the signal as

$$B^2 = \frac{\displaystyle\int_{-\infty}^{\infty} \left|\frac{dx(t)}{dt}\right|^2 dt}{\displaystyle\int_{-\infty}^{\infty} |x(t)|^2\,dt} \tag{4.5.14}$$

Combining Equation (4.5.14) with Equation (4.5.9) gives

$$(TB)^2 = \frac{4\displaystyle\int_{-\infty}^{\infty} t^2|x(t)|^2\,dt \int_{-\infty}^{\infty} |x'(t)|^2\,dt}{\left[\displaystyle\int_{-\infty}^{\infty} |x(t)|^2\,dt\right]^2} \tag{4.5.15}$$

We apply Schwarz's inequality to the numerator of Equation (4.5.15) to obtain

$$TB \geq 2\frac{\left|\displaystyle\int_{-\infty}^{\infty} tx(t)x'(t)\,dt\right|}{\displaystyle\int_{-\infty}^{\infty} |x(t)|^2\,dt} \tag{4.5.16}$$

But the fraction on the right in Equation (4.5.16) is identically equal to 1/2 (as can be seen by integrating the numerator by parts and noting that $x(t)$ must vanish faster than $1/\sqrt{t}$ as $t \to \pm\infty$), which gives the desired result.

To obtain equality in Schwarz's inequality, we must have

$$ktx(t) = \frac{dx(t)}{dt}$$

or

$$\frac{x'(t)}{x(t)} = kt$$

Integrating, we have

$$\ln[x(t)] = \frac{kt^2}{2} + \text{constant}$$

or

$$x(t) = c\exp[kt^2] \tag{4.5.17}$$

If k is a negative real number, $x(t)$ is an acceptable pulselike signal and is referred to as the Gaussian pulse. Thus, among all signals, the Gaussian pulse has the smallest duration-bandwidth product in the sense of Equations (4.5.8) and (4.5.9).

Example 4.5.5

Writing the Fourier transform in the polar form

$$X(\omega) = A(\omega)\exp[j\phi(\omega)]$$

we show that, among all signals with the same amplitude $A(\omega)$, the one that minimizes the duration of $x(t)$ has zero (linear) phase. From Equation (4.3.23), we obtain

$$(-jt)x(t) \leftrightarrow \frac{dX(\omega)}{d\omega} = \left[\frac{dA(\omega)}{d\omega} + jA(\omega)\frac{d\phi(\omega)}{d\omega}\right]\exp[j\phi(\omega)] \tag{4.5.18}$$

From Equations (4.3.14) and (4.5.18), we have

$$\int_{-\infty}^{\infty} t^2|x(t)|^2\,dt = \frac{1}{2\pi}\int_{-\infty}^{\infty}\left\{\left[\frac{dA(\omega)}{d\omega}\right]^2 + A^2(\omega)\left[\frac{d\phi(\omega)}{d\omega}\right]^2\right\}d\omega \tag{4.5.19}$$

Since the left-hand side of Equation (4.5.19) measures the duration of $x(t)$, we conclude that a high ripple in the amplitude spectrum or in the phase angle of $X(\omega)$ results in signals with long duration. A high ripple results in large absolute values of the derivatives of both the amplitude and phase spectrum, and among all signals with the same amplitude $A(\omega)$, the one that minimizes the left-hand side of Equation (4.5.19) has zero (linear) phase.

Example 4.5.6

A convenient measure of the duration of $x(t)$ is the quantity

$$T = \frac{1}{x(0)}\int_{-\infty}^{\infty} x(t)\,dt$$

In this formula, the duration T can be interpreted as the ratio of the area of $x(t)$ to its height. Note that if $x(t)$ represents the impulse response of an LTI system, then T is a measure of the rise time of the system, which is defined as the ratio of the final value of the step response to the slope of the step response at some appropriate point t_0 along the rise ($t_0 = 0$ in this case). If we define the bandwidth of $x(t)$ by

$$B = \frac{1}{X(0)} \int_{-\infty}^{\infty} X(\omega)\, d\omega$$

it is easy to show that

$$BT = 2\pi$$

4.6 SUMMARY

- The Fourier transform of $x(t)$ is defined by

$$X(\omega) = \int_{-\infty}^{\infty} x(t) \exp[-j\omega t]\, dt$$

- The inverse Fourier transform of $X(\omega)$ is defined by

$$x(t) = \frac{1}{2\pi} \int_{-\infty}^{\infty} X(\omega) \exp[j\omega t]\, d\omega$$

- $X(\omega)$ exists if $x(t)$ is "well behaved" and is absolutely integrable. These conditions are sufficient, but not necessary.
- The magnitude of $X(\omega)$ plotted against ω is called the magnitude spectrum of $x(t)$, and $|X(\omega)|^2$ is called the energy spectrum.
- The angle of $X(\omega)$ plotted versus ω is called the phase spectrum.
- Parseval's theorem states that

$$\int_{-\infty}^{\infty} |x(t)|^2\, dt = \frac{1}{2\pi} \int_{-\infty}^{\infty} |X(\omega)|^2\, d\omega$$

- The energy of $x(t)$ within the frequency band $\omega_1 < \omega < \omega_2$ is given by

$$\Delta E = \frac{2}{2\pi} \int_{\omega_1}^{\omega_2} |X(\omega)|^2\, d\omega$$

- The total energy of the aperiodic signal $x(t)$ is

$$E = \frac{1}{2\pi} \int_{-\infty}^{\infty} |X(\omega)|^2\, d\omega$$

- The power-density spectrum of $x(t)$ is defined by

$$S(\omega) = \lim_{\tau \to \infty} \left[\frac{|X_\tau(\omega)|^2}{2\tau} \right]$$

where

$$x_\tau(t) \leftrightarrow X_\tau(\omega)$$

and

$$x_\tau(t) = x(t)\, \text{rect}(t/2\tau)$$

- The convolution property of the Fourier transform states that

$$y(t) = x(t) * h(t) \leftrightarrow Y(\omega) = X(\omega)H(\omega)$$

- If $X(\omega)$ is the Fourier transform of $x(t)$, then the duality property of the transform is expressed as

$$X(t) \leftrightarrow 2\pi x(-\omega)$$

- Amplitude modulation, multiplexing, filtering, and sampling are among the important applications of the Fourier transform.
- If $x(t)$ is a band-limited signal such that

$$X(\omega) = 0, \qquad |\omega| > \omega_B$$

then $x(t)$ is uniquely determined by its values (samples) at equally spaced points in time, provided that $T < \pi/\omega_B$. The radian frequency $\omega_s = 2\pi/T$ is called the sampling frequency. The minimum sampling rate is $2\omega_B$ and is known as the Nyquist rate.
- The bandwidth B of $x(t)$ is a measure of the frequency content of the signal.
- There are several definitions for the bandwidth of the signal $x(t)$ that are useful in particular applications.
- The duration T of $x(t)$ is a measure of the extent of $x(t)$ in time.
- The product of the bandwidth and the duration of $x(t)$ is greater than or equal to a constant that depends on the definition of both B and T.

4.7 CHECKLIST OF IMPORTANT TERMS

Aliasing
Amplitude modulation
Band-pass filter
Bandwidth
Duality
Duration
Energy spectrum
Equivalent bandwidth
Half-power (3-dB) bandwidth
High-pass filter
Low-pass filter
Magnitude spectrum
Multiplexing
Nyquist rate
Parseval's theorem
Periodic pulse train
Phase spectrum
Power-density spectrum
Rectangular pulse
RMS bandwidth
Sampling frequency
Sampling function
Sampling theorem
Signal filtering
Sinc function
Triangular pulse
Two-sided exponential
Uncertainty principle

4.8 PROBLEMS

4.1. Find the Fourier transform of the following signals in terms of $X(\omega)$, the Fourier transform of $x(t)$.

(a) $x(-t)$

(b) $x_e(t) = \dfrac{x(t) + x(-t)}{2}$

(c) $x_o(t) = \dfrac{x(t) - x(-t)}{2}$

(d) $x^*(t)$

(e) $\text{Re}\{x(t)\} = \dfrac{x(t) + x^*(t)}{2}$

(f) $\text{Im}\{x(t)\} = \dfrac{x(t) - x^*(t)}{2j}$

4.2. Determine which of the following signals have a Fourier transform. Why?

(a) $x(t) = \exp[-2t]u(t)$

(b) $x(t) = |t|\,u(t)$

(c) $x(t) = \cos(\pi/t)$

(d) $x(t) = \dfrac{1}{t}$

(e) $x(t) = t^2 \exp[-2t]u(t)$

4.3. Show that the Fourier transform of $x(t)$ can be written as

$$X(\omega) = \sum_{n=0}^{\infty} (-j)^n m_n \frac{\omega^n}{n!}$$

where

$$m_n = \int_{-\infty}^{\infty} t^n x(t)\,dt, \qquad n = 0, 1, 2, \ldots$$

(*Hint:* Expand $\exp[-j\omega t]$ around $t = 0$ and integrate termwise.)

4.4. Using Equation (4.2.12), show that

$$\int_{-\infty}^{\infty} \cos\omega t\,d\omega = 2\pi\,\delta(t)$$

Use the result to prove Equation (4.2.13).

4.5. Let $X(\omega) = \text{rect}[(\omega - 1)/2]$. Find the transform of the following functions, using the properties of the Fourier transform:

(a) $x(-t)$

(b) $tx(t)$

(c) $x(t+1)$

(d) $x(-2t+4)$

(e) $(t - 1)x(t + 1)$

(f) $\dfrac{dx(t)}{dt}$

(g) $t\dfrac{dx(t)}{dt}$

(h) $x(2t - 1)\exp[-j2t]$

(i) $x(t)\exp[-j2t]$

(j) $tx(t)\exp[-j2t]$

(k) $(t-1)x(t-1)\exp[-j2t]$

(l) $\int_{-\infty}^{t} x(\tau)\,d\tau$

4.6. Let $x(t) = \exp[-2t]u(t)$ and let $y(t) = x(t+1) + x(t-1)$. Find $Y(\omega)$.

4.7. Using Euler's form,

$$\exp[j\omega t] = \cos\omega t + j\sin\omega t$$

interpret the integral

$$\frac{1}{2\pi}\int \exp[j\omega t]\,d\omega = \delta(t)$$

(*Hint:* Think of the integral as the sum of a large number of cosines and sines of various frequencies.)

4.8. Consider the two related signals shown in Figure P4.8. Use linearity, time-shifting, and integration properties, along with the transform of a rectangular signal, to find $X(\omega)$ and $Y(\omega)$.

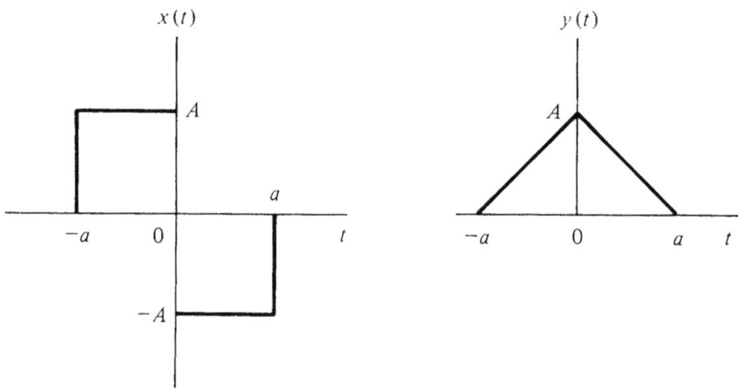

Figure P4.8

4.9. Find the energy of the following signals, using Parseval's theorem.

(a) $x(t) = \exp[-2t]u(t)$

(b) $x(t) = u(t) - u(t-5)$

(c) $x(t) = \Delta(t/4)$

(d) $x(t) = \dfrac{\sin(\pi t)}{\pi t}$

4.10. A Gaussian-shaped signal

$$x(t) = A\exp[-at^2]$$

is applied to a system whose input/output relationship is

$$y(t) = x^2(t)$$

(a) Find the Fourier transform of the output $y(t)$.

(b) If $y(t)$ is applied to a second system, which is LTI with impulse response

$$h(t) = B\exp[-bt^2]$$

find the output of the second system.

(c) How does the output change if we interchange the order of the two systems?

4.11. The two formulas

$$X(0) = \int_{-\infty}^{\infty} x(t)\,dt \qquad \text{and} \qquad x(0) = \frac{1}{2\pi}\int_{-\infty}^{\infty} X(\omega)\,d\omega$$

are special cases of the Fourier-transform pair. These two formulas can be used to evaluate some definite integrals. Choose the appropriate $x(t)$ and $X(\omega)$ to verify the following identities.

(a) $\displaystyle\int_{-\infty}^{\infty} \frac{\sin\theta}{\pi\theta}\,d\theta = 1$

(b) $\displaystyle\int_{-\infty}^{\infty} \exp[-\pi\theta^2]\,d\theta = 1$

(c) $\displaystyle\frac{1}{2\pi}\int_{-\infty}^{\infty} \frac{2a}{a^2+\theta^2}\,d\theta = 1$

(d) $\displaystyle\frac{1}{2\pi}\int_{-\infty}^{\infty} T\frac{\sin^2(\theta T/2)}{(\theta T/2)^2}\,d\theta = 1$

4.12. Use the relation

$$\int_{-\infty}^{\infty} x(t)y^*(t)\,dt = \frac{1}{2\pi}\int_{-\infty}^{\infty} X(\omega)Y^*(\omega)\,d\omega$$

and a known transform pair to show that

(a) $\displaystyle\int_{0}^{\infty} \frac{1}{(a^2+t^2)^2}\,dt = \frac{\pi}{4a^3}$

(b) $\displaystyle\int_{0}^{\infty} \frac{\operatorname{sinc} t}{a^2+t^2}\,dt = \frac{1-\exp[-\pi a]}{2a^2}$

(c) $\displaystyle\int_{-\infty}^{\infty} \frac{\sin^3 t}{t^3}\,dt = \frac{3\pi}{4}$

(d) $\displaystyle\int_{-\infty}^{\infty} \frac{\sin^4 t}{t^4}\,dt = \frac{2\pi}{3}$

4.13. Consider the signal

$$x(t) = \exp[-\varepsilon t]u(t)$$

(a) Find $X(\omega)$.

(b) Write $X(\omega)$ as $X(\omega) = R(\omega) + jI(\omega)$, where $R(\omega)$ and $I(\omega)$ are the real and imaginary parts of $X(\omega)$, respectively.

(c) Take the limit as $\varepsilon \to 0$ of part (b), and show that

$$\mathcal{F}\{\lim_{\varepsilon\to 0} \exp[-\varepsilon t]u(t)\} = \pi\delta(\omega) + \frac{1}{j\omega}$$

Hint: Note that

$$\lim_{\varepsilon \to 0} \frac{\varepsilon}{\varepsilon^2 + \omega^2} = \begin{cases} 0, & \omega \neq 0 \\ \infty, & \omega = 0 \end{cases}$$

and

$$\int_{-\infty}^{\infty} \frac{\varepsilon}{\varepsilon^2 + \omega^2} \, d\omega = \pi$$

4.14. The signal $x(t) = \exp[-\alpha t]u(t)$ is input into a system with impulse response $h(t) = \sin(2t)/(\pi t)$.

(a) Find the Fourier transform $Y(\omega)$ of the output.

(b) For what value of α does the energy in the output signal equal one-half the input-signal energy?

4.15. A signal has Fourier transform

$$X(\omega) = \frac{\omega^2 + j4\omega + 2}{-\omega^2 + j4\omega + 3}$$

Find the transforms of each of the following signals:

(a) $x(-2t + 1)$

(b) $x(t) \exp[-jt]$

(c) $\dfrac{dx(t)}{dt}$

(d) $x(t) \sin(\pi t)$

(e) $x(t) * \delta(t - 1)$

(f) $x(t) * x(t - 1)$

4.16. **(a)** Show that if $x(t)$ is band limited, that is,

$$X(\omega) = 0, \quad \text{for} \quad |\omega| > \omega_c$$

then

$$x(t) * \frac{\sin \alpha t}{\pi t} = x(t), \quad \alpha > \omega_c$$

(b) Use part (a) to show that

$$\frac{1}{\pi} \int_{-\infty}^{\infty} \frac{\sin \alpha\tau}{\tau} \frac{\sin(t - \tau)}{t - \tau} \, d\tau = \begin{cases} \dfrac{\sin t}{t}, & \alpha \geq 1 \\ \dfrac{\sin \alpha t}{t}, & |\alpha| \leq 1 \end{cases}$$

4.17. Show that with current as input and voltage as output, the frequency response of an inductor with inductance L is $j\omega L$ and that of a capacitor of capacitance C is $1/j\omega C$.

4.18. Consider the system shown in Figure P4.18 with $RC = 10$.

(a) Find $H(\omega)$ if the output is the voltage across the capacitor, and sketch $|H(\omega)|$ as a function of ω.

(b) Repeat Part (a) if the output is the resistor voltage. Comment on your results.

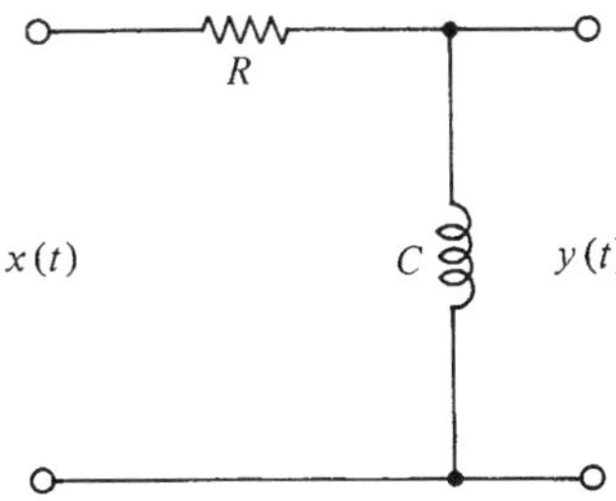

Figure P4.18

4.19. The input to the system shown in Figure P4.19 has the spectrum shown. Let

$$p(t) = \cos \omega_0 t, \qquad \omega_0 \gg \omega_m$$

Find the spectrum $Y(\omega)$ of the output if

$$h_2(t) = \frac{\sin \omega_B t}{\pi t}$$

Consider the cases $\omega_m > \omega_B$ and $\omega_m \leq \omega_B$.

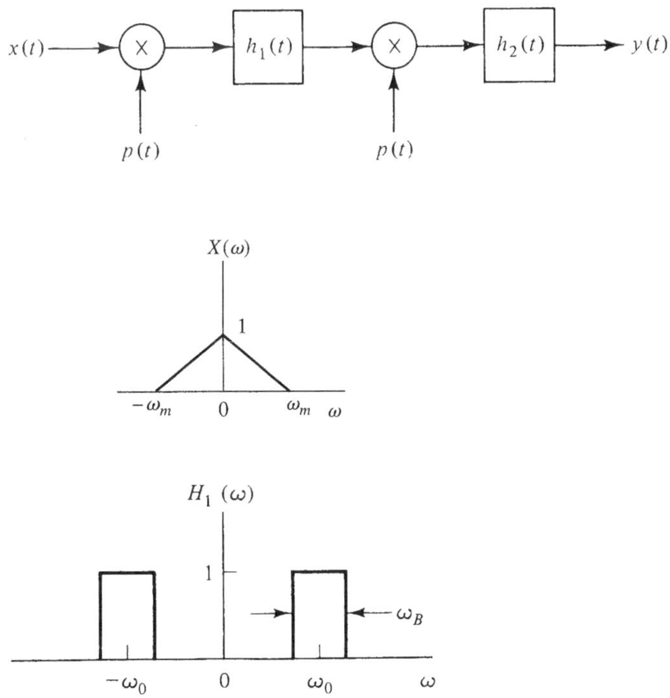

Figure P4.19

4.20. Repeat Problem 4.19 if

$$x(t) = \text{Sa}\,\frac{3\omega_B t}{2}$$

4.21. **(a)** The Hilbert transform of a signal $x(t)$ is obtained by passing the signal through an LTI system with impulse response $h(t) = 1/\pi t$. What is $H(\omega)$?

(b) What is the Hilbert transform of the signal $x(t) = \cos \pi t$?

4.22. The autocorrelation function of signal $x(t)$ is defined as

$$R_x(t) = \int_{-\infty}^{\infty} x^*(\tau)x(\tau + t)\,d\tau$$

Show that

$$R_x(t) \leftrightarrow |X(\omega)|^2$$

4.23. Suppose that $x(t)$ is the input to a linear system with impulse response $h(t)$.

(a) Show that

$$R_y(t) = R_x(t) * (t) * h(-t)$$

where $y(t)$ is the output of the system.

(b) Show that

$$R_y(t) \leftrightarrow |X(\omega)|^2 |H(\omega)|^2$$

4.24. For the system shown in Figure P4.24, assume that

$$x(t) = \frac{\sin \omega_1 t}{t} + \frac{\sin \omega_2 t}{t}, \quad \omega_1 < \omega_2$$

(a) Find $y(t)$ if $0 < \omega_f < \omega_1$.

(b) Find $y(t)$ if $\omega_1 < \omega_f < \omega_2$.

(c) Find $y(t)$ if $\omega_2 < \omega_f$.

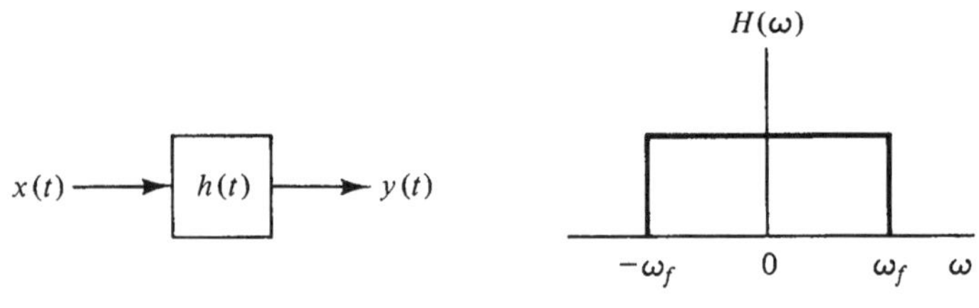

Figure P4.24

4.25. Consider the standard amplitude-modulation system shown in Figure P4.25.

(a) Sketch the spectrum of $x(t)$.

(b) Sketch the spectrum of $y(t)$ for the following cases:

(i) $0 \leq \omega_c < \omega_0 - \omega_m$

(ii) $\omega_0 - \omega_m \leq \omega_c < \omega_0$

(iii) $\omega_c > \omega_0 + \omega_m$

(c) Sketch the spectrum of $z(t)$ for the following cases:

(i) $0 \leq \omega_c < \omega_0 - \omega_m$

(ii) $\omega_0 - \omega_m \leq \omega_c < \omega_0$

(iii) $\omega_c > \omega_0 + \omega_m$

(d) Sketch the spectrum of $v(t)$ if $\omega_c > \omega_0 + \omega_m$ and

(i) $\omega_f < \omega_m$

(ii) $\omega_f < 2\omega_0 - \omega_m$

(iii) $\omega_f > 2\omega_0 + \omega_m$

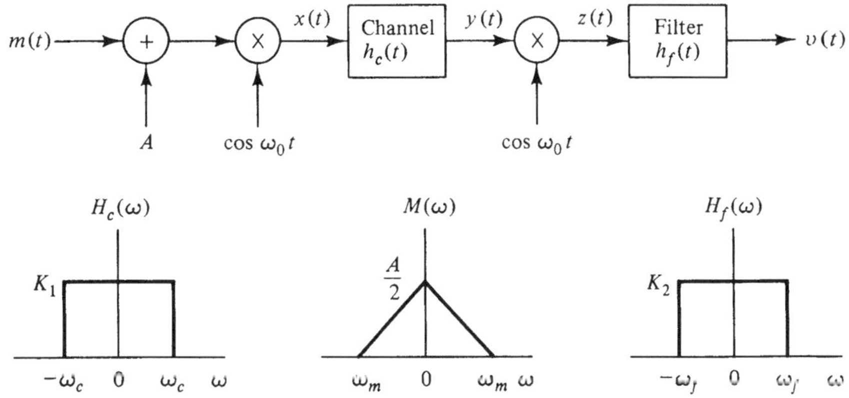

Figure P4.25

4.26. As discussed in Section 4.4.1, AM demodulation consists of multiplying the received signal $y(t)$ by a replica, $A\cos\omega_0 t$, of the carrier and low-pass filtering the resulting signal $z(t)$. Such a scheme is called synchronous demodulation and assumes that the phase of the carrier is known at the receiver. If the carrier phase is not known, $z(t)$ becomes

$$z(t) = y(t)A\cos(\omega_0 t + \theta)$$

where θ is the assumed phase of the carrier.

(a) Assume that the signal $x(t)$ is band limited to ω_m, and find the output $\hat{x}(t)$ of the demodulator.

(b) How does $\hat{x}(t)$ compare with the desired output $x(t)$?

4.27. A single-sideband, amplitude-modulated signal is generated using the system shown in Figure P4.27.

(a) Sketch the spectrum of $y(t)$ for $\omega_f = \omega_m$.

(b) Write a mathematical expression for $h_f(t)$. Is it a realizable filter?

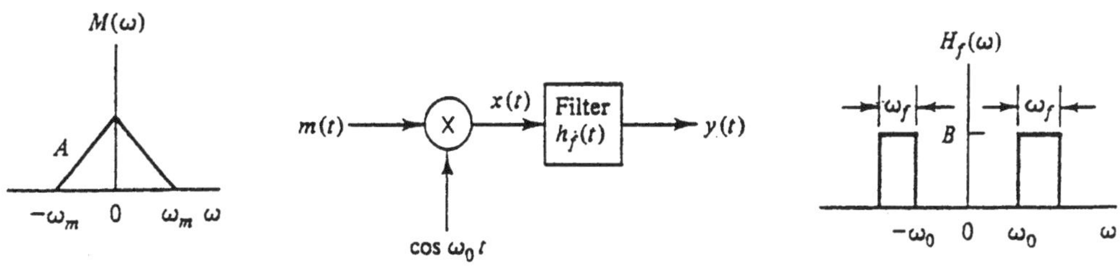

Figure P4.27

4.28. Consider the system shown in Figure P4.28(a). The systems $h_1(t)$ and $h_2(t)$ respectively have frequency responses

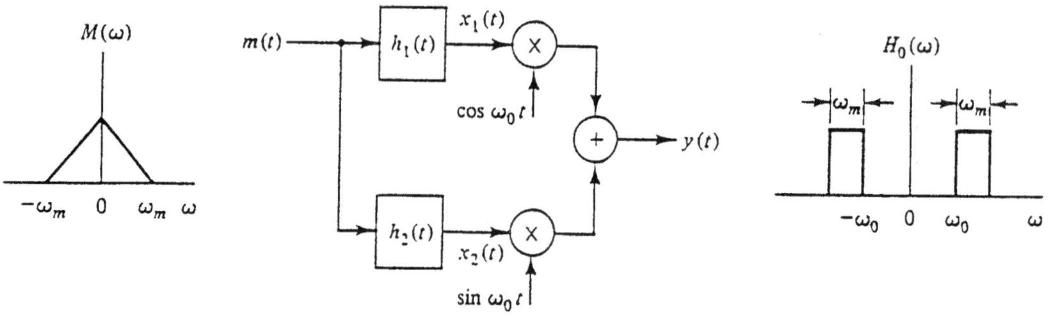

(a)

Figure P4.28(a)

$$H_1(\omega) = \frac{1}{2}[H_0(\omega - \omega_0) + H_0(\omega + \omega_0)]$$

and

$$H_2(\omega) = \frac{-1}{2j}[H_0(\omega - \omega_0) - H_0(\omega + \omega_0)]$$

(a) Sketch the spectrum of $y(t)$.

(b) Repeat part (a) for the $H_0(\omega)$ shown in Figure P4.28(b).

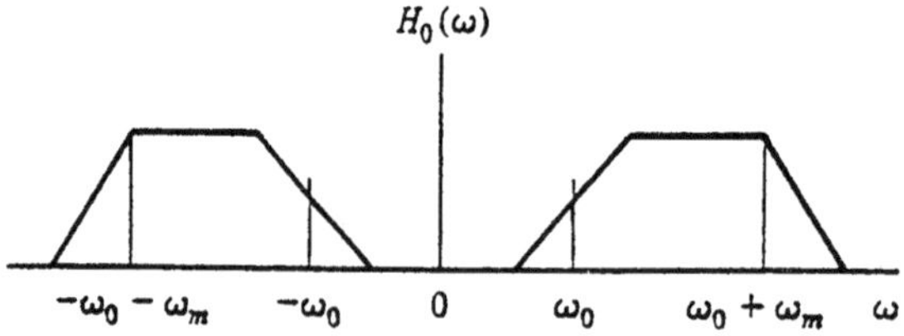

(b)

Figure P4.28(b)

4.29. Let $x(t)$ and $y(t)$ be low-pass signals with bandwidths of 150 Hz and 350 Hz, respectively, and let $z(t) = x(t)y(t)$. The signal $z(t)$ is sampled using an ideal sampler at intervals of T_s secs.

(a) What is the maximum value that T_s can take without introducing aliasing?

(b) If

$$x(t) = \sin\left(\frac{300\pi t}{\pi t}\right), \ y(t) = \sin\left(\frac{700\pi t}{\pi t}\right)$$

sketch the spectrum of the sampled signal for (i) $T_s = 0.5$ ms and (ii) $T_s = 2$ ms.

4.30. In natural sampling, the signal $x(t)$ is multiplied by a train of rectangular pulses, as shown in Figure P4.30.

(a) Find and sketch the spectrum of $x_s(t)$.

(b) Can $x(t)$ be recovered without any distortion?

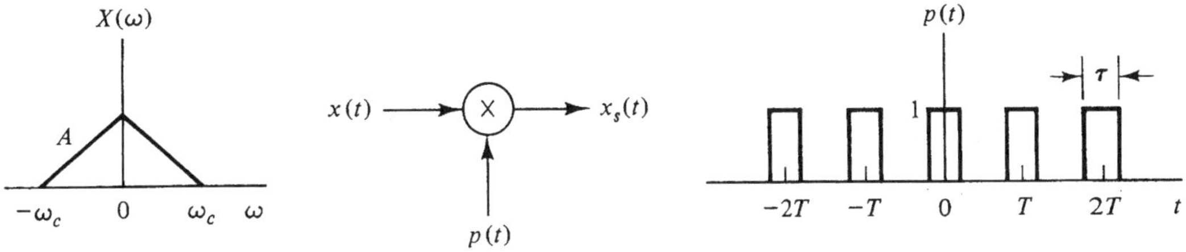

Figure P4.30

4.31. In flat-top sampling, the amplitude of each pulse in the pulse train $x_s(t)$ is constant during the pulse, but is determined by an instantaneous sample of $x(t)$, as illustrated in Figure P4.31(a). The instantaneous sample is chosen to occur at the center of the pulse for convenience. This choice is not necessary in general.

(a) Write an expression for $x_s(t)$.

(b) Find $X_s(\omega)$.

(c) How is this result different from the result in part (a) of Problem 4.30?

(d) Using only a low-pass filter, can $x(t)$ be recovered without any distortion?

(e) Show that you can recover $x(t)$ without any distortion if another filter, $H_{eq}(\omega)$, is added, as shown in Figure P4.31(b), where

$$H(\omega) = \begin{cases} 1, & |\omega| < \omega_c \\ 0, & \text{elsewhere} \end{cases}$$

$$\begin{aligned} H_{eq}(\omega) &= \frac{\omega\tau/2}{\sin(\omega\tau/2)}, && |\omega| < \omega_c \\ &= \text{arbitrary}, && \text{elsewhere} \end{aligned}$$

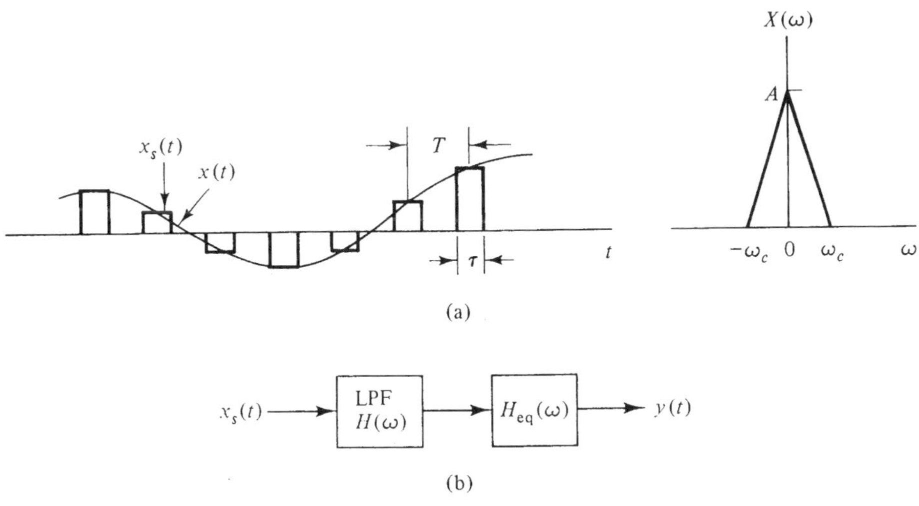

Figure P4.31 Flat-top sampling of $x(t)$.

4.32. Figure P4.32 diagrams the FDM system that generates the baseband signal for FM stereophonic broadcasting. The left-speaker and right-speaker signals are processed to produce $x_L(t) + x_R(t)$ and $x_L(t) - x_R(t)$, respectively.

(a) Sketch the spectrum of $y(t)$

(b) Sketch the spectrum of $z(t)$, $v(t)$, and $w(t)$.

(c) Show how to recover both $x_L(t)$ and $x_R(t)$.

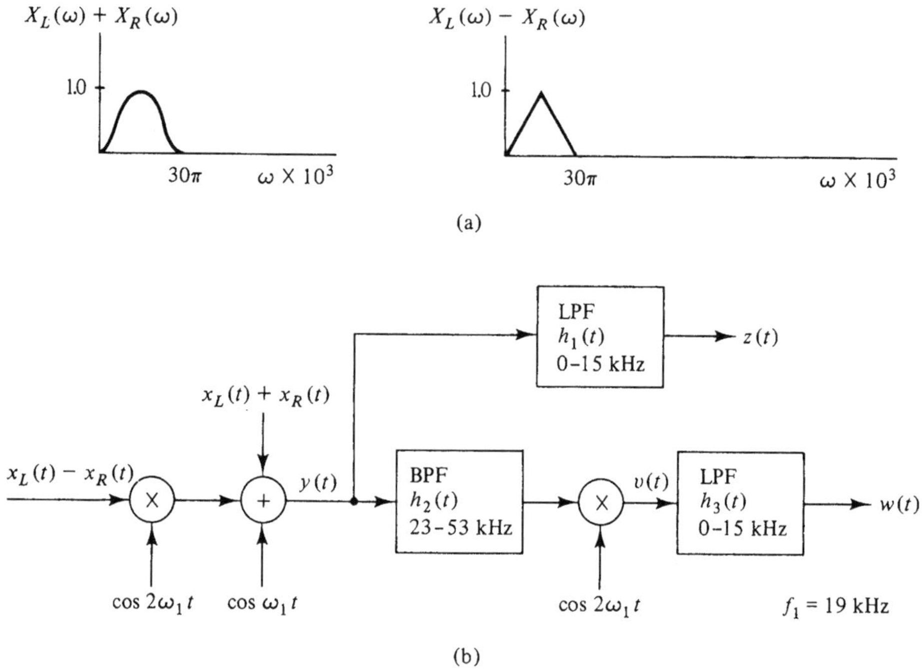

Figure P4.32

4.33. Show that the transform of a train of doublets is a train of impulses:

$$\sum_{n=-\infty}^{\infty} \delta'(t - nT) \leftrightarrow j\omega_0^2 \sum_{n=-\infty}^{\infty} n\,\delta(\omega - n\omega_0), \qquad \omega_0 = \frac{2\pi}{T}$$

4.34. Find the indicated bandwidth of the following signals:

(a) $\text{sinc}\,\dfrac{3Wt}{2}$ (absolute bandwidth)

(b) $\exp[-3t]u(t)$ (3-dB bandwidth)

(c) $\exp[-3t]u(t)$ (95% bandwidth)

(d) $\sqrt{\dfrac{\alpha}{\pi}}\exp[-\alpha t^2]$ (rms bandwidth)

4.35. The signal $X_s(\omega)$ shown in Figure 4.4.10(b) is a periodic signal in ω.

(a) Find the Fourier-series coefficients of this signal.

(b) Show that

$$X_s(\omega) = \sum_{n=-\infty}^{\infty} \frac{2\pi}{\omega_s} x(-nT) \exp\left[\frac{jn2\pi}{\omega_s}\omega\right]$$

(c) Using Equation (4.4.1) along with the shifting property of the Fourier transform, show that

$$x(t) = \sum_{n=-\infty}^{\infty} \frac{2W_B}{\omega_s} x(nT) \frac{\sin[W_B(t - nT)]}{W_B(t - nT)}$$

4.36. Calculate the time-bandwidth product for the following signals:

(a) $x(t) = \dfrac{1}{\sqrt{2\pi\Delta^2}} \exp\left[\dfrac{-t^2}{2\Delta^2}\right]$

(Use the radius of gyration measure for T and the equivalent bandwidth measure for B.)

(b) $x(t) = \dfrac{\sin 2\pi Wt}{\pi t}$

(Use the distance between zeros as a measure of T and the absolute bandwidth as a measure of B.)

(c) $x(t) = A\exp[-at]u(t)$. (Use the time at which $x(t)$ drops to $1/e$ of its value at $t = 0$ as a measure of T and the 3-dB bandwidth as a measure of B.)

Chapter 5

The Laplace Transform

5.1 INTRODUCTION

In Chapters 3 and 4, we saw how frequency-domain methods are extremely useful in the study of signals and LTI systems. In those chapters, we demonstrated that Fourier analysis reduces the convolution operation required to compute the output of LTI systems to just the product of the Fourier transform of the input signal and the frequency response of the system. One of the problems we can run into is that many of the input signals we would like to use do not have Fourier transforms. Examples are $\exp[\alpha t]u(t)$, $\alpha > 0$; $\exp[-\alpha t]$, $-\infty < t < \infty$; $tu(t)$; and other time signals that are not absolutely integrable. If we are confronted, say, with a system that is driven by a ramp-function input, is there any method of solution other than the time-domain techniques of Chapter 2? The difficulty could be resolved by extending the Fourier transform so that the signal $x(t)$ is expressed as a sum of complex exponentials, $\exp[st]$, where the frequency variable is $s = \sigma + j\omega$ and thus is not restricted to the imaginary axis only. This is equivalent to multiplying the signal by an exponential convergent factor. For example, $\exp[-\sigma t]\exp[\alpha t]u(t)$ satisfies Dirichlet's conditions for $\sigma > \alpha$ and, therefore, should have a generalized or extended Fourier transform. Such an extended transform is known as the bilateral Laplace transform, named after the French mathematician Pierre Simon de Laplace. In this chapter, we define the bilateral Laplace transform (section 5.2) and use the definition to determine a set of bilateral transform pairs for some basic signals.

As mentioned in Chapter 2, any signal $x(t)$ can be written as the sum of causal and noncausal signals. The causal part of $x(t)$, $x(t)u(t)$, has a special Laplace transform that we refer to as the unilateral Laplace transform or, simply, the Laplace transform. The unilateral Laplace transform is more often used than the bilateral Laplace transform, not only because most of the signals occurring in practice are causal signals, but also because the response of a causal LTI system to a causal input is causal. In Section 5.3, we define the unilateral Laplace transform and provide some examples to illustrate how to evaluate such transforms. In Section 5.4, we demonstrate how to evaluate the bilateral Laplace transform using the unilateral Laplace transform.

As with other transforms, the Laplace transform possesses a set of valuable properties that are used repeatedly in various applications. Because of their importance, we devote Section 5.5 to the development of the properties of the Laplace transform and give examples to illustrate their use.

Finding the inverse Laplace transform is as important as finding the transform itself. The inverse Laplace transform is defined in terms of a contour integral. In general, such an integral is not easy to evaluate and requires the use of some theorems from the subject of complex variables that are beyond the scope of this text. In Section 5.6, we use the technique of partial fractions to find the inverse Laplace transform for the class of signals that have rational transforms (i.e., that can be expressed as the ratio of two polynomials)

In Section 5.7, we develop techniques for determining the simulation diagrams of continuous-time systems. In Section 5.8, we discuss some applications of the Laplace transform, such as in the solution of differential equations, applications to circuit analysis, and applications to control systems. In Section 5.9, we cover the solution of the state equations in the frequency domain. Finally, in Section 5.10, we discuss the stability of LTI systems in the s domain.

5.2 THE BILATERAL LAPLACE TRANSFORM

The bilateral, or two-sided, Laplace transform of the real-valued signal $x(t)$ is defined as

$$X_B(s) \triangleq \int_{-\infty}^{\infty} x(t) \exp[-st]\, dt \tag{5.2.1}$$

where the complex variable s is, in general, of the form $s = \sigma + j\omega$, with σ and ω the real and imaginary parts, respectively. When $\sigma = 0$, $s = j\omega$, and Equation (5.2.1) becomes the Fourier transform of $x(t)$, while with $\sigma \neq 0$, the bilateral Laplace transform is the Fourier transform of the signal $x(t) \exp[-\sigma t]$. For convenience, we sometimes denote the bilateral Laplace transform in operator form as $\mathscr{L}_B\{x(t)\}$ and denote the transform relationship between $x(t)$ and $X_B(s)$ as

$$x(t) \leftrightarrow X_B(s) \tag{5.2.2}$$

Let us now evaluate a number of bilateral Laplace transforms to illustrate the relationship between them and Fourier transforms.

Example 5.2.1

Consider the signal $x(t) = \exp[-at]u(t)$. From the definition of the bilateral Laplace transform,

$$X_B(s) = \int_{-\infty}^{\infty} \exp[-at]\exp[-st]u(t)\,dt$$

$$= \int_0^{\infty} \exp[-(s+a)t]\,dt$$

$$= \frac{1}{s+a}$$

As stated earlier, we can look at this bilateral Laplace transform as the Fourier transform of the signal $\exp[-at]\exp[-\sigma t]u(t)$. This signal has a Fourier transform only if $\sigma > -a$. Thus, $X_B(s)$ exists only if $\text{Re}\{s\} > -a$.

In general, the bilateral Laplace transform converges for some values of $\text{Re}\{s\}$ and not for others. The values of s for which it converges, i.e.,

$$\int_{-\infty}^{\infty} |x(t)| \exp[-\text{Re}\{s\}t]\,dt < \infty \tag{5.2.3}$$

is called the region of absolute convergence or, simply, the region of convergence and is abbreviated as ROC. It should be stressed that the region of convergence depends on the given signal $x(t)$. For instance, in the preceding example, ROC is defined by $\text{Re}\{s\} > -a$ whether a is positive or negative. Note also that even though the bilateral Laplace transform exists for all values of a, the Fourier transform exists only if $a > 0$.

If we restrict our attention to time signals whose Laplace transforms are rational functions of s, i.e., $X_B(s) = N(s)/D(s)$, then clearly, $X_B(s)$ does not converge at the zeros of the polynomial $D(s)$ (poles of $X_B(s)$), which leads us to conclude that for rational Laplace transforms, the ROC should not contain any poles.

Example 5.2.2

In this example, we show that two signals can have the same algebraic expression for their bilateral Laplace transform, but different ROCs. Consider the signal

$$x(t) = -\exp[-at]u(-t)$$

Its bilateral Laplace transform is

$$X_B(s) = -\int_{-\infty}^{\infty} \exp[-(s+a)t]u(-t)\,dt$$

$$= -\int_{-\infty}^{0} \exp[-(s+a)t]\,dt$$

For this integral to converge, we require that $\text{Re}\{s + a\} < 0$ or $\text{Re}\{s\} < -a$, and the bilateral Laplace transform is

$$X_B(s) = \frac{1}{s+a}$$

In spite of the fact that the algebraic expressions for the bilateral Laplace transform of the two signals in Examples 5.2.1 and 5.2.2 are identical, the two transforms have different ROCs. From these examples, we can conclude that, for signals that exist for positive time only, the behavior of the signal puts a lower bound on the allowable values of $\text{Re}\{s\}$, whereas for signals that exist for negative time, the behavior of the signal puts an upper bound on the allowable values of $\text{Re}\{s\}$. Thus, for a given $X_B(s)$, there can be more than one corresponding $x(t)$, depending on the ROC; in other words, the correspondence between $x(t)$ and $X_B(s)$ is not one to one unless the ROC is specified.

A convenient way to display the ROC is in the complex s plane, as shown in Figure 5.2.1. The horizontal axis is usually referred to as the σ axis, and the vertical axis is normally referred to as the $j\omega$ axis. The shaded region in Figure 5.2.1(a) represents the set of points in the s plane corresponding to the region of convergence for the signal in Example 5.2.1, and the shaded region in Figure 5.2.1(b) represents the region of convergence for the signal in Example 5.2.2.

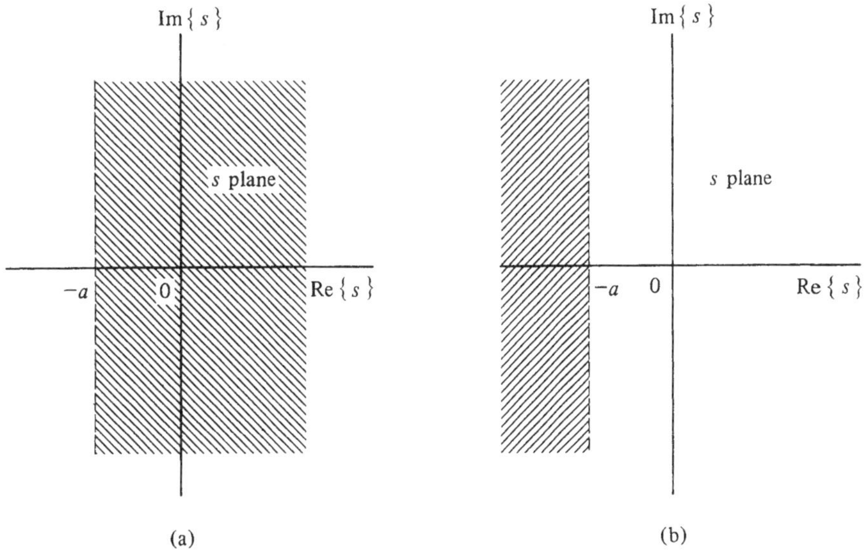

Figure 5.2.1 s-plane representation of the bilateral Laplace transform.

The ROC can also provide us with information about whether $x(t)$ is Fourier transformable or not. Since the Fourier transform is obtained from the bilateral Laplace transform by setting $\sigma = 0$, the region of convergence in this case is a single line (the $j\omega$ axis). Therefore, if the ROC for $X_B(s)$ includes the $j\omega$ axis, $x(t)$ is Fourier transformable, and $X_B(\omega)$ can be obtained by replacing s in $X_B(s)$ by $j\omega$.

Example 5.2.3

Consider the sum of two real exponentials:

$$x(t) = 3\exp[-2t]u(t) + 4\exp[t]u(-t)$$

Note that for signals that exist for both positive and negative time, the behavior of the signal for negative time puts an upper bound on the allowable values of Re$\{s\}$, and the behavior for positive time puts a lower bound on the allowable Re$\{s\}$. Therefore, we expect to obtain a strip as the ROC for such signals. The bilateral Laplace transform of $x(t)$ is

$$X_B(t) = \int_0^\infty 3\exp[-(s+2)t]\,dt + \int_{-\infty}^0 4\exp[-(s-1)t]\,dt$$

The first integral converges for Re$\{s\} > -2$, the second integral converges for Re$\{s\} < 1$, and the algebraic expression for the bilateral Laplace transform is

$$\begin{aligned} X_B(s) &= \frac{3}{s+2} - \frac{4}{s-1} \\ &= \frac{-s-11}{(s+2)(s-1)}, \qquad -2 < \text{Re}\{s\} < 1 \end{aligned}$$

5.3 THE UNILATERAL LAPLACE TRANSFORM

Similar to our definition of Equation (5.2.1), we can define a unilateral or one-sided transform of a signal $x(t)$ as

$$X(s) = \int_{0^-}^\infty x(t)\exp[-st]\,dt \tag{5.3.1}$$

Some texts use $t = 0^+$ or $t = 0$ as a lower limit. All three lower limits are equivalent if $x(t)$ does not contain a singularity function at $t = 0$. This is because there is no contribution to the area under the function $x(t)\exp[-st]$ at $t = 0$ even if $x(t)$ is discontinuous at the origin.

The unilateral transform is of particular interest when we are dealing with causal signals. Recall from our definition in Chapter 2 that if the signal $x(t)$ is causal, we have $x(t) = 0$ for $t < 0$. Thus, the bilateral transform $X_B(s)$ of a causal signal is the same as the unilateral transform of the signal. Our discussion in Section 5.2 showed that, given a transform $X_B(s)$, the corresponding time function $x(t)$ is not uniquely specified and depends on the ROC. For causal signals, however, there is a unique correspondence between the signal $x(t)$ and its unilateral transform $X(s)$. This makes for considerable simplification in analyzing causal signals and systems. In what follows, we will omit the word "unilateral" and simply refer to $X(s)$ as the Laplace transform of $x(t)$, except when it is not clear from the context which transform is being used.

Example 5.3.1

In this example, we find the unilateral Laplace transforms of the following signals:

$$x_1(t) = A, \qquad x_2(t) = \delta(t), \qquad x_3(t) = \exp[j2t], \qquad x_4(t) = \cos 2t, \qquad x_5(t) = \sin 2t$$

From Equation (5.3.1),

$$X_1(s) = \int_{0^-}^\infty A\exp[-st]\,dt = \frac{A}{s}, \qquad \text{Re}\{s\} > 0$$

$$X_2(s) = \int_{0^-}^{\infty} \delta(t)\exp[-st]\,dt = 1, \qquad \text{for all } s$$

$$X_3(s) = \int_{0^-}^{\infty} \exp[j2t]\exp[-st]\,dt$$

$$= \frac{1}{s - j2}$$

$$= \frac{s}{s^2 + 4} + j\frac{2}{s^2 + 4}, \qquad \text{Re}\{s\} > 0$$

Since $\cos 2t = \text{Re}\{\exp[j2t]\}$ and $\sin 2t = \text{Im}\{\exp[j2t]\}$, using the linearity of the integral operation, we have

$$X_4(s) = \text{Re}\left\{\frac{1}{s - j2}\right\} = \frac{s}{s^2 + 4}, \qquad \text{Re}\{s\} > 0$$

$$X_5(s) = \text{Im}\left\{\frac{1}{s - j2}\right\} = \frac{2}{s^2 + 4}, \qquad \text{Re}\{s\} > 0$$

Table 5.1 lists some of the important unilateral Laplace-transform pairs. These are used repeatedly in applications.

5.4 BILATERAL TRANSFORMS USING UNILATERAL TRANSFORMS

The bilateral Laplace transform can be evaluated using the unilateral Laplace transform if we express the signal $x(t)$ as the sum of two signals. The first part represents the behavior of $x(t)$ in the interval $(-\infty, 0)$, and the second part represents the behavior of $x(t)$ in the interval $[0, \infty)$. In general, any signal that does not contain any singularities (a delta function or its derivatives) at $t = 0$ can be written as the sum of a causal part $x_+(t)$ and a noncausal part $x_-(t)$, i.e.,

$$x(t) = x_+(t)u(t) + x_-(t)u(-t) \tag{5.4.1}$$

Taking the bilateral Laplace transform of both sides, we have

$$X_B(s) = X_+(s) + \int_{-\infty}^{0^-} x_-(t)\exp[-st]\,dt$$

Using the substitution $t = -\tau$ yields

$$X_B(s) = X_+(s) + \int_{0^+}^{\infty} x_-(-\tau)\exp[s\tau]\,d\tau$$

If $x(t)$ does not have any singularities at $t = 0$, then the lower limit in the second term can be replaced by 0^-, and the bilateral Laplace transform becomes

$$X_B(s) = X_+(s) + \mathcal{L}\{x_-(-t)u(t)\}_{s\to -s} \tag{5.4.2}$$

TABLE 5.1
Some Selected Unilateral Laplace-Transform Pairs

Signal	Transform	ROC
1. $u(t)$	$\frac{1}{s}$	$\text{Re}\{s\} > 0$
2. $u(t) - u(t-a)$	$\frac{1 - \exp[-as]}{s}$	$\text{Re}\{s\} > 0$
3. $\delta(t)$	1	for all s
4. $\delta(t-a)$	$\exp[-as]$	for all s
5. $t^n u(t)$	$\frac{n!}{s^{n+1}}, \quad n = 1, 2, \ldots$	$\text{Re}\{s\} > 0$
6. $\exp[-at]u(t)$	$\frac{1}{s+a}$	$\text{Re}\{s\} > -a$
7. $t^n \exp[-at]u(t)$	$\frac{n!}{(s+a)^{n+1}}$	$\text{Re}\{s\} > -a$
8. $\cos \omega_0 t\, u(t)$	$\frac{s}{s^2 + \omega_0^2}$	$\text{Re}\{s\} > 0$
9. $\sin \omega_0 t\, u(t)$	$\frac{\omega_0}{s^2 + \omega_0^2}$	$\text{Re}\{s\} > 0$
10. $\cos^2 \omega_0 t\, u(t)$	$\frac{s^2 + 2\omega_0^2}{s(s^2 + 4\omega_0^2)}$	$\text{Re}\{s\} > 0$
11. $\sin^2 \omega_0 t\, u(t)$	$\frac{2\omega_0^2}{s(s^2 + 4\omega_0^2)}$	$\text{Re}\{s\} > 0$
12. $\exp[-at] \cos \omega_0 t\, u(t)$	$\frac{s+a}{(s+a)^2 + \omega_0^2}$	$\text{Re}\{s\} > -a$
13. $\exp[-at] \sin \omega_0 t\, u(t)$	$\frac{\omega_0}{(s+a)^2 + \omega_0^2}$	$\text{Re}\{s\} > -a$
14. $t \cos \omega_0 t\, u(t)$	$\frac{s^2 - \omega_0^2}{(s^2 + \omega_0^2)^2}$	$\text{Re}\{s\} > 0$
15. $t \sin \omega_0 t\, u(t)$	$\frac{2\omega_0 s}{(s^2 + \omega_0^2)^2}$	$\text{Re}\{s\} > 0$

where $\mathcal{L}\{\cdot\}$ stand for the unilateral Laplace transform. Note that if $x_-(-t)u(t)$ has an ROC defined by $\text{Re}\{s\} > \sigma$, then $x_-(t)u(-t)$ should have an ROC defined by $\text{Re}\{s\} < -\sigma$.

Example 5.4.1

The bilateral Laplace transform of the signal $x(t) = \exp[at]u(-t)$, $a > 0$, is

$$X_B(s) = \mathcal{L}\{\exp[-at]u(t)\}_{s\to -s}$$
$$= \left(\frac{1}{s+a}\right)_{s\to -s} = \frac{-1}{s-a}, \qquad \text{Re}\{s\} < a$$

Note that the unilateral Laplace transform of $\exp[at]u(-t)$ is zero.

Example 5.4.2

According to Equation (5.4.2), the bilateral Laplace transform of

$$x(t) = A\exp[-at]u(t) + Bt^2\exp[-bt]u(-t), \qquad a \text{ and } b > 0$$

is

$$X_B(s) = \frac{A}{s+a} + \mathcal{L}\{B(-t)^2\exp[bt]u(t)\}_{s\to -s}$$
$$= \frac{A}{s+a} + \left(B\frac{2!}{(s-b)^3}\right)_{s\to -s}, \qquad \text{Re}\{s\} > -a \cap \text{Re}\{s\} < -b$$
$$= \frac{A}{s+a} - \frac{2B}{(s+b)^3}, \qquad -a < \text{Re}\{s\} < -b$$

where $\mathcal{L}\{B(-t)^2\exp[bt]u(t)\}$ follows from entry 7 in Table 5.1.

Not all signals possess a bilateral Laplace transform. For example, the periodic exponential $\exp[j\omega_0 t]$ does not have a bilateral Laplace transform because

$$\mathcal{L}_B\{\exp[j\omega_0 t]\} = \int_{-\infty}^{\infty}\exp[-(s-j\omega_0)t]\,dt$$
$$= \int_{-\infty}^{0}\exp[-(s-j\omega_0)t]\,dt + \int_{0}^{\infty}\exp[-(s-j\omega_0)t]\,dt$$

For the first integral to converge, we need $\text{Re}\{s\} < 0$, and for the second integral to converge, we need $\text{Re}\{s\} > 0$. These two restrictions are contradictory, and there is no value of s for which the transform converges.

In the remainder of this chapter, we restrict our attention to the unilateral Laplace transform, which we simply refer to as the Laplace transform.

5.5 PROPERTIES OF THE UNILATERAL LAPLACE TRANSFORM

There are a number of useful properties of the unilateral Laplace transform that will allow some problems to be solved almost by inspection. In this section we summarize many of these properties, some of which may be more or less obvious to the reader. By using these properties, it is possible to derive many of the transform pairs in Table 5.1. In this section, we list several of these properties and provide outlines of their proof.

5.5.1 Linearity

If

$$x_1(t) \leftrightarrow X_1(s)$$
$$x_2(t) \leftrightarrow X_2(s)$$

then

$$ax_1(t) + bx_2(t) \leftrightarrow aX_1(s) + bX_2(s) \tag{5.5.1}$$

where a and b are arbitrary constants. This property is the direct result of the linear operation of integration. The linearity property can be easily extended to a linear combination of an arbitrary number of components and simply means that the Laplace transform of a linear combination of an arbitrary number of signals is the same linear combination of the transforms of the individual components. The ROC associated with a linear combination of terms is the intersection of the ROCs for the individual terms.

Example 5.5.1

Suppose we want to find the Laplace transform of

$$(A + B\exp[-bt])u(t)$$

From Table 5.1, we have the transform pair

$$u(t) \leftrightarrow \frac{1}{s} \qquad \text{and} \qquad \exp[-bt]u(t) \leftrightarrow \frac{1}{s+b}$$

Thus, using linearity, we obtain the transform pair

$$Au(t) + B\exp[-bt]u(t) \leftrightarrow \frac{A}{s} + \frac{B}{s+b} = \frac{(A+B)s + Ab}{s(s+b)}$$

The ROC is the intersection of $\text{Re}\{s\} > -b$ and $\text{Re}\{s\} > 0$, and, hence, is given by $\text{Re}\{s\} > \max(-b, 0)$.

5.5.2 Time Shifting

If $x(t) \leftrightarrow X(s)$, then for any positive real number t_0,

$$x(t - t_0)u(t - t_0) \leftrightarrow \exp[-t_0 s]X(s) \tag{5.5.2}$$

The signal $x(t - t_0)u(t - t_0)$ is a t_0-second right shift of $x(t)u(t)$. Therefore, a shift in time to the right corresponds to multiplication by $\exp[-t_0 s]$ in the Laplace-transform domain. The proof follows from Equation (5.3.1) with $x(t - t_0)u(t - t_0)$ substituted for $x(t)$, to obtain

$$\mathcal{L}\{x(t - t_0)u(t - t_0)\} = \int_{0^-}^{\infty} x(t - t_0)u(t - t_0)\exp[-st]\,dt$$

$$= \int_{t_0^-}^{\infty} x(t - t_0)\exp[-st]\,dt$$

Using the transformation of variables $t = \tau + t_0$, we have

$$\mathcal{L}\{x(t - t_0)u(t - t_0)\} = \int_{0^-}^{\infty} x(\tau)\exp[-s(\tau + t_0)]\,d\tau$$

$$= \exp[-t_0 s]\int_{0^-}^{\infty} x(\tau)\exp[-s\tau]\,d\tau$$

$$= \exp[-t_0 s]X(s)$$

Note that all values s in the ROC of $x(t)$ are also in the ROC of $x(t - t_0)$. Therefore, the ROC associated with $x(t - t_0)$ is the same as the ROC associated with $x(t)$.

Example 5.5.2

Consider the rectangular pulse $x(t) = \text{rect}((t - a)/2a)$. This signal can be written as

$$\text{rect}((t - a)/2a) = u(t) - u(t - 2a)$$

Using linearity and time shifting, we find that the Laplace transform of $x(t)$ is

$$X(s) = \frac{1}{s} - \exp[-2as]\frac{1}{s} = \frac{1 - \exp[-2as]}{s}, \qquad \text{Re}\{s\} > 0$$

It should be clear that the time shifting property holds for a right shift only. For example, the Laplace transform of $x(t + t_0)$, for $t_0 > 0$, cannot be expressed in terms of the Laplace transform of $x(t)$. (Why?)

5.5.3 Shifting in the *s* Domain

If

$$x(t) \leftrightarrow X(s)$$

then

$$\exp[s_0 t]x(t) \leftrightarrow X(s - s_0) \tag{5.5.3}$$

The proof follows directly from the definition of the Laplace transform. Since the new transform is a shifted version of $X(s)$, for any s that is in the ROC of $x(t)$, the values $s + \text{Re}\{s_0\}$ are in the ROC of $\exp[s_0 t]x(t)$.

Example 5.5.3

From entry 8 in Table 5.1 and Equation (5.5.3), the Laplace transform of

$$x(t) = A\exp[-at]\cos(\omega_0 t + \theta)u(t)$$

is

$$\begin{aligned}
X(s) &= \mathscr{L}\{A \exp[-at](\cos\omega_0 t \cos\theta - \sin\omega_0 t \sin\theta)u(t)\} \\
&= \mathscr{L}\{A \exp[-at] \cos\omega_0 t \cos\theta\, u(t)\} - \mathscr{L}\{A \exp[-at] \sin\omega_0 t \sin\theta\, u(t)\} \\
&= \frac{A(s+a)\cos\theta}{(s+a)^2+\omega_0^2} - \frac{A\,\omega_0 \sin\theta}{(s+a)^2+\omega_0^2} \\
&= \frac{A[(s+a)\cos\theta - \omega_0\sin\theta]}{(s+a)^2+\omega_0^2}, \qquad \text{Re}\{s\} > -a
\end{aligned}$$

5.5.4 Time Scaling

If

$$x(t) \leftrightarrow X(s), \qquad \text{Re}\{s\} > \sigma_1$$

then for any positive real number α,

$$x(\alpha t) \leftrightarrow \frac{1}{\alpha} X\left(\frac{s}{\alpha}\right), \qquad \text{Re}\{s\} > \alpha\sigma_1 \tag{5.5.4}$$

The proof follows directly from the definition of the Laplace transform and the appropriate substitution of variables.

Aside from the amplitude factor of $1/\alpha$, linear scaling in time by a factor of α corresponds to linear scaling in the s plane by a factor of $1/\alpha$. Also, for any value of s in the ROC of $x(t)$, the value s/α will be in the ROC of $x(\alpha t)$; that is, the ROC associated with $x(\alpha t)$ is a compressed ($\alpha > 1$) or expanded ($\alpha < 1$) version of the ROC of $x(t)$.

Example 5.5.4

Consider the time-scaled unit-step signal $u(\alpha t)$, where α is an arbitrary positive number. The Laplace transform of $u(\alpha t)$ is

$$\mathscr{L}\{u(\alpha t)\} = \frac{1}{\alpha}\frac{1}{s/\alpha} = \frac{1}{s}, \qquad \text{Re}\{s\} > 0.$$

This result is anticipated, since $u(\alpha t) = u(t)$ for $\alpha > 0$.

5.5.5 Differentiation in the Time Domain

If

$$x(t) \leftrightarrow X(s)$$

then

$$\frac{dx(t)}{dt} \leftrightarrow s\,X(s) - x(0^-) \tag{5.5.5}$$

The proof of this property is obtained by computing the transform of $dx(t)/dt$. This transform is

$$\mathcal{L}\left\{\frac{dx(t)}{dt}\right\} = \int_{0^-}^{\infty} \frac{dx(t)}{dt} \exp[-st]\,dt$$

Integrating by parts yields

$$\begin{aligned}\mathcal{L}\left\{\frac{dx(t)}{dt}\right\} &= \exp[-st]\,x(t)\Big|_{0^-}^{\infty} - \int_{0^-}^{\infty} x(t)(-s)\exp[-st]\,dt \\ &= \lim_{t\to\infty}\,[\exp[-st]\,x(t)] - x(0^-) + s\,X(s)\end{aligned}$$

The assumption that $X(s)$ exists implies that

$$\lim_{t\to\infty}\,[\exp[-st]\,x(t)] = 0$$

for s in the ROC. Thus,

$$\mathcal{L}\left\{\frac{dx(t)}{dt}\right\} = s\,X(s) - x(0^-)$$

Therefore, differentiation in the time domain is equivalent to multiplication by s in the s domain. This permits us to replace operations of calculus by simple algebraic operations on transforms.

The differentiation property can be extended to yield

$$\frac{d^n x(t)}{dt^n} \leftrightarrow s^n X(s) - s^{n-1}x(0^-) - \ldots - s x^{(n-2)}(0^-) - x^{(n-1)}(0^-) \tag{5.5.6}$$

Generally speaking, differentiation in the time domain is the most important property (next to linearity) of the Laplace transform. It makes the Laplace transform useful in applications such as solving differential equations. Specifically, we can use the Laplace transform to convert any linear differential equation with constant coefficients into an algebraic equation.

As mentioned earlier, for rational Laplace transforms, the ROC does not contain any poles. Now, if $X(s)$ has a first-order pole at $s = 0$, multiplying by s, as in Equation (5.5.5), may cancel that pole and result in a new ROC that contains the ROC of $x(t)$. Therefore, in general, the ROC associated with $dx(t)/dt$ normally contains the ROC associated with $x(t)$ and can be larger if $X(s)$ has a first-order pole at $s = 0$.

Example 5.5.5

The unit step function $x(t) = u(t)$ has the transform $X(s) = 1/s$, with an ROC defined by $\mathrm{Re}\{s\} > 0$. The derivative of $u(t)$ is the unit-impulse function, whose Laplace transform is unity for all s with associated ROC extending over the entire s plane.

Example 5.5.6

Let $x(t) = \sin^2 \omega t\, u(t)$, for which $x(0^-) = 0$. Note that

$$x'(t) = 2\omega \sin \omega t \cos \omega t\, u(t) = \omega \sin 2\omega t\, u(t)$$

From Table 5.1,

$$\mathcal{L}\{\sin 2\omega t\, u(t)\} = \frac{2\omega}{s^2 + 4\omega^2}$$

and therefore,

$$\mathcal{L}\{\sin^2 \omega t\, u(t)\} = \frac{2\omega^2}{s(s^2 + 4\omega^2)}$$

Example 5.5.7

One of the important applications of the Laplace transform is in solving differential equations with specified initial conditions. As an example, consider the differential equation

$$y''(t) + 3y'(t) + 2y(t) = 0, \qquad y(0^-) = 3, \qquad y'(0^-) = 1$$

Let $Y(s) = \mathcal{L}\{y(t)\}$ be the Laplace transform of the (unknown) solution $y(t)$. Using the differentiation-in-time property, we have

$$\mathcal{L}\{y'(t)\} = sY(s) - y(0^-) = sY(s) - 3$$

$$\mathcal{L}\{y''(t)\} = s^2Y(s) - sy(0^-) - y'(0^-) = s^2Y(s) - 3s - 1$$

If we take the Laplace transform of both sides of the differential equation and use the last two expressions, we obtain

$$s^2Y(s) + 3sY(s) + 2Y(s) = 3s + 10$$

Solving algebraically for $Y(s)$, we get

$$Y(s) = \frac{3s + 10}{(s+2)(s+1)} = \frac{7}{s+1} - \frac{4}{s+2}$$

From Table 5.1, we see that

$$y(t) = 7\exp[-t]u(t) - 4\exp[-2t]u(t)$$

Example 5.5.8

Consider the RC circuit shown in Figure 5.5.1(a). The input is the rectangular signal shown in Figure 5.5.1(b). The circuit is assumed initially relaxed (zero initial condition).

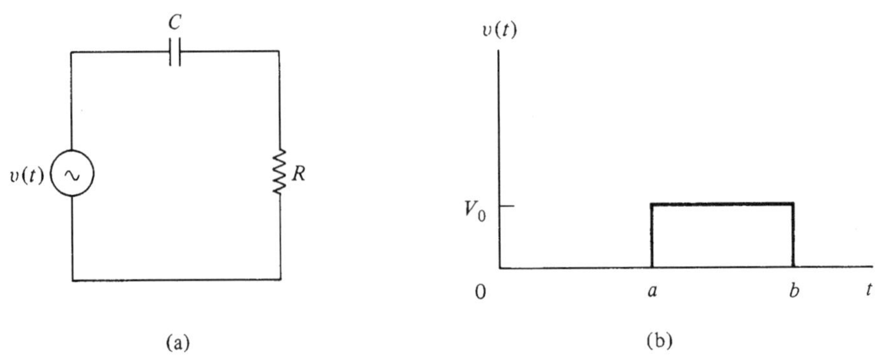

Figure 5.5.1 Circuit for Example 5.5.8.

The differential equation governing the circuit is

$$Ri(t) + \frac{1}{C}\int_0^t i(\tau)\,d\tau = v(t)$$

The input $v(t)$ can be represented in terms of unit-step functions as

$$v(t) = V_0[u(t-a) - u(t-b)]$$

Taking the Laplace transform of both sides of the differential equation yields

$$RI(s) + \frac{I(s)}{Cs} = \frac{V_0}{s}[\exp[-as] - \exp[-bs]]$$

Solving for $I(s)$, we obtain

$$I(s) = \frac{V_0/R}{s + 1/RC}[\exp[-as] - \exp[-sb]]$$

By using the time shift property, we obtain the current

$$i(t) = \frac{V_0}{R}\left[\exp\left[\frac{-(t-a)}{RC}\right]u(t-a) - \exp\left[\frac{-(t-b)}{RC}\right]u(t-b)\right]$$

The solution is shown in Figure 5.5.2.

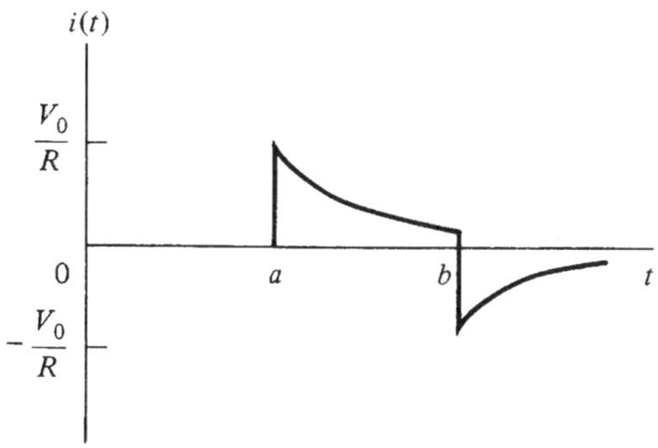

Figure 5.5.2 The current waveform in Example 5.5.8.

5.5.6 Integration in the Time Domain

Since differentiation in the time domain corresponds to multiplication by s in the s domain, one might conclude that integration in the time domain should involve division by s. This is always true if the integral of $x(t)$ does not grow faster than an exponential of the form $A\exp[-at]$, that is, if

$$\lim_{t\to\infty} \exp[-st]\int_{0^-}^t x(\tau)\,d\tau = 0$$

for all s such that $\mathrm{Re}\{s\} > a$.

The integration property can be stated as follows: For any causal signal $x(t)$, if

$$y(t) = \int_{0^-}^{t} x(\tau)\, d\tau$$

then

$$Y(s) = \frac{1}{s} X(s) \tag{5.5.7}$$

To prove this result, we start with

$$X(s) = \int_{0^-}^{\infty} x(t) \exp[-st]\, dt$$

Dividing both sides by s yields

$$\frac{X(s)}{s} = \int_{0^-}^{\infty} x(t) \frac{\exp[-st]}{s}\, dt$$

Integrating the right-hand side by parts, we have

$$\frac{X(s)}{s} = y(t) \frac{\exp[-st]}{s} \bigg|_{0^-}^{\infty} + \int_{0^-}^{\infty} y(t) \exp[-st]\, dt$$

The first term on the right-hand side evaluates to zero at both limits (at the upper limit by assumption and at the lower limit because $y(0^-) = 0$), so that

$$\frac{X(s)}{s} = \mathcal{L}\{y(t)\}$$

Thus, integration in the time domain is equivalent to division by s in the s domain. Integration and differentiation in the time domain are two of the most commonly used properties of the Laplace transform. They can be used to convert the integration and differentiation operations into division or multiplication by s, respectively, which are algebraic operations and, hence, much easier to perform.

5.5.7 Differentiation in the s Domain

Differentiating both sides of Equation (5.3.1) with respect to s, we have

$$\frac{dX(s)}{ds} = \int_{0^-}^{\infty} (-t)x(t) \exp[-st]\, dt$$

Consequently,

$$-t\,x(t) \leftrightarrow \frac{dX(s)}{ds} \tag{5.5.8}$$

Since differentiating $X(s)$ does not add new poles (it may increase the order of some existing poles), the ROC associated with $-tx(t)$ is the same as the the ROC associated with $x(t)$.

By repeated application of Equation (5.5.8), it follows that

$$(-t)^n\, x(t) \leftrightarrow \frac{d^n X(s)}{ds^n} \tag{5.5.9}$$

Example 5.5.9

The Laplace transform of the unit ramp function $r(t) = tu(t)$ can be obtained using Equation (5.5.8) as

$$\begin{aligned} R(s) &= -\frac{d}{ds}\mathcal{L}\{u(t)\} \\ &= -\frac{d}{ds}\frac{1}{s} = \frac{1}{s^2} \end{aligned}$$

Applying Equation (5.5.9), we have, in general,

$$t^n\, u(t) \leftrightarrow \frac{n!}{s^{n+1}} \tag{5.5.10}$$

5.5.8 Modulation

If

$$x(t) \leftrightarrow X(s)$$

then for any real number ω_0,

$$x(t)\cos\omega_0 t \leftrightarrow \tfrac{1}{2}\left[X(s + j\omega_0) + X(s - j\omega_0)\right] \tag{5.5.11}$$

$$x(t)\sin\omega_0 t \leftrightarrow \frac{j}{2}\left[X(s + j\omega_0) - X(s - j\omega_0)\right] \tag{5.5.12}$$

The proof follows from Euler's formula,

$$\exp[j\omega_0 t] = \cos\omega_0 t + j\sin\omega_0 t$$

and the application of the shifting property in the s domain.

Example 5.5.10

The Laplace transform of $(\cos\omega_0 t)u(t)$ is obtained from the Laplace transform of $u(t)$ using the modulation property as follows:

$$\begin{aligned} \mathcal{L}\{(\cos\omega_0 t)u(t)\} &= \frac{1}{2}\left(\frac{1}{s + j\omega_0} + \frac{1}{s - j\omega_0}\right) \\ &= \frac{s}{s^2 + \omega_0^2} \end{aligned}$$

Similarly, the Laplace transform of $\exp[-at]\sin\omega_0 t\, u(t)$ is obtained from the Laplace transform of $\exp[-at]\,u(t)$ and the modulation property as

$$\mathscr{L}\{\exp[-at](\sin\omega_0 t)u(t)\} = \frac{j}{2}\left(\frac{1}{s + j\omega_0 + a} - \frac{1}{s - j\omega_0 + a}\right)$$

$$= \frac{\omega_0}{(s + a)^2 + \omega_0^2}$$

5.5.9 Convolution

This property is one of the most widely used properties in the study and analysis of linear systems. Its use reduces the complexity of evaluating the convolution integral to simple multiplication. The convolution property states that if

$$x(t) \leftrightarrow X(s)$$

$$h(t) \leftrightarrow H(s)$$

then

$$x(t) * h(t) \leftrightarrow X(s)H(s) \tag{5.5.13}$$

where the convolution of $x(t)$ and $h(t)$ is

$$x(t) * h(t) = \int_{-\infty}^{\infty} x(\tau)h(t - \tau)\, d\tau$$

Since both $h(t)$ and $x(t)$ are causal signals, the convolution in this case can be reduced to

$$x(t) * h(t) = \int_{0^-}^{\infty} x(\tau)h(t - \tau)\, d\tau$$

Taking the Laplace transform of both sides results in the transform pair

$$x(t) * h(t) \leftrightarrow \int_{0^-}^{\infty} \left[\int_{0^-}^{\infty} x(\tau)h(t - \tau)\, d\tau\right] \exp[-st]\, dt$$

Interchanging the order of the integrals, we have

$$x(t) * h(t) \leftrightarrow \int_{0^-}^{\infty} x(\tau)\left[\int_{0^-}^{\infty} h(t - \tau) \exp[-st]\, dt\right] d\tau$$

Using the change of variables $\mu = t - \tau$ in the second integral and noting that $h(\mu) = 0$ for $\mu < 0$ yields

$$x(t) * h(t) \leftrightarrow \int_{0^-}^{\infty} x(\tau) \exp[-s\tau]\left[\int_{0^-}^{\infty} h(\mu) \exp[-s\mu]\, d\mu\right] d\tau$$

or

$$x(t) * h(t) \leftrightarrow X(s)H(s)$$

The ROC associated with $X(s)H(s)$ is the intersection of the ROCs of $X(s)$ and $H(s)$. However, because of the multiplication process involved, a zero-pole cancellation can occur that results in a larger ROC than the intersection of the ROCs of $X(s)$

and $H(s)$. In general, the ROC of $X(s)H(s)$ includes the intersection of the ROCs of $X(s)$ and $H(s)$ and can be larger if zero-pole cancellation occurs in the process of multiplying the two transforms.

Example 5.5.11

The integration property can be proved using the convolution property, since

$$\int_{-\infty}^{t} x(\tau)\, d\tau = x(t) * u(t)$$

Therefore, the transform of the integral of $x(t)$ is the product of $X(s)$ and the transform of $u(t)$, which is $1/s$.

Example 5.5.12

Let $x(t)$ be the rectangular pulse $\text{rect}((t-a)/2a)$ centered at $t = a$ and with width $2a$. The convolution of this pulse with itself can be obtained easily with the help of the convolution property.

From Example 5.5.2, the transform of $x(t)$ is

$$X(s) = \frac{1 - \exp[-2as]}{s}$$

The transform of the convolution is

$$\begin{aligned} Y(s) = X^2(s) &= \left[\frac{1 - \exp[-2as]}{s}\right]^2 \\ &= \frac{1}{s^2} - \frac{2\exp[-2as]}{s^2} + \frac{\exp[-4as]}{s^2} \end{aligned}$$

Taking the inverse Laplace transform of both sides and recognizing that $1/s^2$ is the transform of $tu(t)$ yields

$$\begin{aligned} y(t) = x(t) * x(t) &= tu(t) - 2(t-2a)u(t-2a) + (t-4a)u(t-4a) \\ &= r(t) - 2r(t-2a) + r(t-4a) \end{aligned}$$

This signal is illustrated in Figure 5.5.3 and is a triangular pulse, as expected.

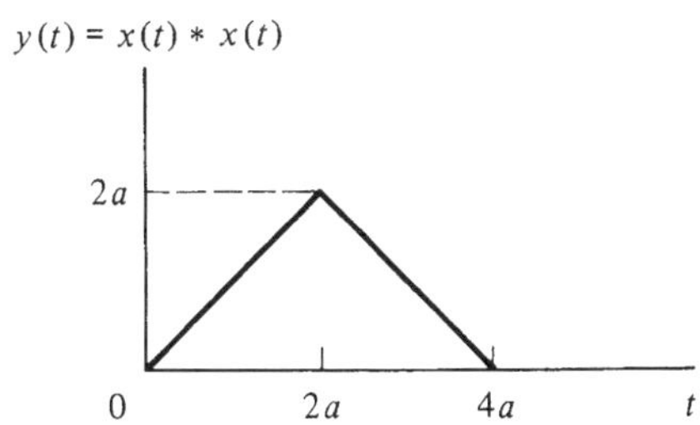

Figure 5.5.3 Convolution of two rectangular signals.

In Equation (5.5.13), $H(s)$ is called the transfer function of the system whose impulse response is $h(t)$. This function is the s-domain representation of the LTI system and describes the "transfer" from the input in the s domain, $X(s)$, to the output in the s domain, $Y(s)$, assuming no initial energy in the system at $t = 0^-$. Dividing both sides of Equation (5.5.13) by $X(s)$, provided that $X(s) \neq 0$, gives

$$H(s) = \frac{Y(s)}{X(s)} \tag{5.5.14}$$

That is, the transfer function is equal to the ratio of the transform $Y(s)$ of the output to the transform $X(s)$ of the input. Equation (5.5.14) allows us to determine the impulse response of the system from a knowledge of the response $y(t)$ to any nonzero input $x(t)$.

Example 5.5.13

Suppose that the input $x(t) = \exp[-2t]u(t)$ is applied to a relaxed (zero initial conditions) LTI system. The output of the system is

$$y(t) = \frac{2}{3}(\exp[-t] + \exp[-2t] - \exp[-3t])u(t)$$

Then

$$X(s) = \frac{1}{s+2}$$

and

$$Y(s) = \frac{2}{3(s+1)} + \frac{2}{3(s+2)} - \frac{2}{3(s+3)}$$

Using Equation (5.5.14), we conclude that the transfer function $H(s)$ of the system is

$$\begin{aligned} H(s) &= \frac{2}{3} + \frac{2(s+2)}{3(s+1)} - \frac{2(s+2)}{3(s+3)} \\ &= \frac{2(s^2 + 6s + 7)}{3(s+1)(s+3)} \\ &= \frac{2}{3}\left[1 + \frac{1}{s+1} + \frac{1}{s+3}\right] \end{aligned}$$

from which it follows that

$$h(t) = \frac{2}{3}\delta(t) + \frac{2}{3}[\exp[-t] + \exp[-3t]]u(t)$$

Example 5.5.14

Consider the LTI system described by the differential equation

$$y'''(t) + 2y''(t) - y'(t) + 5y(t) = 3x'(t) + x(t)$$

Assuming that the system was initially relaxed, and taking the Laplace transform of both sides, we obtain

$$s^3Y(s) + 2s^2Y(s) - sY(s) + 5Y(s) = 3sX(s) + X(s)$$

Solving for $H(s) = Y(s)/X(s)$, we have

$$H(s) = \frac{3s + 1}{s^3 + 2s^2 - s + 5}$$

5.5.10 Initial-Value Theorem

Let $x(t)$ be infinitely differentiable on an interval around $x(0^+)$ (an infinitesimal interval); then

$$x(0^+) = \lim_{s\to\infty} s\,X(s) \tag{5.5.15}$$

Equation (5.5.15) implies that the behavior of $x(t)$ for small t is determined by the behavior of $X(s)$ for large s. This is another aspect of the inverse relationship between time- and frequency-domain variables. To establish this result, we expand $x(t)$ in a Maclaurin series (a Taylor series about $t = 0^+$) to obtain

$$x(t) = \left[x(0^+) + x'(0^+)t + \cdots + x^{(n)}(0^+)\frac{t^n}{n!} + \cdots\right]u(t)$$

where $x^{(n)}(0^+)$ denotes the nth derivative of $x(t)$ evaluated at $t = 0^+$. Taking the Laplace transform of both sides yields

$$\begin{aligned} X(s) &= \frac{x(0^+)}{s} + \frac{x'(0^+)}{s^2} + \cdots + \frac{x^{(n)}(0^+)}{s^{n+1}} + \cdots \\ &= \sum_{n=0}^{\infty} x^{(n)}(0^+)\frac{1}{s^{n+1}} \end{aligned}$$

Multiplying both sides by s and taking the limit as $s \to \infty$ proves the initial-value theorem. As a generalization, multiplying by s^{n+1} and taking the limit as $s \to \infty$ yields

$$x^{(n)}(0^+) = \lim_{s\to\infty} [s^{n+1}X(s) - s^n x(0^+) - s^{n-1}x'(0^+) - \cdots - sx^{(n-1)}(0^+)] \tag{5.5.16}$$

This more general form of the initial-value theorem is simplified if $x^{(n)}(0^+) = 0$ for $n < N$. In that case,

$$x^{(N)}(0^+) = \lim_{s\to\infty} s^{N+1}X(s) \tag{5.5.17}$$

This property is useful, since it allows us to compute the initial value of the signal $x(t)$ and its derivatives directly from the Laplace transform $X(s)$ without having to find the inverse $x(t)$. Note that the right-hand side of Equation (5.5.15) can exist without the existence of $x(0^+)$. Therefore, the initial-value theorem should be applied only when $x(0^+)$ exists. Note also that the initial-value theorem produces $x(0^+)$, not $x(0^-)$.

Example 5.5.15

The initial value of the signal whose Laplace transform is given by

$$X(s) = \frac{cs + d}{(s - a)(s - b)}, \qquad a \neq b$$

is

$$x(0^+) = \lim_{s\to\infty} s\frac{cs + d}{(s - a)(s - b)} = c$$

The result can be verified by determining $x(t)$ first and then substituting $t = 0^+$. For this example, the inverse Laplace transform of $X(s)$ is

$$x(t) = \frac{c}{a - b}[a \exp[at] - b \exp[bt]]u(t) + \frac{d}{a - b}[\exp[at] - \exp[bt]]u(t)$$

so that $x(0^+) = c$. Note that $x(0^-) = 0$.

5.5.11 Final-Value Theorem

The final-value theorem allows us to compute the limit of signal $x(t)$ as $t \to \infty$ from its Laplace transform as follows:

$$\lim_{t\to\infty} x(t) = \lim_{s\to 0} s\,X(s) \tag{5.5.18}$$

The final-value theorem is useful in some applications, such as control theory, where we may need to find the final value (steady-state value) of the output of the system without solving for the time-domain function. Equation (5.5.18) can be proved using the differentiation-in-time property. We have

$$\int_{0^-}^{\infty} x'(t) \exp[-st]dt = s\,X(s) - x(0^-) \tag{5.5.19}$$

Taking the limit as $s \to 0$ of both sides of Equation (5.5.19) yields

$$\lim_{s\to 0}\int_{0^-}^{\infty} x'(t) \exp[-st]dt = \lim_{s\to 0}[s\,X(s) - x(0^-)]$$

or

$$\int_{0^-}^{\infty} x'(t)\,dt = \lim_{s\to 0}[s\,X(s) - x(0^-)]$$

Assuming that $\lim_{t\to 0} x(t)$ exists, this becomes

$$\lim_{t\to 0} x(t) - x(0^-) = \lim_{s\to 0} s\,X(s) - x(0^-)$$

which, after simplification, results in Equation (5.5.18). One must be careful in using the final-value theorem, since $\lim_{s\to 0} s\,X(s)$ can exist even though $x(t)$ does not have a limit as $t \to \infty$. Hence, it is important to know that $\lim_{t \to 0} x(t)$ exists before applying the final-value theorem. For example, if

$$X(s) = \frac{s}{s^2 + \omega^2}$$

then

$$\lim_{s \to 0} s\,X(s) = \lim_{s \to 0} \frac{s^2}{s^2 + \omega^2} = 0$$

But $x(t) = \cos \omega t$, which does not have a limit as $t \to \infty$ ($\cos \omega t$ oscillates between $+1$ and -1). Why do we have a discrepancy? To use the final-value theorem, we need the point $s = 0$ to be in the ROC of $sX(s)$. (Otherwise we cannot substitute $s = 0$ in $sX(s)$.) We have seen earlier that for rational-function Laplace transforms, the ROC does not contain any poles. Therefore, to use the final-value theorem, all the poles of $sX(s)$ must be in the left-hand side of the s plane. In our example, $sX(s)$ has two poles on the imaginary axis.

Example 5.5.16

The input $x(t) = Au(t)$ is applied to an automatic position-control system whose transfer function is

$$H(s) = \frac{c}{s(s + b) + c}$$

The final value of the output $y(t)$ is obtained as

$$\begin{aligned} \lim_{t \to \infty} y(t) &= \lim_{s \to 0} s\,Y(s) = \lim_{s \to 0} s\,X(s)H(s) \\ &= \lim_{s \to 0} s \left[\frac{A}{s} \frac{c}{s(s + b) + c} \right] \\ &= A \end{aligned}$$

assuming that the zeros of $s^2 + bs + c$ are in the left half plane. Thus, after a sufficiently long time, the output follows (tracks) the input $x(t)$.

Example 5.5.17

Suppose we are interested in the value of the integral

$$\int_{0^-}^{\infty} t^n \exp[-at]\, dt$$

Consider the integral

$$y(t) = \int_{0^-}^{t} \tau^n \exp[-a\tau]\, d\tau = \int_{0^-}^{t} x(\tau)\, d\tau$$

Note that the final value of $y(t)$ is the quantity of interest; that is,

TABLE 5.2
Some Selected Properties of the Laplace Transform

1. Linearity	$\sum_{n=1}^{N} \alpha_n x_n(t)$	$\sum_{n=1}^{N} \alpha_n X_n(s)$	(5.5.1)
2. Time shift	$x(t - t_0)u(t - t_0)$	$X(s) \exp(-st_0)$	(5.5.2)
3. Frequency shift	$\exp(s_0 t) x(t)$	$X(s - s_0)$	(5.5.3)
4. Time scaling	$x(\alpha t),\ \alpha > 0$	$1/\alpha\, X(s/\alpha)$	(5.5.4)
5. Differentiation	$dx(t)/dt$	$s\,X(s) - x(0^-)$	(5.5.5)
6. Integration	$\int_{0^-}^{t} x(\tau)\, d\tau$	$\frac{1}{s} X(s)$	(5.5.7)
7. Multiplication by t	$t\,x(t)$	$-\frac{dX(s)}{ds}$	(5.5.8)
8. Modulation	$x(t) \cos \omega_0 t$	$\frac{1}{2}[X(s - j\omega_0) + X(s + j\omega_0)]$	(5.5.11)
	$x(t) \sin \omega_0 t$	$\frac{1}{2j}[X(s - j\omega_0) - X(s + j\omega_0)]$	(5.5.12)
9. Convolution	$x(t) * h(t)$	$X(s)H(s)$	(5.5.13)
10. Initial value	$x(0^+)$	$\lim_{s\to\infty} s\,X(s)$	(5.5.15)
11. Final value	$\lim_{t\to\infty} x(t)$	$\lim_{s\to 0} s\,X(s)$	(5.5.18)

$$\lim_{t\to\infty} y(t) = \lim_{s\to 0} s \frac{1}{s} X(s) = \lim_{s\to 0} X(s)$$

From Table 5.1,

$$X(s) = \frac{n!}{(s + a)^{n+1}}$$

Therefore,

$$\int_{0^-}^{\infty} t^n \exp[-at]\, dt = \frac{n!}{a^{n+1}}$$

Table 5.2 summarizes the properties of the Laplace transform. These properties, along with the transform pairs in Table 5.1, can be used to derive other transform pairs.

5.6 THE INVERSE LAPLACE TRANSFORM

We saw in Section 5.2 that with $s = \sigma + j\omega$ such that $\mathrm{Re}\{s\}$ is inside the ROC, the Laplace transform of $x(t)$ can be interpreted as the Fourier transform of the exponentially weighted signal $x(t) \exp[-\sigma t]$; that is,

$$X(\sigma + j\omega) = \int_{-\infty}^{\infty} x(t) \exp[-\sigma t] \exp[-j\omega t]\, dt$$

Using the inverse Fourier-transform relationship given in Equation (4.2.5), we can find $x(t) \exp[-\sigma t]$ as

$$x(t) \exp[-\sigma t] = \frac{1}{2\pi} \int_{-\infty}^{\infty} X(\sigma + j\omega) \exp[j\omega t]\, d\omega$$

Multiplying by $\exp[\sigma t]$, we obtain

$$x(t) = \frac{1}{2\pi} \int_{-\infty}^{\infty} X(\sigma + j\omega) \exp[(\sigma + j\omega)t]\, d\omega$$

Using the change of variables $s = \sigma + j\omega$, we get the inverse Laplace-transform equation

$$x(t) = \frac{1}{2\pi j} \int_{\sigma - j\infty}^{\sigma + j\infty} X(s) \exp[st]\, ds \tag{5.6.1}$$

The integral in Equation (5.6.1) is evaluated along the straight line $\sigma + j\omega$ in the complex plane from $\sigma - j\infty$ to $\sigma + j\infty$, where σ is any fixed real number for which $\mathrm{Re}\{s\} = \sigma$ is a point in the ROC of $X(s)$. Thus, the integral is evaluated along a straight line that is parallel to the imaginary axis and at a distance σ from it.

Evaluation of the integral in Equation (5.6.1) requires the use of contour integration in the complex plane, which is not only difficult, but also outside of the scope of this text; hence, we will avoid using Equation (5.6.1) to compute the inverse Laplace transform. In many cases of interest, the Laplace transform can be expressed in the form

$$X(s) = \frac{N(s)}{D(s)} \tag{5.6.2}$$

where $N(s)$ and $D(s)$ are polynomials in s given by

$$N(s) = b_m s^m + b_{m-1} s^{s-1} + \cdots + b_1 s + b_0$$
$$D(s) = a_n s^n + a_{n-1} s^{n-1} + \cdots + a_1 s + a_0, \quad a_n \neq 0$$

The function $X(s)$ given by Equation (5.6.2) is said to be a rational function of s, since it is a ratio of two polynomials. We assume that $m < n$; that is, the degree of $N(s)$ is strictly less than the degree of $D(s)$. In this case, the rational function is proper in s. If $m = n$ i.e., when the rational function is improper, we can use long division to reduce it to a proper rational function. For proper rational transforms, the inverse Laplace transform can be determined by utilizing partial-fraction expansion techniques. Actually, this is what we did in some simple cases, ad hoc and without difficulty. Appendix D is devoted to the subject of partial fractions. We recommend that the reader not familiar with partial fractions review that appendix before studying the following examples.

Example 5.6.1

To find the inverse Laplace transform of

$$X(s) = \frac{2s + 1}{(s^3 + 3s^2 - 4s)}$$

we factor the polynomial $D(s) = s^3 + 3s^2 - 4s$ and use the partial-fractions form

$$X(s) = \frac{A_1}{s} + \frac{A_2}{s+4} + \frac{A_3}{s-1}$$

Using Equation (D.2), we find that the coefficients A_i, $i = 1, 2, 3$, are

$$A_1 = -\frac{1}{4}$$

$$A_2 = \frac{7}{20}$$

$$A_3 = \frac{3}{5}$$

and the inverse Laplace transform is

$$x(t) = -\frac{1}{4}u(t) + \frac{7}{20}\exp[-4t]u(t) + \frac{3}{5}\exp[t]u(t)$$

Example 5.6.2

In this example, we consider the case where we have repeated factors. Suppose the Laplace transform is given by

$$X(s) = \frac{2s^2 - 3s}{s^3 - 4s^2 + 5s - 2}$$

The denominator $D(s) = s^3 - 4s^2 + 5s - 2$ can be factored as

$$D(s) = (s-2)(s-1)^2$$

Since we have a repeated factor of order 2, the corresponding partial-fraction form is

$$X(s) = \frac{B}{s-2} + \frac{A_2}{(s-1)^2} + \frac{A_1}{s-1}$$

The coefficient B can be found using Equation (D.2); we obtain

$$B = 2$$

The coefficients A_i, $i = 1, 2$, are found using Equations (D.3) and (D.4); we get

$$A_2 = 1$$

and

$$A_1 = \frac{d}{ds}\left(\frac{2s^2 - 3s}{s-2}\right)\Bigg|_{s=1}$$

$$= \frac{(s-2)(4s-3) - (2s^2 - 3s)}{(s-2)^2}\Bigg|_{s=1} = 0$$

so that

$$X(s) = \frac{2}{s-2} + \frac{1}{(s-1)^2}$$

The inverse Laplace transform is therefore

$$x(t) = 2\exp[2t]u(t) + t\exp[t]u(t)$$

Example 5.6.3

In this example, we treat the case of complex conjugate poles (irreducible second-degree factors). Let

$$X(s) = \frac{s+3}{s^2+4s+13}$$

Since we cannot factor the denominator, we complete the square as follows:

$$D(s) = (s+2)^2 + 3^2$$

Then

$$X(s) = \frac{s+2}{(s+2)^2+3^2} + \frac{1}{(s+2)^2+3^2}$$

By using the shifting property of the transform, or alternatively, by using entries 12 and 13 in Table 5.1, we find the inverse Laplace transform to be

$$x(t) = \exp[-2t](\cos 3t)u(t) + \tfrac{1}{3}\exp[-2t](\sin 3t)u(t)$$

Example 5.6.4

As an example of repeated complex conjugate poles, consider the rational function

$$X(s) = \frac{5s^3 - 3s^2 + 7s - 3}{(s^2+1)^2}$$

Writing $X(s)$ in partial-fraction form, we have

$$X(s) = \frac{A_1 s + B_1}{s^2+1} + \frac{A_2 s + B_2}{(s^2+1)^2}$$

and therefore,

$$5s^3 - 3s^2 + 7s - 3 = (A_1 s + B_1)(s^2+1) + A_2 s + B_2$$

Comparing the coefficients of the different powers of s, we obtain

$$A_1 = 5, \qquad B_1 = -3, \qquad A_2 = 2, \qquad B_2 = 0$$

Thus,

$$X(s) = \frac{5s}{s^2+1} - \frac{3}{s^2+1} + \frac{2s}{(s^2+1)^2}$$

and the inverse Laplace transform can be determined from Table 5.1 to be

$$x(t) = (5\cos t - 3\sin t + t\sin t)u(t)$$

5.7 SIMULATION DIAGRAMS FOR CONTINUOUS-TIME SYSTEMS

In Section 2.5.3, we introduced two canonical forms to simulate (realize) LTI systems and showed that, since simulation is basically a synthesis problem, there are several ways to simulate LTI systems, but all are equivalent. Now consider the Nth order system described by the differential equation

$$\left(D^N + \sum_{i=0}^{N-1} a_i D^i\right) y(t) = \left(\sum_{i=0}^{M} b_i D^i\right) x(t) \tag{5.7.1}$$

Assuming that the system is initially relaxed, and taking the Laplace transform of both sides, we obtain

$$\left(s^N + \sum_{i=0}^{N-1} a_i s^i\right) Y(s) = \left(\sum_{i=0}^{M} b_i s^i\right) X(s) \tag{5.7.2}$$

Solving for $Y(s)/X(s)$, we get the transfer function of the system:

$$H(s) = \frac{\sum_{i=0}^{M} b_i s^i}{s^N + \sum_{i=0}^{N-1} a_i s^i} \tag{5.7.3}$$

Assuming that $N = M$, we can express Equation (5.7.2) as

$$s^N[Y(s) - b_N X(s)] + s^{N-1}[a_{N-1}Y(s) - b_{N-1}X(s)] + \cdots + a_0 Y(s) - b_0 X(s) = 0$$

Dividing through by s^N and solving for $Y(s)$ yields

$$\begin{aligned} Y(s) = b_N X(s) &+ \frac{1}{s}[b_{N-1}X(s) - a_{N-1}Y(s)] + \cdots + \\ &+ \frac{1}{s^{N-1}}[b_1 X(s) - a_1 Y(s)] + \frac{1}{s^N}[b_0 X(s) - a_Y(s)] \end{aligned} \tag{5.7.4}$$

Thus, $Y(s)$ can be generated by adding all the components on the right-hand side of Equation (5.7.4). Figure 5.7.1 demonstrates how $H(s)$ is simulated using this technique. Notice that the figure is similar to Figure 2.5.4, except that each integrator is replaced by its transfer function $1/s$.

The transfer function in Equation (5.7.3) can also be realized in the second canonical form if we express Equation (5.7.2) as

$$\begin{aligned} Y(s) &= \frac{\sum_{i=0}^{M} b_i s^i}{s^N + \sum_{i=0}^{N-1} a_i s^i} X(s) \\ &= \left(\sum_{i=0}^{M} b_i s^i\right) V(s) \end{aligned} \tag{5.7.5}$$

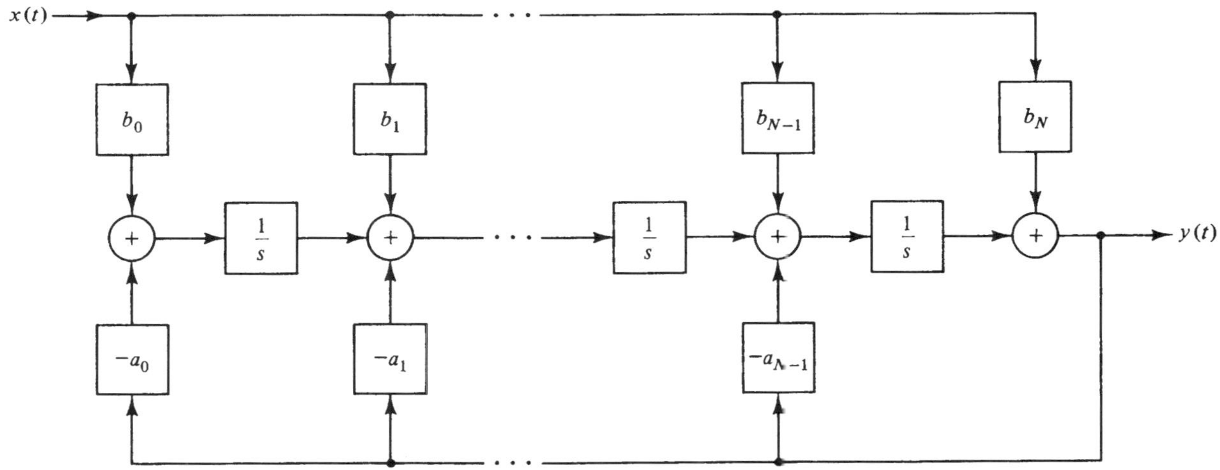

Figure 5.7.1 Simulation diagram using the first canonical form.

where

$$V(s) = \frac{1}{s^N + \sum_{i=0}^{N-1} a_i s^i} X(s) \tag{5.7.6}$$

or

$$\left(s^N + \sum_{i=0}^{N-1} a_i s^i\right) V(s) = X(s) \tag{5.7.7}$$

Therefore, we can generate $Y(s)$ in two steps: First, we generate $V(s)$ from Equation (5.7.7) and then use Equation (5.7.5) to generate $Y(s)$ from $V(s)$. The result is shown in Figure 5.7.2. Again, this figure is similar to Figure 2.5.5, except that each integrator is replaced by its transfer function $1/s$.

Example 5.7.1

The two canonical realization forms for the system with the transfer function

$$H(s) = \frac{s^2 - 3s + 2}{s^3 + 6s^2 + 11s + 6}$$

are shown in Figures 5.7.3 and 5.7.4.

As we saw earlier, the Laplace transform is a useful tool for computing the system transfer function if the system is described by its differential equation or if the output is expressed explicitly in terms of the input. The situation changes considerably in cases where a large number of components or elements are interconnected to form the complete system. In such cases, it is convenient to represent the system by suitably interconnected subsystems, each of which can be separately and easily analyzed. Three of the most common such subsystems involve series (cascade), parallel, and feedback interconnections.

In the case of cascade interconnections, as shown in Figure 5.7.5,

$$Y_1(s) = H_1(s)X(s)$$

and

$$\begin{aligned} Y_2(s) &= H_2(s)Y_1(s) \\ &= [H_2(s)H_1(s)]X(s) \end{aligned}$$

which shows that the combined transfer function is given by

$$H(s) = H_1(s)H_2(s) \tag{5.7.8}$$

We note that Equation (5.7.8) is valid only if there is no initial energy in either system. It is also implied that connecting the second system to the first does not affect the output of the latter. In short, the transfer function of first subsystem, $H_1(s)$, is computed under the assumption that the second subsystem with transfer function $H_2(s)$ is not connected. In other words, the input/output relationship of the first subsystem

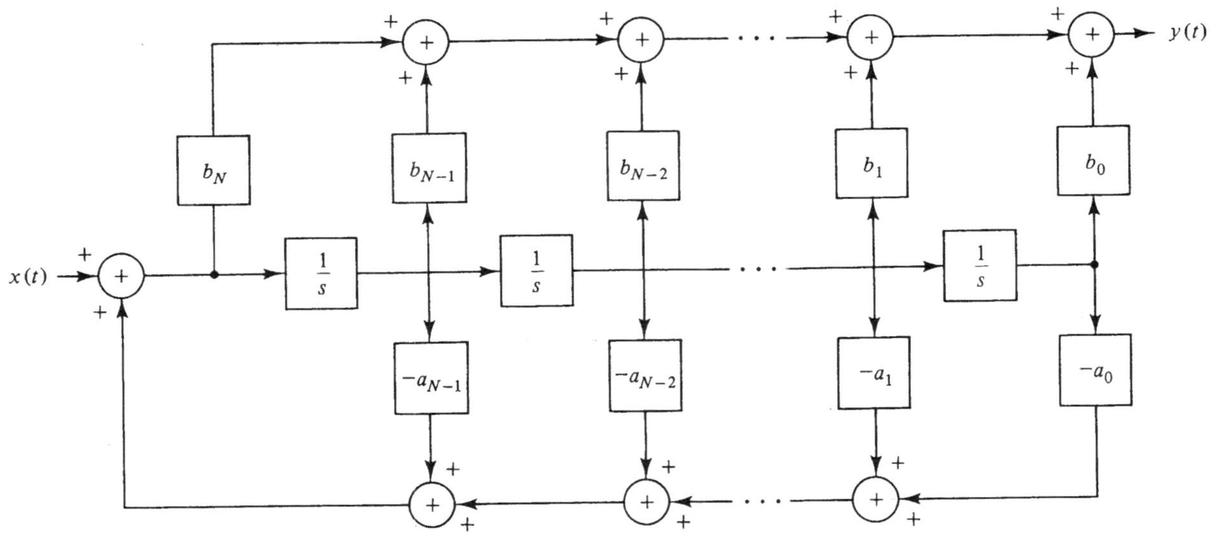

Figure 5.7.2 Simulation diagram using the second canonical form.

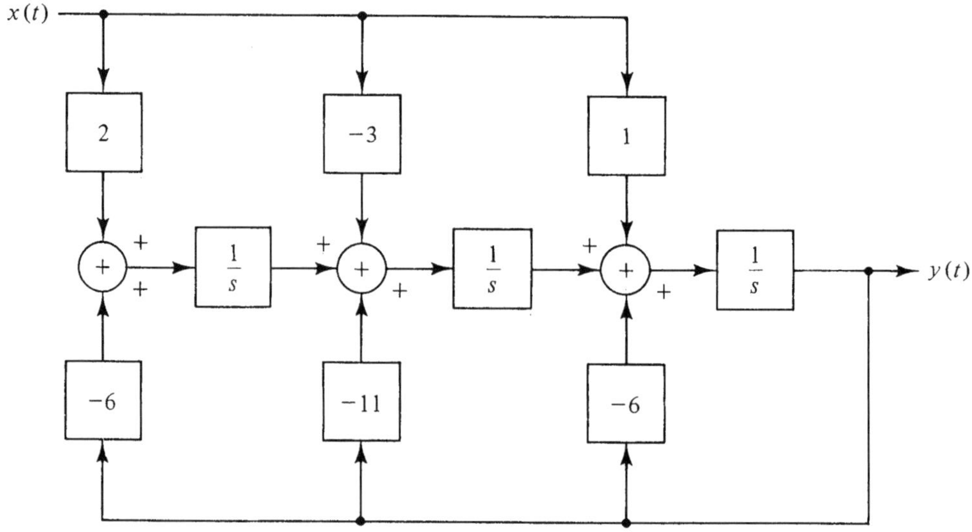

Figure 5.7.3 Simulation diagram using first canonical form for Example 5.7.1.

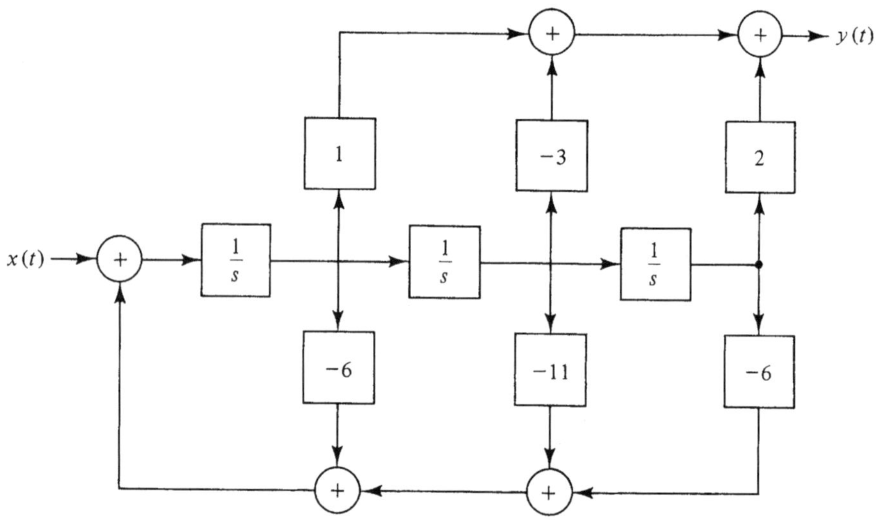

Figure 5.7.4 Simulation diagram using second canonical form for Example 5.7.1.

remains unchanged, regardless of whether $H_2(s)$ is connected to it. If this assumption is not satisfied, $H_1(s)$ must be computed under loading conditions, i.e., when $H_2(s)$ is connected to it.

If there are N systems connected in cascade, then their overall transfer function is

$$H(s) = H_1(s)H_2(s) \ldots H_N(s) \tag{5.7.9}$$

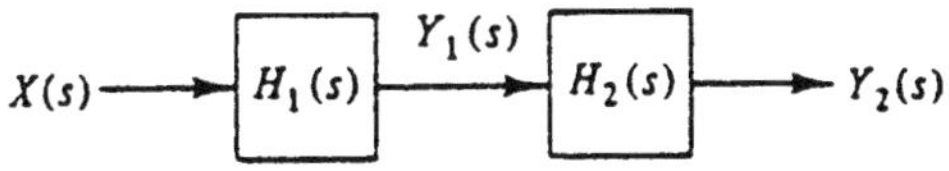

Figure 5.7.5 Cascade interconnection of two subsystems.

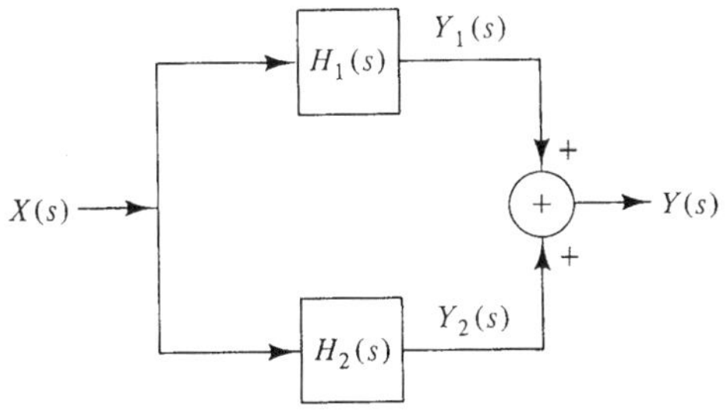

Figure 5.7.6 Parallel interconnection of two subsystems.

Using the convolution property, the impulse response of the overall system is

$$h(t) = h_1(t) * h_2(t) * \dots * h_N(t) \tag{5.7.10}$$

If two subsystems are connected in parallel, as shown in Figure 5.7.6, and each subsystem has no initial energy, then the output

$$\begin{aligned} Y(s) &= Y_1(s) + Y_2(s) \\ &= H_1(s)X(s) + H_2(s)X(s) \\ &= [H_1(s) + H_2(s)]X(s) \end{aligned}$$

and the overall transfer function is

$$H(s) = H_1(s) + H_2(s) \tag{5.7.11}$$

For N subsystems connected in parallel, the overall transfer function is

$$H(s) = H_1(s) + H_2(s) + \dots + H_N(s) \tag{5.7.12}$$

From the linearity of the Laplace transform, the impulse response of the overall system is

$$h(t) = h_1(t) + h_2(t) + \dots + h_N(t) \tag{5.7.13}$$

These two results are consistent with those obtained in Chapter 2 for the same interconnections.

Example 5.7.2

The transfer function of the system described in Example 5.7.1 also can be written as

$$H(s) = \frac{s-1}{s+1}\,\frac{s-2}{s+2}\,\frac{1}{s+3}$$

This system can be realized as a cascade of three subsystems, as shown in Figure 5.7.7. Each subsystem is composed of a pole-zero combination. The same system can be realized in parallel, too. This can be done by expanding $H(s)$ using the method of partial fractions as follows:

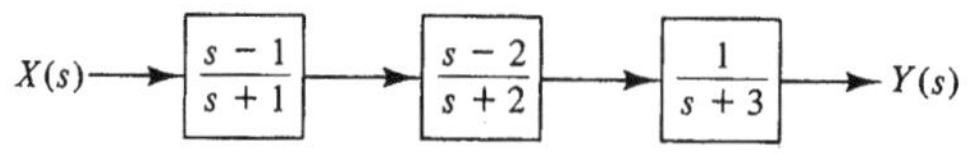

Figure 5.7.7 Cascade-form simulation for Example 5.7.2.

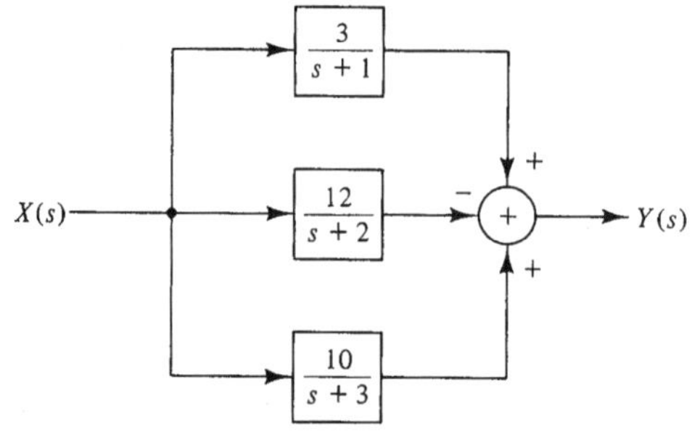

Figure 5.7.8 Parallel-form simulation for Example 5.7.2.

$$H(s) = \frac{3}{s+1} - \frac{12}{s+2} + \frac{10}{s+3}$$

A parallel interconnection is shown in Figure 5.7.8.

The connection in Figure 5.7.9 is called a positive feedback system. The output of the first system $H_1(s)$ is fed back to the summer through the system $H_2(s)$; hence the name "feedback connection." Note that if the feedback loop is disconnected, the transfer function from $X(s)$ to $Y(s)$ is $H_1(s)$, and hence $H_1(s)$ is called the open-loop transfer function. The system with transfer function $H_2(s)$ is called a feedback system. The whole system is called a closed-loop system.

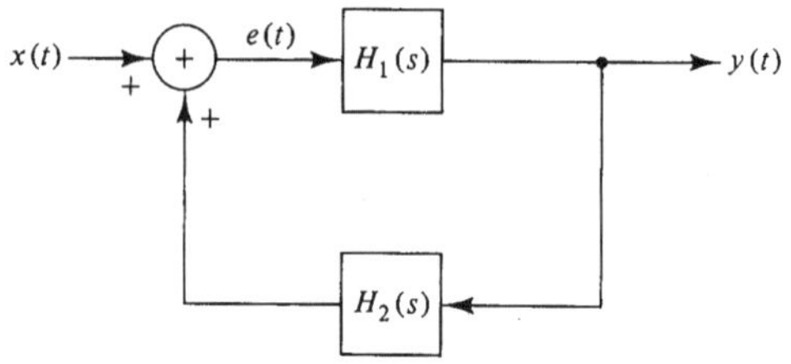

Figure 5.7.9 Feedback connection.

We assume that each system has no initial energy and that the feedback system does not load the open-loop system. Let $e(t)$ be the input signal to the system with transfer function $H_1(s)$. Then

$$Y(s) = E(s)H_1(s)$$
$$E(s) = X(s) + H_2(s)Y(s)$$

so that

$$Y(s) = H_1(s)\,[X(s) + H_2(s)Y(s)]$$

Solving for the ratio $Y(s)/X(s)$ yields the transfer function of the closed-loop system:

$$H(s) = \frac{Y(s)}{X(s)} = \frac{H_1(s)}{1 - H_1(s)H_2(s)} \tag{5.7.14}$$

Thus, the closed-loop transfer function is equal to the open-loop transfer function divided by 1 minus the product of the transfer functions of the open-loop and feedback systems. If the adder in Figure 5.7.9 is changed to a subtractor, the system is called a negative feedback system, and the closed-loop transfer function changes to

$$H(s) = \frac{H_1(s)}{1 + H_1(s)H_2(s)} \tag{5.7.15}$$

5.8 APPLICATIONS OF THE LAPLACE TRANSFORM

The Laplace transform can be applied to a number of problems in system analysis and design. These applications depend on the properties of the Laplace transform, especially those associated with differentiation, integration, and convolution.

In this section, we discuss three applications, beginning with the solution of differential equations.

5.8.1 Solution of Differential Equations

One of the most common uses of the Laplace transform is to solve linear, constant-coefficient differential equations. As we saw in Section 2.5, such equations are used to model continuous-time LTI systems. Solving these equations depends on the differentiation property of the Laplace transform. The procedure is straightforward and systematic, and we summarize it in the following steps:

1. For a given set of initial conditions, take the Laplace transform of both sides of the differential equation to obtain an algebraic equation in $Y(s)$.
2. Solve the algebraic equation for $Y(s)$.
3. Take the inverse Laplace transform to obtain $y(t)$.

Example 5.8.1

Consider the second-order, linear, constant-coefficient differential equation

$$y''(t) + 5y'(t) + 6y(t) = \exp[-t]u(t), \qquad y'(0^-) = 1 \text{ and } y(0^-) = 2$$

Taking the Laplace transform of both sides results in

$$[s^2Y(s) - 2s - 1] + 5[sY(s) - 2] + 6Y(s) = \frac{1}{s+1}$$

Solving for $Y(s)$ yields

$$Y(s) = \frac{2s^2 + 13s + 12}{(s+1)(s^2 + 5s + 6)}$$

$$= \frac{1}{2(s+1)} + \frac{6}{s+2} - \frac{9}{2(s+3)}$$

Taking the inverse Laplace transform, we obtain

$$y(t) = \left(\frac{1}{2}\exp[-t] + 6\exp[-2t] - \frac{9}{2}\exp[-3t]\right)u(t)$$

Higher order differential equations can be solved using the same procedure.

5.8.2 Application to RLC Circuit Analysis

In the analysis of circuits, the Laplace transform can be carried one step further by transforming the circuit itself rather than the differential equation. The s-domain current-voltage equivalent relations for arbitrary R, L, and C are as follows:

Resistors. The s domain current-voltage characterization of a resistor with resistance R is obtained by taking the Laplace transform of the current-voltage relationship in the time domain, $Ri_R(t) = v_R(t)$. This yields

$$V_R(s) = RI_R(s) \tag{5.8.1}$$

Inductors. For an inductor with inductance L and time-domain current-voltage relationship $Ldi_L(t)/dt = v_L(t)$, the s-domain characterization is

$$V_L(s) = sLI_L(s) - Li_L(0^-) \tag{5.8.2}$$

That is, an energized inductor (an inductor with nonzero initial conditions) at $t = 0^-$ is equivalent to an unenergized inductor at $t = 0^-$ in series with an impulsive voltage source with strength $Li_L(0^-)$. This impulsive source is called an initial-condition generator. Alternatively, Equation (5.8.2) can be written as

$$I_L(s) = \frac{1}{sL}V_L(s) + \frac{i_L(0^-)}{s} \tag{5.8.3}$$

That is, an energized inductor at $t = 0^-$ is equivalent to an unenergized inductor at $t = 0^-$ in parallel with a step-function current source. The height of the step function is $i_L(0^-)$.

Capacitors. For a capacitor with capacitance C and time-domain current-voltage relationship $Cdv_C(t)/dt = i_C(t)$, the s-domain characterization is

$$I_C(s) = sC\,V_C(s) - Cv_C(0^-) \tag{5.8.4}$$

That is, a charged capacitor (a capacitor with nonzero initial conditions) at $t = 0^-$ is equivalent to an uncharged capacitor at $t = 0^-$ in parallel with an impulsive current source. The strength of the impulsive source is $Cv(0^-)$, and the source itself is called an initial-condition generator. Equation (5.8.4) also can be written as

$$V_C(s) = \frac{1}{sC}I_C(s) + \frac{v_c(0^-)}{s} \tag{5.8.5}$$

Thus, a charged capacitor can be replaced by an uncharged capacitor in series with a step-function voltage source. The height of the step function is $v_c(0^-)$.

We can similarly write Kirchhoff's laws in the s domain. The equivalent statement of the current law is that at any node of an equivalent circuit, the algebraic sum of the currents in the s domain is zero; i.e.,

$$\sum_k I_k(s) = 0 \tag{5.8.6}$$

The voltage law states that around any loop in an equivalent circuit, the algebraic sum of the voltages in the s domain is zero; i.e.,

$$\sum_k V_k(s) = 0 \tag{5.8.7}$$

Caution must be exercised when assigning the polarity of the initial-condition generators.

Example 5.8.2

Consider the circuit shown in Figure 5.8.1(a) with $i_L(0^-) = -2$, $v_C(0^-) = 2$, and $x(t) = u(t)$. The equivalent s-domain circuit is shown in Figure 5.8.1(b).

Writing the node equation at node 1, we obtain

$$2 - \frac{Y(s) - 1/s - 2}{2 + s} - sY(s) - Y(s) = 0$$

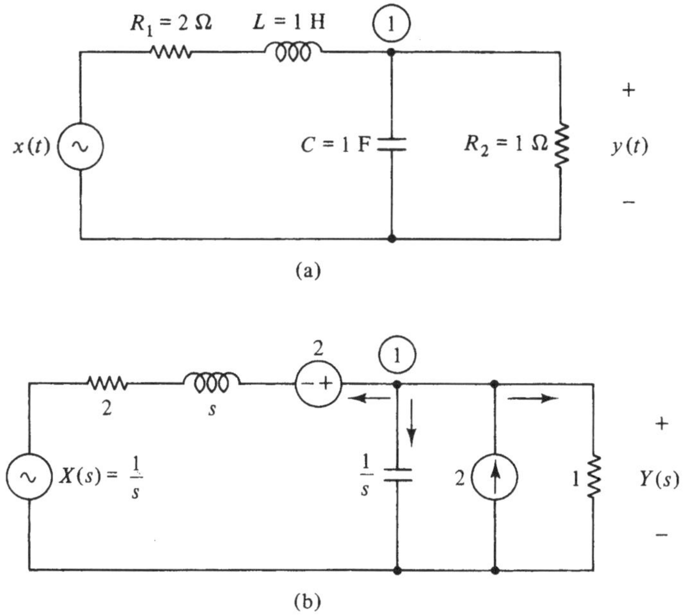

Figure 5.8.1 Circuit for Example 5.8.2.

Solving for $Y(s)$ yields

$$Y(s) = \frac{2s^2 + 6s + 1}{s(s^2 + 3s + 3)}$$

$$= \frac{1}{3s} + \frac{5s/3 + 5}{(s + 3/2)^2 + (\sqrt{3}/2)^2}$$

$$= \frac{1}{3s} + \frac{5}{3}\frac{s + 3/2}{(s + 3/2)^2 + (\sqrt{3}/2)^2} + \frac{5}{\sqrt{3}}\frac{\sqrt{3}/2}{(s + 3/2)^2 + (\sqrt{3}/2)^2}$$

Taking the inverse Laplace transform of both sides results in

$$y(t) = \frac{1}{3}u(t) + \frac{5}{3}\exp\left[\frac{-3t}{2}\right]\left(\cos\frac{\sqrt{3}}{2}t\right)u(t) + \frac{5}{\sqrt{3}}\exp\left[\frac{-3t}{2}\right]\left(\sin\frac{\sqrt{3}}{2}t\right)u(t)$$

The analysis of any circuit can be carried out using this procedure.

5.8.3 Application to Control

One of the major applications of the Laplace transform is in the study of control systems. Many important and practical problems can be formulated as control problems. Examples can be found in many areas, such as communications systems, radar systems, and speed control.

Consider the control system shown in Figure 5.8.2. The system is composed of two subsystems. The first subsystem is called the plant and has a known transfer function $H(s)$. The second subsystem is called the controller and is designed to achieve a certain system performance. The input to the system is the reference signal $r(t)$. The signal $w(t)$ is introduced to model any disturbance (noise) in the system. The difference between the reference and the output signals is an error signal

$$e(t) = r(t) - y(t)$$

The error signal is applied to the controller, whose function is to force $e(t)$ to zero as $t \to \infty$; that is,

$$\lim_{t\to\infty} e(t) = 0$$

This condition implies that the system output follows the reference signal $r(t)$. This type of system performance is called tracking in the presence of the disturbance $w(t)$.

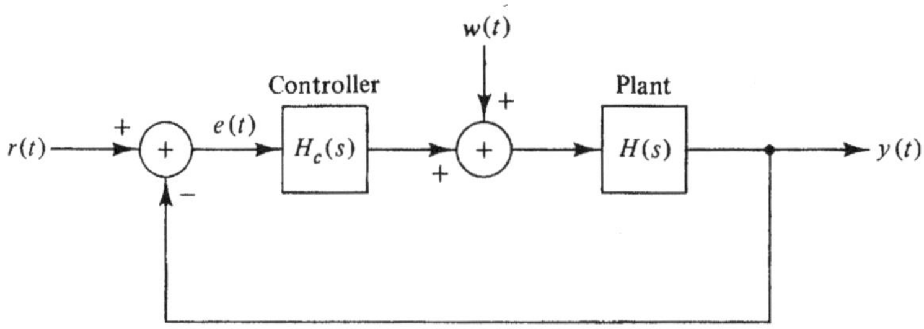

Figure 5.8.2 Block diagram of a control system.

The following example demonstrates how to design the controller to achieve the tracking effect.

Example 5.8.3

Suppose that the LTI system we have to control has the transfer function

$$H(s) = \frac{N(s)}{D(s)} \tag{5.8.8}$$

Let the input be $r(t) = Au(t)$ and the disturbance be $w(t) = Bu(t)$, where A and B are constants. Because of linearity, we can divide the problem into two simpler problems, one with input $r(t)$ and the other with input $w(t)$. That is, the output $y(t)$ is expressed as the sum of two components. The first component is due to $r(t)$ when $w(t) = 0$ and is labeled $y_1(t)$. It can be easily verified that

$$Y_1(s) = \frac{H_c(s)H(s)}{1 + H_c(s)H(s)} R(s)$$

where $R(s)$ is the Laplace transform of $r(t)$. The second component is due to $w(t)$ when $r(t) = 0$ and has the Laplace transform

$$Y_2(s) = \frac{H(s)}{1 + H_c(s)H(s)} W(s)$$

where $W(s)$ is the Laplace transform of the disturbance $w(t)$. The complete output has the Laplace transform

$$\begin{aligned} Y(s) &= Y_1(s) + Y_2(s) \\ &= \frac{H_c(s)H(s)}{1 + H_c(s)H(s)} R(s) + \frac{H(s)}{1 + H_c(s)H(s)} W(s) \\ &= \frac{H(s)[H_c(s)A + B]}{s[1 + H_c(s)H(s)]} \end{aligned} \tag{5.8.9}$$

We have to design $H_c(s)$ such that $r(t)$ tracks $y(t)$; that is,

$$\lim_{t\to\infty} y(t) = A$$

Let $H_c(s) = N_c(s)/D_c(s)$. Then we can write

$$Y(s) = \frac{N(s)[N_c(s)A + D_c(s)B]}{s[D(s)D_c(s) + N(s)N_c(s)]}$$

Let us assume that the real parts of all the zeros of $D(s)D_c(s) + N(s)N_c(s)$ are strictly negative. Then by using the final-value theorem, it follows that

$$\begin{aligned} \lim_{t\to\infty} y(t) &= \lim_{s\to 0} sY(s) \\ &= \lim_{s\to 0} \frac{N(s)[N_c(s)A + D_c(s)B]}{D(s)D_c(s) + N(s)N_c(s)} \end{aligned} \tag{5.8.10}$$

For this to be equal to A, one needs that $\lim_{s\to 0} D_c(s) = 0$ or $D_c(s)$ has a zero at $s = 0$. Substituting in the expression for $Y(s)$, we obtain

$$\lim_{t\to\infty} y(t) = \frac{N(0)N_c(0)A}{N(0)N_c(0)} = A$$

Example 5.8.4

Consider the control system shown in Figure 5.8.3. This system represents an automatic position-control system that can be used in a tracking antenna or in an antiaircraft gun mount. The input $r(t)$ is the desired angular position of the object to be tracked, and the output is the position of the antenna.

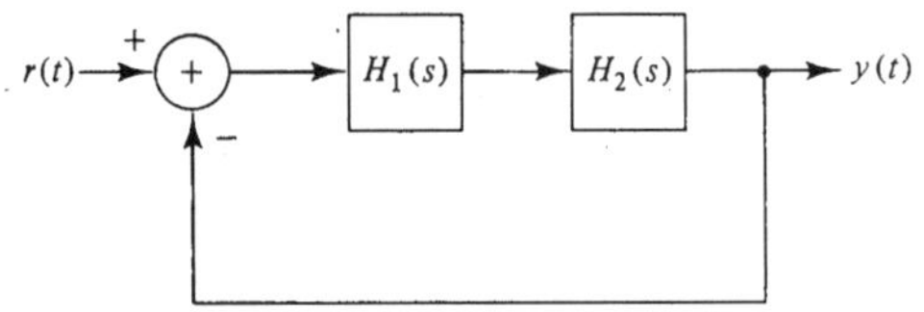

Figure 5.8.3 Block diagram of a tracking antenna.

The first subsystem is an amplifier with transfer function $H_1(s) = 8$, and the second subsystem is a motor with transfer function $H_2(s) = 1/[s(s + \alpha)]$, where $0 < \alpha < \sqrt{32}$. Let us investigate the step response of the system as the parameter α changes. The output $Y(s)$ is

$$\begin{aligned} Y(s) &= \frac{1}{s}H(s) = \frac{H_1(s)H_2(s)}{s[1 + H_1(s)H_2(s)]} \\ &= \frac{8}{s(s^2 + \alpha s + 8)} \\ &= \frac{1}{s} - \frac{s + \alpha}{s^2 + \alpha s + 8} \end{aligned}$$

The restriction $0 < \alpha < \sqrt{32}$ is chosen to ensure that the roots of the polynomial $s^2 + \alpha s + 8$ are complex numbers and lie in the left half plane. The reason for this will become clear in Section 5.10.

The step response of this system is obtained by taking the inverse Laplace transform of $Y(s)$ to yield

$$\begin{aligned} y(t) = \Bigg(1 - \exp\left[\frac{-\alpha t}{2}\right]\Bigg\{\cos\left[\sqrt{8 - \left(\frac{\alpha}{2}\right)^2}\,t\right] \\ + \frac{\alpha}{2\sqrt{8 - \left(\frac{\alpha}{2}\right)^2}}\sin\left[\sqrt{8 - \left(\frac{\alpha}{2}\right)^2}\,t\right]\Bigg\}\Bigg)u(t) \end{aligned}$$

The step response $y(t)$ for two values of α, namely, $\alpha = 2$ and $\alpha = 3$, is shown in Figure 5.8.4. Note that the response is oscillatory with overshoots of 30% and 14%, respectively. The time required for the response to rise from 10% to 90% of its final value is called the rise time. The first system has a rise time of 0.48 s, and the second system has a rise time of 0.60 s. Systems with longer rise times are inferior (sluggish) to those with shorter rise times. Reducing the rise time increases the overshoot, however, and high overshoots may not be acceptable in some applications.

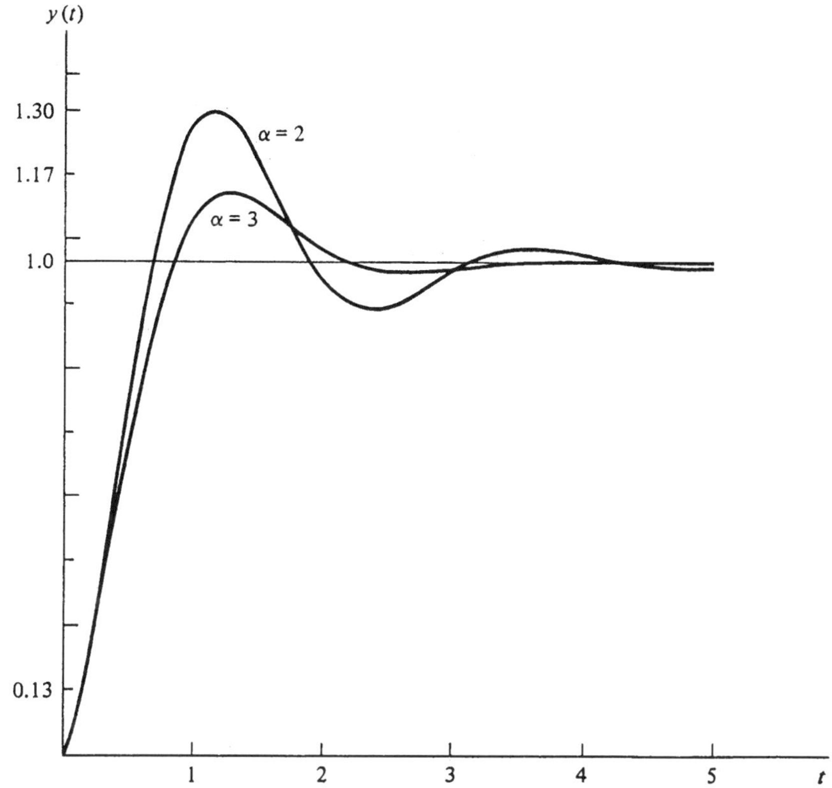

Figure 5.8.4 Step response of an antenna tracking system.

5.9 STATE EQUATIONS AND THE LAPLACE TRANSFORM

We saw that the Laplace transform is an efficient and convenient tool in solving differential equations. In Chapter 2, we introduced the concept of state variables and demonstrated that any LTI system can be described by a set of first-order differential equations called state equations.

Using the time-domain differentiation property of the Laplace transform, we can reduce this set of differential equations to a set of algebraic equations. Consider the LTI system described by

$$\mathbf{v}'(t) = \mathbf{A}\mathbf{v}(t) + \mathbf{b}x(t) \tag{5.9.1}$$

$$y(t) = \mathbf{c}\mathbf{v}(t) + dx(t) \tag{5.9.2}$$

Taking the Laplace transform of Equation (5.9.1), we obtain

$$s\mathbf{V}(s) - \mathbf{v}(0^-) = \mathbf{A}\mathbf{V}(s) + \mathbf{b}X(s)$$

which can be written as

$$(s\mathbf{I} - \mathbf{A})\mathbf{V}(s) = \mathbf{v}(0^-) + \mathbf{b}X(s)$$

where $\mathbf{I}$ is the unit matrix. Left multiplying both sides by the inverse of $(s\mathbf{I} - \mathbf{A})$, we obtain

$$\mathbf{V}(s) = (s\mathbf{I} - \mathbf{A})^{-1}\mathbf{v}(0^-) + (s\mathbf{I} - \mathbf{A})^{-1}\mathbf{b}X(s) \tag{5.9.3}$$

The Laplace transform of the output equation is

$$Y(s) = \mathbf{c}\mathbf{V}(s) + d\,X(s)$$

Substituting for $\mathbf{V}(s)$ from Equation (5.9.3), we obtain

$$Y(s) = \mathbf{c}(s\mathbf{I} - \mathbf{A})^{-1}\mathbf{v}(0^-) + [\mathbf{c}(s\mathbf{I} - \mathbf{A})^{-1}\mathbf{b} + d]X(s) \tag{5.9.4}$$

The first term is the transform of the output when the input is set to zero and is identified as the Laplace transform of the zero-input component of $y(t)$. The second term is the Laplace transform of the output when the initial state vector is zero and is identified as the Laplace transform of the zero-state component of $y(t)$. In Chapter 2, we saw that the solution to Equation (5.9.1) is given by

$$\mathbf{v}(t) = \exp[\mathbf{A}t]\,\mathbf{v}(0^-) + \int_{0^-}^{t} \exp[\mathbf{A}(t-\tau)]\,\mathbf{b}x(\tau)d\tau \tag{5.9.5}$$

(See Equation (2.6.13) with $t_0 = 0^-$.) The integral on the right side of Equation (5.9.5) represents the convolution of the signals exp $[\mathbf{A}t]$ and $\mathbf{b}x(t)$. Thus, the Laplace transformation of Equation (5.9.5) yields

$$\mathbf{V}(s) = \mathscr{L}\{\exp[\mathbf{A}t]\}\,\mathbf{v}(0^-) + \mathscr{L}\{\exp[\mathbf{A}t]\}\,\mathbf{b}X(s) \tag{5.9.6}$$

A comparison of Equations (5.9.3) and (5.9.6) shows that

$$\mathscr{L}\{\exp[\mathbf{A}t]\} = (s\mathbf{I} - \mathbf{A})^{-1} = \mathbf{\Phi}(s) \tag{5.9.7}$$

where $\mathbf{\Phi}(s)$ represents the Laplace transform of the state-transition matrix $\exp[\mathbf{A}t]$. $\mathbf{\Phi}(s)$ is usually referred to as the resolvent matrix.

Equation (5.9.7) gives us a convenient alternative method for determining $\exp[\mathbf{A}t]$: We first form the matrix $s\mathbf{I} - \mathbf{A}$ and then take the inverse Laplace transform of $(s\mathbf{I} - \mathbf{A})^{-1}$.

With zero initial conditions, Equation (5.9.4) becomes

$$Y(s) = [\mathbf{c}(s\mathbf{I} - \mathbf{A})^{-1}\mathbf{b} + d]X(s) \tag{5.9.8}$$

and hence, the transfer function of the system can be written as

$$H(s) = \mathbf{c}[s\mathbf{I} - \mathbf{A}]^{-1}\mathbf{b} + d = \mathbf{c}\mathbf{\Phi}(s)\mathbf{b} + d \tag{5.9.9}$$

Example 5.9.1

Consider the system described by

$$\mathbf{v}'(t) = \begin{bmatrix} -3 & 4 \\ -2 & 3 \end{bmatrix}\mathbf{v}(t) + \begin{bmatrix} 1 \\ 3 \end{bmatrix}x(t)$$

$$y(t) = [-1 \quad -1]\,\mathbf{v}(t) + 2x(t)$$

with

$$\mathbf{v}(0^-) = \begin{bmatrix} -1 \\ 3 \end{bmatrix}$$

The resolvent matrix of this system is

$$\mathbf{\Phi}(s) = \begin{bmatrix} s+3 & -4 \\ 2 & s-3 \end{bmatrix}^{-1}$$

Using Appendix C, we obtain

$$\mathbf{\Phi}(s) = \frac{\begin{bmatrix} s-3 & 4 \\ -2 & s+3 \end{bmatrix}}{(s+3)(s-3)+8}$$

$$= \begin{bmatrix} \dfrac{s-3}{(s+1)(s-1)} & \dfrac{4}{(s+1)(s-1)} \\ \dfrac{-2}{(s+1)(s-1)} & \dfrac{s+3}{(s+1)(s-1)} \end{bmatrix}$$

The transfer function is obtained using Equation (5.9.9):

$$H(s) = [-1 \quad -1] \begin{bmatrix} \dfrac{s-3}{(s+1)(s-1)} & \dfrac{4}{(s+1)(s-1)} \\ \dfrac{-2}{(s+1)(s-1)} & \dfrac{s+3}{(s+1)(s-1)} \end{bmatrix} \begin{bmatrix} 1 \\ 3 \end{bmatrix} + 2$$

$$= \frac{2s^2 - 4s - 18}{(s+1)(s-1)}$$

Taking the inverse Laplace transform, we obtain

$$h(t) = 2[\delta(t) + 3\exp[-t]u(t) - 5\exp[t]u(t)]$$

The zero-input response of the system is

$$Y_1(s) = \mathbf{c}(s\mathbf{I} - \mathbf{A})^{-1}\mathbf{v}(0^-)$$

$$= \frac{-2(s+13)}{(s+1)(s-1)}$$

and the zero-state response is

$$Y_2(s) = c(s\mathbf{I} - \mathbf{A})^{-1}\mathbf{b}X(s) + 2X(s)$$

$$= \frac{2s^2 - 4s - 18}{(s+1)(s-1)} X(s)$$

The overall response is

$$Y(s) = \frac{-2(s+13)}{(s+1)(s-1)} + \frac{2s^2 - 4s - 18}{(s+1)(s-1)} X(s)$$

The step response of this system is obtained by substituting $X(s) = 1/s$, so that

$$
\begin{aligned}
Y(s) &= \frac{-2(s+13)}{(s+1)(s-1)} + \frac{2s^2 - 4s - 18}{s(s+1)(s-1)} \\
&= \frac{-30s - 18}{s(s+1)(s-1)} \\
&= \frac{18}{s} + \frac{6}{s+1} - \frac{24}{s-1}
\end{aligned}
$$

Taking the inverse Laplace transform of both sides yields

$$y(t) = [18 + 6\exp[-t] - 24\exp[t]]u(t)$$

Example 5.9.2

Let us find the state-transition matrix of the system in Example 5.9.1. The resolvent matrix is

$$\mathbf{\Phi}(s) = \begin{bmatrix} \dfrac{s-3}{(s+1)(s-1)} & \dfrac{4}{(s+1)(s-1)} \\ \dfrac{-2}{(s+1)(s-1)} & \dfrac{s+3}{(s+1)(s-1)} \end{bmatrix}$$

The various elements of $\mathbf{\Phi}(t)$ are obtained by taking the inverse Laplace transform of each entry in the matrix $\mathbf{\Phi}(s)$. Doing so, we obtain

$$\mathbf{\Phi}(t) = \begin{bmatrix} 2\exp[-t] - \exp[t] & -2\exp[-t] + 2\exp[t] \\ \exp[-t] - \exp[t] & -\exp[-t] + 2\exp[t] \end{bmatrix} u(t)$$

5.10 STABILITY IN THE *s* DOMAIN

Stability is an important issue in system design. In Chapter 2, we showed that the stability of a system can be examined either through the impulse response of the system or through the eigenvalues of the state-transition matrix. Specifically, we demonstrated that for a stable system, the output, as well as all internal variables, should remain bounded for any bounded input. A system that satisfies this condition is called a bounded-input, bounded-output (BIBO) stable system.

Stability can also be examined in the s domain through the transfer function $H(s)$. The transfer function for any LTI system of the type we have been discussing always has the form of a ratio of two polynomials in s. Since any polynomial can be factored in terms of its roots, the rational transfer function can always be written in the following form (assuming that the degree of $N(s)$ is less than the degree of $D(s)$):

$$H(s) = \frac{A_1}{s - s_1} + \frac{A_2}{s - s_2} + \cdots + \frac{A_N}{s - s_N} \tag{5.10.1}$$

The zeros s_k of the denominator are the poles of $H(s)$ and, in general, may be complex numbers. If the coefficients of the governing differential equation are real, then the complex roots occur in conjugate pairs. If all the poles are distinct, then they are simple poles. If one of the poles corresponds to a repeated factor of the form $(s - s_k)^m$, then it is a multiple-order pole with order m. The impulse response of the system, $h(t)$, is obtained by taking the inverse Laplace transform of Equation (5.10.1). From entry 6 in Table 5.1, the kth pole contributes the term $h_k(t) = A_k \exp[s_k t]$ to $h(t)$. Thus, the behavior of the system depends on the location of the pole in the s plane. A pole can be in the left half of the s plane, on the imaginary axis, or in the right half of the s plane. Also, it may be a simple or multiple-order pole. The following is a discussion of the effects of the location and order of the pole on the stability of LTI systems.

1. *Simple Poles in the Left Half Plane.* In this case, the pole has the form

$$s_k = \sigma_k + j\,\omega_k, \qquad \sigma_k < 0$$

and the impulse-response component of the system, $h_k(t)$, corresponding to this pole is

$$\begin{aligned} h_k(t) &= A_k \exp[(\sigma_k + j\omega_k)t] + A_k^* \exp[(\sigma_k - j\omega_k)t] \\ &= |A_k| \exp[\sigma_k t](\exp[j(\omega_k t + \beta_k)] + \exp[-j(\omega_k t + \beta_k)]) \\ &= 2|A_k| \exp[\sigma_k t] \cos(\omega_k t + \beta_k), \qquad \sigma_k < 0 \end{aligned} \tag{5.10.2}$$

where

$$A_k = |A_k| \exp[j\beta_k]$$

As t increases, this component of the impulse response decays to zero and thus results in a stable system. Therefore, systems with only simple poles in the left half plane are stable.

2. *Simple Poles on the Imaginary Axis.* This case can be considered a special case of Equation (5.10.2) with $\sigma_k = 0$. The kth component in the impulse response is then

$$h_k(t) = 2|A_k| \cos(\omega_k t + \beta_k)$$

Note that there is no exponential damping; that is, the response does not decay as time progresses. It may appear that the response to the bounded input is also bounded. This is not true if the system is excited by a cosine function with the same frequency ω_k. In that case, a multiple-order pole of the form

$$\frac{B_s}{(s^2 + \omega_k^2)^2}$$

appears in the Laplace transform of the output. This term gives rise to a time response

$$\frac{B}{2\omega_k} t \sin \omega_k t$$

that increases without bound as t increases. Physically, ω_k is the natural frequency of the system. If the input frequency matches the natural frequency, the system resonates and the output grows without bound. An example is the lossless (nonresistive) LC circuit. A system with poles on the imaginary axis is sometimes called a marginally stable system.

3. *Simple Poles in the Right Half Plane.* If the system function has poles in the right half plane, then the system response is of the form

$$h_k(t) = 2|A_k| \exp[\sigma_k t] \cos(\omega_k t + \beta_k), \qquad \sigma_k > 0$$

Because of the increasing exponential term, the output of the system increases without bound, even for bounded input. Systems for which poles are in the right half plane are unstable.

4. *Multiple-order Poles in the Left Half Plane.* A pole of order m in the left half plane gives rise to a response of the form (see entry 7 in Table 5.1)

$$h_k = |A_k|\, t^m \exp[\sigma_k t] \cos(\omega_k t + \beta_k), \qquad \sigma_k < 0$$

For negative values of σ_k, the exponential function decreases faster than the polynomial t^m. Thus, the response decays as t increases, and a system with such poles is stable.

5. *Multiple-order Poles on the Imaginary Axis.* In this case, the response of the system takes the form

$$h_k = |A_k|\, t^m \cos(\omega_k t + \beta_k)$$

This term increases with time, and therefore, the system is unstable.

6. *Multiple-order Poles in the Right Half Plane.* The system response is

$$h_k = |A_k|\, t^m \exp[\sigma_k t] \cos(\omega_k t + \beta_k), \qquad \sigma_k > 0$$

Because $\sigma_k > 0$, the response increases with time, and therefore, the system is unstable.

In sum, a LTI (causal) system is stable if all its poles are in the open left half plane (the region of the complex plane consisting of all points to the left of, but not including, the $j\omega$-axis). A LTI system is marginally stable if it has simple poles on the $j\omega$-axis. An LTI system is unstable if it has poles in the right half plane or multiple poles on the $j\omega$-axis.

5.11 SUMMARY

- The bilateral Laplace transform of $x(t)$ is defined by

$$X_B(s) = \int_{-\infty}^{\infty} x(t) \exp[-st]\, dt$$

- The values of s for which $X(s)$ converges ($X(s)$ exists) constitute the region of convergence (ROC).
- The transformation $x(t) \leftrightarrow X(s)$ is not one to one unless the ROC is specified.
- The unilateral Laplace transform is defined as

$$X(s) = \int_{0^-}^{\infty} x(t) \exp[-st]\, dt$$

The bilateral and the unilateral Laplace transforms are related by

$$X_B(s) = X_+(s) + \mathcal{L}\{x_-(-t)u(t)\}_{s \to -s}$$

where $X_+(s)$ is the unilateral Laplace transform of the causal part of $x(t)$ and $x_-(t)$ is the noncausal part of $x(t)$.

- Differentiation in the time domain is equivalent to multiplication by s in the s domain; that is,

$$\mathcal{L}\left\{\frac{dx(t)}{dt}\right\} = sX(s) - x(0^-)$$

- Integration in the time domain is equivalent to division by s in the s domain; that is,

$$\mathcal{L}\left\{\int_{-\infty}^{t} x(\tau)\, d\tau\right\} = \frac{X(s)}{s}$$

- Convolution in the time domain is equivalent to multiplication in the s domain; that is,

$$y(t) = x(t) * h(t) \leftrightarrow Y(s) = X(s)H(s)$$

- The initial-value theorem allows us to compute the initial value of the signal $x(t)$ and its derivatives directly from $X(s)$:

$$x^{(n)}(0^+) = \lim_{s\to\infty} [s^{n+1}X(s) - s^n x(0^+) - s^{n-1}x'(0^+) - \cdots - sx^{(n-1)}(0^+)]$$

- The final-value theorem enables us to find the final value of $x(t)$ from $X(s)$:

$$\lim_{t\to\infty} x(t) = \lim_{s\to 0} sX(s)$$

- Partial-fraction expansion can be used to find the inverse Laplace transform of signals whose Laplace transforms are rational functions of s.
- There are many applications of the Laplace transform; among them are the solution of differential equations, the analysis of electrical circuits, and the design and analysis of control systems.
- If two subsystems with transfer functions $H_1(s)$ and $H_2(s)$ are connected in parallel, then the overall transfer function $H(s)$ is

$$H(s) = H_1(s) + H_2(s)$$

- If two subsystems with transfer functions $H_1(s)$ and $H_2(s)$ are connected in series, then the overall transfer function $H(s)$ is

$$H(s) = H_1(s)H_2(s)$$

- The closed-loop transfer function of a negative-feedback system with open-loop transfer function $H_1(s)$ and feedback transfer function $H_2(s)$ is

$$H(s) = \frac{H_1(s)}{1 + H_1(s)H_2(s)}$$

- Simulation diagrams for LTI systems can be obtained in the frequency domain. These diagrams can be used to obtain representations of state variables.
- The solution to the state equation can be written in the s domain as

$$\mathbf{V}(s) = \mathbf{\Phi}(s)\mathbf{v}(0^-) + \mathbf{\Phi}(s)\mathbf{b}X(s)$$

$$Y(s) = \mathbf{c}\mathbf{V}(s) + dX(s)$$

- The matrix

$$\mathbf{\Phi}(s) = (s\mathbf{I} - \mathbf{A})^{-1} = \mathscr{L}\{\exp[\mathbf{A}t]\}$$

is called the resolvent matrix.

- The transfer function of a system can be written as

$$H(s) = \mathbf{c\Phi}(s)\mathbf{b} + d$$

- An LTI system is stable if and only if all its poles are in the open left half plane. An LTI system is marginally stable if it has only simple poles on the $j\omega$-axis; otherwise it is unstable.

5.12 CHECKLIST OF IMPORTANT TERMS

Bilateral Laplace transform
Cascade interconnection
Causal part of $x(t)$
Controller
Convolution property
Feedback interconnection
Final-value theorem
Initial-conditions generator
Initial-value theorem
Inverse Laplace transform
Kirchhoff's current law
Kirchhoff's voltage law
Left half plane
Multiple-order pole
Negative feedback
Noncausal part of $x(t)$
Parallel Interconnection
Partial-fraction expansion
Plant
Poles of $D(s)$
Positive feedback
Rational function
Region of convergence
Simple pole
Simulation diagram
s plane
Transfer function
Unilateral Laplace transform
Zero-input response
Zero-state response

5.13 PROBLEMS

5.1. Find the bilateral Laplace transform and the ROC of the following functions:

(a) $\exp[t + 1]$
(b) $\exp[bt]u(-t)$
(c) $|t|$
(d) $(1 - |t|)$
(e) $\exp[-2|t|]$
(f) $t^n \exp[-t]u(-t)$
(g) $(\cos at)u(-t)$
(h) $(\sinh at)u(-t)$

5.2. Use the definition in Equation (5.3.1) to determine the unilateral Laplace transforms of the following signals:

(i) $x_1(t) = t\,\mathrm{rect}[(t - 1)/2]$

(ii) $x_2(t) = x_1(t) + \frac{1}{2}\delta(t)$

(iii) $x(t) = \text{rect}\left[\dfrac{t}{2}\right]$

5.3. Use Equation (5.4.2) to evaluate the bilateral Laplace transform of the signals in Problem 5.1.

5.4. The Laplace transform of a signal $x(t)$ that is zero for $t < 0$ is

$$X(s) = \frac{s^3 + 2s^2 + 3s + 2}{s^4 + 2s^3 + 2s^2 + 2s + 2}$$

Determine the Laplace transform of the following signals:

(a) $y(t) = 3x\left(\dfrac{t}{3}\right)$

(b) $y(t) = tx(t)$

(c) $y(t) = tx(t - 1)$

(d) $y(t) = \dfrac{dx(t)}{dt}$

(e) $y(t) = (t - 1)x(t - 1) + \dfrac{dx(t)}{dt}$

(f) $y(t) = \displaystyle\int_0^t x(\tau)\,d\tau$

5.5. Derive entry 5 in Table 5.1.

5.6. Show that

$$\mathcal{L}\{t^a u(t)\} = \frac{\Gamma(a + 1)}{s^a + 1}, \; a > 0$$

where

$$\Gamma(v) = \int_0^\infty t^{v-1} \exp[-t]\, dt$$

5.7. Use the property

$$\Gamma(v + 1) = v\Gamma(v)$$

to show that the result in Problem 5.6 reduces to entry 5 in Table 5.1.

5.8. Derive formulas 8 and 9 in Table 5.1 using integration by parts.

5.9. Use entries 8 and 9 in Table 5.1 to find the Laplace transforms of $\sinh(\omega_0 t)u(t)$ and $\cosh(\omega_0 t)u(t)$.

5.10. Determine the initial and final values of each of the signals whose unilateral Laplace transforms are as follows without computing the inverse Laplace transform. If there is no final value, state why not.

(a) $\dfrac{1}{s + a}$

(b) $\dfrac{1}{(s + a)^n}$

(c) $\dfrac{6}{s(s^2 + 25)}$

(d) $\dfrac{s+2}{s+3}$

(e) $\dfrac{s^2+s+3}{s^3+4s^2+2s+2}$

(f) $\dfrac{s}{s^2-2s-3}$

5.11. Find $x(t)$ for the following Laplace tranforms:

(a) $\dfrac{s+2}{s^2-s-2}$

(b) $\dfrac{s^2+8}{s(s^2+16)}$

(c) $\dfrac{2s^3+3s^2+6s+4}{(s^2+4)(s^2+2s+2)}$

(d) $\dfrac{s^2}{s^2+3s+2}$

(e) $\dfrac{s^2-s+1}{s^3-2s^2+s}$

(f) $\dfrac{(s+2)e^{-s}}{s^2+2s+1}$

(g) $\dfrac{2s^2-6s+3}{s^2-3s+2}$

(h) $\dfrac{2(s^2+2s+4)}{(s^2+4)^2}$

(i) $\dfrac{2}{(2s+1)^3}$

(j) $\dfrac{(s^2+8)e^{-2s}}{s(s^2+16)}$

5.12. Find the following convolutions using Laplace transforms:

(a) $\exp[at]u(t) * \exp[bt]u(t),\ a \neq b$

(b) $\exp[at]u(t) * \exp[at]u(t)$

(c) $\text{rect}(t/2) * u(t)$

(d) $tu(t) * \exp[at]u(t)$

(e) $\exp[-bt]u(t) * u(t)$

(f) $\sin(at)u(t) * \cos(bt)u(t)$

(g) $\exp[-2t]u(t) * \text{rect}[(t-1)/2]$

(h) $[\exp(-2t)u(t) + \delta(t)] * u(t-1)$

5.13. **(a)** Use the convolution property to find the time signals corresponding to the following Laplace transforms:

$$\text{(i)}\ \frac{1}{(s-a)^2} \qquad \text{(ii)}\ \frac{1}{(s-a)^3}$$

(b) Can you infer the inverse Laplace transform of $1/(s-a)^n$ from your answers in Part (a)?

5.14. We have seen that the output of an LTI system can be determined as $Y(s) = H(s)X(s)$, where the system transfer function $H(s)$ is the Laplace transform of the system impulse response $h(t)$. Let $H(s) = N(s)/D(s)$, where $N(s)$ and $D(s)$ are polynomials in s. The roots of $N(s)$ are the zeros of $H(s)$, while the roots of $D(s)$ are the poles.

(a) For the transfer function

$$H(s) = \frac{s^2 + 3s + 2}{s^3 - s^2 + 9s - 9}$$

plot the locations of the poles and zeros in the complex s plane.

(b) What is $h(t)$ for this system? Is $h(t)$ real?

(c) Show that if $h(t)$ is real, $H(s^*) = H^*(s)$. Hence show that if $s = s_0$ is a pole (zero) of $H(s)$, so is $s = s_0{}^*$. That is poles and zeros occur in complex conjugate pairs.

(d) Verify that the given $H(s)$ satisfies (c).

5.15. Find the system transfer functions for each of the systems in Figure P5.15. (*Hint:* You may have to move the pickoff, or summation, point.)

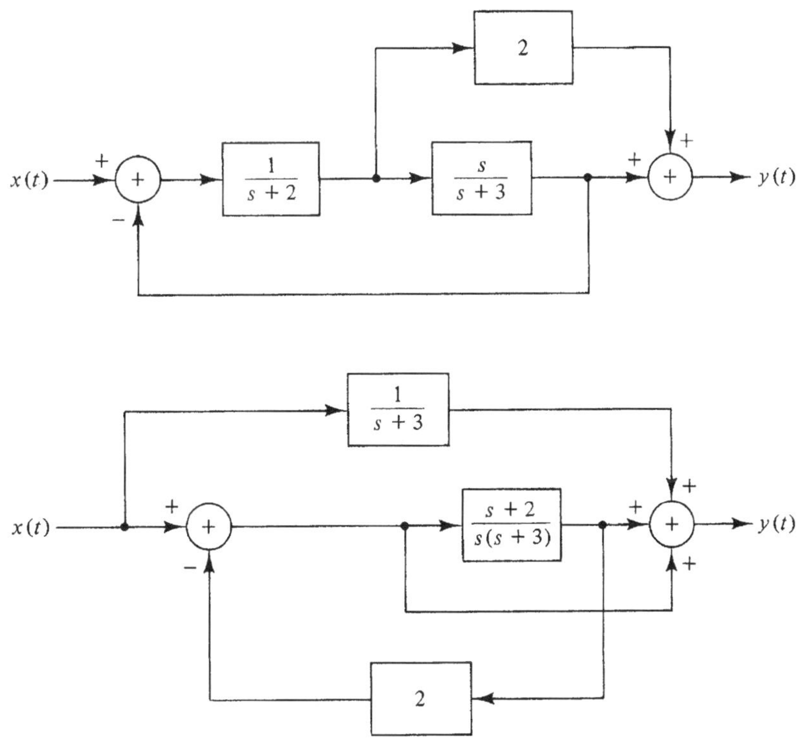

Figure P5.15

5.16. Draw the simulation diagrams in the first and second canonical forms for the LTI system described by the transfer function

$$H(s) = \frac{s^2 + s + 1}{s^3 + 4s^2 + 3s + 2}$$

5.17. Repeat Problem 5.16 for the system described by

$$H(s) = \frac{s^3 + 3s + 1}{s^3 + 3s^2 + s}$$

5.18. Find the transfer function of the system described by

$$2y''(t) + 3y'(t) + 4y(t) = 2x'(t) - x(t)$$

(Assume zero initial conditions.) Find the system impulse response.

5.19. Find the transfer function of the system shown in Figure P5.19.

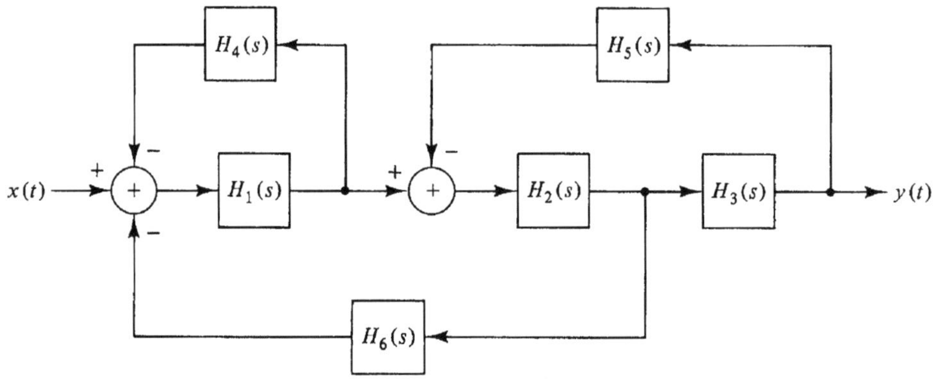

Figure P5.19

5.20. Solve the following differential equations:

(a) $y'(t) + 2y(t) = u(t), \quad y(0^-) = 1$

(b) $y'(t) + 2y(t) = (\cos t)u(t), \quad y(0^-) = 1$

(c) $y'(t) + 2y(t) = \exp[-3t]u(t), \quad y(0^-) = 1$

(d) $y''(t) + 4y'(t) + 3y(t) = u(t), \quad y(0^-) = 2, y(0^-) = 1$

(e) $y''(t) + 4y'(t) + 3y(t) = \exp[-3t]u(t), \quad y(0^-) = 0, y'(0^-) = 1$

(f) $y'''(t) + 3y'(t) + 2y'(t) - 6y(t) = \exp[-2t]u(t), \quad y(0^-) = y'(0^-) = y''(0^-) = 0$

5.21. Find the impulse response $h(t)$ for the systems described by the following differential equations:

(a) $y'(t) + 5y(t) = x(t) + 2x'(t)$

(b) $y''(t) + 4y'(t) + 3y(t) = 2x(t) - 3x'(t)$

(c) $y'''(t) + y'(t) - 2y(t) = x''(t) + x'(t) + 2x(t)$

5.22. One major problem in systems theory is system identification. Observing the output of an LTI system in response to a known input can provide us with the impulse response of the system. Find the impulse response of the systems whose input and output are as follows:

(a) $x(t) = 2\exp[-2t]u(t)$
$y(t) = (1 - t + \exp[-t] + \exp[-2t])u(t)$

(b) $x(t) = 2u(t)$
$y(t) = tu(t) - \exp[-2t]u(t)$

(c) $x(t) = \exp[-2t]u(t)$
$y(t) = \exp[-t] - 3\exp[-2t])u(t)$

(d) $x(t) = tu(t)$
$y(t) = (t^2 - 2\exp[-3t])u(t)$

(e) $x(t) = 2u(t)$
$y(t) = \exp[-2t]\cos(4t + 135°)u(t)$

(f) $x(t) = 3tu(t)$
$y(t) = \exp[-4t][\cos(4t + 135°) - 2\sin(4t + 135°)]u(t)$

5.23. For the circuit shown in Figure P5.23, let $v_C(0^-) = 1$ volt, $i_L(0^-) = 2$ amperes, and $x(t) = u(t)$. Find $y(t)$. (Incorporate the initial energy for the inductor and the capacitor in your transformed model.)

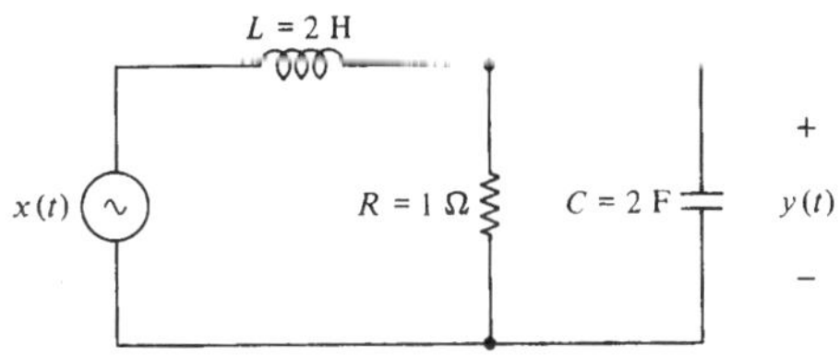

Figure P5.23

5.24. For the circuit shown in Figure P5.24, let $v_C(0^-) = 1$ volt, $i_L(0^-) = 2$ amperes, and $x(t) = u(t)$. Find $y(t)$. (Incorporate the initial energy for the inductor and the capacitor in your transformed model.)

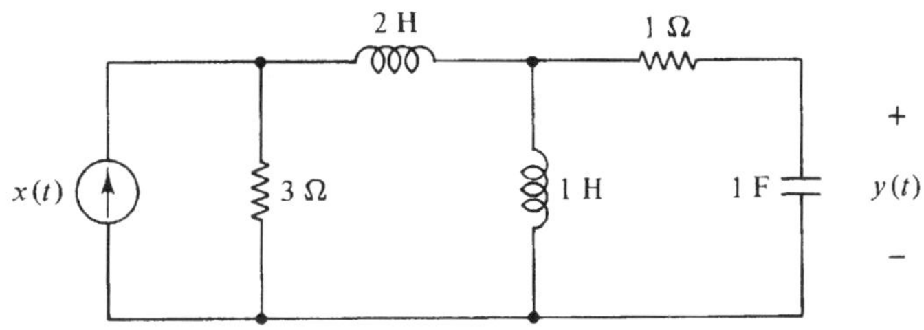

Figure P5.24

5.25. Repeat Problem 5.23 for $x(t) = (\cos t)u(t)$.

5.26. Repeat Problem 5.24 for $x(t) = (\sin 2t)u(t)$.

5.27. Repeat Problem 5.23 for the circuit shown in Figure P5.27.

5.28. Consider the control system shown in Figure P5.28. For $x(t) = u(t)$, $H_1(s) = K$, and $H_2(s) = 1/(s(s + a))$, find the following:

(a) $Y(s)$

(b) $y(t)$ for $K = 29$, $a = 5$, $a = 3$, and $a = 1$

5.29. Consider the control system shown in Figure P5.29.
Let

$$x(t) = Au(t)$$

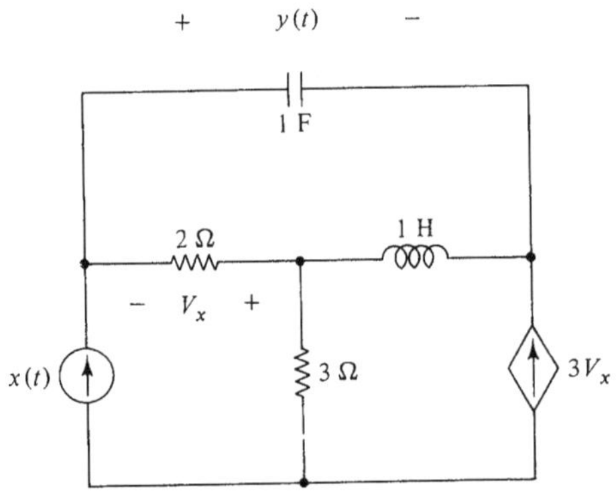

Figure P5.27

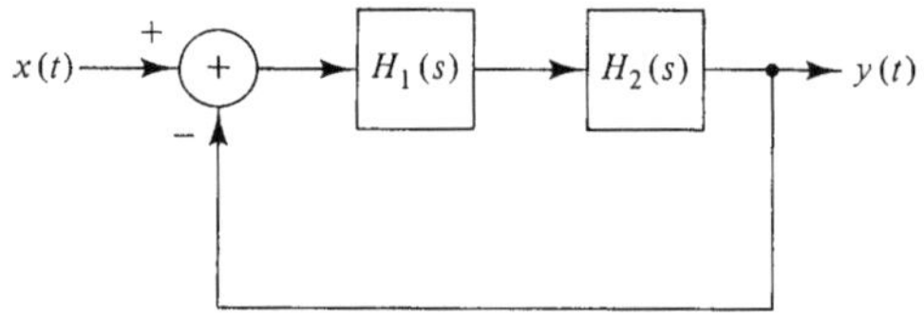

Figure P5.28

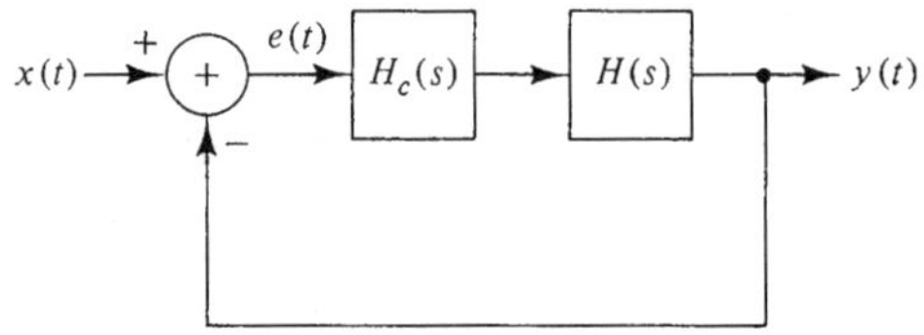

Figure P5.29

$$H_c(s) = \frac{s+1}{s}$$

$$H(s) = \frac{1}{s+2}$$

(a) Show that $\lim_{t\to\infty} y(t) = A$.

(b) Determine the error signal $e(t)$.

(c) Does the system track the input if $H_c(s) = \dfrac{1}{(s-2)}$? If not, why?

(d) Does the system work if $H_c(s) = \dfrac{(s+1)}{(s+3)}$? Why?

5.30. Find $\exp[\mathbf{A}t]$ using the Laplace transform for the following matrices:

(a) $\mathbf{A} = \begin{bmatrix} 0 & -1 \\ 1 & 2 \end{bmatrix}$ **(b)** $\mathbf{A} = \begin{bmatrix} 1 & -1 \\ 2 & 0 \end{bmatrix}$

(c) $\mathbf{A} = \begin{bmatrix} 1 & -2 \\ 0 & 1 \end{bmatrix}$ **(d)** $\mathbf{A} = \begin{bmatrix} 3 & -2 \\ 1 & 1 \end{bmatrix}$

(e) $\mathbf{A} = \begin{bmatrix} 1 & 0 & 0 \\ -1 & 1 & 1 \\ -1 & 0 & 0 \end{bmatrix}$ **(f)** $\mathbf{A} = \begin{bmatrix} 2 & 1 & 1 \\ 0 & 3 & 1 \\ 0 & -1 & 1 \end{bmatrix}$

5.31. Consider the circuit shown in Figure P5.31. Select the capacitor voltage and the inductor current as state variables. Assume zero initial conditions.

(a) Write the state equations in the transform domain.

(b) Find $Y(s)$ if the input $x(t)$ is the unit step.

(c) What is $y(t)$?

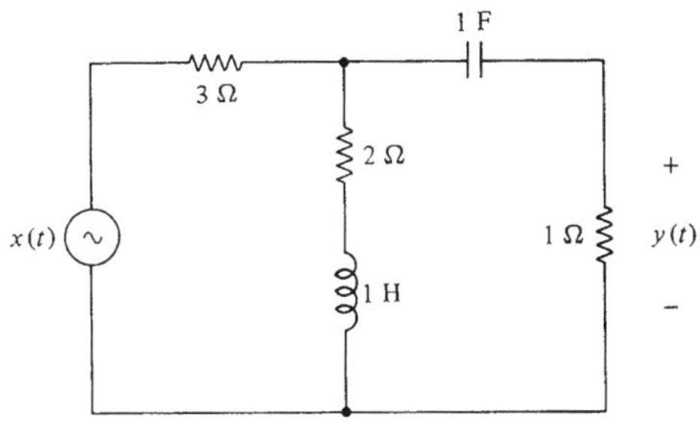

Figure P5.31

5.32. Use the Laplace-transform method to find the solution of the following state equations:

(a) $\begin{bmatrix} v_1'(t) \\ v_2'(t) \end{bmatrix} = \begin{bmatrix} -1 & 0 \\ -3 & -2 \end{bmatrix} \begin{bmatrix} v_1(t) \\ v_2(t) \end{bmatrix}, \quad \begin{bmatrix} v_1(0^-) \\ v_2(0^-) \end{bmatrix} = \begin{bmatrix} 1 \\ 2 \end{bmatrix}$

(b) $\begin{bmatrix} v_1'(t) \\ v_2'(t) \end{bmatrix} = \begin{bmatrix} 2 & -7 \\ 1 & -3 \end{bmatrix} \begin{bmatrix} v_1(t) \\ v_2(t) \end{bmatrix}, \quad \begin{bmatrix} v_1(0^-) \\ v_2(0^-) \end{bmatrix} = \begin{bmatrix} 0 \\ -1 \end{bmatrix}$

5.33. Check the stability of the systems shown in Figure P5.33.

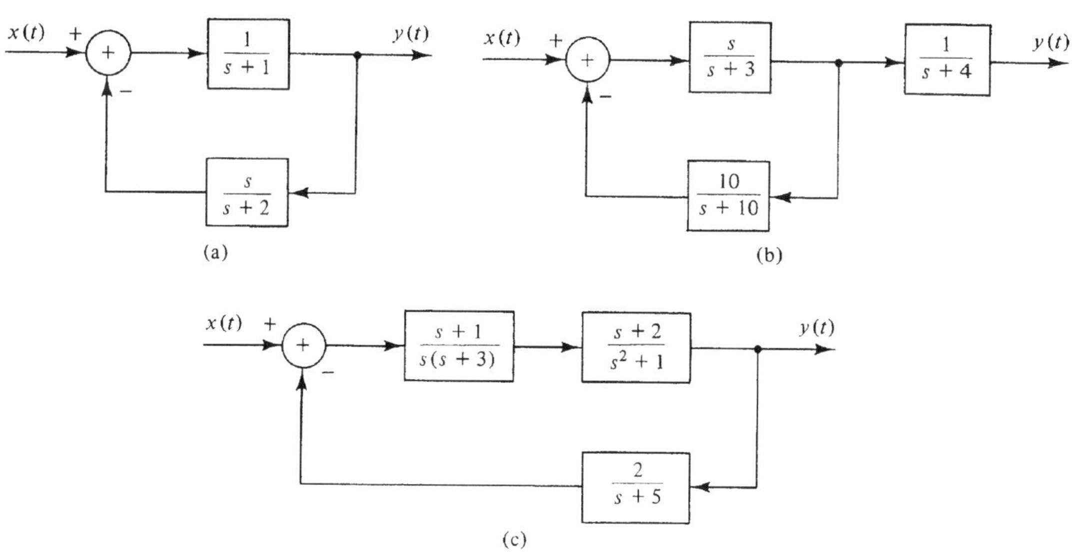

Figure P5.33

Chapter 6

Discrete-Time Systems

6.1 INTRODUCTION

In the preceding chapters, we discussed techniques for the analysis of analog or continuous-time signals and systems. In this and subsequent chapters, we consider corresponding techniques for the analysis of discrete-time signals and systems.

Discrete-time signals, as the name implies, are signals that are defined only at discrete instants of time. Examples of such signals are the number of children born on a specific day in a year, the population of the United States as obtained by a census, the interest on a bank account, etc. A second type of discrete-time signal occurs when an analog signal is converted into a discrete-time signal by the process of *sampling*. (We will have more to say about sampling later.) An example is the digital recording of audio signals. Another example is a telemetering system in which data from several measurement sensors are transmitted over a single channel by *time-sharing*.

In either case, we represent the discrete-time signal as a sequence of values $x(t_n)$, where the t_n correspond to the instants at which the signal is defined. We can also write the sequence as $x(n)$, with n assuming only integer values.

As with continuous-time signals, we usually represent discrete-time signals in functional form—for example,

$$x(n) = \frac{1}{2}\cos 3n \tag{6.1.1}$$

Alternatively, if a signal is nonzero only over a finite interval, we can list the values of the signal as the elements of a sequence. Thus, the function shown in Figure 6.1.1 can be written as

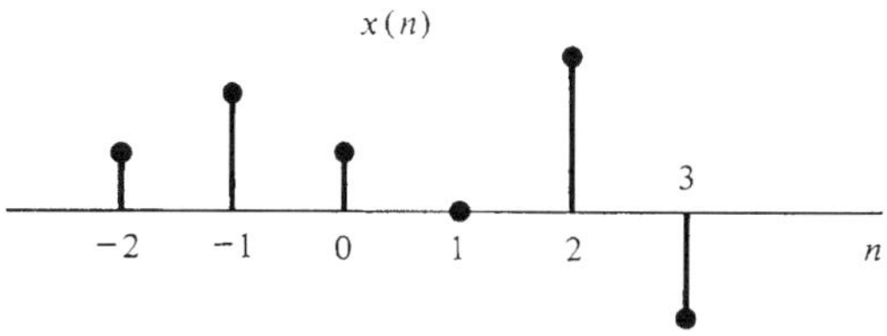

Figure 6.1.1 Example of a discrete-time sequence.

$$x(n) = \left\{\frac{1}{4}, \frac{1}{2}, \frac{1}{3}, 0, \frac{3}{4}, -\frac{1}{2}\right\} \tag{6.1.2}$$
$$\uparrow$$

where the arrow indicates the value for $n = 0$. In this notation, it is assumed that all values not listed are zero. For causal sequences, in which the first entry represents the value at $n = 0$, we omit the arrow.

The sequence shown in Equation (6.1.2) is an example of a *finite-length* sequence. The length of the sequence is given by the number of terms in the sequence. Thus, Equation (6.1.2) represents a six-point sequence.

6.1.1 Classification of Discrete-Time Signals

As with continuous-time signals, discrete-time signals can be classified into different categories. For example, we can define the energy of a discrete-time signal $x(n)$ as

$$E = \lim_{N\to\infty} \sum_{n=-N}^{N} |x(n)|^2 \tag{6.1.3}$$

The average power of the signal is

$$P = \lim_{N\to\infty} \frac{1}{2N+1} \sum_{n=-N}^{N} |x(n)|^2 \tag{6.1.4}$$

The signal $x(n)$ is an energy signal if E is finite. It is a power signal if E is not finite, but P is finite. Since $P = 0$ when E is finite, all energy signals are also power signals. However, if P is finite, E may or may not be finite. Thus, not all power signals are energy signals. If neither E nor P is finite, the signal is neither an energy nor a power signal.

The signal $x(n)$ is periodic if, for some integer $N > 0$,

$$x(n + N) = x(n) \qquad \text{for all } n \tag{6.1.5}$$

The smallest value of N that satisfies this relation is the fundamental period of the signal.

If there is no integer N that satisfies Equation (6.1.5), $x(n)$ is an aperiodic signal.

Example 6.1.1

Consider the signal

$$x(n) = A \sin(2\pi f_0 n + \phi_0)$$

Then

$$x(n + N) = A \sin(2\pi f_0(n + N) + \phi_0)$$
$$= A \sin(2\pi f_0 n + \phi_0)\cos(2\pi f_0 N) + A\cos(2\pi f_0 n + \phi_0)\sin(2\pi f_0 N)$$

Clearly, $x(n + N)$ will be equal to $x(n)$ if

$$N = \frac{m}{f_0}$$

where m is some integer. The fundamental period is obtained by choosing m as the smallest integer that yields an integer value for N. For example, if $f_0 = 3/5$, we can choose $m = 3$ to get $N = 5$.

On the other hand, if $f_0 = \sqrt{2}$, N will not be an integer, and thus, $x(n)$ is aperiodic.

Let $x(n)$ be the sum of two periodic sequences $x_1(n)$ and $x_2(n)$, with periods N_1 and N_2 respectively. Let p and q be two integers such that

$$pN_1 = qN_2 = N \tag{6.1.6}$$

Then $x(n)$ is periodic with period N, since

$$x(n + N) = x_1(n + pN_1) + x_2(n + qN_2) = x_1(n) + x_2(n)$$

Because we can always find integers p and q to satisfy Equation (6.1.6), it follows that the sum of two discrete-time periodic sequences is also periodic.

Example 6.1.2

Let

$$x(n) = \cos\left(\frac{\pi n}{9}\right) + \sin\left(\frac{\pi n}{7} + \frac{1}{2}\right)$$

It can be easily verified, as in Example 6.1.1, that the two terms in $x(n)$ are both periodic with periods $N_1 = 18$ and $N_2 = 14$, respectively, so that $x(n)$ is periodic with period $N = 126$.

The signal $x(n)$ is even if

$$x(n) = x(-n) \qquad \text{for all } n \tag{6.1.7}$$

and is odd if

$$x(n) = -x(-n) \qquad \text{for all } n \tag{6.1.8}$$

The even part of $x(n)$ can be determined as

$$x_e(n) = \frac{1}{2}[x(n) + x(-n)] \tag{6.1.9}$$

whereas its odd part is given by

$$x_0(n) = \frac{1}{2}[x(n) - x(-n)] \tag{6.1.10}$$

6.1.2 Tranformations of the Independent Variable

For integer values of k, the sequence $x(n-k)$ represents the sequence $x(n)$ shifted by k samples. The shift is to the right if $k > 0$ and to the left if $k < 0$. Similarly, the signal $x(-n)$ corresponds to reflecting $x(n)$ around the time origin $n = 0$. As in the continuous-time case, the operations of shifting and reflecting are not commutative.

While amplitude scaling is no different than in the continuous-time case, time scaling must be interpreted with care in the discrete-time case, since the signals are defined only for integer values of the time variable. We illustrate this by a few examples.

Example 6.1.3

Let

$$x(n) = e^{-n/2}u(n)$$

and suppose we want to find (i) $2x(5n/3)$ and (ii) $x(2n)$.

With $y(n) = 2x(5n/3)$, we have

$$y(0) = 2x(0) = 2,\ y(1) = 2x(5/3) = 0,\ y(2) = 2x(10/3) = 0,$$

$$y(3) = 2x(5) = 2\exp(-5/2),\ y(4) = 2x(10/3) = 0, \text{ etc.}$$

Here we have assumed that $x(n)$ is zero if n is not an integer. It is clear that the general expression for $y(n)$ is

$$y(n) = \begin{cases} 2\exp\left[\dfrac{-5n}{6}\right], & n = 0, 3, 6, \text{ etc.} \\ 0, & \text{otherwise} \end{cases}$$

Similarly, with $z(n) = x(2n)$, we have

$$z(0) = x(0) = 1,\ z(1) = x(2) = \exp[-1],\ z(3) = x(6) = \exp[-3], \text{ etc.}$$

The general expression for $z(n)$ is therefore

$$z(n) = \begin{cases} \exp[-n], & n \geq 0 \\ 0, & n < 0 \end{cases}$$

The preceding example shows that for discrete-time signals, time scaling does not yield just a stretched or compressed version of the original signal, but may give a totally different waveform.

Example 6.1.4

Let

$$x(n) = \begin{cases} 1, & n \text{ even} \\ -1, & n \text{ odd} \end{cases}$$

Then

$$y(n) = x(2n) = 1 \qquad \text{for all } n$$

Example 6.1.5

Consider the waveform shown in Fig. (6.1.2a), and let

$$y(n) = x\left(-\frac{n}{3} + \frac{2}{3}\right)$$

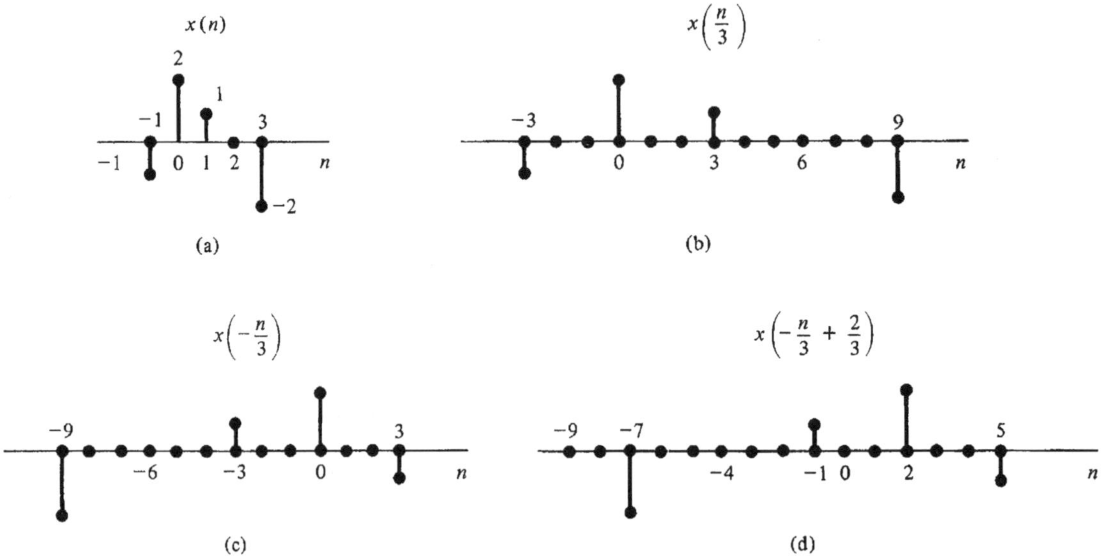

Figure 6.1.2 Signals for Example 6.1.5 (a) $x(n)$, (b) $x(n/3)$, (c) $x(-n/3)$, and (d) $x(-n/3 + 2/3)$.

We determine $y(n)$ by writing it as

$$y(n) = x\left[-\frac{(n-2)}{3}\right]$$

We first scale $x(.)$ by a factor of 1/3 to obtain $x(n/3)$ and then reflect this about the vertical axis to obtain $x(-n/3)$. The result is shifted to the right by two samples to obtain $y(n)$. These steps are illustrated in Figs. (6.1.2b)–(6.1.2d). The resulting sequence is

$$y(n) = [-2, 0, 0, 0, 0, 0, 1, 0, 0, 2, 0, 0, -1]$$
$$\uparrow$$

6.2 ELEMENTARY DISCRETE-TIME SIGNALS

Thus far, we have seen that continuous-time signals can be represented in terms of elementary signals such as the delta function, unit-step function, exponentials, and sine and cosine waveforms. We now consider the discrete-time equivalents of these signals. We will see that these discrete-time signals have characteristics similar to those of their continuous-time counterparts, but with some significant differences. As with continuous-time systems, the analysis of the responses of discrete-time linear systems to arbitrary inputs is considerably simplified by expressing the inputs in terms of elementary time functions.

6.2.1 Discrete Impulse and Step Functions

We define the unit-impulse function in discrete time as

$$\delta(n) = \begin{cases} 1, & n = 0 \\ 0, & n \neq 0 \end{cases} \tag{6.2.1}$$

as shown in Figure 6.2.1. We refer to $\delta(n)$ as the unit sample occurring at $n = 0$ and the shifted function $\delta(n - k)$ as the unit sample occurring at $n = k$. That is,

$$\delta(n - k) = \begin{cases} 1, & n = k \\ 0, & n \neq k \end{cases} \tag{6.2.2}$$

Whereas $\delta(n)$ is somewhat similar to the continuous-time impulse function $\delta(t)$, we note that the magnitude of the discrete impulse is always finite. Thus, there are no analytical difficulties in defining $\delta(n)$.

The unit-step sequence shown in Figure 6.2.2 is defined as

$$u(n) = \begin{cases} 1, & n \geq 0 \\ 0, & n < 0 \end{cases} \tag{6.2.3}$$

The discrete-time delta and step functions have properties somewhat similar to their continuous-time counterparts. For example, the *first difference* of the unit-step function is

$$u(n) - u(n - 1) = \delta(n) \tag{6.2.4}$$

If we compute the sum from $-\infty$ to n of the δ function, as can be seen from Figure 6.2.3, we get the unit step function:

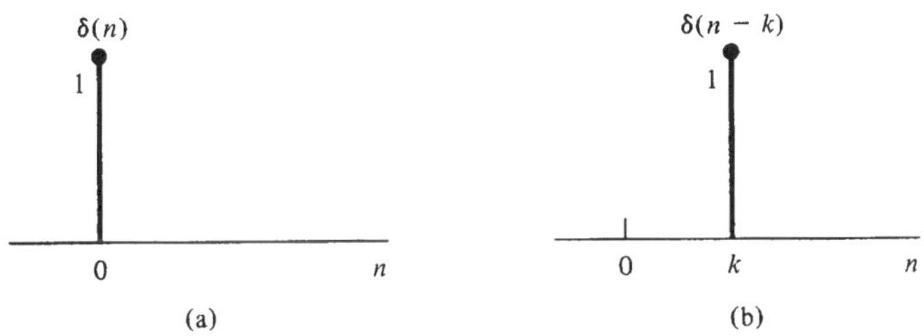

Figure 6.2.1 (a) The unit sample of δ function. (b) The shifted δ function.

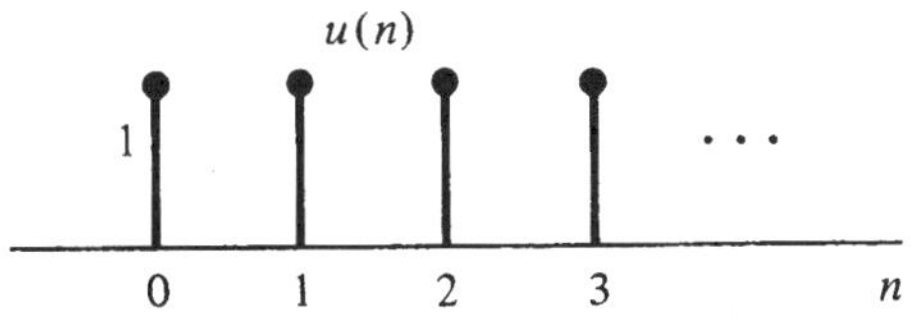

Figure 6.2.2 The unit step function.

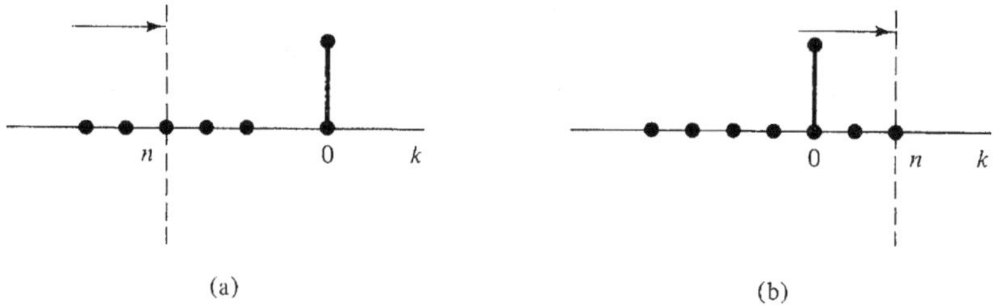

Figure 6.2.3 Summing the δ function. (a) $n < 0$. (b) $n \geq 0$.

$$\sum_{k=-\infty}^{n} \delta(k) = \begin{cases} 0, & n < 0 \\ 1, & n \geq 0 \end{cases} \tag{6.2.5}$$
$$= u(n)$$

By replacing k by $n - k$, we can write Equation (6.2.5) as

$$\sum_{k=0}^{\infty} \delta(n-k) = u(n) \tag{6.2.6}$$

From Equations (6.2.4) and (6.2.5), we see that in discrete-time systems, the first difference, in a sense, takes the place of the first derivative in continuous-time systems, and the sum operator replaces the integral.

Other analogous properties of the δ function follow easily. For any arbitrary sequence $x(n)$, we have

$$x(n)\,\delta(n-k) = x(k)\,\delta(n-k) \tag{6.2.7}$$

Since we can write $x(n)$ as

$$x(n) = \cdots + x(-1)\,\delta(n+1) + x(0)\,\delta(n) + x(1)\,\delta(n-1) + \cdots$$

it follows that

$$x(n) = \sum_{k=-\infty}^{\infty} x(k)\,\delta(n-k) \tag{6.2.8}$$

Thus, Equation (6.2.6) is a special case of Equation (6.2.8).

6.2.2 Exponential Sequences

The exponential sequence in discrete time is given by

$$x(n) = C\,\alpha^n \tag{6.2.9}$$

where, in general, C and α are complex numbers. The fact that this is a direct analog of the exponential function in continuous time can be seen by writing $\alpha = e^{\beta}$, so that

$$x(n) = C\,e^{\beta n} \tag{6.2.10}$$

For C and α real, $x(n)$ increases with increasing n if $|\alpha| > 1$. Similarly, if $|\alpha| < 1$, we have a decreasing exponential.

Consider the complex exponential signal in continuous time,

$$x(t) = C \exp[j\omega_0 t] \tag{6.2.11}$$

Suppose we sample $x(t)$ at equally spaced intervals nT to get the discrete-time signal

$$x(n) = C \exp[j\omega_0 Tn] \tag{6.2.12}$$

By replacing $\omega_0 T$ in this equation by Ω_0, we obtain the complex exponential in discrete time,

$$x(n) = C \exp[j\Omega_0 n] \tag{6.2.13}$$

Recall that ω_0 is the frequency of the continuous-time signal $x(t)$. Correspondingly, we will refer to Ω_0 as the frequency of the discrete-time signal $x(n)$. It can be seen, however, that whereas the continuous-time or analog frequency ω_0 has units of radians per second, the discrete-time frequency Ω_0 has units of radians.

Furthermore, while the signal $x(t)$ is periodic with period $T = 2\pi/\omega_0$ for any ω_0, in the discrete-time case, since the period is constrained to be an integer, not all values of Ω_0 correspond to a periodic signal. To see this, suppose $x(n)$ in Equation (6.2.13) is periodic with period N. Then, since $x(n) = x(n + N)$, we must have

$$e^{j\Omega_0 N} = 1$$

For this to hold, $\Omega_0 N$ must be an integer multiple of 2π, so that

$$\Omega_0 N = m\,2\pi,\ m = 0, \pm 1, \pm 2, \text{etc.}$$

or

$$\frac{\Omega_0}{2\pi} = \frac{m}{N}$$

for m any integer. Thus, $x(n)$ will be periodic only if $\Omega_0/2\pi$ is a rational number. The period is given by $N = 2\pi m/\Omega_0$, with the fundamental period corresponding to the smallest possible value for m.

Example 6.2.1

Let

$$x(n) = \exp\left[j\frac{7\pi}{9}n\right]$$

so that

$$\frac{\Omega_0}{2\pi} = \frac{7}{18} = \frac{m}{N}$$

Thus, the sequence is periodic, and the fundamental period, obtained by choosing $m = 7$, is given by $N = 18$.

Example 6.2.2

For the sequence

$$x(n) = \exp\left[j\frac{7n}{9}\right]$$

we have

$$\frac{\Omega_0}{2\pi} = \frac{7}{18\pi}$$

which is not rational. Thus, the sequence is not periodic.

Let $x_k(n)$ define the set of functions

$$x_k(n) = e^{jk\Omega_0 n} \qquad k = 0, \pm 1, \pm 2, \ldots \tag{6.2.14}$$

with $\Omega_0 = 2\pi/N$, so that $x_k(n)$ represents the kth harmonic of fundamental signal $x_1(n)$. In the case of continuous-time signals, we saw that the set of harmonics $\exp[jk(2\pi/T)t]$, $k = 0, \pm 1, \pm 2, \ldots$ are all distinct, so that we have an infinite number of harmonics. However, in the discrete-time case, since

$$x_{k+N}(n) = e^{j(k+N)\frac{2\pi}{n}n} = e^{j2\pi n}\, e^{jk\frac{2\pi}{N}n} = x_k(n) \tag{6.2.15}$$

there are only N distinct waveforms in the set given by Equation (6.2.14). These correspond to the frequencies $\Omega_k = 2\pi k/N$ for $k = 0, 1, \ldots, N-1$. Since $\Omega_{k+N} = \Omega_k + 2\pi$, waveforms separated in frequency by 2π radians are identical. As we shall see later, this has implications in the Fourier analysis of discrete-time, periodic signals.

Example 6.2.3

Consider the continuous-time signal

$$x(t) = \sum_{k=-2}^{2} c_k e^{jk\frac{2\pi}{3}t}$$

where $c_0 = 1$, $c_1 = (1 + j1) = c_{-1}^*$, and $c_2 = c_{-2}^* = 3/2$.

Let us sample $x(t)$ uniformly at a rate $T = 4$ to get the sampled signal

$$x(n) = \sum_{k=-2}^{2} c_k e^{jk\frac{2\pi}{3}4n} = \sum_{k=-2}^{2} c_k e^{jk\Omega_0 n}$$

where $\Omega_0 = 4(2\pi/3)$. Thus, $x(n)$ represents a sum of harmonic signals with fundamental period $N = 2\pi m/\Omega_0$. Choosing $m = 4$ then yields $N = 3$. It follows, therefore, that there are only three distinct harmonics, and hence, the summation can be reduced to one consisting only of three terms.

To see this, we note that, from Equation (6.2.15), we have $\exp(j2\Omega_0 n) = \exp(-j\Omega_0 n)$ and $\exp(j(-2\Omega_0 n)) = \exp(j\Omega_0 n)$, so that grouping like terms together gives

$$x(n) = \sum_{k=-1}^{1} d_k e^{jk\frac{2\pi}{3}n}$$

where

$$d_0 = c_0 = 1, d_1 = c_1 + c_{-2} = -1 + j\frac{1}{2}, d_{-1} = c_{-1} + c_2 = -1 - j\frac{1}{2} = d_1^*$$

6.3 DISCRETE-TIME SYSTEMS

A discrete-time system is a system in which all the signals are discrete-time signals. That is, a discrete-time system transforms discrete-time inputs into discrete-time outputs. Such concepts as linearity, time invariance, causality, etc., which we defined for continuous-time systems carry over to discrete-time systems. As in our discussion of continuous-time systems, we consider only linear, time-invariant (or *shift-invariant*) systems in discrete time.

Again, as with continuous-time systems, we can use either a time-domain or a frequency-domain characterization of a discrete-time system. In this section, we examine the time-domain characterization of discrete-time systems using (a) the impulse-response and (b) the difference equation representations.

Consider a linear, shift-invariant, discrete-time system with input $x(n)$. We saw in Section 6.2.1 that any arbitrary signal $x(n)$ can be written as the weighted sum of shifted unit-sample functions:

$$x(n) = \sum_{k=-\infty}^{\infty} x(k)\,\delta(n-k) \tag{6.3.1}$$

It follows, therefore, that we can use the linearity property of the system to determine its response to $x(n)$ in terms of its response to a unit-sample input. Let $h(n)$ denote the response of the system measured at time n to a unit impulse applied at time zero. If we apply a shifted impulse $\delta(n-k)$ occurring at time k, then, by the assumption of shift invariance, the response of the system at time n is given by $h(n-k)$. If the input is amplitude scaled by a factor $x(k)$, then, again, by linearity, so is the output. If we now fix n, let k vary from $-\infty$ to ∞, and take the sum, it follows from Equation (6.3.1) that the output of the system at time n is given in terms of the input as

$$y(n) = \sum_{k=-\infty}^{\infty} x(k)\,h(n-k) \tag{6.3.2}$$

As in the case of continuous-time systems, the impulse response is determined assuming that the system has no initial energy; otherwise the linearity property does not hold (why?), so that $y(n)$, as determined by using Equation (6.3.2), corresponds to only the forced response of the system.

The right-hand side of Equation (6.3.2) is referred to as the *convolution sum* of the two sequences $x(n)$ and $h(n)$ and is represented symbolically as $x(n) * h(n)$. By replacing k by $n-k$ in the equation, the output can also be written as

$$y(n) = \sum_{k=-\infty}^{\infty} x(n-k)\,h(k)$$
$$= h(n) * x(n) \tag{6.3.3}$$

Thus, the convolution operation is commutative.

For causal systems, it is clear that

$$h(n) = 0, \qquad n < 0 \tag{6.3.4}$$

so that Equation (6.3.2) can be written as

$$y(n) = \sum_{k=-\infty}^{n} x(k)\,h(n-k) \tag{6.3.5}$$

or, in the equivalent form,

$$y(n) = \sum_{k=0}^{\infty} x(n-k)\,h(k) \tag{6.3.6}$$

For continuous-time systems, we saw that the impulse response is, in general, the sum of several complex exponentials. Consequently, the impulse response is nonzero over any finite interval of time (except, possibly, at isolated points) and is generally referred to as an *infinite impulse response* (IIR). With discrete-time systems, on the other hand, the impulse response can become identically zero after a few samples. Such systems are said to have a *finite impulse response* (FIR). Thus, discrete-time systems can be either IIR or FIR.

We can interpret Equation (6.3.2) in a manner similar to the continuous-time case. For a fixed value of n, we consider the product of the two sequences $x(k)$ and $h(n-k)$, where $h(n-k)$ is obtained from $h(k)$ by first reflecting $h(k)$ about the origin and then shifting to the right by n if n is positive or to the left by $|n|$ if n is negative. This is illustrated in Figure (6.3.1). The output $y(n)$ for this value of n is determined by summing the values of the sequence $x(k)h(n-k)$.

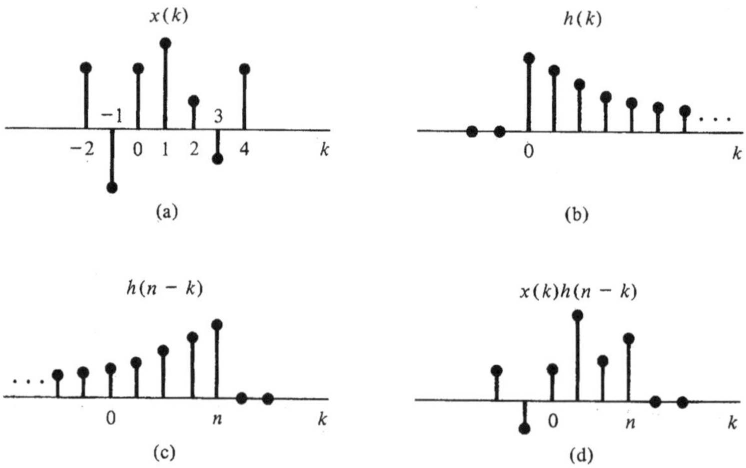

Figure 6.3.1 Convolution operation of Equation (6.3.2). (a) $x(k)$, (b) $h(k)$, (c) $h(n-k)$, and (d) $x(k)h(n-k)$.

We note that the convolution of $h(n)$ with $\delta(n)$ is, by definition, equal to $h(n)$. That is, the convolution of any function with the δ function gives back the original function.

We now consider a few examples.

Example 6.3.1

When an input $x(n) = 3\delta(n-2)$ is applied to a causal, linear time-invariant system, the output is found to be

$$y(n) = 2\left(-\frac{1}{2}\right)^n + 8\left(\frac{1}{4}\right)^n, \qquad n \geq 2$$

Find the impulse response $h(n)$ of the system.

By definition, $h(n)$ is the response of the system to the input $\delta(n)$. Since the system is LTI, it follows that

$$h(n) = \frac{1}{3}y(n+2)$$

We note that the output can be written as

$$y(n) = \left[\frac{1}{2}\left(-\frac{1}{2}\right)^{(n-2)} + \frac{1}{2}\left(\frac{1}{4}\right)^{n-2}\right]u(n-2)$$

so that

$$h(n) = \frac{1}{6}\left[\left(-\frac{1}{2}\right)^n + \left(\frac{1}{4}\right)^n\right]u(n)$$

Example 6.3.2

Let

$$x(n) = \alpha^n u(n)$$

$$h(n) = \beta^n u(n)$$

Then

$$y(n) = \sum_{k=-\infty}^{\infty} \alpha^k u(k)\, \beta^{n-k} u(n-k)$$

Since $u(k) = 0$ for $k < 0$, and $u(n-k) = 0$ for $k > n$, we can rewrite the summation as

$$y(n) = \sum_{k=0}^{n} \alpha^k \beta^{n-k} = \beta^n \sum_{k=0}^{n} (\alpha\beta^{-1})^k$$

Clearly, $y(n) = 0$ if $n < 0$.

For $n \geq 0$, if $\alpha = \beta$, we have

$$y(n) = \beta^n \sum_{k=0}^{n} (1) = (n+1)\,\beta^n$$

If $\alpha \neq \beta$, the sum can be put in closed form by using the formula (see Problem 6.6)

$$\sum_{k=n_1}^{n_2} a^k = \frac{a^{n_1} - a^{n_2+1}}{1-a}, \qquad a \neq 1 \tag{6.3.7}$$

Assuming that $\alpha\beta^{-1} \neq 1$, we can write

$$y(n) = \beta^n \frac{1 - (\alpha\beta^{-1})^{n+1}}{1 - \alpha\beta^{-1}} = \frac{\alpha^{n+1} - \beta^{n+1}}{\alpha - \beta}$$

As a special case of this example, let $\alpha = 1$, so that $x(n)$ is the unit step. The step response of this system obtained by setting $\alpha = 1$ in the last expression for $y(n)$ is

$$y(n) = \frac{1 - \beta^{n+1}}{1 - \beta}$$

In general, as can be seen by letting $x(n) = u(n)$ in Equation (6.3.3), the step response of a system whose impulse response is $h(n)$ is given by

$$s(n) = \sum_{k=-\infty}^{n} h(k) \tag{6.3.8}$$

For a causal system, this reduces to

$$s(n) = \sum_{k=0}^{n} h(k) \tag{6.3.9}$$

It follows that, given the step response $s(n)$ of a system, we can find the impulse response as

$$h(n) = s(n) - s(n-1) \tag{6.3.10}$$

Example 6.3.3

We want to find the step response of the system with impulse response

$$h(n) = 2\left(\frac{1}{2}\right)^n \cos\left(\frac{2\pi}{3}n\right)u(n)$$

By writing $h(n)$ as

$$h(n) = \left[\left(\frac{1}{2}e^{j\frac{2\pi}{3}}\right)^n + \left(\frac{1}{2}e^{-j\frac{2\pi}{3}}\right)^n\right]u(n)$$

it follows from the last equation in Example 6.3.2 that the step response is equal to

$$s(n) = \left[\frac{1 - \left(\frac{1}{2}e^{j\frac{2\pi}{3}}\right)^{n+1}}{1 - \frac{1}{2}e^{j\frac{2\pi}{3}}} + \frac{1 - \left(\frac{1}{2}e^{-j\frac{2\pi}{3}}\right)^{n+1}}{1 - \frac{1}{2}e^{-j\frac{2\pi}{3}}}\right]u(n)$$

which can be simplified as

$$s(n) = 2 + \frac{2}{\sqrt{3}}\left(\frac{1}{2}\right)^n \sin\left(\frac{2\pi}{3}n\right), \qquad n \geq 0.$$

We can use Equation (6.3.10) to verify that this result is correct by finding the impulse response as

$$h(n) = s(n) - s(n-1)$$
$$= \frac{2}{\sqrt{3}}\left(\frac{1}{2}\right)^n \sin\left(\frac{2\pi}{3}n\right) - \frac{2}{\sqrt{3}}\left(\frac{1}{2}\right)^{n-1} \sin\left(\frac{2\pi}{3}(n-1)\right)$$

which simplifies to the expression for $h(n)$ in the problem statement.

The following examples consider the convolution of two finite-length sequences.

Example 6.3.4

Let $x(n)$ be a finite sequence that is nonzero for $n \in [N_1, N_2]$ and $h(n)$ be a finite sequence that is nonzero for $n \in [N_3, N_4]$. Then for fixed n, $h(n-k)$ is nonzero for $k \in [n - N_4, n - N_3]$, whereas $x(k)$ is nonzero only for $k \in [N_1, N_2]$, so that the product $x(k)h(n-k)$ is zero if $n - N_3 < N_1$ or if $n - N_4 > N_2$. Thus, $y(n)$ is nonzero only for $n \in [N_1 + N_3, N_2 + N_4]$.

Let $M = N_2 - N_1 + 1$ be the length of the sequence $x(n)$ and $N = N_4 - N_3 + 1$ be the length of the sequence $h(n)$. The length of the sequence $y(n)$, which is $(N_2 + N_4) - (N_1 + N_3) + 1$ is thus equal to $M + N - 1$. That is, the convolution of an M-point sequence and an N-point sequence results in an $(M + N - 1)$-point sequence.

Example 6.3.5

Let $h(n) = \{1, 2, 0, -1, 1\}$ and $x(n) = \{1, 3, -1, -2\}$ be two causal sequences. Since $h(n)$ is a five-point sequence and $x(n)$ is a four-point sequence, from the results of Example 6.3.3, $y(n)$ is an eight-point sequence that is zero for $n < 0$ or $n > 7$.

Since both sequences are finite, we can perform the convolution easily by setting up a table of values of $h(k)$ and $x(n-k)$ for the relevant values of n and using

$$y(n) = \sum_{k=0}^{n} h(k)x(n-k)$$

as shown in Table 6.1 The entries for $x(n-k)$ in the table are obtained by first reflecting $x(k)$ about the origin to form $x(-k)$ and successively shifting the resulting sequence by 1 to the right. All entries not explicitly shown are assumed to be zero. The output $y(n)$ is determined by multiplying the entries in the rows corresponding to $h(k)$ and $x(n-k)$ and summing the results. Thus, to find $y(0)$, multiply the entries in rows 2 and 4; for $y(1)$, multiply rows 2 and 5; and so on. The last two columns list n and $y(n)$, respectively.

From the last column in the table, we see that

$$y(n) = \{1, 5, 5, -5, -6, 4, 1, -2\}$$

Example 6.3.6

We can use an alternative tabular form to determine $y(n)$ by noting that

$$y(n) = h(0)x(n) + h(1)x(n-1) + h(2)x(n-2) + \cdots$$
$$+ h(-1)x(n+1) + h(-2)x(n+2) + \cdots$$

TABLE 6.1
Convolution Table for Example 6.3.4.

k	−3	−2	−1	0	1	2	3	4	5	6	7	n	$y(n)$
$h(k)$				1	2	0	−1	1					
$x(k)$				1	3	−1	−2						
$x(-k)$	−2	−1	3	1								0	1
$x(1-k)$		−2	−1	3	1							1	5
$x(2-k)$			−2	−1	3	1						2	5
$x(3-k)$				−2	−1	3	1					3	−5
$x(4-k)$					−2	−1	3	1				4	−6
$x(5-k)$						−2	−1	3	1			5	4
$x(6-k)$							−2	−1	3	1		6	1
$x(7-k)$								−2	−1	3	1	7	−2

We consider the convolution of sequences

$$h(n) = \{-2, \underset{\uparrow}{2}, 0, -1, 1\} \quad \text{and} \quad x(n) = \{-1, \underset{\uparrow}{3}, -1, -2\}$$

The convolution table is shown in Table 6.2. Rows 2 through 5 list $x(n-k)$ for the relevant values of k, namely, $k = -1, 0, 1, 2$, and 3. Values of $h(k)x(n-k)$ are shown in rows 7 through 11, and $y(n)$ for each n is obtained by summing these entries in each column.

TABLE 6.2
Convolution Table for Example 6.3.5.

n	−2	−1	0	1	2	3	4	5
$x(n+1)$	−1	3	−1	−2				
$x(n)$		−1	3	−1	−2			
$x(n-1)$			−1	3	−1	−2		
$x(n-2)$				−1	3	−1	−2	
$x(n-3)$					−1	3	−1	−2
$h(-1)x(n+1)$	2	−6	2	4				
$h(0)x(n)$		−2	6	−2	−4			
$h(1)x(n-1)$			0	0	0	0	0	
$h(2)x(n-2)$				1	−3	1	2	
$h(3)x(n-3)$					−1	3	−1	−2
$y(n)$	2	−8	8	3	−8	4	1	−2

Finally, we note that just as with the convolution integral, the convolution sum defined in Equation (6.3.2) is additive, distributive, and commutative. This enables us to determine the impulse response of series or parallel combinations of systems in terms of their individual impulse responses, as shown in Figure 6.3.2.

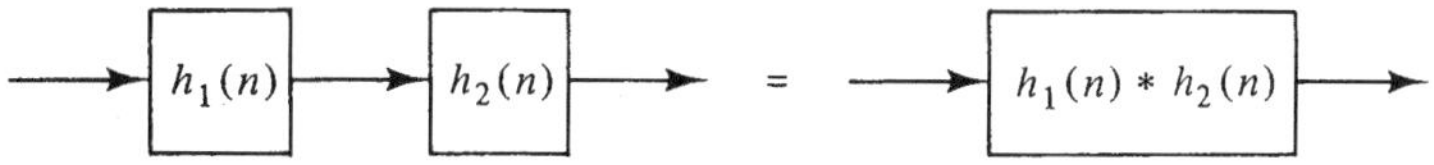

(a)

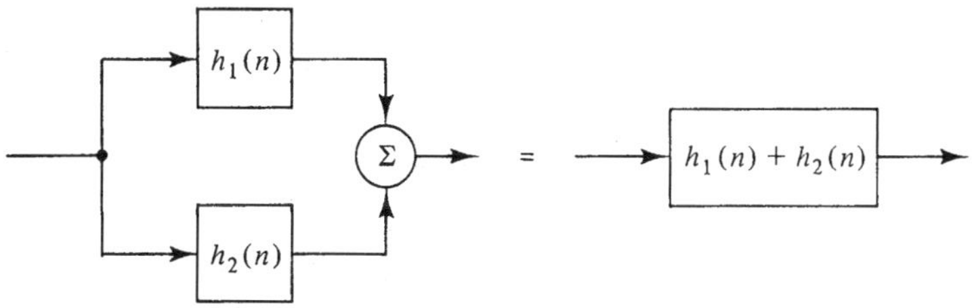

(b)

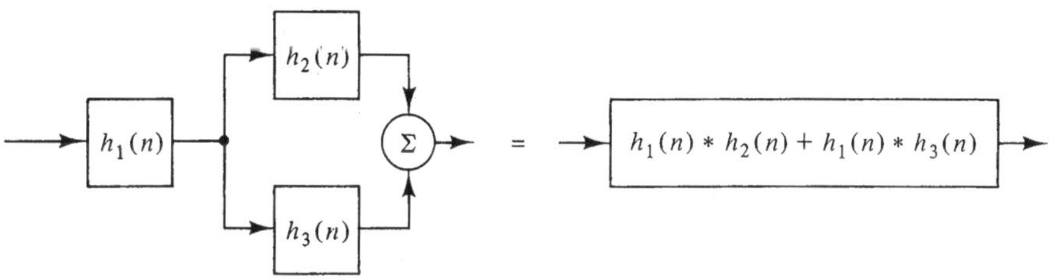

(c)

Figure 6.3.2 Impulse responses of series and parallel combinations.

Example 6.3.7

Consider the system shown in Figure 6.3.3 with

$$h_1(n) = \delta(n) - a\delta(n-1)$$

$$h_2(n) = \left(\frac{1}{2}\right)^n u(n)$$

$$h_3(n) = a^n u(n)$$

$$h_4(n) = (n-1)u(n)$$

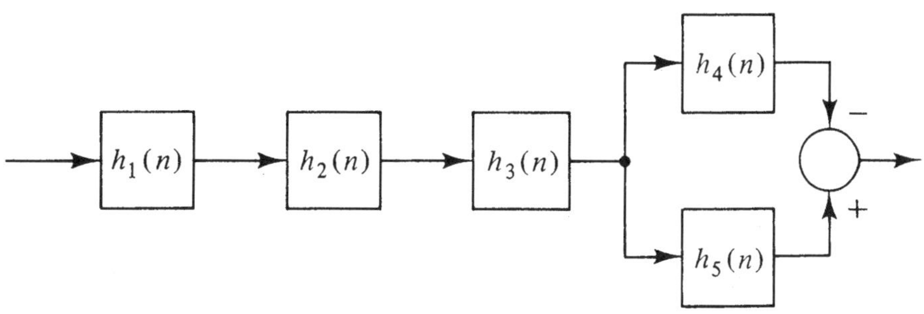

Figure 6.3.3 System for Example 6.3.7.

and

$$h_5(n) = \delta(n) + n\,u(n-1) + \delta(n-2)$$

It is clear from the figure that

$$h(n) = h_1(n) * h_2(n) * h_3(n) * [h_5(n) - h_4(n)]$$

To evaluate $h(n)$, we first form the convolution $h_1(n) * h_3(n)$ as

$$\begin{aligned} h_1(n) * h_3(n) &= [\delta(n) - a\,\delta(n-1)] * a^n u(n) \\ &= a^n u(n) - a^n u(n-1) = \delta(n) \end{aligned}$$

Also,

$$\begin{aligned} h_5(n) - h_4(n) &= \delta(n) + n\,u(n-1) + \delta(n-2) - (n-1)u(n) \\ &= \delta(n) + \delta(n-2) + u(n) \end{aligned}$$

so that

$$\begin{aligned} h(n) &= \delta(n) * h_2(n) * [\delta(n) + \delta(n-2) + u(n)] \\ &= h_2(n) + h_2(n-2) + s_2(n) \end{aligned}$$

where $s_2(n)$ represents the step response corresponding to $h_2(n)$. (See Equation (6.3.9).) We have, therefore,

$$h(n) = \left(\frac{1}{2}\right)^n u(n) + \left(\frac{1}{2}\right)^{n-2} u(n-2) + \sum_{k=0}^{n} \left(\frac{1}{2}\right)^k$$

which can be put in closed form, using Equation (6.3.7), as

$$h(n) = \left(\frac{1}{2}\right)^{n-2} u(n-2) + 2\,u(n)$$

6.4 PERIODIC CONVOLUTION

In certain applications, it is desirable to consider the convolution of two periodic sequences $x_1(n)$ and $x_2(n)$, with common period N. However, the convolution of two periodic sequences in the sense of Equation (6.3.2) or (6.3.3) does not converge. This can be seen by letting $k = rN + m$ in Equation (6.3.2) and rewriting the sum over k as a double sum over r and m:

$$y(n) = \sum_{k=-\infty}^{\infty} x_1(k)x_2(n-k) = \sum_{r=-\infty}^{\infty} \sum_{m=0}^{N-1} x_1(rN+m)x_2(n-rN-m)$$

Since both sequences on the right side are periodic with period N, we have

$$y(n) = \sum_{r=-\infty}^{\infty} \sum_{m=0}^{N-1} x_1(m)x_2(n-m)$$

For a fixed value of n, the inner sum is a constant; thus, the infinite sum on the right does not converge.

In order to get around this problem, as in continuous time, we define a different form of convolution for periodic signals, namely, periodic convolution:

$$y(n) = \sum_{k=0}^{N-1} x_1(k)x_2(n-k) \tag{6.4.1}$$

Note that the sum on the right has only N terms. We denote this operation as

$$y(n) = x_1(n) \circledast x_2(n) \tag{6.4.2}$$

By replacing k by $n - k$ in Equation (6.4.1), we obtain the equivalent form,

$$y(n) = \sum_{k=0}^{N-1} x_1(n-k)x_2(k) \tag{6.4.3}$$

We emphasize that periodic convolution is defined only for sequences with the same period. Recall that, since the convolution of Equation (6.3.2) represents the output of a linear system, it is usual to call it a *linear convolution* in order to distinguish it from the convolution of Equation (6.4.1).

It is clear that $y(n)$ as defined in Equation (6.4.1) is periodic, since

$$y(n+N) = \sum_{k=0}^{N-1} x_2(n+N-k)x_1(k) = y(n) \tag{6.4.4}$$

so that $y(n)$ has to be evaluated only for $0 \le n \le N-1$. It can also be easily verified that the sum can be taken over any one period. (See Problem 6.12). That is,

$$y(n) = \sum_{k=N_0}^{N_0+N-1} x_1(k)x_2(n-k) \tag{6.4.5}$$

The convolution operation of Equation (6.4.1) involves the shifted sequence $x_2(n-k)$, which is obtained from $x_2(n)$ by successive shifts to the right. However, we are interested only in values of n in the range $0 \le n \le N-1$. On each successive shift, the first value in this range is replaced by the value at -1. Since the sequence is periodic, this is the same as the value at $N-1$, as shown in the example in Figure 6.4.1. We can assume, therefore, that on each successive shift, each entry in the sequence moves one place to the right, and the last entry moves into the first place. Such a shift is known as a *periodic*, or *circular*, shift.

From Equation (6.4.1), $y(n)$ can be explicitly written as

$$y(n) = x_1(0)x_2(n) + x_1(1)x_2(n-1) + \cdots + x_1(N-1)x_2(n-N+1)$$

We can use the tabular form of Example 6.3.6 to calculate $y(n)$. However, since the sum is taken only over values of n from 0 to $N-1$, the table has to have only N columns. We present an example to illustrate this.

Example 6.4.1

Consider the convolution of the *periodic extensions* of two sequences:

$$x(n) = \{1, 2, 0, -1\} \qquad \text{and} \qquad h(n) = \{1, 3, -1, -2\}$$

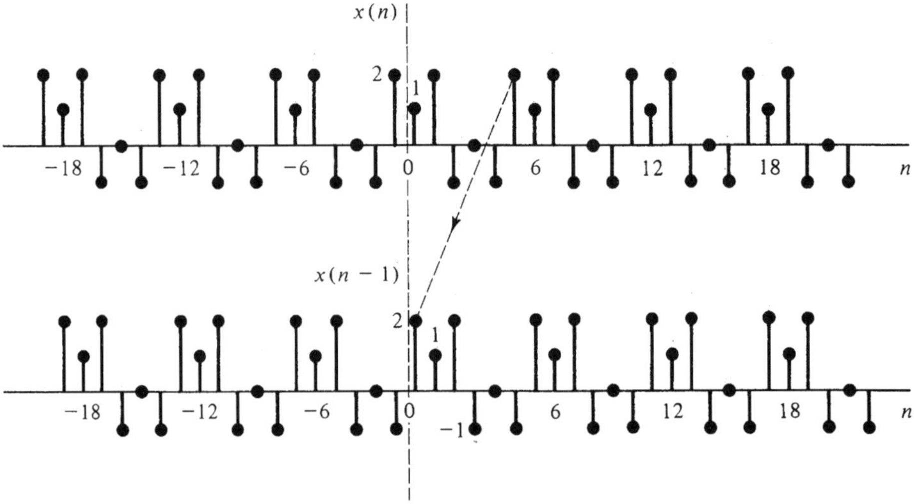

Figure 6.4.1 Shifting of periodic sequences.

It follows that $y(n)$ is periodic with period $N = 4$. The convolution table of Table 6.3 illustrates the steps involved in determining $y(n)$. For $n = 0, 1, 2, 3$, rows 2 through 5 list the values of $x(n-k)$ obtained by circular shifts $x(n)$. Rows 6 through 9 list the values of $h(k)x(n-k)$. The oputput $y(n)$ is determined by summing the entries in each column corresponding to these rows.

TABLE 6.3
Periodic Convolution of Example 6.4.1

n	0	1	2	3
$x(n)$	1	2	0	−1
$x(n-1)$	−1	1	2	0
$x(n-2)$	0	−1	1	2
$x(n-3)$	2	0	−1	1
$h(0)x(n)$	1	2	0	−1
$h(1)x(n-1)$	−3	3	6	0
$h(2)x(n-2)$	0	1	−1	−2
$h(3)x(n-3)$	−4	0	2	−2
$y(n)$	−6	6	7	−5

While we have defined periodic convolution in terms of periodic sequences, given two finite-length sequences, we can define a periodic convolution of the two sequences in a similar manner. Thus, given two N-point sequences $x_1(n)$ and $x_2(n)$, we define their N-point periodic convolution as

$$y_p(n) = \sum_{k=0}^{N-1} x_1(k)x_2(n-k) \tag{6.4.6}$$

where $x_2(n-k)$ denotes that the shift is periodic.

In order to distinguish $y(n)$ discussed in the previous section from $y_p(n)$, $y(n)$ is usually referred to as the *linear* convolution of the sequences $x_1(n)$ and $x_2(n)$, since it corresponds to the output of a linear system driven by an input.

It is clear that $y_p(n)$ in Equation (6.4.6) is the same as the periodic convolution of the periodic extensions of the signals $x_1(n)$ and $x_2(n)$, so that $y_p(n)$ can also be considered periodic with period N. If the two sequences are not of the same length, we can still define their convolution by augmenting the shorter sequence with zeros to make the two sequences the same length. This is known as *zero-padding* or zero-augmentation. Since zero-augmentation of a finite-length sequence does not change the sequence, given two sequences of length N_1 and N_2, we can define their periodic convolution, of arbitrary length M, denoted $[y_p(n)]_M$, provided that $M \geq \text{Max}\{N_1, N_2\}$. We illustrate this in the following example.

Example 6.4.2

Consider the periodic convolution of the sequences $h(n) = \{1, 2, 0, -1, 1\}$ and $x(n) = \{1, 3, -1, -2\}$ of Example 6.3.5. We can find the M-point periodic convolution of the two sequences for $M \geq 5$ by zero-padding the sequences appropriately and following the procedure of Example 6.4.1. Thus, for $M = 5$, we form

$$x_a(n) = \{1, 3, -1, -2, 0\}$$

so that both $h(n)$ and $x_a(n)$ are five points long. It can then easily be verified that

$$[y_p(n)]_5 = \{5, 6, 3, -5, -6\}$$

Comparing this result with $y(n)$ obtained in Example 6.3.4, we note that while the first three values of $y(n)$ and $[y_p(n)]_5$ are different, the next two values are the same. In fact,

$$[y_p(0)]_5 = y(0) + y(6), \; [y_p(1)]_5 = y(1) + y(7), \; [y_p(2)]_5 = y(2) + y(8)$$

It can similarly be verified that the eight-point circular convention of $x(n)$ and $h(n)$ obtained by considering the augmented sequences

$$x_a(n) = \{1, 3, -1, -2, 0, 0, 0, 0\}$$

and

$$h_a(n) = \{1, 2, 0, -1, 1, 0, 0, 0\}$$

is given by

$$[y_p(n)]_8 = \{1, 5, 5, -5, -6, 4, 1, -2\}$$

which is exactly the same as $y(n)$ obtained in Example 6.3.5.

The preceding example shows that the periodic convolution $y_p(n)$ of two finite-length sequences is related to their linear convolution $y_l(n)$. We will explore this relationship further in Section 9.4.

6.5 DIFFERENCE-EQUATION REPRESENTATION OF DISCRETE-TIME SYSTEMS

Earlier, we saw that we can characterize a continuous-time system in terms of a differential equation relating the output and its derivatives to the input and its derivatives. The discrete-time counterpart of this characterization is the difference equation, which, for linear, time-invariant systems, is of the form

$$\sum_{k=0}^{N} a_k y(n-k) = \sum_{k=0}^{M} b_k x(n-k), \qquad n \geq 0 \tag{6.5.1}$$

where a_k and b_k are known constants.

By defining the operator

$$D^k y(n) = y(n-k) \tag{6.5.2}$$

we can write Equation (6.5.1) in operator notation as

$$\sum_{k=0}^{N} a_k D^k y(n) = \sum_{k=0}^{M} b_k D^k x(n) \tag{6.5.3}$$

Note that an alternative form of Equation (6.5.1) is sometimes given as

$$\sum_{k=0}^{N} a_k y(n+k) = \sum_{k=0}^{M} b_k x(n+k), \qquad n \geq 0 \tag{6.5.4}$$

In this form, if the system is causal, we must have $M \leq N$.

The solution to either Equation (6.5.1) or Equation (6.5.4) can be determined, by analogy with the differential equation, as the sum of two components: the homogeneous solution, which depends on the initial conditions that are assumed to be known, and the particular solution, which depends on the input.

Before we explore this approach to finding the solution to Equation (6.5.1), let us consider an alternative approach by rewriting that equation as

$$y(n) = \frac{1}{a_0}\left[\sum_{k=0}^{M} b_k x(n-k) - \sum_{k=1}^{N} a_k y(n-k)\right] \tag{6.5.5}$$

In this equation, $x(n-k)$ are known. If $y(n-k)$ are also known, then $y(n)$ can be determined. Setting $n = 0$ in Equation (6.5.5) yields

$$y(0) = \frac{1}{a_0}\left[\sum_{k=0}^{M} b_k x(-k) - \sum_{k=1}^{N} a_k y(-k)\right] \tag{6.5.6}$$

The quantities $y(-k)$, for $k = 1, 2, \ldots, N$, represent the initial conditions for the difference equation and are therefore assumed to be known. Thus, since all the terms on the right-hand side are known, we can determine $y(0)$.

We now let $n = 1$ in Equation (6.5.5) to get

$$y(1) = \frac{1}{a_0}\left[\sum_{k=0}^{M} b_k x(1-k) - \sum_{k=1}^{N} a_k y(1-k)\right]$$

and use the value of $y(0)$ determined earlier to solve for $y(1)$. This process can be repeated for successive values of n to determine $y(n)$ by iteration.

Using an argument similar to the previous one, we can see that the initial conditions needed to solve Equation (6.5.4) are $y(0), y(1), \ldots, y(N-1)$. Starting with these initial conditions, Equation (6.5.4) can be solved iteratively in a similar manner.

Example 6.5.1

Consider the difference equation

$$y(n) - \frac{3}{4}y(n-1) + \frac{1}{8}y(n-2) = \left(\frac{1}{2}\right)^n, \qquad n \geq 0$$

with

$$y(-1) = 1 \qquad \text{and} \qquad y(-2) = 0$$

Then

$$y(n) = \frac{3}{4}y(n-1) - \frac{1}{8}y(n-2) + \left(\frac{1}{2}\right)^n$$

so that

$$y(0) = \frac{3}{4}y(-1) - \frac{1}{8}y(-2) + 1 = \frac{7}{4}$$

$$y(1) = \frac{3}{4}y(0) - \frac{1}{8}y(-1) + \frac{1}{2} = \frac{27}{16}$$

$$y(2) = \frac{3}{4}y(1) - \frac{1}{8}y(0) + \frac{1}{4} = \frac{83}{64}$$

etc.

Whereas we can use the iterative procedure described before to obtain $y(n)$ for several values of n, the procedure does not, in general, yield an analytical expression for evaluating $y(n)$ for any arbitrary n. The procedure, however, is easily implemented on a digital computer. We now consider the analytical solution of the difference equation by determining the homogeneous and particular solutions of Equation (6.5.1).

6.5.1 Homogeneous Solution of the Difference Equation

The homogeneous equation corresponding to Equation (6.5.1) is

$$\sum_{k=0}^{N} a_k y(n-k) = 0 \tag{6.5.7}$$

By analogy with our discussion of the continuous-time case, we assume that the solution to this equation is given by the exponential function

$$y_h(n) = A\alpha^n$$

Substituting into the difference equation yields

$$\sum_{k=0}^{N} a_k A\alpha^{n-k} = 0$$

Thus, any homogeneous solution must satisfy the algebraic equation

$$\sum_{k=0}^{N} a_k \alpha^{-k} = 0 \tag{6.5.8}$$

Equation (6.5.8) is the characteristic equation for the difference equation, and the values of α that satisfy this equation are the characteristic values. It is clear that there are N characteristic roots $\alpha_1, \alpha_2, \ldots, \alpha_N$, and these roots may or may not be distinct. If they are distinct, the corresponding characteristic solutions are independent, and we can obtain the homogeneous solution $y_h(n)$ as a linear combination of terms of the type α_i^n, so that

$$y_h(n) = A_1\alpha_1^n + A_2\alpha_2^n + \cdots + A_N\alpha_N^n \tag{6.5.9}$$

If any of the roots are repeated, then we generate N independent solutions by multiplying the corresponding characteristic solution by the appropriate power of n. For example, if α_1 has a multiplicity of P_1, while the other $N - P_1$ roots are distinct, we assume a homogeneous solution of the form

$$\begin{aligned} y_h(n) = {} & A_1\alpha_1^n + A_2 n\alpha_1^n + \cdots + A_{P_1} n^{P_1 - 1}\alpha_1^n \\ & + A_{P_1+1}\alpha_{P_1+1}^n + \cdots + A_N\alpha_N^n \end{aligned} \tag{6.5.10}$$

Example 6.5.2

Consider the equation

$$y(n) - \frac{13}{12}y(n-1) + \frac{3}{8}y(n-2) - \frac{1}{24}y(n-3) = 0$$

with

$$y(-1) = 6, \qquad y(-2) = 6 \qquad y(-3) = -2$$

The characteristic equation is

$$1 - \frac{13}{12}\alpha^{-1} + \frac{3}{8}\alpha^{-2} - \frac{1}{24}\alpha^{-3} = 0$$

or

$$\alpha^3 - \frac{13}{12}\alpha^2 + \frac{3}{8}\alpha - \frac{1}{24} = 0$$

which can be factored as

$$\left(\alpha - \frac{1}{2}\right)\left(\alpha - \frac{1}{3}\right)\left(\alpha - \frac{1}{4}\right) = 0$$

so that the characteristic roots are

$$\alpha_1 = \frac{1}{2}, \qquad \alpha_2 = \frac{1}{3}, \qquad \alpha_3 = \frac{1}{4}$$

Since these roots are distinct, the homogeneous solution is of the form

$$y_h(n) = A_1\left(\frac{1}{2}\right)^n + A_2\left(\frac{1}{3}\right)^n + A_3\left(\frac{1}{4}\right)^n$$

Substitution of the initial conditions then gives the following equations for the unknown constants A_1, A_2, and A_3:

$$2A_1 + 3A_2 + 4A_3 = 6$$
$$4A_1 + 9A_2 + 16A_3 = 6$$
$$8A_1 + 27A_2 + 64A_3 = -2$$

The simultaneous solution of these equations yields

$$A_1 = 7, \qquad A_2 = -\frac{10}{3}, \qquad A_3 = \frac{1}{2}$$

The homogeneous solution, therefore, is equal to

$$y_h(n) = 7\left(\frac{1}{2}\right)^n - \frac{10}{3}\left(\frac{1}{3}\right)^n - \frac{1}{2}\left(\frac{1}{4}\right)^n$$

Example 6.5.3

Consider the equation

$$y(n) - \frac{5}{4}y(n-1) + \frac{1}{2}y(n-2) - \frac{1}{16}y(n-3) = 0$$

with the same initial conditions as in the previous example. The characteristic equation is

$$1 - \frac{5}{4}\alpha^{-1} + \frac{1}{2}\alpha^{-2} - \frac{1}{16}\alpha^{-3} = 0$$

with roots

$$\alpha_1 = \frac{1}{2}, \qquad \alpha_2 = \frac{1}{2}, \qquad \alpha_3 = \frac{1}{4}$$

Therefore, we write the homogeneous solution as

$$y_h(n) = A_1\left(\frac{1}{2}\right)^n + A_2 n\left(\frac{1}{2}\right)^n + A_3\left(\frac{1}{4}\right)^n$$

Substituting the initial conditions and solving the resulting equations gives

$$A_1 = \frac{9}{2}, \qquad A_2 = \frac{5}{4}, \qquad A_3 = -\frac{1}{8}$$

so that the homogeneous solution is

$$y_h(n) = \frac{9}{2}\left(\frac{1}{2}\right)^n + \frac{5n}{4}\left(\frac{1}{2}\right)^n - \frac{1}{8}\left(\frac{1}{4}\right)^n$$

6.5.2 The Particular Solution

We now consider the determination of the particular solution for the difference equation

$$\sum_{k=0}^{N} a_k y(n-k) = \sum_{k=0}^{M} b_k x(n-k) \tag{6.5.11}$$

We note that the right side of this equation is the weighted sum of the input $x(n)$ and its delayed versions. Therefore, we can obtain $y_p(n)$, the particular solution to Equation (6.5.11), by first determining $\bar{y}(n)$, the particular solution to the equation

$$\sum_{k=0}^{N} a_k \bar{y}(n-k) = x(n) \tag{6.5.12}$$

Use of the principle of superposition then enables us to write

$$y_p(n) = \sum_{k=0}^{M} b_k \bar{y}(n-k) \tag{6.5.13}$$

To find $\bar{y}(n)$, we assume that it is a linear combination of $x(n)$ and its delayed versions $x(n-1)$, $x(n-2)$, etc. For example, if $x(n)$ is a constant, so is $x(n-k)$ for any k. Therefore, $\bar{y}(n)$ is also a constant. Similarly, if $x(n)$ is an exponential function of the form β^n, $\bar{y}(n)$ is an exponential of the same form. If

$$x(n) = \sin \Omega_0 n$$

then

$$x(n-k) = \sin \Omega_0 (n-k) = \cos \Omega_0 k \sin \Omega_0 n - \sin \Omega_0 k \cos \Omega_0 n$$

Correspondingly, we have

$$\bar{y}(n) = A \sin \Omega_0 n + B \cos \Omega_0 n$$

We get the same form for $\bar{y}(n)$ when

$$x(n) = \cos \Omega_0 n$$

We can determine the unknown constants in the assumed solution by substituting into the difference equation and equating like terms.

As in the solution of differential equations, the assumed form for the particular solution has to be modified by multiplying by an appropriate power of n if the forcing function is of the same form as one of the characteristic solutions.

Example 6.5.4

Consider the difference equation

$$y(n) - \frac{3}{4}y(n-1) + \frac{1}{8}y(n-2) = 2\sin\frac{n\pi}{2}$$

with initial conditions

$$y(-1) = 2 \qquad \text{and} \qquad y(-2) = 4$$

We assume the particular solution to be

$$y_p(n) = A \sin\frac{n\pi}{2} + B \cos\frac{n\pi}{2}$$

Then

$$y_p(n-1) = A \sin\frac{(n-1)\pi}{2} + B \cos\frac{(n-1)\pi}{2}$$

By using trigonometric identities, it can easily be verified that

$$\sin\frac{(n-1)\pi}{2} = -\cos\frac{n\pi}{2} \qquad \text{and} \qquad \cos\frac{(n-1)\pi}{2} = \sin\frac{n\pi}{2}$$

so that

$$y_p(n-1) = -A \cos\frac{n\pi}{2} + B \sin\frac{n\pi}{2}$$

Similarly, $y_p(n-2)$ can be shown to be

$$y_p(n-2) = -A \cos\frac{(n-1)\pi}{2} + B \sin\frac{(n-1)\pi}{2}$$

$$= -A \sin\frac{n\pi}{2} - B \cos\frac{n\pi}{2}$$

Substitution into the difference equation yields

$$\left(A - \frac{3}{4}B - \frac{1}{8}A\right) \sin\frac{n\pi}{2} + \left(B + \frac{3}{4}A - \frac{1}{8}B\right) \cos\frac{n\pi}{2} = 2 \sin\frac{n\pi}{2}$$

Equating like terms gives the following equations for the unknown constants A and B:

$$A - \frac{3}{4}B - \frac{1}{8}A = 2$$

$$B + \frac{3}{4}A - \frac{1}{8}B = 0$$

Solving these equations simultaneously, we obtain

$$A = \frac{112}{85} \qquad \text{and} \qquad B = -\frac{96}{85}$$

so that the particular solution is

$$y_p(n) = \frac{112}{85} \sin\frac{n\pi}{2} - \frac{96}{85} \cos\frac{n\pi}{2}$$

To find the homogeneous solution, we write the characteristic equation for the difference equation as

$$1 - \frac{3}{4}\alpha^{-1} + \frac{1}{8}\alpha^{-2} = 0$$

Since the characteristic roots are

$$\alpha_1 = \frac{1}{4} \quad \text{and} \quad \alpha_2 = \frac{1}{2}$$

the homogeneous solution is

$$y_h(n) = A_1\left(\frac{1}{4}\right)^n + A_2\left(\frac{1}{2}\right)^n$$

so that the total solution is

$$y(n) = A_1\left(\frac{1}{4}\right)^n + A_2\left(\frac{1}{2}\right)^n + \frac{112}{85}\sin\frac{n\pi}{2} - \frac{96}{85}\cos\frac{n\pi}{2}$$

We can now substitute the given initial conditions to solve for the constants A_1 and A_2 as

$$A_1 = -\frac{8}{17} \quad \text{and} \quad A_2 = \frac{13}{5}$$

so that

$$y(n) = -\frac{8}{17}\left(\frac{1}{4}\right)^n + \frac{13}{5}\left(\frac{1}{2}\right)^n + \frac{112}{85}\sin\frac{n\pi}{2} - \frac{96}{85}\cos\frac{n\pi}{2}$$

Example 6.5.5

Consider the difference equation

$$y(n) - \frac{3}{4}y(n-1) + \frac{1}{8}y(n-2) = x(n) + \frac{1}{2}x(n-1)$$

with

$$x(n) = 2\sin\frac{n\pi}{2}$$

From our earlier discussion, we can determine the particular solution for this equation in terms of the particular solution $y_p(n)$ of Example 6.5.4 as

$$y(n) = y_p(n) + \frac{1}{2}y_p(n-1)$$

$$= \frac{112}{85}\sin\frac{n\pi}{2} - \frac{96}{85}\cos\frac{n\pi}{2} + \frac{56}{85}\sin\frac{(n-1)\pi}{2} - \frac{48}{85}\cos\frac{(n-1)\pi}{2}$$

$$= \frac{74}{85}\sin\frac{n\pi}{2} - \frac{152}{85}\cos\frac{n\pi}{2}$$

6.5.3 Determination of the Impulse Response

We conclude this section by considering the determination of the impulse response of systems described by the difference equation, Equation (6.5.1). Recall that the impulse response is the response of the system to a unit sample input with zero initial conditions, so that the impulse response is just the particular solution to the difference equation when the input $x(n)$ is a δ function. We thus consider the equation

$$\sum_{k=0}^{N} a_k y(n-k) = \sum_{k=0}^{M} b_k \delta(n-k) \tag{6.5.14}$$

with $y(-1)$, $y(-2)$, etc., set equal to zero.

Clearly, for $n > M$, the right side of Equation (6.5.14) is zero, so that we have a homogeneous equation. The N initial conditions required to solve this equation are $y(M), y(M-1), \ldots, y(M-N+1)$. Since $N \geq M$ for a causal system, we have to determine only $y(0)$, $y(1)$, $\ldots$, $y(M)$. By successively letting n take on the values $0, 1, 2 \ldots,$ M in Equation (6.5.14) and using the fact that $y(k)$ is zero if $k < 0$, we get the set of $M + 1$ equations;

$$\sum_{k=0}^{j} a_k y(n-k) = b_j, \qquad j = 0, 1, 2, \cdots, M \tag{6.5.15}$$

or equivalently, in matrix form,

$$\begin{bmatrix} a_0 & 0 & \cdot & \cdot & 0 \\ a_1 & a_0 & \cdot & \cdot & 0 \\ a_2 & a_1 & a_0 & \cdot & 0 \\ \cdot & \cdot & \cdot & \cdot & \cdot \\ \cdot & \cdot & \cdot & \cdot & \cdot \\ a_M & a_{M-1} & \cdot & \cdot & a_0 \end{bmatrix} \begin{bmatrix} y(0) \\ y(1) \\ \cdot \\ \cdot \\ \cdot \\ y(M) \end{bmatrix} = \begin{bmatrix} b_0 \\ b_1 \\ \cdot \\ \cdot \\ \cdot \\ b_M \end{bmatrix} \tag{6.5.16}$$

The initial conditions obtained by solving these equations are now used to determine the impulse response as the solution to the homogeneous equation

$$\sum_{k=0}^{N} a_k y(n-k) = 0, \qquad n > M \tag{6.5.17}$$

Example 6.5.6

Consider the system

$$y(n) - \frac{5}{4}y(n-1) + \frac{1}{2}y(n-2) - \frac{1}{16}y(n-3) = x(n) + \frac{1}{3}x(n-1)$$

so that $N = 3$ and $M = 1$. It follows that the impulse response is determined as the solution to the equation

$$y(n) - \frac{5}{4}y(n-1) + \frac{1}{2}y(n-2) - \frac{1}{16}y(n-3) = 0, \qquad n \geq 2$$

and is therefore of the form (see Example 6.5.3)

$$h(n) = A_1\left(\frac{1}{2}\right)^n + A_2 n\left(\frac{1}{2}\right)^n + A_3\left(\frac{1}{4}\right)^n, \qquad n \geq 2$$

The initial conditions needed to determine the constants A_1, A_2, and A_3 are $y(-1)$, $y(0)$, and $y(1)$. By assumption, $y(-1) = 0$. We can determine $y(0)$ and $y(1)$ by using Equation (6.5.16) to get

$$\begin{bmatrix} 1 & 0 \\ -\frac{5}{4} & 1 \end{bmatrix} \begin{bmatrix} y(0) \\ y(1) \end{bmatrix} = \begin{bmatrix} 1 \\ \frac{1}{3} \end{bmatrix}$$

so that $y(0) = 1$ $y(1) = 19/12$. Use of these initial conditions gives the impulse response as

$$h(n) = -\frac{4}{3}\left(\frac{1}{2}\right)^n + \frac{10}{3}n\left(\frac{1}{2}\right)^n + \frac{7}{3}\left(\frac{1}{4}\right)^n, \qquad n \geq 0$$

This is an infinite impulse response as defined in Section 6.3.

Example 6.5.7

Consider the following special case of Equation (6.5.1) in which all the coefficients on the left-hand side are zero except for a_0, which is assumed to be unity:

$$y(n) = \sum_{k=0}^{M} b_k x(n-k) \tag{6.5.18}$$

We let $x(n) = \delta(n)$ and solve for $y(n)$ iteratively to get

$$y(0) = b_0$$
$$y(1) = b_1$$
$$\vdots$$
$$y(M) = b_M$$

Clearly, $y(n) = 0$ for $n > M$, so that

$$h(n) = \{b_0, b_1, b_2, \ldots, b_M\} \tag{6.5.19}$$

This result can be confirmed by comparing Equation (6.5.18) with Equation (6.3.3), which yields $h(k) = b_k$. The impulse response becomes identically zero after M values, so that the system is a finite-impulse-reponse system as defined in Section 6.3.

6.6 SIMULATION DIAGRAMS FOR DISCRETE-TIME SYSTEMS

We can obtain simulation diagrams for discrete-time systems by developing such diagrams in a manner similar to that for continuous-time systems. The simulation diagram in this case is obtained by using summers, coefficient multipliers, and unit delays. The first two are the same as in the continuous-time case, and the unit delay takes the place of the integrator. As in the case of continuous-time systems, we can obtain several different simulation diagrams for the same system. We illustrate this by considering two approaches to obtaining the diagrams, similar to the two approaches we used for continuous-time systems in Chapter 2. In Chapter 8, we explore other methods for deriving simulation diagrams.

Example 6.6.1

We obtain a simulation diagram for the system described by the difference equation

$$y(n) - 0.25\,y(n-1) - 0.25\,y(n-2) + 0.0625\,y(n-3)$$
$$= x(n) + 0.5\,x(n-1) - x(n-2) + 0.25\,x(n-3) \tag{6.6.1}$$

If we now solve for $y(n)$ and group like terms together, we can write

$$y(n) = x(n) + D[0.5\,x(n) + 0.25\,y(n)] + D^2[-x(n) + 0.25\,y(n)]$$
$$+ D^3[0.25\,x(n) - 0.0625\,y(n)]$$

where D represents the unit-delay operator defined in Equation (6.5.2). To obtain the simulation diagram for this system, we assume that $y(n)$ is available and first form the signal

$$v_4(n) = 0.25\,x(n) - 0.0625\,y(n)$$

We pass this signal through a unit delay and add $-x(n) + 0.25\,y(n)$ to form

$$v_3(n) = D[0.25\,x(n) - 0.0625\,y(n)] + [-x(n) + 0.25\,y(n)]$$

We now delay this signal and add $0.5\,x(n) + 0.25\,y(n)$ to it to get

$$v_2(n) = D^2[0.25\,x(n) - 0.0625\,y(n)] + D[-x(n) + 0.25\,y(n)] + [0.5\,x(n) + 0.25\,y(n)]$$

If we now pass $v_2(n)$ through a unit delay and add $x(n)$, we get

$$v_1(n) = D^3[0.25\,x(n) - 0.0625\,y(n)] + D^2[-x(n) + 0.25\,y(n)]$$
$$+ D[0.5\,x(n) + 0.25\,y(n)] + x(n)$$

Clearly, $v_1(n)$ is the same as $y(n)$, so that we can complete the simulation diagram by equating the two expressions. The simulation diagram is shown in Figure 6.6.1.

Consider the general Nth-order difference equation

$$y(n) + a_1 y(n-1) + \dots + a_N y(n-N)$$
$$= b_0 x(n) + b_1 x(n-1) + \dots + b_N x(n-N) \tag{6.6.2}$$

By following the approach given in the last example, we can construct the simulation diagram shown in Figure 6.6.2.

To derive an alternative simulation diagram for the system of Equation (6.6.2), we rewrite the equation in terms of a new variable $v(n)$ as

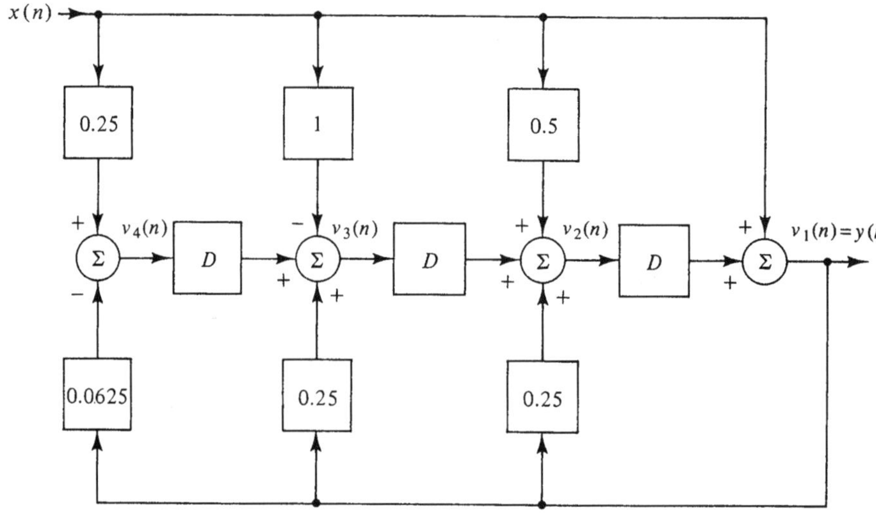

Figure 6.6.1 Simulation diagram for Example 6.6.1.

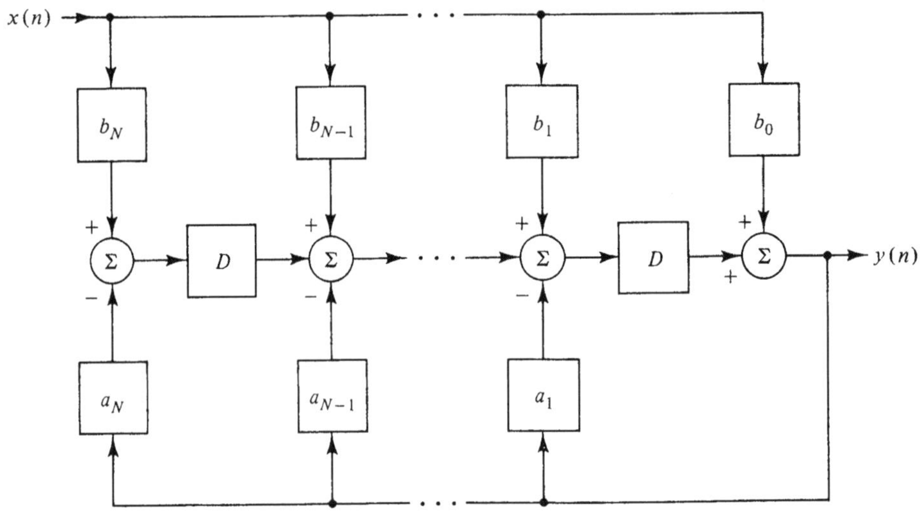

Figure 6.6.2 Simulation diagram for discrete-time system of order N.

$$v(n) + \sum_{j=1}^{N} a_j v(n-j) = x(n) \tag{6.6.3a}$$

$$y(n) = \sum_{m=0}^{N} b_m v(n-m) \tag{6.6.3b}$$

Note that the left side of Equation (6.6.3a) is of the same form as the left side of Equation (6.6.2), and the right side of Equation (6.6.3b) is of the form of the right side of Equation (6.6.2).

To verify that these two equations are equivalent to Equation (6.6.2), we substitute Equation (6.6.3b) into the left side of Equation (6.6.2) to obtain

$$\begin{aligned} y(n) + \sum_{j=1}^{N} a_j y(n-j) &= \sum_{m=0}^{N} b_m v(n-m) + \sum_{j=1}^{N} a_j \left[\sum_{m=0}^{N} b_m v(n-m-j) \right] \\ &= \sum_{m=0}^{N} b_m \left[v(n-m) + \sum_{j=1}^{N} a_j v(n-m-j) \right] \\ &= \sum_{m=0}^{N} b_m x(n-m) \end{aligned}$$

where the last step follows from Equation (6.6.3a).

To generate the simulation diagram, we first determine the diagram for Equation (6.6.3a). If we have $v(n)$ available, we can generate $v(n-1)$, $v(n-2)$, etc., by passing $v(n)$ through successive unit delays. To generate $v(n)$, we note from Equation (6.6.3a) that

$$v(n) = x(n) - \sum_{j=1}^{N} a_j v(n-j) \tag{6.6.4}$$

To complete the simulation diagram, we generate $y(n)$ as in Equation (6.6.3b) by suitably combining $v(n)$, $v(n-1)$, etc., The complete diagram is shown in Figure 6.6.3.

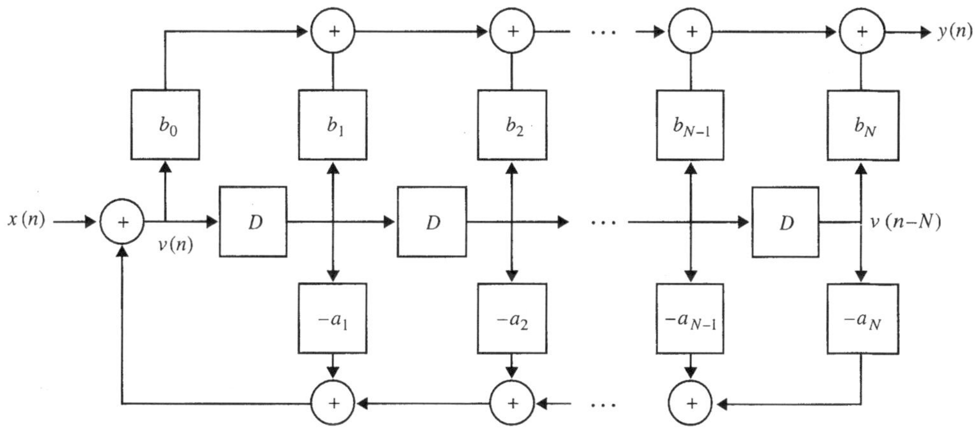

Figure 6.6.3 Alternative simulation diagram for Nth-order system.

Note that both simulation diagrams can be obtained in a straightforward manner from the corresponding difference equations.

Example 6.6.2

The alternative simulation diagram for the system of Equation (6.6.1) is obtained by writing the equation as

$$v(n) - 0.25v(n-1) - 0.25v(n-2) + 0.0625v(n-3) = x(n)$$

and

$$y(n) = v(n) + 0.5v(n-1) - v(n-2) + 0.25v(n-3)$$

Figure 6.6.4 gives the simulation diagram using these two equations.

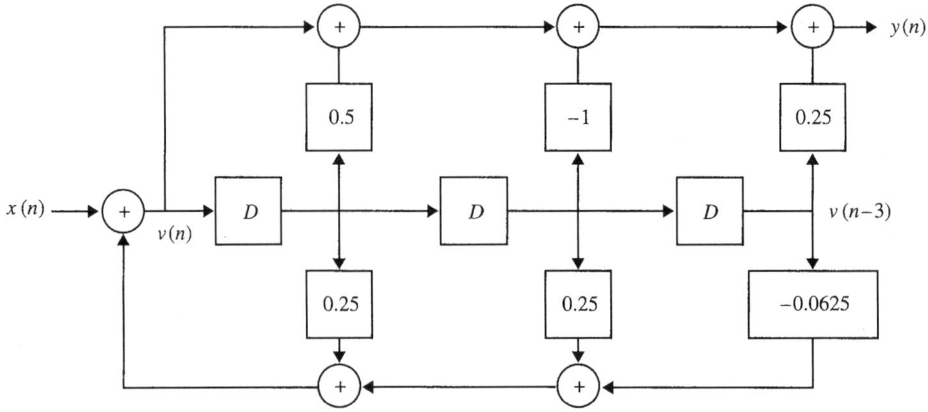

Figure 6.6.4 Alternative simulation diagram for Example 6.6.2.

6.7 STATE-VARIABLE REPRESENTATION OF DISCRETE-TIME SYSTEMS

As with continuous-time systems, the use of state variables permits a more complete description of a system in discrete time. We define the state of a discrete-time system as the minimum amount of information needed to determine the future output states of the system. If we denote the state by the N-dimensional vector

$$\mathbf{v}(n) = [v_1(n)\ v_2(n) \cdots v_N(n)]^T \tag{6.7.1}$$

then the state-space description of a single-input, single-output, time-invariant, discrete-time system with input $x(n)$ and output $y(n)$ is given by the vector-matrix equations

$$\mathbf{v}(n+1) = \mathbf{A}\mathbf{v}(n) + \mathbf{b}\,x(n) \tag{6.7.2a}$$

$$y(n) = \mathbf{c}\mathbf{v}(n) + d\,x(n) \tag{6.7.2b}$$

where $\mathbf{A}$ is an $N \times N$ matrix, $\mathbf{b}$ is an $N \times 1$ column vector, $\mathbf{c}$ is a $1 \times N$ row vector, and d is a scalar.

As with continuous-time systems, in deriving a set of state equations for a system, we can start with a simulation diagram of the system and use the outputs of the delays as the states. We illustrate this in the following example.

Example 6.7.1

Consider the problem of Example 6.6.1, and use the simulation diagrams that we obtained (Figures 6.6.1 and 6.6.4) to derive two state descriptions. For convenience, the two diagrams are repeated in Figures 6.7.1(a) and 6.7.1(b). For our first description, we use the outputs of the delays in Figure 6.7.1(a) as states to get

$$y(n) = v_1(n) + x(n) \tag{6.7.3a}$$

$$\begin{aligned} v_1(n+1) &= v_2(n) + 0.25y(n) + 0.5x(n) \\ &= 0.25v_1(n) + v_2(n) + 0.75x(n) \end{aligned} \tag{6.7.3b}$$

$$\begin{aligned} v_2(n+1) &= v_3(n) + 0.25y(n) - x(n) \\ &= 0.25v_1(n) + v_3(n) - 0.75x(n) \end{aligned} \tag{6.7.3c}$$

$$\begin{aligned} v_3(n+1) &= -0.0625y(n) + 0.25x(n) \\ &= -0.0625v_1(n) + 0.1875x(n) \end{aligned} \tag{6.7.3d}$$

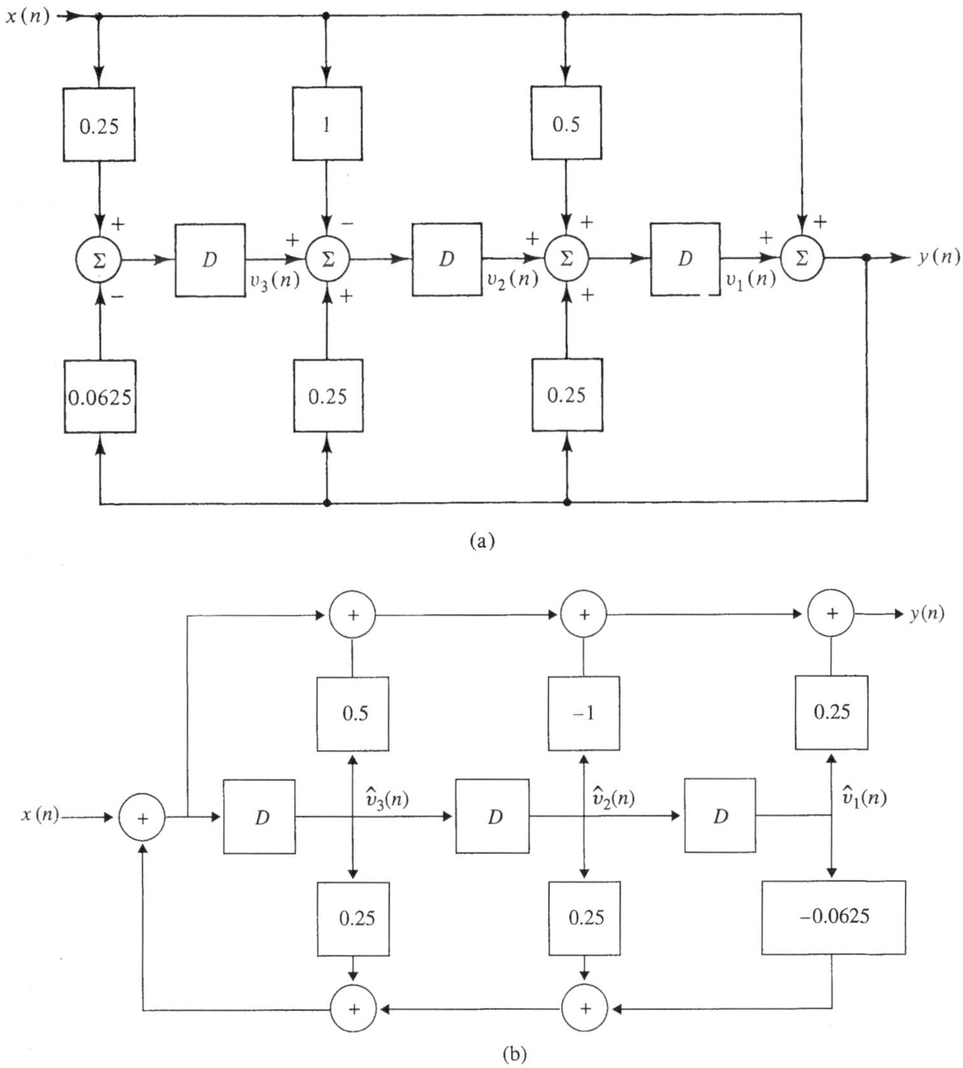

Figure 6.7.1 Simulation diagrams for Example 6.7.1.

In vector-matrix format, these equations can be written as

$$\mathbf{v}(n+1) = \begin{bmatrix} 0.25 & 1 & 0 \\ 0.25 & 0 & 1 \\ -0.0625 & 0 & 0 \end{bmatrix} \mathbf{v}(n) + \begin{bmatrix} 0.75 \\ -0.75 \\ 0.1875 \end{bmatrix} x(n) \tag{6.7.4}$$

$$y(n) = [1 \quad 0 \quad 0]\,\mathbf{v}(n) + x(n)$$

so that

$$\mathbf{A} = \begin{bmatrix} 0.25 & 1 & 0 \\ 0.25 & 0 & 1 \\ -0.0625 & 0 & 0 \end{bmatrix} \quad \mathbf{b} = \begin{bmatrix} 0.75 \\ -0.75 \\ 0.1875 \end{bmatrix}, \qquad \mathbf{c} = [1 \quad 0 \quad 0], \qquad d = 1 \tag{6.7.5}$$

As in continuous time, we refer to this form as the first canonical form. For our second representation, we have, from Figure 6.7.1(b),

$$\hat{v}_1(n+1) = \hat{v}_2(n) \tag{6.7.6a}$$

$$\hat{v}_2(n+1) = \hat{v}_3(n) \tag{6.7.6b}$$

$$\hat{v}_3(n+1) = -0.0625\hat{v}_1(n) + 0.25\hat{v}_2(n) + 0.25\hat{v}_3(n) + x(n) \tag{6.7.6c}$$

$$\begin{aligned} y(n) &= 0.25\hat{v}_1(n) - \hat{v}_2(n) + 0.5\hat{v}_3(n) + \hat{v}_3(n+1) \\ &= -0.1875\hat{v}_1(n) - 0.75\hat{v}_2(n) + 0.75\hat{v}_3(n) + x(n) \end{aligned} \tag{6.7.6d}$$

where the last step follows by substituting Equation (6.7.6c). In vector-matrix format, we can write

$$\hat{\mathbf{v}}(n+1) = \begin{bmatrix} 0 & 1 & 0 \\ 0 & 0 & 1 \\ -0.0625 & 0.25 & 0.25 \end{bmatrix} \hat{\mathbf{v}}(n) + \begin{bmatrix} 0 \\ 0 \\ 1 \end{bmatrix} x(n) \tag{6.7.7}$$

$$y(n) = [-0.1875 \quad -0.75 \quad 0.75]\,\hat{\mathbf{v}}(n) + x(n)$$

so that

$$\hat{\mathbf{A}} = \begin{bmatrix} 0 & 1 & 0 \\ 0 & 0 & 1 \\ -0.0625 & 0.25 & 0.25 \end{bmatrix}, \qquad \mathbf{b} = \begin{bmatrix} 0 \\ 0 \\ 1 \end{bmatrix},$$

$$\mathbf{c} = [-0.1875 \quad -0.75 \quad 0.75], \qquad d = 1 \tag{6.7.8}$$

This is the second canonical form of the state equations.

By generalizing the results of the last example to the system of Equation (6.6.2), we can show that the first form of the state equations yields

$$\mathbf{A} = \begin{bmatrix} -a_1 & 1 & \cdots & 0 \\ -a_2 & 0 & \cdots & 0 \\ \cdot & \cdot & \cdots & \cdot \\ \cdot & \cdot & \cdots & \cdot \\ \cdot & \cdot & \cdots & \cdot \\ -a_N & 0 & \cdots & 0 \end{bmatrix}, \quad \mathbf{b} = \begin{bmatrix} b_1 - a_1 b_0 \\ b_2 - a_2 b_0 \\ \cdot \\ \cdot \\ \cdot \\ b_N - a_N b_0 \end{bmatrix}, \quad \mathbf{c} = \begin{bmatrix} 1 \\ 0 \\ 0 \\ \cdot \\ \cdot \\ 0 \end{bmatrix}^T, \quad d = b_0 \tag{6.7.9}$$

whereas for the second form, we get

$$\hat{\mathbf{A}} = \begin{bmatrix} 0 & 1 & 0 & \cdot & 0 \\ 0 & 0 & 1 & \cdot & 0 \\ \cdot & \cdot & \cdot & \cdot & \cdot \\ 0 & \cdot & \cdot & \cdot & 1 \\ -a_N & -a_{N-1} & \cdot & \cdot & -a_1 \end{bmatrix}, \qquad \hat{\mathbf{b}} = \begin{bmatrix} 0 \\ 0 \\ \cdot \\ \cdot \\ 1 \end{bmatrix},$$

$$\hat{\mathbf{c}} = \begin{bmatrix} b_N - a_N b_0 \\ b_{N-1} - a_{N-1} b_0 \\ \cdot \\ \cdot \\ b_1 - a_1 b_0 \end{bmatrix}^T, \qquad \hat{d} = b_0 \tag{6.7.10}$$

These two forms can be directly obtained by inspection of the difference equation or Figures 6.6.2 and 6.6.3. Let

$$\begin{aligned} \mathbf{v}(n+1) &= \mathbf{A}\mathbf{v}(n) + \mathbf{b}x(n) \\ y(n) &= \mathbf{c}\mathbf{v}(n) + dx(n) \end{aligned} \tag{6.7.11}$$

and

$$\begin{aligned} \hat{\mathbf{v}}(n+1) &= \hat{\mathbf{A}}\hat{\mathbf{v}}(n) + \hat{\mathbf{b}}x(n) \\ y(n) &= \hat{\mathbf{c}}\hat{\mathbf{v}}(n) + \hat{d}x(n) \end{aligned} \tag{6.7.12}$$

be two alternative state-space descriptions of a system. Then there exists a nonsingular matrix $\mathbf{P}$ of dimension $N \times N$ such that

$$\mathbf{v}(n) = \mathbf{P}\hat{\mathbf{v}}(n) \tag{6.7.13}$$

It can easily be verified that the following relations hold:

$$\hat{\mathbf{A}} = \mathbf{P}\mathbf{A}\mathbf{P}^{-1}, \qquad \hat{\mathbf{b}} = \mathbf{P}^{-1}\mathbf{b}, \qquad \hat{\mathbf{c}} = \mathbf{c}\mathbf{P}, \qquad \hat{d} = d \tag{6.7.14}$$

6.7.1 Solution of State-Space Equations

We can find the solution of the equation

$$\mathbf{v}(n+1) = \mathbf{A}\mathbf{v}(n) + \mathbf{b}x(n), \qquad n \geq 0; \qquad \mathbf{v}(0) = \mathbf{v}_0 \tag{6.7.15}$$

by iteration. Thus, setting $n = 0$ in Equation (6.7.15) gives

$$\mathbf{v}(1) = \mathbf{A}\mathbf{v}(0) + \mathbf{b}x(0)$$

For $n = 1$, we have

$$\begin{aligned} \mathbf{v}(2) &= \mathbf{A}\mathbf{v}(1) + \mathbf{b}x(1) \\ &= \mathbf{A}[\mathbf{A}\mathbf{v}(0) + \mathbf{b}x(0)] + \mathbf{b}x(1) \\ &= \mathbf{A}^2\mathbf{v}(0) + \mathbf{A}\mathbf{b}x(0) + \mathbf{b}x(1) \end{aligned}$$

which can be written as

$$\mathbf{v}(2) = \mathbf{A}^2\mathbf{v}(0) + \sum_{j=0}^{1} \mathbf{A}^{2-j-1}\mathbf{b}x(j)$$

By continuing this procedure, it is clear that the solution for general n is

$$\mathbf{v}(n) = \mathbf{A}^n\mathbf{v}(0) + \sum_{j=0}^{n-1} \mathbf{A}^{n-j-1}\mathbf{b}x(j) \tag{6.7.16}$$

The first term in the solution corresponds to the initial-condition response, and the second term, which is the convolution sum of $\mathbf{A}^{n-1}$ and $\mathbf{b}x(n)$, corresponds to the forced response of the system. The quantity $\mathbf{A}^n$, which defines how the state changes as time progresses, represents the state-transition matrix for the discrete-time system $\boldsymbol{\Phi}(n)$. In terms of $\boldsymbol{\Phi}(n)$, Equation (6.7.16) can be written as

$$\mathbf{v}(n) = \boldsymbol{\Phi}(n)\,\mathbf{v}(0) + \sum_{j=0}^{n-1} \boldsymbol{\Phi}(n-j-1)\,\mathbf{b}x(j) \tag{6.7.17}$$

Clearly, the first step in obtaining the solution of the state equations is the determination of $\mathbf{A}^n$. We can use the Cayley-Hamilton theorem for this purpose.

Example 6.7.2

Consider the system

$$v_1(n+1) = v_2(n) \tag{6.7.18}$$

$$v_2(n+1) = \frac{1}{8}v_1(n) - \frac{1}{4}v_2(n) + x(n)$$

$$y(n) = v_1(n)$$

By using the Cayley-Hamilton theorem as in Chapter 2, we can write

$$\mathbf{A}^n = \alpha_0(n)\mathbf{I} + \alpha_1(n)\mathbf{A} \tag{6.7.19}$$

Replacing $\mathbf{A}$ in Equation (6.7.19) by its eigenvalues, $-\frac{1}{2}$ and $\frac{1}{4}$, leads to the equations

$$\alpha_0(n) - \frac{1}{2}\alpha_1(n) = \left(-\frac{1}{2}\right)^n$$

and

$$\alpha_1(n) + \frac{1}{4}\alpha_1(n) = \left(\frac{1}{4}\right)^n$$

so that

$$\alpha_0(n) = \frac{2}{3}\left(\frac{1}{4}\right)^n + \frac{1}{3}\left(-\frac{1}{2}\right)^n$$

$$\alpha_1(n) = \frac{4}{3}\left(\frac{1}{4}\right)^n - \frac{4}{3}\left(-\frac{1}{2}\right)^n$$

Substituting into Equation (6.7.16) gives

$$\mathbf{A}^n = \begin{bmatrix} \frac{2}{3}\left(\frac{1}{4}\right)^n + \frac{1}{3}\left(-\frac{1}{2}\right)^n & \frac{4}{3}\left(\frac{1}{4}\right)^n - \frac{4}{3}\left(-\frac{1}{2}\right)^n \\ \frac{1}{6}\left(\frac{1}{4}\right)^n - \frac{1}{6}\left(-\frac{1}{2}\right)^n & \frac{1}{3}\left(\frac{1}{4}\right)^n + \frac{2}{3}\left(-\frac{1}{2}\right)^n \end{bmatrix} \tag{6.7.20}$$

Example 6.7.3

Let us determine the unit-step response of the system of Example 6.7.2 for the case when $\mathbf{v}(0) = [1 \ \ -1]^T$. Substituting into Equation (6.7.16) gives

$$\mathbf{v}(n) = \mathbf{A}^n \begin{bmatrix} 1 \\ -1 \end{bmatrix} + \sum_{j=0}^{n-1} \mathbf{A}^{n-j-1} \begin{bmatrix} 0 \\ 1 \end{bmatrix} (1)$$

$$= \begin{bmatrix} -\frac{2}{3}\left(\frac{1}{4}\right)^n + \frac{5}{3}\left(-\frac{1}{2}\right)^n \\ -\frac{1}{6}\left(\frac{1}{4}\right)^n - \frac{5}{6}\left(-\frac{1}{2}\right)^n \end{bmatrix} + \sum_{j=0}^{n-1} \begin{bmatrix} \frac{4}{3}\left(\frac{1}{4}\right)^{n-j-1} - \frac{4}{3}\left(-\frac{1}{2}\right)^{n-j-1} \\ \frac{1}{3}\left(\frac{1}{4}\right)^{n-j-1} + \frac{2}{3}\left(-\frac{1}{2}\right)^{n-j-1} \end{bmatrix}$$

Putting the second term in closed form yields

$$\mathbf{v}(n) = \begin{bmatrix} -\frac{2}{3}\left(\frac{1}{4}\right)^n + \frac{5}{3}\left(-\frac{1}{2}\right)^n \\ -\frac{1}{6}\left(\frac{1}{4}\right)^n - \frac{5}{6}\left(-\frac{1}{2}\right)^n \end{bmatrix} + \begin{bmatrix} \frac{8}{9} + \frac{8}{9}\left(-\frac{1}{2}\right)^n - \frac{16}{9}\left(\frac{1}{4}\right)^n \\ \frac{8}{9} - \frac{4}{9}\left(-\frac{1}{2}\right)^n - \frac{4}{9}\left(\frac{1}{4}\right)^n \end{bmatrix}$$

The first term corresponds to the initial-condition response and the second term to the forced response. Combining the two terms yields the total response:

$$\mathbf{v}(n) = \begin{bmatrix} v_1(n) \\ v_2(n) \end{bmatrix} = \begin{bmatrix} \frac{8}{9} + \frac{23}{9}\left(-\frac{1}{2}\right)^n - \frac{22}{9}\left(\frac{1}{4}\right)^n \\ \frac{8}{9} - \frac{23}{18}\left(-\frac{1}{2}\right)^n - \frac{11}{18}\left(\frac{1}{4}\right)^n \end{bmatrix}, \qquad n \geq 0 \tag{6.7.21}$$

The output is given by

$$y(n) = v_1(n) = \frac{8}{9} + \frac{23}{9}\left(-\frac{1}{2}\right)^n - \frac{22}{9}\left(\frac{1}{4}\right)^n, \qquad n \geq 0 \tag{6.7.22}$$

We conclude this section with a brief summary of the properties of the state-transition matrix. These properties, which are easily verified, are somewhat similar to the corresponding ones in continuous time:

1.

$$\mathbf{\Phi}(n+1) = \mathbf{A}\mathbf{\Phi}(n) \tag{6.7.23a}$$

2.

$$\mathbf{\Phi}(0) = \mathbf{I} \tag{6.7.23b}$$

3. Transition property:

$$\mathbf{\Phi}(n-k) = \mathbf{\Phi}(n-j)\mathbf{\Phi}(j-k) \tag{6.7.23c}$$

4. Inversion property:

$$\mathbf{\Phi}^{-1}(n) = \mathbf{\Phi}(-n) \text{ if the inverse exists.} \tag{6.7.23d}$$

Note that unlike the situation in continuous time, the transition matrix can be singular and hence noninvertible. Clearly, $\mathbf{\Phi}^{-1}(n)$ is nonsingular (invertible) only if $\mathbf{A}$ is nonsingular.

6.7.2 Impulse Response of Systems Described by State Equations

We can find the impulse response of the system described by Equation (6.7.2) by setting $\mathbf{v}_0 = 0$ and $x(n) = \delta(n)$ in the solution to the state equation, Equation (6.7.16), to get

$$\mathbf{v}(n) = \mathbf{A}^{n-1}\mathbf{b} \tag{6.7.24}$$

The impulse response is then obtained from Equation (6.7.2b) as

$$h(n) = \mathbf{c}\mathbf{A}^{n-1}\,\mathbf{b} + d\,\delta(n) \tag{6.7.25}$$

Example 6.7.4

The impulse response of the system of Example 6.7.2 easily follows from our previous results as

$$h(n) = [1 \quad 0]\begin{bmatrix} \frac{2}{3}\left(\frac{1}{4}\right)^{n-1} - \frac{1}{3}\left(-\frac{1}{2}\right)^{n-1} & \frac{4}{3}\left(\frac{1}{4}\right)^{n-1} - \frac{4}{3}\left(-\frac{1}{2}\right)^{n-1} \\ \frac{1}{6}\left(\frac{1}{4}\right)^{n-1} - \frac{1}{6}\left(-\frac{1}{2}\right)^{n-1} & \frac{1}{3}\left(\frac{1}{4}\right)^{n-1} + \frac{2}{3}\left(-\frac{1}{2}\right)^{n-1} \end{bmatrix}\begin{bmatrix} 0 \\ 1 \end{bmatrix}$$

$$= \frac{4}{3}\left(\frac{1}{4}\right)^{n-1} - \frac{4}{3}\left(-\frac{1}{2}\right)^{n-1}, \qquad n \geq 0 \tag{6.7.26}$$

6.8 STABILITY OF DISCRETE-TIME SYSTEMS

As with continuous-time systems, an important property associated with discrete-time systems is system stability. We can extend our definition of stability to the discrete-time case by saying that a discrete-time system is input/output stable if a bounded input produces a bounded output. That is, if

$$|x(n)| \leq M < \infty \tag{6.8.1}$$

then

$$|y(n)| \leq L < \infty$$

By using a similar procedure as in the case of continuous-time systems, we can obtain a condition for stability in terms of the system impulse response. Given a system with impulse response $h(n)$, let $x(n)$ be such that $|x(n)| \leq M$. Then the output $y(n)$ is given by the convolution sum:

$$y(n) = \sum_{k=-\infty}^{\infty} h(k)\, x(n-k) \tag{6.8.2}$$

so that

$$\begin{aligned} |y(n)| &= \Big|\sum_{k=-\infty}^{\infty} h(k)\, x(n-k)\Big| \\ &\leq \sum_{k=-\infty}^{\infty} |h(k)|\, |x(n-k)| \\ &\leq M \sum_{k=-\infty}^{\infty} |h(k)| \end{aligned}$$

Thus, a sufficient condition for the system to be stable is that the impulse response must be absolutely summable; that is,

$$\sum_{k=-\infty}^{\infty} |h(k)| < \infty \tag{6.8.3}$$

That it is also a necessary condition can be seen by considering as input the bounded signal $x(k) = \text{sgn}[h(n-k)]$, or equivalently, $x(n-k) = \text{sgn}[h(k)]$, with corresponding output

$$y(n) = \sum_{k=-\infty}^{\infty} h(k)\, \text{sgn}[h(k)] = \sum_{k=-\infty}^{\infty} |h(k)|$$

Clearly, if $h(n)$ is not absolutely summable, $y(n)$ will be unbounded.

For causal systems, the condition for stability becomes

$$\sum_{k=0}^{\infty} |h(k)| < \infty \tag{6.8.4}$$

We can obtain equivalent conditions in terms of the locations of the characteristic values of the system. Recall that for a causal system described by a difference equation, the solution consists of terms of the form $n^k \alpha^n$, $k = 0, 1, \ldots, M$, where α denotes a characteristic value of multiplicity M. It is clear that if $|\alpha| \geq 1$, the response is not bounded for all inputs. Thus, for a system to be stable, all the characteristic values must have magnitude less than 1. That is, they must all lie inside a circle of unity radius in the complex plane.

For the state-variable representation, we saw that the solution depends on the state-transition matrix $\mathbf{A}^n$. The form of $\mathbf{A}^n$ is determined by the eigenvalues or characteristic values of the matrix $\mathbf{A}$. Suppose we obtain the difference equation relating the output $y(n)$ to the input $x(n)$ by eliminating the state variables from Equations (6.7.2a) and (6.7.2b). It can be verified that the characteristic values of this equation are exactly the same as those of the matrix **A**. (We leave the proof of this relation as an exercise for the reader; see Problem 6.31.) It follows, therefore, that a system described by state equations is stable if the eigenvalues of **A** lie inside the unit circle in the complex plane.

Example 6.8.1

Determine if the following causal, time-invariant systems are stable:

(i) System with impulse response

$$h(n) = \left[n^2\left(-\frac{1}{2}\right)^n + 2\left(\frac{1}{4}\right)^n\right]u(n)$$

(ii) System described by the difference equation

$$y(n) - \frac{11}{6}y(n-1) - \frac{1}{2}y(n-2) + \frac{1}{3}y(n-3) = x(n) + 2x(n-2)$$

(iii) System described by the state equations

$$\mathbf{v}(n+1) = \begin{bmatrix} \frac{3}{4} & -\frac{1}{4} \\ \frac{13}{4} & \frac{9}{4} \end{bmatrix}\mathbf{v}(n) + \begin{bmatrix} \frac{1}{2} \\ \frac{2}{5} \end{bmatrix}\mathbf{x}(n)$$

$$y(n) = \begin{bmatrix} 1 & -\frac{2}{3} \end{bmatrix}\mathbf{v}(n)$$

For the first system, we have

$$\sum_{n=-\infty}^{\infty} |h(n)| = \sum_{n=0}^{\infty} n^2\left(\frac{1}{2}\right)^n + 2\left(\frac{1}{4}\right)^n = 6 + \frac{8}{3}$$

so that the systems is stable.

For the second system, the characteristic equation is

$$\alpha^3 - \frac{11}{6}\alpha^2 - \frac{1}{2}\alpha + \frac{1}{3} = 0$$

and the characteristic roots are $\alpha_1 = 2$, $\alpha_2 = -1/2$ and $\alpha_3 = 1/3$. Since $|\alpha_1| > 1$, this system is unstable.

It can easily be verified that the eigenvalues of the **A** matrix in the last system are equal to $3/2 \pm j\,1/2$. Since both have a magnitude greater than 1, it follows that the system is unstable.

6.9 SUMMARY

- A discrete-time (DT) signal is defined only at discrete instants of time.
- A DT signal is usually represented as a sequence of values $x(n)$ for integer values of n.
- A DT signal $x(n)$ is periodic with period N if $x(n + N) = x(n)$ for some integer N.

- The DT unit-step and impulse functions are related as

$$u(n) = \sum_{k=-\infty}^{n} \delta(k)$$

$$\delta(n) = u(n) - u(n-1)$$

- Any DT signal $x(n)$ can be expressed in terms of shifted impulse functions as

$$x(n) = \sum_{k=-\infty}^{\infty} x(k)\,\delta(n-k)$$

- The complex exponential $x(n) = \exp[j\Omega_0 n]$ is periodic only if $\Omega_0/2\pi$ is a rational number.
- The set of harmonic signals $x_k(n) = \exp[jk\Omega_0 n]$ consists of only N distinct waveforms.
- Time scaling of DT signals may yield a signal that is completely different from the original signal.
- Concepts such as linearity, memory, time invariance, and causality in DT systems are similar to those in continuous-time (CT) systems.
- A DT LTI system is completely characterized by its impulse response.
- The output $y(n)$ of an LTI DT system is obtained as the convolution of the input $x(n)$ and the system impulse response $h(n)$:

$$y(n) = h(n) * x(n) = \sum_{m=-\infty}^{\infty} h(m)x(n-m)$$

- The convolution sum gives only the forced response of the system.
- The periodic convolution of two periodic sequences $x_1(n)$ and $x_2(n)$ is

$$x_1(n) \circledast x_2(n) = \sum_{k=0}^{N-1} x_1(n-k)\,x_2(k)$$

- An alternative representation of a DT system is in terms of the difference equation (DE)

$$\sum_{k=0}^{N} a_k y(n-k) = \sum_{k=0}^{M} b_k x(n-k), \qquad n \geq 0$$

- The DE can be solved either analytically or by iterating from known initial conditions. The analytical solution consists of two parts: the homogeneous (zero-input) solution and the particular (zero-state) solution. The homogeneous solution is determined by the roots of the characteristic equation. The particular solution is of the same form as the input $x(n)$ and its delayed versions.
- The impulse response is obtained by solving the system DE with input $x(n) = \delta(n)$ and all initial conditions zero.
- The simulation diagram for a DT system can be obtained from the DE using summers, coefficient multipliers, and delays as building blocks.
- The state equations for an LTI DT system can be obtained from the simulation diagram by assigning a state to the output of each delay. The equations are of the form

$$\mathbf{v}(n+1) = \mathbf{A}\mathbf{v}(n) + \mathbf{b}x(n)$$
$$y(n) = \mathbf{c}\mathbf{v}(n) + dx(n)$$

- As in the CT case, for a given DT system, we can obtain several equivalent simulation diagrams and, hence, several equivalent state representations.
- The solution of the state equation is

$$\mathbf{v}(n) = \mathbf{\Phi}(n)\,\mathbf{v}(n) + \sum_{j=0}^{n-1} \mathbf{\Phi}(n-j-1)\,\mathbf{b}x(j)$$
$$y(n) = \mathbf{c}\mathbf{v}(n) + dx(n)$$

where

$$\mathbf{\Phi}(n) = \mathbf{A}^n$$

is the state-transition matrix and can be evaluated using the Cayley-Hamilton theorem.

- The following conditions for the BIBO stability of a DT LTI system are equivalent:
  (a) $\sum_{k=-\infty}^{\infty} |h(k)| < \infty$
  (b) The roots of the characteristic equation are inside the unit circle.
  (c) The eigenvalues of $\mathbf{A}$ are inside the unit circle.

6.10 CHECKLIST OF IMPORTANT TERMS

Cayley-Hamilton theorem
Characteristic equation
Coefficient multiplier
Complex exponential
Convolution sum
Delay
Difference equation
Discrete-time signal
Homogeneous solution
Impulse response
Particular solution
Periodic convolution
Periodic signal
Simulation diagram
State equations
State variables
Summer
Transition matrix
Unit-impulse function
Unit-step function

6.11 PROBLEMS

6.1. For the discrete-time signal shown in Figure P6.1, sketch each of the following:
(a) $x(2-n)$
(b) $x(3n-4)$
(c) $x(\frac{2}{3}n+1)$
(d) $x\left(-\frac{n+8}{4}\right)$
(e) $x(n^3)$

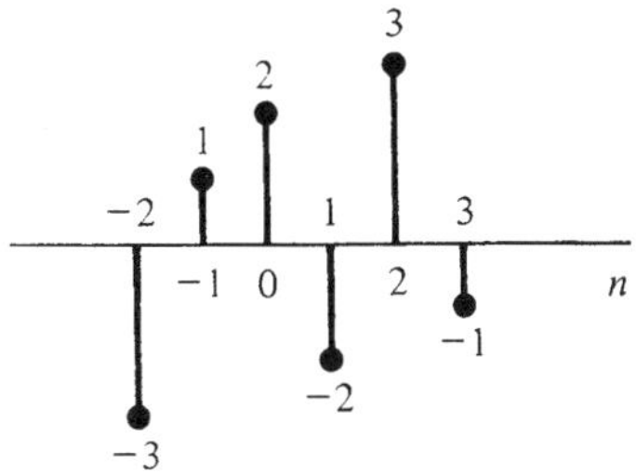

Figure P6.1 $x(n)$ for Problem 6.1.

(f) $x_e(n)$

(g) $x_0(n)$

(h) $x(2-n) + x(3n-4)$

6.2. Repeat Problem 6.1 if

$$x(n) = \left\{-1, \underset{\uparrow}{\frac{1}{3}}, 0, 1, -\frac{3}{2}, -1\right\}$$

6.3. Determine whether each of the following signals is periodic, and if it is, find its period.

(a) $x(n) = \sin\left(\frac{\pi n}{4} + \frac{\pi}{8}\right)$

(b) $x(n) = \sin\left(\frac{3\pi n}{4}\right) + \sin\left(\frac{\pi}{3} n\right)$

(c) $x(n) = \sin\left(\frac{3\pi n}{4}\right) \sin\left(\frac{\pi}{3} n\right)$

(d) $x(n) = \exp\left[\frac{6\pi}{5} n\right]$

(e) $x(n) = \exp\left[j\frac{5\pi}{6} n\right]$

(f) $x(n) = \sum_{n=-\infty}^{\infty} [\delta(n-2m) + 2\delta(n-3m)]$

(g) $x(n) = \sin\left(\frac{3\pi n}{4}\right) + \cos\left(\frac{\pi}{3} n\right)$

6.4. The signal $x(t) = 5\cos(120t - \pi/3)$ is sampled to yield uniformly spaced samples T seconds apart. What values of T cause the resulting discrete-time sequence to be periodic? What is the period?

6.5. Repeat Problem 6.4 if $x(t) = 3\sin 100\pi t + 4\cos 120t$.

6.6. The following equalities are used several places in the text. Prove their validity.

(a) $$\sum_{n=0}^{N-1} \alpha^n = \begin{cases} \dfrac{1-\alpha^N}{1-\alpha} & \alpha \neq 1 \\ N & \alpha = 1 \end{cases}$$

(b) $$\sum_{n=0}^{\infty} \alpha^n = \frac{1}{1-\alpha}, \qquad |\alpha| < 1$$

(c) $\sum_{n=n_0}^{n_f} \alpha^n = \frac{\alpha^{n_0} - \alpha^{n_f+1}}{1-\alpha}, \qquad \alpha \neq 1$

6.7. Such concepts from continuous-time systems as memory, time invariance, linearity, and causality carry over to discrete-time systems. In the following, $x(n)$ refers to the input to a system and $y(n)$ refers to the output. Determine whether the systems are (i) linear, (ii)memoryless, (iii) shift invariant, and (iv) causal. Justify your answer in each case.

(a) $y(n) = \log[x(n)]$

(b) $y(n) = x(n)x(n-2)$

(c) $y(n) = 3nx(n)$

(d) $y(n) = nx(n) + 3$

(e) $y(n) = x(n-1)$

(f) $y(n) = x(n) + 2x(n-1)$

(g) $y(n) = \sum_{k=0}^{\infty} x(k)$

(h) $y(n) = \sum_{k=0}^{n} x(n)$

(i) $y(n) = \frac{1}{N}\sum_{k=0}^{N-1} x(n-k)$

(j) $y(n) = \frac{1}{2N+1}\sum_{k=-N}^{N} x(n-k)$

(k) $y(n) = \text{median}\{x(n-1), x(n), x(n+1)\}$

(l) $y(n) = \begin{bmatrix} x(n), & n \geq 0 \\ -x(n), & n < 0 \end{bmatrix}$

(m) $y(n) = \begin{bmatrix} x(n), & x(n) \geq 0 \\ -x(n), & x(n) < 0 \end{bmatrix}$

6.8. **(a)** Find the convolution $y(n) = h(n) * x(n)$ of the following signals:

(i) $x(n) = \begin{cases} -1 & -5 \leq n \leq -1 \\ 1 & 0 \leq n \leq 4 \end{cases}$

$h(n) = 2u(n)$

(ii) $x(n) = \left(\frac{1}{2}\right)^n u(n)$

$h(n) = \delta(n) + \delta(n-1) + \left(\frac{1}{3}\right)^n u(n)$

(iii) $x(n) = u(n)$

$h(n) = 1 \qquad 0 \leq n \leq 9$

(iv) $x(n) = \left(\frac{1}{3}\right)^n u(n)$

$h(n) = \delta(n) + \left(\frac{1}{3}\right)^n u(n)$

(v) $x(n) = \left(\frac{1}{3}\right)^n u(n)$

$h(n) = \delta(n) + \left(\frac{1}{2}\right)^n u(n)$

(vi) $x(n) = nu(n)$

$h(n) = u(n) - u(n - 10)$

(b) Use any mathematical software package to verify your results.

(c) Plot $y(n)$ vs n.

6.9. Find the convolution $y(n) = h(n) * x(n)$ for each of the following pairs of finite sequences:

(a) $x(n) = \left\{1, -\frac{1}{2}, \underset{\uparrow}{\frac{1}{4}}, -\frac{1}{8}, \frac{1}{16}\right\}, \qquad h(n) = \{1, -1, 1, -1\}$

(b) $x(n) = \{1, 2, 3, 0, -1,\}, \qquad h(n) = \{2, -1, 3, 1, -2\}$

(c) $x(n) = \left\{3, \frac{1}{2}, \underset{\uparrow}{-\frac{1}{4}}, 1, 4\right\}, \qquad h(n) = \left\{2, -1, \frac{1}{2}, -\frac{1}{2}\right\}$

(d) $x(n) = \left\{-1, \frac{1}{2}, \frac{3}{4}, -\frac{1}{5}, 1\right\}, \qquad h(n) = \{1, 1, 1, 1, 1\}$

(e) Verify your results using any mathematical software package.

(f) Plot the resulting $y(n)$.

6.10. **(a)** Find the impulse response of the system shown in Figure P6.10. Assume that

$$h_1(n) = h_2(n) = \left(\frac{1}{3}\right)^n u(n)$$

$$h_3(n) = u(n)$$

$$h_4(n) = \left(\frac{1}{2}\right)^n u(n)$$

(b) Find the response of the system to a unit-step input.

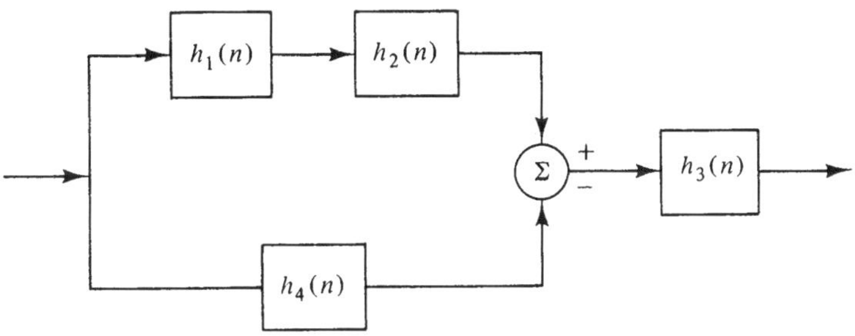

Figure P6.10 System for Problem 6.10.

6.11. **(a)** Repeat Problem 6.10 if

$$h_1(n) = \left(\frac{1}{2}\right)^n u(n)$$

$$h_2(n) = \delta(n)$$

$$h_3(n) = h_4(n) = \left(\frac{1}{3}\right)^n u(n)$$

(b) Find the response of the system to a unit-step input.

6.12. Let $x_1(n)$ and $x_2(n)$ be two periodic sequences with period N. Show that

$$\sum_{k=0}^{N-1} x_1(k)\, x_2(n-k) = \sum_{k=n_0}^{N+n_0-1} x_1(k)\, x_2(n-k)$$

6.13. **(a)** Find the periodic convolution $y_p(n)$ of the finite-length sequences of Problem 6.9 by zero-padding the shorter sequence. How is $y_p(n)$ related to the $y(n)$ that you determined in Problem 6.9.

(b) Verify your results using any mathematical software package.

6.14. **(a)** In solving differential equations on a computer, we can approximate the derivatives of successive order by the corresponding differences at discrete time increments T. That is, we replace

$$y(t) = \frac{dx(t)}{dt}$$

with

$$y(nT) = \frac{x(nT) - x((n-1)T)}{T}$$

and

$$z(t) = \frac{dy(t)}{dt} = \frac{d^2x(t)}{dt^2}$$

with

$$z(nT) = \frac{y(nT) - y((n-1)T)}{T} = \frac{x(nT) - 2x((n-1)T) + x((n-2)T)}{T^2}, \qquad \text{etc.}$$

Use this approximation to derive the equation you would employ to solve the differential equation

$$2\frac{dy(t)}{dt} + y(t) = x(t)$$

(b) Repeat part (a) using the forward-difference approximation

$$\frac{dx(t)}{dt} \simeq \frac{x((n+1)T) - x(nt)}{T}$$

6.15. We can use a similar procedure as in Problems 6.13 and 6.14 to evaluate the integral of continuous-time functions. That is, if we want to find

$$y(t) = \int_{t_0}^{t} x(\tau)\, d\tau + x(t_0)$$

we can write

$$\frac{dy(t)}{dt} = x(t)$$

If we use the backward-difference approximation for $y(t)$, we get

$$y(nT) = Tx(nT) + y((n-1)T), \qquad y(0) = x(t_0)$$

whereas the forward-difference approximation gives

$$y((n+1)T) = Tx(nT) + y(nT), \qquad y(0) = x(t_0)$$

(a) Use these approximation to determine the integral of the continuous-time function shown in Figure P6.15 for t in the range [0, 3]. Assume that $T = 0.02$ s. What is the error at (i) 1 s, (ii) 2 s, and (iii) 3 s?

(b) Repeat part (a) for $T = 0.01$ s. Comment on your results.

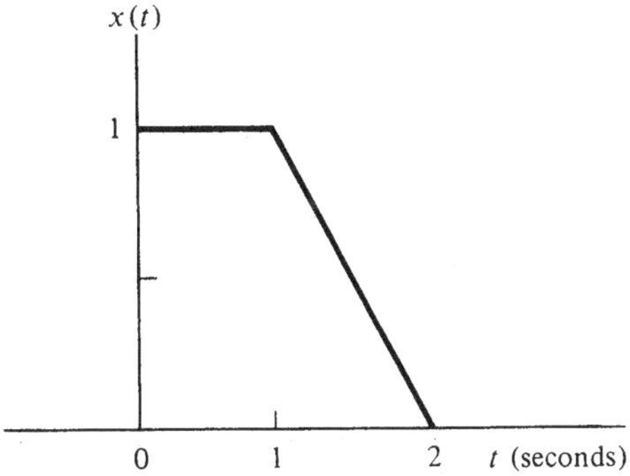

Figure P6.15 Input for Problem 6.15.

6.16. A better approximation to the integral in Problem 6.15 can be obtained by the trapezoidal rule

$$y(nT) = \frac{T}{2}(x(nT) + x(n-1)T) + y((n-1)T)$$

Determine the integral of the function in Problem 6.15 using this rule.

6.17. **(a)** Solve the following difference equations by iteration:

(i) $y(n) + y(n-1) + \frac{1}{4}y(n-2) = x(n), \qquad n \geq 0$

$y(-1) = 0, \quad y(-2) = 1, \quad x(n) = u(n)$

(ii) $y(n) + \frac{3}{4}y(n-1) + \frac{1}{8}y(n-2) = x(n), \qquad n \geq 0$

$y(-1) = 1, \quad y(-2) = 0, \quad x(n) = \left(\frac{1}{3}\right)^n u(n)$

(iii) $y(n) + \frac{3}{4}y(n-1) + \frac{1}{8}y(n-2) = x(n), \qquad n \geq 0$

$y(-1) = 0, \quad y(-2) = 0, \quad x(n) = \left(\frac{1}{2}\right)^n u(n)$

(iv) $y(n+1) + \frac{1}{2}y(n-1) = x(n) - \frac{1}{2}x(n-1), \qquad n \geq 0$

$y(0) = 1, \quad x(n) = \left(\frac{1}{2}\right)^n u(n)$

(v) $y(n) = x(n) + \frac{1}{3}x(n-1) + 2x(n-2), \qquad n \geq 0$

$x(n) = u(n)$

(b) Using any mathematical software package, verify your results for n in the range 0 to 20. Obtain a plot of $y(n)$ vs. n.

6.18. Determine tha characteristic roots and the homogeneous solutions of the following difference equations:

(i) $y(n) - y(n-1) + \frac{1}{2}y(n-2) = x(n), \qquad n \geq 0$

$y(-1) = 0, \quad y(-2) = 1$

(ii) $y(n) - \frac{1}{6}y(n-1) - \frac{1}{6}y(n-2) = x(n) + \frac{1}{2}x(n-1), \qquad n \geq 0$

$y(-1) = 1, \quad y(-2) = 0$

(iii) $y(n) - y(n-1) + \frac{1}{4}y(n-2) = x(n), \qquad n \geq 0$

$y(-1) = 1, \quad y(-2) = 1$

(iv) $y(n) - \frac{3}{2}y(n-1) + \frac{1}{2}y(n-2) = x(n), \qquad n \geq 0$

$y(-1) = 2, \quad y(-2) = 0$

(v) $y(n) - \frac{1}{4}y(n-1) - \frac{1}{8}y(n-2) = x(n), \qquad n \geq 0$

$y(-1) = 1, \quad y(-2) = -1$

6.19. Find the total solution to the following difference equations:

(a) $y(n) + \frac{1}{6}y(n-1) - \frac{1}{6}y(n-2) = x(n) + \frac{1}{2}x(n-1), \qquad n \geq 0$

if $y(-1) = 1, \quad y(-2) = 0,$ and $x(n) = 2\cos\frac{3n\pi}{4}$

(b) $y(n) + \frac{1}{2}y(n-1) = x(n)$ if $y(-1) = 0$ and $x(n) = \left(-\frac{1}{2}\right)^n u(n)$

(c) $y(n) + \frac{1}{2}y(n-1) = x(n)$ if $y(-1) = 1$ and $x(n) = \left(\frac{1}{2}\right)^n u(n)$

(d) $y(n) + \frac{8}{15}y(n-1) + \frac{1}{15}y(n-2) = x(n), \qquad n \geq 0$

if $y(-1) = y(-2) = 1$ and $x(n) = \left(\frac{1}{3}\right)^n$

(e) $y(n+2) - \frac{1}{12}y(n+1) - \frac{1}{12}y(n) = x(n) + \frac{1}{4}x(n+1), \qquad n \geq 0$

if $y(1) = y(0) = 0$ and $x(n) = \left(\frac{1}{4}\right)^n$

6.20. Find the impulse response of the systems in Problem 6.18.

6.21. Find the impulse responses of the systems in Problem 6.19.

6.22. We can find the difference-equation characterization of a system wu=ith a given impulse respnse $h(n)$ by assuming a difference equation of appropriate order with unknown coefficients. We can substitute the given $h(n)$ in the difference equation and solve for the coefficients. Use this procedure to find the difference-equation representation of the system with impulse response $h(n) = (-\frac{1}{2})^n u(n) + (\frac{1}{2})^n u(n)$ by assuming that

$$y(n) + ay(n-1) + by(n-2) = cx(n) + dx(n-1)$$

and find a, b, c, and d.

6.23. Repeat Problem 6.22 if

$$h(n) = \left(-\frac{1}{2}\right)^n u(n) - \left(-\frac{1}{2}\right)^n u(n-1)$$

6.24. We can find the impulse response $h(n)$ of the system of Equation (6.5.11) by first finding the impulse response $h_0(n)$ of the system

$$\sum_{k=0}^{N} a_k y(n-k) = x(n) \tag{P6.1}$$

and then using superposition to find

$$h(n) = \sum_{k=0}^{M} b_k h_0(n-k)$$

(a) Verify that the initial condition for solving Equation (P6.1) is

$$y(0) = \frac{1}{a_0}$$

(b) Use this method to find the impulse response of the system of Example 6.5.6.

(c) Find the impulse responses of the systems of Problem 6.17 by using this method.

6.25. Find the two canonical simulation diagrams for the systems of Problem 6.17.

6.26. Find the corresponding forms of the state equations for the systems of Problem 6.17 by using the simulation diagrams that you determined in Problem 6.25.

6.27. Repeat Problems 6.25 and 6.26 for the systems of Problem 6.18.

6.28. **(a)** Find an appropriate set of state equations for the systems described by the following difference equations:

(i) $y(n) - \dfrac{13}{12}y(n-1) + \dfrac{9}{24}y(n-2) - \dfrac{1}{24}y(n-3) = x(n)$

(ii) $y(n) + 0.707\,y(n-1) + y(n-2) = x(n) + \dfrac{1}{2}x(n-1) + \dfrac{1}{4}x(n-2)$

(iii) $y(n) - 3y(n-1) + 2y(n-2) = x(n)$

(b) Find $\mathbf{A}^n$ for the preceding systems.

(c) Verify your results using any mathematical software package.

6.29. **(a)** Find the unit-step response of the systems in Problem 6.28 if $\mathbf{v}(0) = \mathbf{0}$.

(b) Verify your results using any mathematical software package

6.30. Consider the state-space systems with

$$\mathbf{A} = \begin{bmatrix} 1 & \frac{1}{2} \\ -\frac{5}{3} & -\frac{7}{6} \end{bmatrix}, \qquad \mathbf{b} = \begin{bmatrix} 1 \\ -1 \end{bmatrix}, \qquad \mathbf{c} = [1 \quad 1], \quad d = 0$$

(a) Verify that the eigenvalues of $\mathbf{A}$ are $\frac{1}{2}$ and $-\frac{2}{3}$.

(b) Let $\hat{\mathbf{v}}(n) = \mathbf{P}\mathbf{v}(n)$. Find $\mathbf{P}$ such that the state representation in terms of $\hat{\mathbf{v}}(n)$ has

$$\hat{\mathbf{A}} = \begin{bmatrix} \frac{1}{2} & 0 \\ 0 & -\frac{2}{3} \end{bmatrix}$$

(This is the diagonal form of the state equations.) Find the corresponding values for $\hat{\mathbf{b}}$, $\hat{\mathbf{c}}$, $\hat{\mathbf{d}}$, and $\hat{\mathbf{v}}(0)$.

(c) Find the unit-step response of the system representation that you obtained in part (b).

(d) Find the unit-step response of the original system.

(e) Verify your results using any mathematical software package.

6.31. By using the second canonical form of the state equations, show that the characteristic values of the difference-equation representation of a system are the same as the eigenvalues of the $\mathbf{A}$ matrix in the state-space characterization.

6.32. Determine which of the following systems are stable:

(a)

$$h(n) = \begin{cases} \left(\frac{1}{2}\right)^n, & n \geq 0 \\ \left(\frac{1}{3}\right)^n, & n < 0 \end{cases}$$

(b)

$$h(n) = \begin{cases} 3^n, & 0 \leq n \leq 100 \\ 0, & \text{otherwise} \end{cases}$$

(c)

$$h(n) = \begin{cases} \left(\frac{1}{3}\right)^n \cos 5n, & n \geq 0 \\ 2^n \cos 3n, & n < 0 \end{cases}$$

(d) $y(n) = x(n) + 2x(n-1) + \frac{1}{2}x(n-2)$

(e) $y(n) - 2y(n-1) + y(n-2) = x(n) + x(n-1)$

(f) $y(n+2) - \frac{2}{3}y(n-1) - \frac{1}{3}y(n-2) = x(n)$

(g) $\mathbf{v}(n+1) = \begin{bmatrix} \frac{1}{2} & \frac{1}{4} \\ -\frac{1}{4} & 2 \end{bmatrix} \mathbf{v}(n) + \begin{bmatrix} 1 \\ -1 \end{bmatrix} x(n) \qquad y(n) = [1 \quad 0]\,\mathbf{v}(n)$

(h) $\mathbf{v}(n+1) = \begin{bmatrix} 0 & 1 \\ -2 & -3 \end{bmatrix} \mathbf{v}(n) + \begin{bmatrix} 0 \\ 1 \end{bmatrix} x(n) \qquad y(n) = [2 \quad 1]\,\mathbf{v}(n)$

Chapter 7

Fourier Analysis of Discrete-Time Systems

7.1 INTRODUCTION

In the previous chapter, we considered techniques for the time-domain analysis of discrete-time systems. Recall that, as in the case of continuous-time systems, the primary characterization of a linear, time-invariant, discrete-time system that we used was in terms of the response of the system to the unit impulse. In this and subsequent chapters, we consider frequency-domain techniques for analyzing discrete-time systems. We start our discussion of these techniques with an examination of the Fourier analysis of discrete-time signals. As we might suspect, the results that we obtain closely parallel those for continuous-time systems.

To motivate our discussion of frequency-domain techniques, let us consider the response of a linear, time-invariant, discrete-time system to a complex exponential input of the form

$$x(n) = z^n \tag{7.1.1}$$

where z is a complex number. If the impulse response of the system is $h(n)$, the output of the system is determined by the convolution sum as

$$\begin{aligned} y(n) &= \sum_{k=-\infty}^{\infty} h(k)x(n-k) \\ &= \sum_{k=-\infty}^{\infty} h(k)z^{n-k} \\ &= z^n \sum_{k=-\infty}^{\infty} h(k)z^{-k} \end{aligned} \tag{7.1.2}$$

For a fixed z, the summation is just a constant, which we denote by $H(z)$; that is,

$$H(z) = \sum_{k=-\infty}^{\infty} h(k)z^{-k} \tag{7.1.3}$$

so that

$$y(n) = H(z)x(n) \tag{7.1.4}$$

As can be seen from Equation (7.1.4), the output $y(n)$ is just the input $x(n)$ multiplied by a scaling factor $H(z)$.

We can extend this result to the case where the input to the system consists of a linear combination of complex exponentials of the form of Equation (7.1.1). Specifically, let

$$x(n) = \sum_{k=1}^{N} a_k z_k^n \tag{7.15}$$

It then follows from the superposition property and Equation (7.1.3) that the ouput is

$$\begin{aligned} y(n) &= \sum_{k=1}^{N} a_k H(z_k) z_k^n \\ &= \sum_{k=1}^{N} b_k z_k^n \end{aligned} \tag{7.1.6}$$

That is, the output is also a linear combination of the complex exponentials in the input. The coefficient b_k associated with the function z_k^n in the output is just the corresponding coefficient a_k multiplied by the scaling factor $H(z_k)$.

Example 7.1.1

Suppose we want to find the output of the system with impulse response

$$h(n) = \left(\frac{1}{2}\right)^n u(n)$$

when the input is

$$x(n) = 2\cos\frac{2\pi}{3}n$$

To find the output, we first find

$$H(z) = \sum_{n=0}^{\infty}\left(\frac{1}{2}\right)^n z^{-n} = \sum_{n=0}^{\infty}\left(\frac{1}{2}z^{-1}\right)^n = \frac{1}{1-\dfrac{1}{2}z^{-1}}, \quad \left|\frac{1}{2}z^{-1}\right| < 1$$

where we have used Equation (6.3.7). The input can be expressed as

$$x(n) = \exp\left[j\frac{2\pi}{3}n\right] + \exp\left[-j\frac{2\pi}{3}n\right]$$

so that we have

$$z_1 = \exp\left[j\frac{2\pi}{3}\right], \quad z_2 = \exp\left[-j\frac{2\pi}{3}\right], \quad a_1 = a_2 = 1$$

and

$$H(z_1) = \frac{1}{1 - \frac{1}{2}\exp[-j(2\pi/3)]} = \frac{2}{\sqrt{7}}\exp[-j\phi],$$

$$H(z_2) = \frac{1}{1 - \frac{1}{2}\exp[j(2\pi/3)]} = \frac{2}{\sqrt{7}}\exp[j\phi], \quad \phi = \tan^{-1}\frac{\sqrt{3}}{5}$$

It follows from Equation (7.1.5) that the output is

$$\begin{aligned} y(n) &= \frac{2}{\sqrt{7}}\exp[j\phi]\exp\left[j\frac{2\pi}{3}n\right] + \frac{2}{\sqrt{7}}\exp\left[-j\frac{2\pi}{3}n\right] \\ &= 3\cos\left(\frac{2\pi}{3}n + \phi\right) \end{aligned}$$

A special case occurs when the input is of the form exp $[j\Omega_k]$, where Ω_k is a real, continuous variable. This corresponds to the case $|z_k| = 1$. For this input, the output is

$$y(n) = H(e^{j\Omega k})\exp[j\Omega_k n] \tag{7.1.7}$$

where, from Equation (7.1.3),

$$H(e^{j\Omega}) = H(\Omega) = \sum_{n=-\infty}^{\infty} h(n)\exp[-j\Omega n] \tag{7.1.8}$$

7.2 FOURIER-SERIES REPRESENTATION OF DISCRETE-TIME PERIODIC SIGNALS

Often, as we have seen in our considerations of continuous-time systems, we are interested in the response of linear systems to periodic inputs. Recall that a discrete time signal $x(n)$ is periodic with period N if

$$x(n) = x(n + N) \tag{7.2.1}$$

for some positive integer N. It follows from our discussion in the previous section that if $x(n)$ can be expressed as the sum of several complex exponentials, the response of the system is easily determined. By analogy with our representation of periodic signals in continuous time, we can expect that we can obtain such a representation in terms of the harmonics corresponding to the fundamental frequency $2\pi/N$. That is, we seek a representation for $x(n)$ of the form

$$x(n) = \sum_k a_k \exp[j\Omega_k n] = \sum_k a_k x_k(n) \tag{7.2.2}$$

where $\Omega_k = 2\pi k/N$. It is clear that the $x_k(n)$ are periodic, since $\Omega_k/2\pi$ is a rational number. Also, from our discussions in Chapter 6, there are only N distinct waveforms in this set, corresponding to $k = 0, 1, 2, \ldots, N-1$, since

$$x_k(n) = x_{k+N}(n), \qquad \text{for all } k \tag{7.2.3}$$

Therefore, we have to include only N terms in the summation on the right side of Equation (7.2.2). This sum can be taken over any N consecutive values of k. We indicate this by expressing the range of summation as $k = \langle N \rangle$. However, for the most part, we consider the range $0 \le k \le N-1$. The representation for $x(n)$ can now be written as

$$x(n) = \sum_{k=\langle N \rangle} a_k \exp\left[j\frac{2\pi}{N}kn\right] \tag{7.2.4}$$

Equation (7.2.4) is the *discrete-time Fourier-series* representation of the periodic sequence $x(n)$ with coefficients a_k.

To determine the coefficients a_k, we replace the summation variable k by m on the right side of Equation (7.2.4) and multiply both sides by $\exp[-j2\pi kn/N]$ to get

$$x(n)\exp\left[-j\frac{2\pi}{N}kn\right] = \sum_{m=\langle N \rangle} a_m \exp\left[j\frac{2\pi}{N}(m-k)n\right] \tag{7.2.5}$$

Then we sum over values of n in $[0, N-1]$ to get

$$\sum_{n=0}^{N-1} x(n)\exp\left[-j\frac{2\pi}{N}kn\right] = \sum_{n=0}^{N-1}\sum_{m=\langle N \rangle} a_m \exp\left[j\frac{2\pi}{N}(m-k)n\right] \tag{7.2.6}$$

By interchanging the order of summation in Equation (7.2.6), we can write

$$\sum_{n=0}^{N-1} x(n)\exp\left[-j\frac{2\pi}{N}kn\right] = \sum_{m=\langle N \rangle}\sum_{n=0}^{N-1} a_m \exp\left[j\frac{2\pi}{N}(m-k)n\right] \tag{7.2.7}$$

From Equation (6.3.7), we note that

$$\sum_{n=0}^{N-1} \alpha^n = \frac{1-\alpha^N}{1-\alpha}, \qquad \alpha \neq 1 \tag{7.2.8}$$

For $\alpha = 1$, we have

$$\sum_{n=0}^{N-1} \alpha^n = N \tag{7.2.9}$$

If $m-k$ is not an integer multiple of N (i.e., $m-k \neq rN$ for $r = 0, \pm 1, \pm 2$, etc.), we can let $\alpha = \exp[j2\pi(m-k)/N]$ in Equation (7.2.8) to get

$$\sum_{n=0}^{N-1} \exp\left[j\frac{2\pi}{N}(m-k)n\right] = \frac{1-\exp[j(2\pi/N)(m-k)N]}{1-\exp[j(2\pi/N)(m-k)]} = 0 \tag{7.2.10}$$

If $m-k$ is an integer multiple of N, we can use Equation (7.2.9), so that we have

$$\sum_{n=0}^{N-1} \exp\left[j\frac{2\pi}{N}(m-k)n\right] = N \tag{7.2.11}$$

Combining Equations (7.2.10) and (7.2.11), we write

$$\sum_{n=0}^{N-1} \exp\left[j\frac{2\pi}{N}(m-k)n\right] = N\,\delta(m-k-rN) \tag{7.2.12}$$

where $\delta(m-k-rN)$ is the unit sample occurring at $m = k + rN$. Substitution into Equation (7.2.7) then yields

$$\sum_{n=0}^{N-1} x(n)\exp\left[-j\frac{2\pi}{N}kn\right] = \sum_{m=\langle N\rangle} Na_m\delta(m-k-rN) \tag{7.2.13}$$

Since the summation on the right is carried out over N consecutive values of m for a fixed value of k, it is clear that the only value that r can take in the range of summation is $r = 0$. Thus, the only nonzero value in the sum corresponds to $m = k$, and the right hand side of Equation (7.2.13) evaluates to Na_k, so that

$$a_k = \frac{1}{N}\sum_{n=0}^{N-1} x(n)\exp\left[-j\frac{2\pi}{N}kn\right] \tag{7.2.14}$$

Because each of the terms in the summation in Equation (7.2.14) is periodic with period N, the summation can be taken over any N sucessive values of n. We thus have the pair of equations

$$x(n) = \sum_{k=\langle N\rangle} a_k\exp\left[j\frac{2\pi}{N}kn\right] \tag{7.2.15}$$

and

$$a_k = \frac{1}{N}\sum_{n=\langle N\rangle} x(n)\exp\left[-j\frac{2\pi}{N}kn\right] \tag{7.2.16}$$

which together form the discrete-time Fourier-series pair.

Since $x_{k+N}(n) = x_k(n)$, it is clear that

$$a_{k+N} = a_k \tag{7.2.17}$$

Because the Fourier series for discrete-time periodic signals is a finite sum defined entirely by the values of the signal over one period, the series always converges. The Fourier series provides an *exact* alternative representation of the time signal, and issues such as convergence or the Gibbs phenomenon do not arise.

Example 7.2.1

Let $x(n) = \exp[jK\Omega_0 n]$ for some K with $\Omega_0 = 2\pi/N$, so that $x(n)$ is periodic with period N. By writing $x(n)$ as

$$x(n) = \exp\left[j\frac{2\pi}{N}Kn\right] \quad 0 \le n \le n-1$$

it follows from Equation (7.2.15) that in the range $0 \le k \le N-1$, only $a_k = 1$, with all other a_i being zero. Since $a_{K+N} = a_K$, the spectrum of $x(n)$ is a line spectrum consisting of discrete impulses of magnitude 1 repeated at intervals $N\Omega_0$, as shown in Figure 7.2.1.

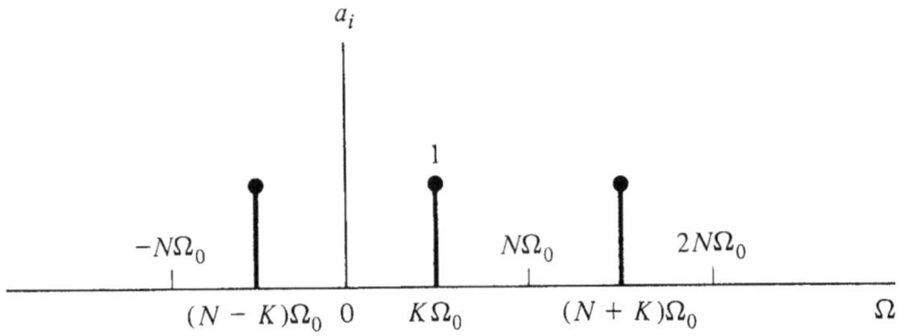

Figure 7.2.1 Spectrum of complex exponential for Example 7.2.1.

Example 7.2.2

Let $x(n)$ be the signal

$$x(n) = \cos\left(\frac{\pi n}{9}\right) + \sin\left(\frac{\pi n}{7} + \frac{1}{2}\right)$$

As we saw in Example 6.1.2, this signal is periodic with period $N = 126$ and fundamental frequency $\Omega_0 = 2\pi/126$, so that $\pi/9$ and $\frac{\pi}{7}$ correspond to $14\Omega_0$ and $18\Omega_0$ respectively. Since $-k\Omega_0$ corresponds to $(N-k)\Omega_0$, it follows that $-\frac{\pi}{9}$ and $-\frac{\pi}{7}$ can be replaced by $108\Omega_0$ and $112\Omega_0$. We can therefore write

$$\begin{aligned} x(n) &= \frac{1}{2}\,[e^{j\frac{\pi n}{9}} + e^{-j\frac{\pi n}{9}}] + \frac{1}{2j}\,[e^{j\frac{1}{2}}e^{j\frac{\pi n}{9}} + e^{-j\frac{1}{2}}e^{-j\frac{\pi n}{9}}] \\ &= \frac{1}{2}\,e^{j14\Omega_0 n} + \frac{e^{j\frac{1}{2}}}{2j}\,e^{j14\Omega_0 n} + \frac{e^{-j\frac{1}{2}}}{2j}\,e^{j108\Omega_0 n} + \frac{1}{2}\,e^{j112\Omega_0 n} \end{aligned}$$

so that

$$a_{14} = \frac{1}{2} = a_{112},\, a_{18} = \frac{e^{j\frac{1}{2}}}{2j} = a_{108}^*, \text{ all other } a_k = 0,\, 0 \le k \le 125$$

Example 7.2.3

Consider the discrete-time periodic square wave shown in Figure 7.2.2. From Equation (7.2.16), the Fourier coefficients can be evaluated as

$$a_k = \frac{1}{N}\sum_{n=-M}^{M} \exp\left[-j\frac{2\pi}{N}kn\right]$$

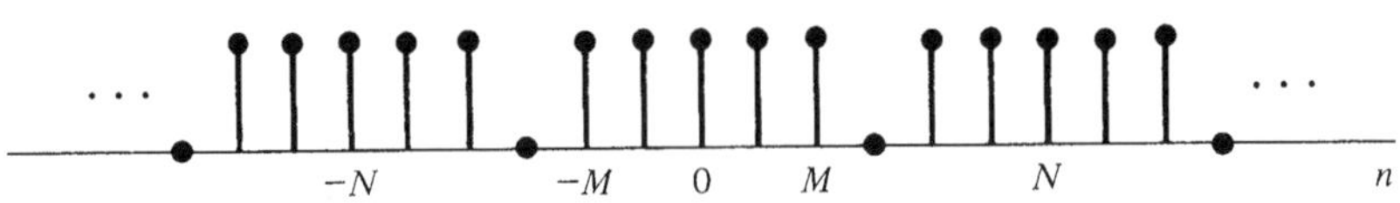

Figure 7.2.2 Periodic square wave of Example 7.2.3.

For $k = 0$,

$$a_0 = \frac{1}{N}\sum_{n=-M}^{M}(1) = \frac{2M+1}{N}$$

For $k \neq 0$, we can use Equation (6.3.7) to get

$$a_k = \frac{1}{N}\frac{\exp[j(2\pi/N)kM] - \exp[-j(2\pi/N)k(M+1)]}{1 - \exp[-j(2\pi/N)k]}$$

$$= \frac{1}{N}\frac{\exp[-j(2\pi/N)(k/2)]\left\{\left[\exp\left[j(2\pi/N)k\left(M+\frac{1}{2}\right)\right] - \exp\left[-j(2\pi/N)k\left(M+\frac{1}{2}\right)\right]\right]\right\}}{\exp[-j(2\pi/N)(k/2)]\{\exp[j(2\pi/N)(k/2)] - \exp[-j(2\pi/N)(k/2)]\}}$$

$$= \frac{1}{N}\frac{\sin\left[\frac{2\pi k}{N}\left(M+\frac{1}{2}\right)\right]}{\sin\left[\left(\frac{2\pi}{N}\right)\left(\frac{k}{2}\right)\right]}, \quad k = 1, 2, \ldots, N-1$$

We can, therefore, write an expression for the coefficients a_k in terms of the sample values of the function

$$f(\Omega) = \frac{\sin[(2M+1)(\Omega/2)]}{\sin(\Omega/2)}$$

as

$$a_k = \frac{1}{N}f\left(\frac{2\pi k}{N}\right)$$

The function $f(\cdot)$ is similar to the sampling function $(\sin x)/x$ that we have encountered in the continuous-time case. Whereas the sinc function is not periodic, the function $f(\Omega)$, being the ratio of two sinusoidal signals with commensurate frequencies, is periodic with period 2π. Figure 7.2.3 shows a plot of the Fourier-series coefficients for $M = 3$, for values of N corresponding to 10, 20, and 30.

Figure 7.2.4 shows the partial sums $x_p(n)$ of the Fourier-series expansion for this example for $N = 11$ and $M = 3$ and for values of $p = 1, 2, 3, 4$, and 5, where

$$x_p(n) = \sum_{k=-p}^{p} a_k \exp\left[j\frac{2\pi}{N}kn\right]$$

As can be seen from the figure, the partial sum is exactly the original sequence for $p = 5$.

Example 7.2.4

Let $x(n)$ be the periodic extension of the sequence

$$\{2, -1, 1, 2\}$$

The period is $N = 4$, so that $\exp[-j2\pi/N] = -j$. The coefficients a_k are therefore given by

$$a_0 = \frac{1}{4}(2 - 1 + 1 + 2) = 1$$

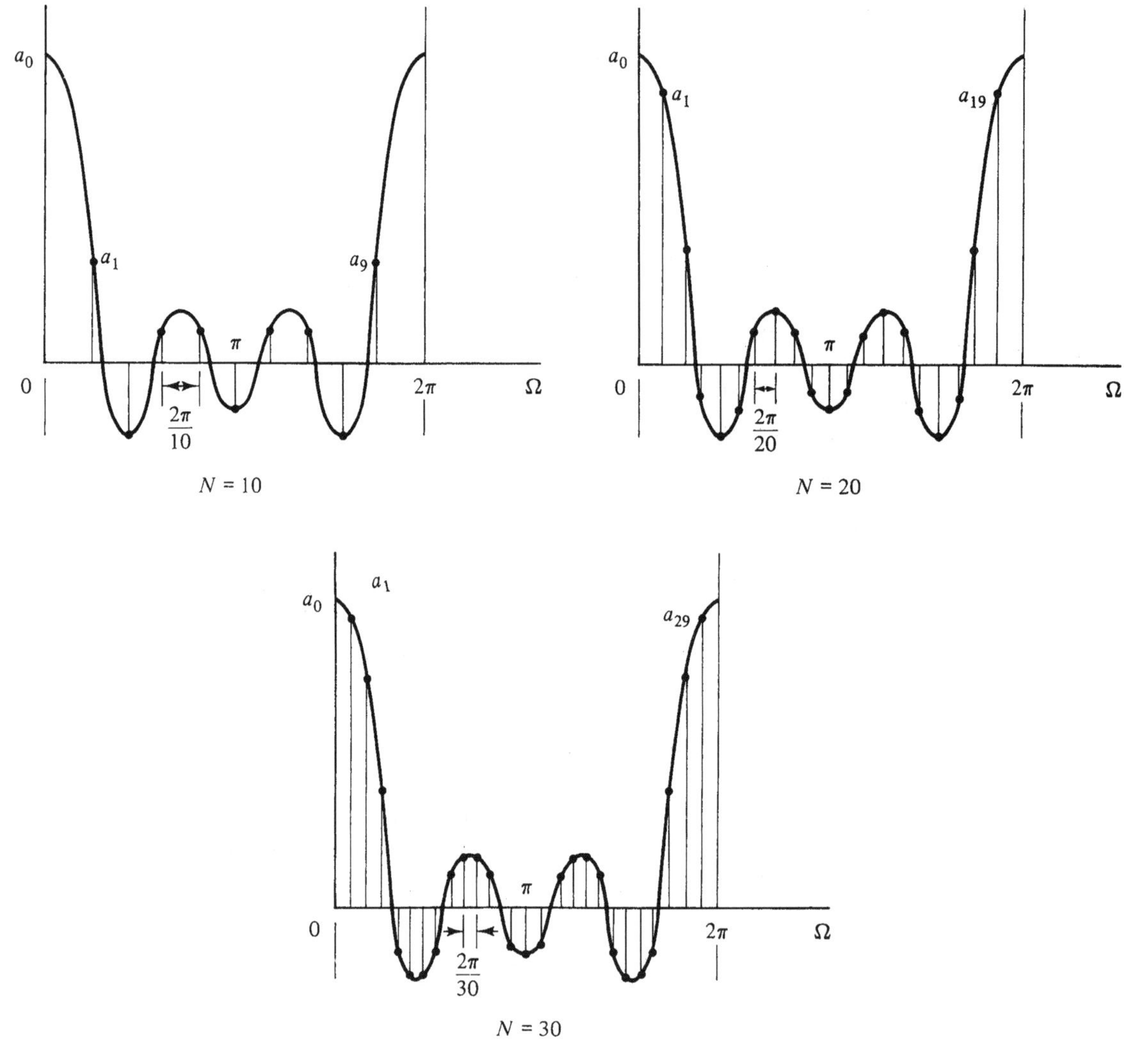

Figure 7.2.3 Fourier series coefficients for the periodic square wave of Example 7.2.3.

$$a_1 = \frac{1}{4}(2 + j - 1 - 2j) = \frac{1}{4} - j\frac{1}{4}$$

$$a_2 = \frac{1}{4}(2 + 1 + 1 - 2) = \frac{1}{2}$$

$$a_3 = \frac{1}{4}(2 - j - 1 + 2j) = \frac{1}{4} + j\frac{1}{4} = a_1^*$$

In general, if $x(n)$ is a real periodic sequence, then

$$a_k = a_{N-k}^* \tag{7.2.18}$$

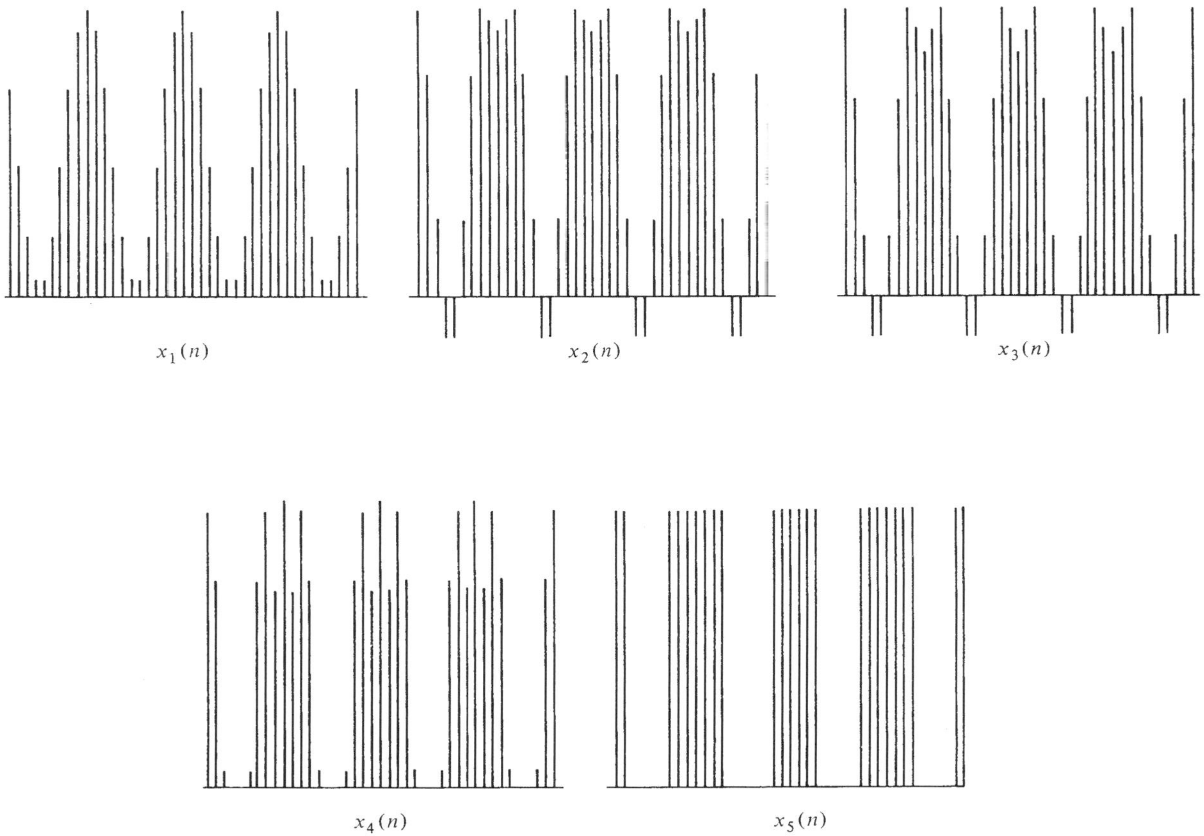

Figure 7.2.4 Partial sums of the Fourier series expansion of a periodic square wave.

Example 7.2.5

Consider the periodic sequence with the following Fourier-series coefficients:

$$a_k = \frac{1}{6}\sin\frac{k\pi}{6} + \frac{1}{12}\cos\frac{k\pi}{2}, \qquad 0 \le k \le 11$$

The signal $x(n)$ can be determined as

$$\begin{aligned}
x(n) &= \sum_{k=0}^{11} a_k \exp\left[j\frac{2\pi}{12}nk\right] \\
&= \sum_{k=0}^{11}\left[\frac{\exp[j(k\pi/6)] - \exp[-j(k\pi/6)]}{12j} + \frac{\exp[j(k\pi/2)] + \exp[-j(k\pi/2)]}{24}\right]\exp\left[j\frac{2\pi}{12}nk\right] \\
&= \sum_{k=0}^{11}\left[\frac{1}{12j}\left\{\exp\left[j\frac{2\pi}{12}k(n+1)\right] - \exp\left[j\frac{2\pi}{12}k(n-1)\right]\right\}\right. \\
&\quad \left. + \frac{1}{24}\left\{\exp\left[j\frac{2\pi}{12}k(n+3)\right] + \exp\left[j\frac{2\pi}{12}k(n-3)\right]\right\}\right]
\end{aligned}$$

Using Equation (7.2.12), we can write this equation as

$$x(n) = \frac{1}{j}\delta(n+1) - \frac{1}{j}\delta(n-1) + \frac{1}{2}\delta(n+3) + \frac{1}{2}\delta(n-3)$$

The values of the sequence $x(n)$ in one period are therefore given by

$$\left\{0, -\frac{1}{j}, 0, \frac{1}{2}, 0, 0, 0, 0, 0, \frac{1}{2}, 0, \frac{1}{j}\right\}$$

where we have used the fact that $x(N + k) = x(k)$.

It follows from the definition of Equation (7.2.16) that, given two sequences $x_1(n)$ and $x_2(n)$, both of period N, with Fourier-series coefficients a_{1k} and a_{2k}, the coefficients for the sequence $Ax_1(n) + Bx_2(n)$ are equal to $Aa_{1k} + Ba_{2k}$.

For a periodic sequence with coefficients a_k, we can find the coefficients b_k corresponding to the shifted sequence $x(n - m)$ as

$$b_k = \frac{1}{N}\sum_{n=\langle N\rangle} x(n-m)\exp\left[-j\frac{2\pi}{N}kn\right] \tag{7.2.19}$$

By replacing $n - m$ by n and noting that the summation is taken over any N successive values of n, we can write

$$b_k = \left(\frac{1}{N}\sum_{n=\langle N\rangle} x(n)\exp\left[-j\frac{2\pi}{N}kn\right]\right)\exp\left[-j\frac{2\pi}{N}km\right] = \exp\left[-j\frac{2\pi}{N}km\right]a_k \tag{7.2.20}$$

Let the periodic sequence $x(n)$, with Fourier coefficients a_k, be the input to a linear system with impulse response $h(n)$, where $h(n)$ is not periodic. [Note that if $h(n)$ is also periodic, the linear convolution of $x(n)$ and $h(n)$ is not defined.] Since, from Equation (7.1.7), the response $y_k(n)$ to input $a_k \exp[j(2\pi/N)kn]$ is

$$y_k(n) = a_k H\left(\frac{2\pi}{N}k\right) \exp\left[j\frac{2\pi}{N}kn\right] \tag{7.2.21}$$

it follows that the response to $x(n)$ can be written as

$$y(n) = \sum_{k=\langle N\rangle} y_k(n) = \sum_{k=\langle N\rangle} a_k H\left(\frac{2\pi}{N}k\right) \exp\left[j\frac{2\pi}{N}kn\right] \tag{7.2.22}$$

where $H(2\pi k/N)$is obtained by evaluating $H(\Omega)$ in Equation (7.1.8) at $\Omega = 2\pi k/N$.

If $x_1(n)$ and $x_2(n)$ are both periodic sequences with the same period and with Fourier coefficients a_{1k} and a_{2k}, it can be shown that the Fourier-series coefficients of their periodic convolution is (see Problem 7.7b) $Na_{1k}a_{2k}$. That is,

$$x_1(n) \circledast x_2(n) \leftrightarrow Na_{1k}a_{2k}$$

Also, the Fourier-series coefficients for their product is (see Problem 7.7a) $a_{1k} \circledast a_{2k}$

$$x_1(n)x_2(n) \leftrightarrow a_{1k} \circledast a_{2k}$$

These properties are summarized in Table 7-1.

TABLE 7-1
Properties of Discrete-Time Fourier Series

1. Fourier coefficients	$x_i(n)$ periodic with period N	$a_{ik} = \frac{1}{N}\sum_{\langle N\rangle} x_i(n) \exp\left[-j\frac{2\pi}{N}nk\right]$	(7.2.14)
2. Linearity	$Ax_1(n) + Bx_2(n)$	$Aa_{1k} + Ba_{2k}$	
3. Time shift	$x(n - m)$	$\exp\left[-j\frac{2\pi}{N}km\right]a_k$	(7.2.20)
4. Convolution	$x(n) * h(n)$; $h(n)$ not periodic	$a_k H\left(\frac{2\pi}{N}k\right)$	(7.2.21)
		$H(\Omega) = \sum_{n=-\infty}^{\infty} h(n) \exp[-j\Omega_n]$	
5. Periodic convolution	$x_1(n) \circledast x_2(n)$	$Na_{1k}a_{2k}$	(7.2.23)
6. Modulation	$x_1(n)x_2(n)$	$a_{1k} \circledast a_{2k}$	(7.2.24)

Example 7.2.6

Consider the system with impulse response $h(n) = (1/3)^n u(n)$. Suppose want to find the Fourier-series representation for the output $y(n)$ when the input $x(n)$ is the periodic extension of the sequence $\{2, -1, 1, 2\}$. From Equation (7.2.22), it follows that we can write $y(n)$ in a Fourier series as

$$y(n) = \sum_{\langle N\rangle} b_k \exp\left[j\frac{2\pi}{N}kn\right]$$

with

$$b_k = a_k H\left(\frac{2\pi}{N}k\right)$$

From Example 7.2.4, we have

$$a_0 = 1, \quad a_1 = a_3^* = \frac{1}{4} - j\frac{1}{4}, \quad a_2 = \frac{1}{2}$$

Using Equation (7.1.8), we can write

$$H(\Omega) = \sum_{n=0}^{\infty}\left(\frac{1}{3}\right)^n \exp[-j\Omega n] = \frac{1}{1 - \frac{1}{3}\exp[-j\Omega]}$$

so that with $N = 4$, we have

$$H\left(\frac{2\pi}{N}k\right) = \frac{1}{1 - \frac{1}{3}\exp\left[-j\frac{\pi}{2}k\right]}$$

It follows that

$$b_0 = H(0)a_0 = \frac{3}{2}$$

$$b_1 = H\left(\frac{\pi}{2}\right)a_1 = \frac{3(1 - j2)}{20}$$

$$b_2 = H(\pi)a_2 = \frac{3}{8}$$

$$b_3 = b_1^* = \frac{3(1 + j2)}{20}$$

7.3 THE DISCRETE-TIME FOURIER TRANSFORM

We now consider the frequency-domain representation of discrete-time signals that are not necessarily periodic. For continuous-time signals, we obtained such a representation by defining the Fourier transform of a signal $x(t)$ as

$$X(\omega) = \mathscr{F}\{x(t)\} = \int_{-\infty}^{\infty} x(t)\exp[-j\omega t]\,dt \tag{7.3.1}$$

with respect to the transform (frequency) variable ω. For discrete-time signals, we consider an analogous definition. To motivate this definition, let us sample $x(t)$ uniformly every T seconds to obtain the samples $x(nT)$. Recall from Equations (4.4.1) and (4.4.2) that the sampled signal can be written as

$$x_s(t) = x(t)\sum_{n=-\infty}^{\infty}\delta(t - nT) \tag{7.3.2}$$

so that its Fourier transform is given by

$$\begin{aligned} X_s(\omega) &= \int_{-\infty}^{\infty} x_s(t)e^{-j\omega t}\,dt \\ &= \int_{-\infty}^{\infty} x(t) \sum_{n=-\infty}^{\infty} \delta(t - nT)e^{-j\omega t}\,dt \\ &= \sum_{n=-\infty}^{\infty} x(nT)e^{-j\omega Tn} \end{aligned} \tag{7.3.3}$$

where the last step follows from the sifting property of the δ function.

If we replace ωT in the previous equation by the discrete-time frequency variable Ω, we get the discrete-time Fourier transform, $X(\Omega)$, of the discrete-time signal $x(n)$, obtained by sampling $x(t)$, as

$$X(\Omega) = \mathcal{F}\{x(n)\} = \sum_{n=-\infty}^{\infty} x(n)\exp[-j\Omega n] \tag{7.3.4}$$

Equation (7.3.4), in fact, defines the discrete-time Fourier transform of *any* discrete-time signal $x(n)$. The transform exists if $x(n)$ satisfies a relation of the type

$$\sum_{n=-\infty}^{\infty} |x(n)| < \infty \qquad \text{or} \qquad \sum_{n=-\infty}^{\infty} |x(n)|^2 < \infty \tag{7.3.5}$$

These conditions are sufficient to guarantee that the sequence has a discrete-time Fourier transform. As in the case of continuous-time signals, there are signals that neither are absolutely summable nor have finite energy, but still have discrete-time Fourier transforms.

We reiterate that although ω has units of radians/second, Ω has units of radians. Since $\exp[j\Omega n]$ is periodic with period 2π, it follows that $X(\Omega)$ is also periodic with the same period, since

$$\begin{aligned} X(\Omega + 2\pi) &= \sum_{n=-\infty}^{\infty} x(n)\exp[-j(\Omega + 2\pi)n] \\ &= \sum_{n=-\infty}^{\infty} x(n)\exp[-j\Omega n] = X(\Omega) \end{aligned} \tag{7.3.6}$$

As a consequence, while in the continuous-time case we have to consider values of ω over the entire real axis, in the discrete-time case we have to consider values of Ω only over the range $[0, 2\pi]$.

To find the inverse relation between $X(\Omega)$ and $x(n)$, we replace the variable n in Equation (7.3.4) by p to get

$$X(\Omega) = \sum_{p=-\infty}^{\infty} x(p)\exp[-j\Omega p] \tag{7.3.7}$$

Next, we multiply both sides of Equation (7.3.7) by $\exp[j\Omega n]$ and integrate over the range $[0, 2\pi]$ to get

$$\int_{\Omega=0}^{2\pi} X(\Omega)\exp[j\Omega n]\,d\Omega = \int_{\Omega=0}^{2\pi}\sum_{p=-\infty}^{\infty} x(p)\exp[j\Omega(n-p)]\,d\Omega \qquad (7.38)$$

Interchanging the order of summation and integration on the right side of Equation (7.3.8) then gives

$$\int_{0}^{2\pi} X(\Omega)\exp[j\Omega n]\,d\Omega = \sum_{p=-\infty}^{\infty} x(p)\int_{0}^{2\pi}\exp[j\Omega(n-p)]\,d\Omega \qquad (7.39)$$

It can be verified (see Problem 7.10) that

$$\int_{0}^{2\pi}\exp[j\Omega(n-p)]\,d\Omega = \begin{cases} 2\pi, & n = p \\ 0, & n \neq p \end{cases} \qquad (7.3.10)$$

so that the right-hand side of Equation (7.3.9) evaluates to $2\pi x(n)$. We can therefore write

$$x(n) = \frac{1}{2\pi}\int_{0}^{2\pi} X(\Omega)\exp[j\Omega n]\,d\Omega \qquad (7.3.11)$$

Again, since the integrand in Equation (7.3.11) is periodic with period 2π, the integration can be carried out over any interval of length 2π. Thus, the discrete-time Fourier-transform relations can be written as

$$X(\Omega) = \sum_{n=-\infty}^{\infty} x(n)\exp[-j\Omega n] \qquad (7.3.12)$$

$$x(n) = \frac{1}{2\pi}\int_{\langle 2\pi\rangle} X(\Omega)\exp[j\Omega n]\,d\Omega \qquad (7.3.13)$$

Example 7.3.1

Consider the sequence

$$x(n) = \alpha^n u(n), \qquad |\alpha| < 1$$

For this sequence,

$$X(\Omega) = \sum_{n=0}^{\infty} \alpha^n \exp[-j\Omega n] = \frac{1}{1 - \alpha\exp[-j\Omega]}$$

The magnitude is given by

$$|X(\Omega)| = \frac{1}{\sqrt{1 + \alpha^2 - 2\alpha\cos\Omega}}$$

and the phase by

$$\operatorname{Arg} X(\Omega) = -\tan^{-1}\frac{\alpha\sin\Omega}{1 - \alpha\cos\Omega}$$

Figure 7.3.1 shows the magnitude and phase spectra of this signal for $\alpha > 0$. Note that these functions are periodic with period 2π.

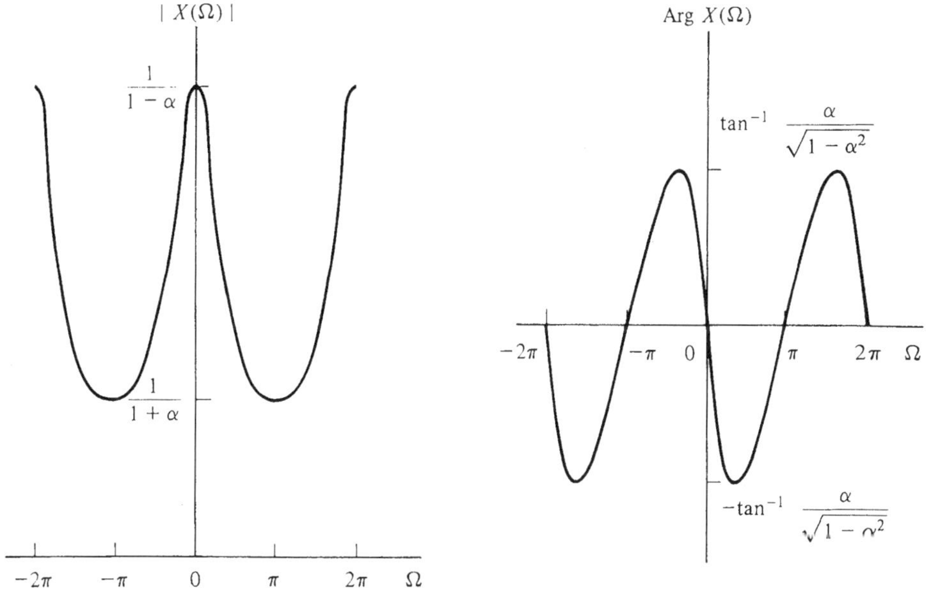

Figure 7.3.1 Fourier spectra of signal for Example 7.3.1.

Example 7.3.2

Let

$$x(n) = \alpha^{|n|}, \qquad |\alpha| < 1$$

We obtain the Fourier tranform of $x(n)$ as

$$X(\Omega) = \sum_{n=-\infty}^{\infty} \alpha^{|n|} \exp[-j\Omega n]$$

$$= \sum_{n=-\infty}^{-1} \alpha^{-n} \exp[-j\Omega n] + \sum_{n=0}^{\infty} \alpha^{n} \exp[-j\Omega n]$$

which can be put in closed form, by using Equation (6.3.7), as

$$X(\Omega) = \frac{1}{1 - \alpha^{-1}\exp[-j\Omega]} + \frac{1}{1 - \alpha \exp[-j\Omega]}$$

$$= \frac{1 - \alpha^2}{1 - 2\alpha \cos\Omega + \alpha^2}$$

In this case, $X(\Omega)$ is real, so that the phase is identically zero. The magnitude is plotted in Figure 7.3.2.

Example 7.3.3

Consider the sequence $x(n) = \exp[j\Omega_0 n]$, with Ω_0 arbitrary. Thus, $x(n)$ is not necessarily a periodic signal. Then

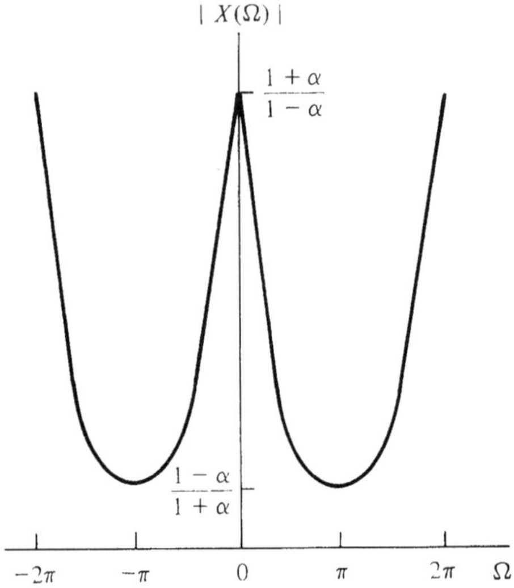

Figure 7.3.2 Magnitude spectrum of signal for Example 7.3.2.

$$X(\Omega) = \sum_{m=-\infty}^{\infty} 2\pi\delta(\Omega - \Omega_0 - 2\pi m) \tag{7.3.14}$$

In the range $[0, 2\pi]$, $X(\Omega)$ consists of a δ function of strength 2π, occurring at $\Omega = \Omega_0$. As can be expected, and as indicated by Equation (7.3.14), $X(\Omega)$ is a periodic extension, with period 2π, of this δ function. (See Figure 7.3.3.) To establish Equation (7.3.14), we use the inverse Fourier relation of Equation (7.3.13) as

$$\begin{aligned} x(n) = \mathcal{F}^{-1}\{x(\Omega)\} &= \frac{1}{2\pi}\int_{-\pi}^{\pi} X(\Omega)\exp[j\Omega n]\,d\Omega \\ &= \frac{1}{2\pi}\int_{-\pi}^{\pi}\left[\sum_{m=-\infty}^{\infty} 2\pi\,\delta(\Omega - \Omega_0 - 2\pi m)\right]\exp[j\Omega n]\,d\Omega \\ &= \exp[j\Omega_0 n] \end{aligned}$$

where the last step follows because the only pemissible value for m in the range of integration is $m = 0$.

We can modify the results of this example to determine the Fourier transform of an exponential signal that is periodic. Thus, let $x(n) = \exp[jk\Omega_0 n]$ be such that $\Omega_0 = 2\pi/N$. We can write the Fourier transform from Equation (7.3.12) as

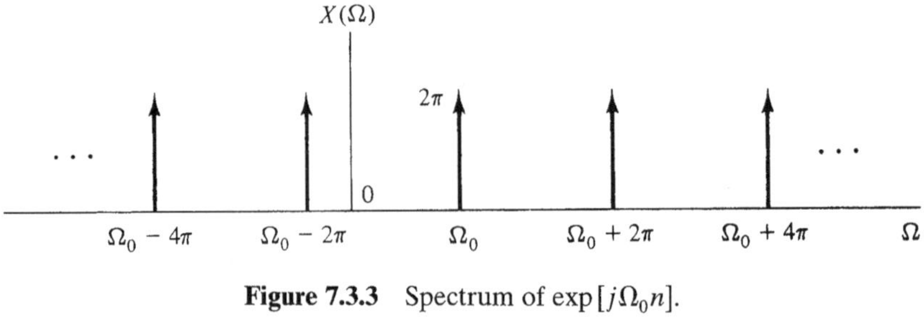

Figure 7.3.3 Spectrum of $\exp[j\Omega_0 n]$.

$$X(\Omega) = \sum_{m=-\infty}^{\infty} 2\pi\,\delta(\Omega - k\Omega_0 - 2\pi m)$$

Replacing 2π by $N\Omega_0$ yields

$$X(\Omega) = \sum_{m=-\infty}^{\infty} 2\pi\,\delta(\Omega - k\Omega_0 - N\Omega_0 m)$$

That is, the spectrum consists of an infinite set of impulses of strength 2π centered at $k\Omega_0$, $(k \pm N)\Omega_0$, $(k \pm 2N)\Omega_0$, etc. This can be compared to the result we obtained in Example 7.2.1, where we considered the Fourier-series representation for $x(n)$. The difference, as in continuous time, is that in the Fourier-series representation the frequency variable takes on only discrete values, whereas in the Fourier transform the frequency variable is continuous.

7.4 PROPERTIES OF THE DISCRETE-TIME FOURIER TRANSFORM

The properties of the discrete-time Fourier transform closely parallel those of the continuous-time transform. These properties can prove useful in the analysis of signals and systems and in simplifying our manipulations of both the forward and inverse transforms. In this section, we consider some of the more useful properties.

7.4.1 Periodicity

We saw that the discrete-time Fourier transform is periodic in Ω with period 2π, so that

$$X(\Omega + 2\pi) = X(\Omega) \tag{7.4.1}$$

7.4.2 Linearity

Let $x_1(n)$ and $x_2(n)$ be two sequences with Fourier transforms $X_1(\Omega)$ and $X_2(\Omega)$, respectively. Then

$$\mathcal{F}[a_1 x_1(n) + a_2 x_2(n)] = a_1 X_1(\Omega) + a_2 X_2(\Omega) \tag{7.4.2}$$

for any constants a_1 and a_2.

7.4.3 Time and Frequency Shifting

By direct substitution into the defining equations of the Fourier transform, it can easily be shown that

$$\mathcal{F}[x(n - n_0)] = \exp[-j\Omega n_0]X(\Omega) \tag{7.4.3}$$

and

$$\mathcal{F}[\exp[j\Omega_0 n]x(n)] = X(\Omega - \Omega_0) \tag{7.4.4}$$

7.4.4 Differentiation in Frequency

Since

$$X(\Omega) = \sum_{n=-\infty}^{\infty} x(n)\exp[-j\Omega n]$$

it follows that if we differentiate both sides with respect to Ω, we get

$$\frac{dX(\Omega)}{d\Omega} = \sum_{n=-\infty}^{\infty} (-jn)x(n)\exp[-j\Omega n]$$

from which we can write

$$\mathcal{F}[nx(n)] = \sum_{n=-\infty}^{\infty} nx(n)\exp[-j\Omega n] = j\frac{dX(\Omega)}{d\Omega} \tag{7.4.5}$$

Example 7.4.1

Let $x(n) = n\alpha^n u(n)$, with $|\alpha| < 1$. Then, by using the results of Example 7.3.1, we can write

$$X(\Omega) = j\frac{d}{d\Omega}\mathcal{F}\{\alpha^n u(n)\} = j\frac{d}{d\Omega}\frac{1}{1-\alpha\exp[-j\Omega]}$$

$$= \frac{\alpha\exp[-j\Omega]}{(1-\alpha\exp[-j\Omega])^2}$$

7.4.5 Convolution

Let $y(n)$ represent the convolution of two discrete-time signals $x(n)$ and $h(n)$; that is,

$$y(n) = h(n) * x(n) \tag{7.4.6}$$

Then

$$Y(\Omega) = H(\Omega)X(\Omega) \tag{7.4.7}$$

This result can easily be established by using the definition of the convolution operation given in Equation (6.3.2) and the definition of the Fourier transform:

$$Y(\Omega) = \sum_{n=-\infty}^{\infty} y(n)\exp[-j\Omega n]$$

$$= \sum_{n=-\infty}^{\infty}\left[\sum_{k=-\infty}^{\infty} h(k)x(n-k)\right]\exp[-j\Omega n]$$

$$= \sum_{k=-\infty}^{\infty} h(k)\left[\sum_{n=-\infty}^{\infty} x(n-k)\exp[-j\Omega n]\right]$$

Here, the last step follows by interchanging the order of summation. Now we replace $n-k$ by n in the inner sum to get

$$Y(\Omega) = \sum_{k=-\infty}^{\infty} h(k)\left[\sum_{n=-\infty}^{\infty} x(n)\exp[-j\Omega n]\right] \exp[-j\Omega k]$$

so that

$$\begin{aligned} Y(\Omega) &= \sum_{k=-\infty}^{\infty} h(k)X(\Omega)\exp[-j\Omega k] \\ &= H(\Omega)X(\Omega) \end{aligned}$$

As in the case of continuous-time systems, this property is extremely useful in the analysis of discrete-time linear systems. The function $H(\Omega)$ is referred to as the *frequency response* of the system.

Example 7.4.2

A pure delay is described by the input/output relation

$$y(n) = x(n - n_0)$$

Taking the Fourier transformation of both sides, using Equation (7.4.3), yields

$$Y(\Omega) = \exp[-j\Omega n_0]X(\Omega)$$

The frequency response of a pure delay is therefore

$$H(\Omega) = \exp[-j\Omega n_0]$$

Since $H(\Omega)$ has unity gain for all frequencies and a linear phase, it is distortionless.

Example 7.4.3

Let

$$h(n) = \left(\frac{1}{2}\right)^n u(n)$$

$$x(n) = \left(\frac{1}{3}\right)^n u(n)$$

Their respective Fourier transforms are given by

$$H(\Omega) = \frac{1}{1 - \frac{1}{2}\exp[-j\Omega]}$$

$$X(\Omega) = \frac{1}{1 - \frac{1}{3}\exp[-j\Omega]}$$

so that

$$Y(\Omega) = H(\Omega)X(\Omega) = \frac{1}{1 - \frac{1}{2}\exp[-j\Omega]} \frac{1}{1 - \frac{1}{3}\exp[-j\Omega]}$$

$$= \frac{3}{1 - \frac{1}{2}\exp[-j\Omega]} - \frac{2}{1 - \frac{1}{3}\exp[-j\Omega]}$$

By comparing the two terms in the previous equation with $X(\Omega)$ in Example 7.3.1, we see that $y(n)$ can be written down as

$$y(n) = 3\left(\frac{1}{2}\right)^n u(n) - 2\left(\frac{1}{3}\right)^n u(n)$$

Example 7.4.4

As a modification of the problem in Example 7.3.2, let

$$h(n) = \alpha^{|n-n_0|}, \qquad -\infty < n < \infty$$

represent the impulse response of a discrete-time system. It is clear that this is a noncausal IIR system. By following the same procedure as in Example 7.3.2, it can easily be verified that the frequency response of the system is

$$H(\Omega) = \frac{1 - \alpha^2}{1 - 2\alpha\cos\Omega + \alpha^2}\exp[-j\Omega n_0]$$

The magnitude function $|H(\Omega)|$ is the same as $X(\Omega)$ in Example 7.3.2 and is plotted in Figure 7.3.2. The phase is given by

$$\operatorname{Arg} H(\Omega) = -n_0\Omega$$

Thus, $H(\Omega)$ represents a linear-phase system, with the associated delay equal to n_0. It can be shown that, in general, a system will have a linear phase, if $h(n)$ satisfies

$$h(n) = h(2n_0 - n), \qquad -\infty < n < \infty$$

If the system is an IIR system, this condition implies that the system is noncausal. Since a continuous-time system is always IIR, we cannot have a linear phase in a continuous-time causal system. For an FIR discrete-time system, for which the impulse response is an N-point sequence, we can find a causal $h(n)$ to satisfy the linear-phase condition by letting delay n_0 be equal to $(N - 1)/2$. It can easily be verified that $h(n)$ then satisfies

$$h(n) = h(N - 1 - n), \qquad 0 \le n \le N - 1$$

Example 7.4.5

Let

$$H(\Omega) = \begin{cases} 1, & 0 \le |\Omega| \le \Omega_c \\ 0, & \Omega_c < |\Omega| < \pi \end{cases}$$

That is, $H(\Omega)$ represents the transfer function of an ideal low-pass discrete-time filter with a cutoff of Ω_c radians. We can find the impulse response of this filter by using Equation (7.3.11):

$$h(n) = \frac{1}{2\pi}\int_{-\Omega_c}^{\Omega_c} \exp[j\Omega n]\,d\Omega$$

$$= \frac{\sin \Omega_c n}{\pi n}$$

Example 7.4.6

We will find the output $y(n)$ of the system with

$$h(n) = \delta(n) - \frac{\sin\left(\dfrac{\pi n}{8}\right)}{\pi n}$$

when the input is

$$x(n) = \cos\left(\frac{\pi n}{9}\right) + \sin\left(\frac{\pi n}{7} + \frac{1}{2}\right)$$

From Example 7.2.2 and Equation (7.4.12), it follows that, with $\Omega_0 = \pi/128$, in the range $0 \le \Omega < 2\pi$

$$X(\Omega) = 2\pi\left[\frac{1}{2}\delta(\Omega - 14\Omega_0) + \frac{e^{j\frac{1}{2}}}{2j}\delta(\Omega - 18\Omega_0) + \frac{e^{-j\frac{1}{2}}}{2j}\delta(\Omega - 108\Omega_0) + \frac{1}{2}\delta(\Omega - 112\Omega_0)\right],$$

Now

$$H(\Omega) = 1 - \operatorname{rect}\left(\frac{\Omega}{\pi/4}\right), \qquad -\pi \le |\Omega| < \pi$$

so that in the range $0 \le \Omega < 2\pi$ we have

$$II(\Omega) - \begin{cases} 1 & \dfrac{\pi}{8} \le \Omega < \dfrac{15\pi}{8} \\ 0 & \text{otherwise} \end{cases}$$

Thus

$$Y(\Omega) = H(\Omega)X(\Omega) = 2\pi\left[\frac{e^{j\frac{1}{2}}}{2j}\delta(\Omega - 18\Omega_0) + \frac{e^{-j\frac{1}{2}}}{2j}\delta(\Omega - 108\Omega_0)\right]$$

and

$$y(n) = \frac{e^{j\frac{1}{2}}}{2j}e^{j18\Omega_0} + \frac{e^{-j\frac{1}{2}}}{2j}e^{j108\Omega_0}$$

which can be simplified as

$$y(n) = \sin\left(\frac{\pi n}{7} + \frac{1}{2}\right)$$

7.4.6 Modulation

Let $y(n)$ be the product of the two sequences $x_1(n)$ and $x_2(n)$ with transforms $X_1(\Omega)$ and $X_2(\Omega)$, respectively. Then

$$Y(\Omega) = \sum_{n=-\infty}^{\infty} x_1(n)x_2(n) \exp[-j\Omega n]$$

If we use the inverse Fourier-transform relation to write $x_1(n)$ in terms of its Fourier transform, we have

$$Y(\Omega) = \sum_{n=-\infty}^{\infty} \left[\frac{1}{2\pi} \int_{\langle 2\pi \rangle} X_1(\theta) \exp[j\theta n]\, d\theta\right] x_2(n) \exp[-j\Omega n]$$

Interchanging the order of summation and integration yields

$$Y(\Omega) = \frac{1}{2\pi} \int_{\langle 2\pi \rangle} X_1(\theta) \left\{ \sum_{n=-\infty}^{\infty} x_2(n) \exp[-j(\Omega - \theta)n] \right\} d\theta$$

so that

$$Y(\Omega) = \frac{1}{2} \int_{\langle 2\pi \rangle} X_1(\theta) X_2(\Omega - \theta)\, d\theta \qquad (7.4.8)$$

7.4.7 Fourier Transform of Discrete-Time Periodic Sequences

Let $x(n)$ be a periodic sequence with period N, so that we can express $x(n)$ in a Fourier-series expansion as

$$x(n) = \sum_{k=0}^{N-1} a_k \exp[jk\Omega_0 n] \qquad (7.4.9)$$

where

$$\Omega_0 = \frac{2\pi}{N} \qquad (7.4.10)$$

Then

$$X(\Omega) = \mathcal{F}[x(n)] = \mathcal{F}\left[\sum_{k=0}^{N-1} a_k \exp[jk\Omega_0 n]\right]$$

$$= \sum_{k=0}^{N-1} a_k \mathcal{F}[\exp[jk\Omega_0 n]]$$

We saw in Example 7.3.3, that, in the range $[0, 2\pi]$,

$$\mathcal{F}[\exp[jk\Omega_0 n]] = 2\pi\,\delta(\Omega - k\Omega_0)$$

so that

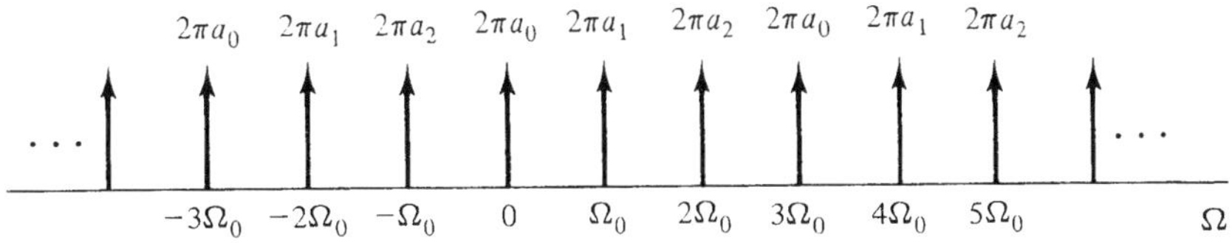

Figure 7.4.1 Spectrum of a periodic signal $N = 3$.

$$X(\Omega) = \sum_{k=0}^{N-1} 2\pi a_k \delta(\Omega - k\Omega_0), \quad 0 \le \Omega < 2\pi \tag{7.4.11}$$

Since the discrete-time Fourier transform is periodic with period 2π, it follows that $X(\Omega)$ consists of a set of N impulses of strength $2\pi a_k$, $k = 0, 1, 2, \ldots, N-1$, repeated at intervals of $N\Omega_0 = 2\pi$. Thus, $X(\Omega)$ can be compactly written as

$$X(\Omega) = \sum_{k=0}^{N-1} 2\pi a_k \delta(\Omega - k\Omega_0), \quad \text{for all } \Omega \tag{7.4.12}$$

This is illustrated in Figure 7.4.1 for the case $N = 3$.

Table 7-2 summarizes the properties of the discrete-time Fourier transform, while Table 7-3 lists the transforms of some common sequences.

TABLE 7-2
Properties of the Discrete-Time Fourier Transform

1. Linearity	$Ax_1(n) + Bx_2(n)$	$AX_1(\Omega) + BX_2(\Omega)$	(7.4.2)
2. Time shift	$x(n - n_0)$	$\exp[-j\Omega n_0]X(\Omega)$	(7.4.3)
3. Frequency shift	$x(n)\exp[j\Omega_0 n]$	$X(\Omega - \Omega_0)$	(7.4.4)
4. Convolution	$x_1(n) * x_2(n)$	$X_1(\Omega)X_2(\Omega)$	(7.4.7)
5. Modulation	$x_1(n)x_2(n)$	$\frac{1}{2\pi}\int_{2\pi} X_1(p)X_2(\Omega - p)\,dp$	(7.4.8)
6. Periodic signals	$x(n)$ periodic with period N	$\sum_{k=-\infty}^{\infty} 2\pi a_k \delta(\Omega - k\Omega_0)$	(7.4.11)
	$\Omega_0 = \frac{2\pi}{N}$	$a_k = \frac{1}{N}\sum_{\langle N\rangle} x(n)\exp[-j\Omega_0 n]$	

7.5 FOURIER TRANSFORM OF SAMPLED CONTINUOUS-TIME SIGNALS

We conclude this chapter with a discussion of the Fourier transform of sampled continuous-time (analog) signals. Recall that we can obtain a discrete-time signal by *sampling* a continuous-time signal. Let $x_a(t)$ be the analog signal that is sampled at equally spaced intervals T:

TABLE 7-3
Some Common Discrete-Time Fourier Transform Pairs

Signal	Fourier Transform (periodic in Ω, period 2π)
$\delta(n)$	1
1	$2\pi\delta(\Omega)$
$\exp[j\Omega_0 n]$, Ω_0 arbitrary	$2\pi\delta(\Omega - \Omega_0)$
$\sum_{k=0}^{N-1} a_k \exp[jk\Omega_0 n]$, $N\Omega_0 = 2\pi$	$\sum_{k=0}^{N-1} 2\pi a_k \delta(\Omega - k\Omega_0)$
$\alpha^n u(n)$, $\lvert\alpha\rvert < 1$	$\dfrac{1}{1 - \alpha \exp[-j\Omega]}$
$\alpha^{\lvert n\rvert}$, $\lvert\alpha\rvert < 1$	$\dfrac{1 - \alpha^2}{1 - 2\alpha \cos\Omega + \alpha^2}$
$n\alpha^n u(n)$, $\lvert\alpha\rvert < 1$	$\dfrac{\alpha \exp[-j\Omega]}{(1 - \alpha \exp[-j\Omega])^2}$
$\text{rect}(n/N_1)$	$\dfrac{\sin\left[\Omega\left(N_1 + \frac{1}{2}\right)\right]}{\sin(\Omega/2)}$
$\dfrac{\sin\Omega_c n}{\pi n}$	$\text{rect}(\Omega/2\Omega_c)$

$$x(n) = x_a(nT) \tag{7.5.1}$$

In some applications, the signals in parts of a system can be discrete-time signals, whereas other signals in the system are analog signals. An example is a system in which a microprocessor is used to process the signals in the system or to provide on-line control. In such *hybrid* or *sampled-data* systems, it is advantageous to consider the sampled signal as being a continuous-time signal so that all the signals in the system can be treated in the same manner. When we consider the sampled signal in this manner, we denote it by $x_s(t)$.

We can write the analog signal $x_a(t)$ in terms of its Fourier transform $X_a(\omega)$ as

$$x_a(t) = \frac{1}{2\pi}\int_{-\infty}^{\infty} X_a(\omega)\exp[j\omega t]\,d\omega \tag{7.5.2}$$

Sample values $x(n)$ can be determined by setting $t = nT$ in Equation (7.5.2), yielding

$$x(n) = x_a(nT) = \frac{1}{2\pi}\int_{-\infty}^{\infty} X_a(\omega)\exp[j\omega nT]\,d\omega \tag{7.5.3}$$

However, since $x(n)$ is a discrete-time signal, we can write it in terms of its *discrete-time* Fourier transform $X(\Omega)$ as

$$x(n) = \frac{1}{2\pi}\int_{-\pi}^{\pi} X(\Omega)\exp[j\Omega n]\,d\Omega \tag{7.5.4}$$

Both Equations (7.5.3) and (7.5.4) represent the same sequence $x(n)$. Hence, the transforms must also be related. In order to find this relation, let us divide the range $-\infty < \omega < \infty$ into equal intervals of length $2\pi/T$ and express the right-hand side of Equation (7.5.3) as a sum of integrals over these intervals:

$$x(n) = \frac{1}{2\pi} \sum_{r=-\infty}^{\infty} \int_{(2r-1)\pi/T}^{(2r+1)\pi/T} X_a(\omega) \exp[j\omega nT]\, d\omega \tag{7.5.5}$$

If we now replace ω by $\omega + 2\pi r/T$, we can write Equation (7.5.5) as

$$x(n) = \frac{1}{2\pi} \sum_{r=-\infty}^{\infty} \int_{-\pi/T}^{\pi/T} X_a\left(\omega + \frac{2\pi}{T} r\right) \exp\left[j\left(\omega + \frac{2\pi r}{T}\right) nT\right] d\omega \tag{7.5.6}$$

Interchanging the orders of summation and integration, and noting that $\exp[j2\pi rnT/T] = 1$, we can write Equation (7.5.6) as

$$x(n) = \frac{1}{2\pi} \int_{-\pi/T}^{\pi/T} \left[\sum_{r=-\infty}^{\infty} X_a\left(\omega + \frac{2\pi}{T} r\right)\right] \exp[j\omega nT]\, d\omega \tag{7.5.7}$$

If we make the change of variable $\omega = \Omega/T$, Equation (7.5.7) becomes

$$x(n) = \frac{1}{2\pi} \int_{-\pi}^{\pi} \left[\frac{1}{T} \sum_{r=-\infty}^{\infty} X_a\left(\frac{\Omega}{T} + \frac{2\pi}{T} r\right)\right] \exp[j\Omega n]\, d\Omega \tag{7.5.8}$$

A comparison of Equations (7.5.8) and (7.5.4) then yields

$$X(\Omega) = \frac{1}{T} \sum_{r=-\infty}^{\infty} X_a\left(\frac{\Omega}{T} + \frac{2\pi}{T} r\right) \tag{7.5.9}$$

We can express this relation in terms of the frequency variable ω by setting

$$\omega = \frac{\Omega}{T} \tag{7.5.10}$$

With this change of variable, the left-hand side of Equation (7.5.9) can be identified as the continuous-time Fourier transform of the sampled signal and is therefore equal to $X_s(\omega)$, the Fourier transform of the signal $x_s(t)$. That is,

$$X_s(\omega) = X(\Omega)\Big|_{\Omega = \omega T} \tag{7.5.11}$$

Also, since the sampling interval is T, the sampling frequency ω_s is equal to $2\pi/T$ rad/s. We can therefore write Equation (7.5.9) as

$$X_s(\omega) = \frac{1}{T} \sum_{r=-\infty}^{\infty} X_a(\omega + r\omega_s) \tag{7.5.12}$$

This is the result that we obtained in Chapter 4 when we were discussing the Fourier transform of sampled signals. It is clear from Equation (7.5.12) that $X_s(\omega)$ is the periodic extension, with period ω_s, of the continuous-time Fourier transform $X_a(\omega)$ of the analog signal $x_a(t)$, amplitude scaled by a factor $1/T$. Suppose that $x_a(t)$ is a low-pass signal, such that its spectrum is zero for $\omega > \omega_0$. Figure 7.5.1 shows the spectra of a typical band-limited analog signal and the corresponding sampled signal. As discussed in

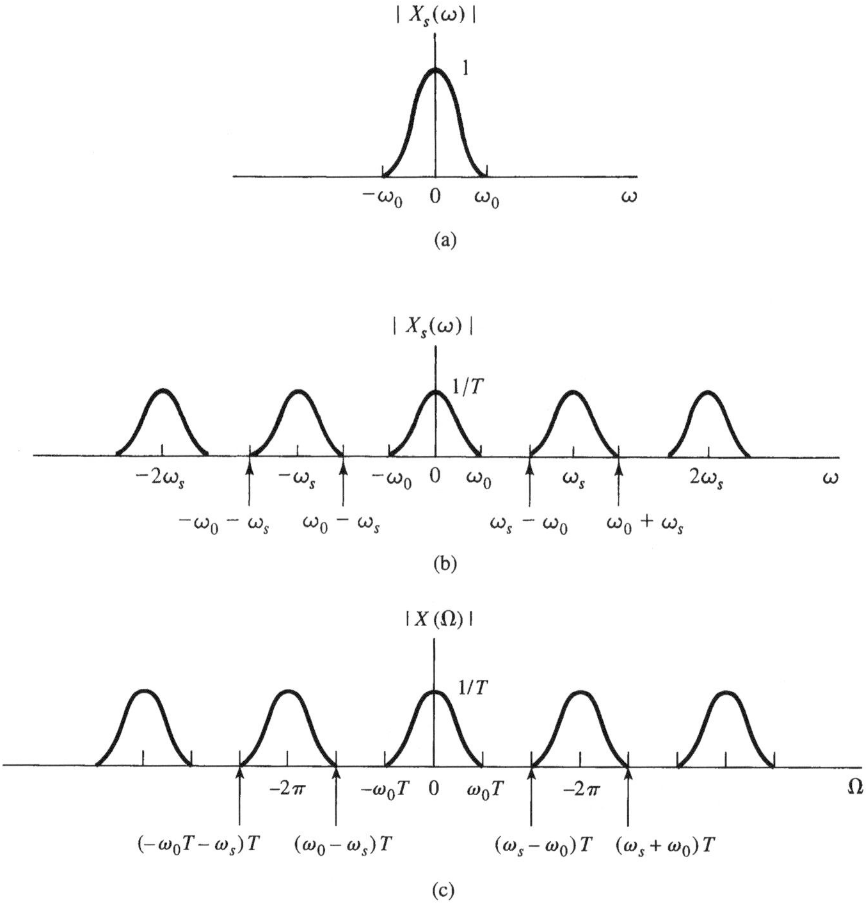

Figure 7.5.1 Spectra of sampled signals. (a) Analog spectrum. (b) Spectrum of $x_s(t)$. (c) Spectrum of $x(n)$.

Chapter 4, and as can be seen from the figure, there is no overlap of the spectral components in $X_s(\omega)$ if $\omega_s - \omega_0 > \omega_0$. We can then recover $x_a(t)$ from the sampled signal $x_s(t)$ by passing $x_s(t)$ through an ideal low-pass filter with a cutoff at ω_0 rad/s and a gain of T. Thus, there is no aliasing distortion if the sampling frequency is such that

$$\omega_s - \omega_0 > \omega_0$$

or

$$\omega_s > 2\omega_0 \tag{7.5.13}$$

This is a restatement of the Nyquist sampling theorem that we encountered in Chapter 4 and specifies the minimum sampling frequency that must be used to recover a continuous-time signal from its samples. Clearly, if $x_a(t)$ is not band limited, there is always an overlap (aliasing).

Equation (7.5.10) describes the mapping between the analog frequency ω and the digital frequency Ω. It follows from this equation that, whereas the units of ω are rad/s, those for Ω are just rad.

From Equation (7.5.11) and the definition of $X(\Omega)$, it follows that the Fourier transform of the signal $x_s(t)$ is

$$X_s(\omega) = \sum_{n=-\infty}^{\infty} x_a(nT)\exp[-j\omega nT] \tag{7.5.14}$$

We can use Equation (7.5.14) to justify the impulse-modulation model for sampled signals that we employed in Chapter 4. From the sifting property of the δ function, we can write Equation (7.5.14) as

$$X_s(\omega) = \frac{1}{2}\int_{-\infty}^{\infty} x_a(t) \sum_{n=-\infty}^{\infty} \delta(t-nT)\exp[-j\omega t]\,dt$$

from which it follows that

$$x_s(t) = x_a(t) \sum_{n=-\infty}^{\infty} \delta(t-nT) \tag{7.5.15}$$

That is, the sampled signal $x_s(t)$ can be considered to be the product of the analog signal $x_a(t)$ and the impulse train $\sum_{n=-\infty}^{\infty} \delta(t-nT)$.

To summarize our discussions so far, when an analog signal $x_a(t)$ is sampled, the sampled signal may be considered to be either a discrete-time signal $x(n)$ or a continuous-time signal $x_s(t)$, as given by Equations (7.5.1) and (7.5.15), respectively. When the sampled signal is considered to be the discrete-time signal $x(n)$, we can find its discrete-time Fourier transform

$$X(\Omega) = \sum_{n=-\infty}^{\infty} x(n)e^{-j\Omega n} \tag{7.5.16}$$

If we consider the sampled signal to be the continuous-time signal $x_s(t)$, we can find its continuous-time Fourier transform by using either Equation (7.5.12) or (7.5.14). However, Equation (7.5.12), being in the form of an infinite sum, is not useful in determining $X_s(\omega)$ in closed form. Still, it enables us to derive the Nyquist sampling theorem, which specifies the minimum sampling frequency ω_s that must be used so that there is no aliasing distortion. From Equation (7.5.11), it follows that, to obtain $X(\Omega)$ from $X_s(\omega)$, we must scale the frequency axis. Therefore, with reference to Figure 7.5.1 (b), to find $X(\Omega)$, we replace ω in Figure 7.5.1 (c) by ωT.

If there is no aliasing, $X_s(\omega)$ is just the periodic repetition of $X_a(\omega)$ at intervals of ω_s, amplitude scaled by the factor $1/T$, so that

$$X_s(\omega) = \frac{1}{T}X_a(\omega) \qquad -\omega_s \le \omega \le \omega_s \tag{7.5.17}$$

Since $X(\Omega)$ is a frequency-scaled version of $X_s(\omega)$ with $\Omega = \omega T$, it follows that

$$X(\Omega) = \frac{1}{T}X_a\left(\frac{\Omega}{T}\right) \qquad -\pi \le \Omega \le \pi \tag{7.5.18}$$

Example 7.5.1

We consider the analog signal $x_a(t)$ with spectrum as shown in Figure 7.5 2 (a). The signal has a one-sided bandwidth $f_0 = 5000$ Hz, or equivalently, $\omega_0 = 2\pi f_0 = 10{,}000\pi$ rad/sec. The minimum sampling frequency that can be used without introducing aliasing is $[\omega_s]_{\min} = 2\omega_0 = 20{,}000\pi$ rad/sec. Thus, the maximum sampling rate that can be used is $T_{\max} = 1/(2f_0) = 100$ μsec.

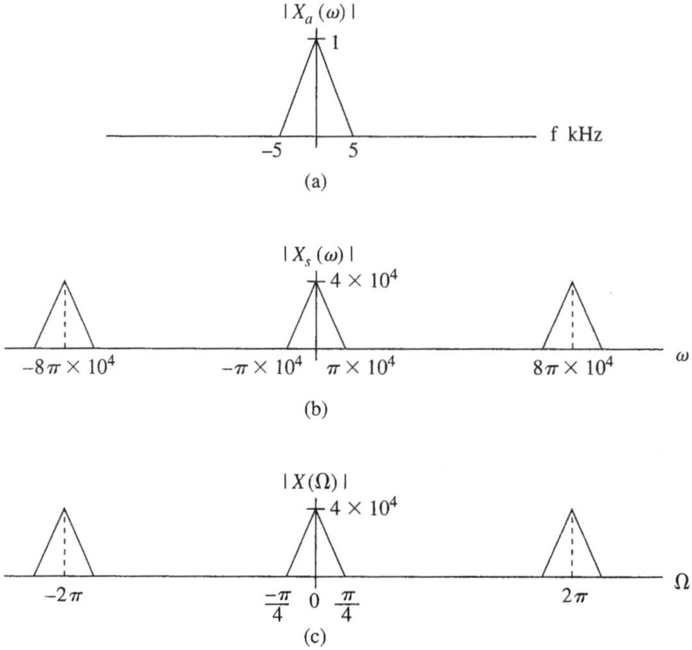

Figure 7.5.2 Spectra for Example 7.5.1.

Suppose we sample the signal at a rate $T = 25$ μsec. Then $\omega_s = 8\pi \times 10^4$ rad/sec. Figure 7.5.2 (b) shows the spectrum $X_s(\omega)$ of the sampled signal. The spectrum is periodic with period ω_s. To get $X(\Omega)$, we simply scale the frequency axis, replacing ω by $\Omega = \omega T$, as shown in Figure 7.5.2 (c). The resulting spectrum is, as expected, periodic with period 2π.

7.5.1 Reconstruction of Sampled Signals

If there is no aliasing, we can recover the analog signal $x_a(t)$ from the samples $x_a(nT)$ by using a *reconstruction filter*. The input to the reconstruction filter is a discrete-time signal, whereas its output is a continuous-time signal. As discussed in Chapter 4, the reconstruction filter is an ideal low-pass filter. Referring to Figure 7.5.1, we see that if we pass $x_s(t)$ through a filter with frequency response function

$$H(\omega) = \begin{cases} T, & |\omega| < \omega_B \\ 0, & \text{otherwise} \end{cases} \tag{7.5.19}$$

with ω_B chosen to lie between ω_0 and $\omega_s - \omega_0$, the spectrum of the filter output will be identical to $X_a(\omega)$, so that the output is equal to $x_a(t)$. For a signal sampled at the Nyquist rate, $\omega_s = 2\omega_0$, so that the bandwidth of the reconstruction filter must be equal to $\omega_B = \omega_s/2 = \pi/T$. In this case, the reconstruction filter is said to be matched to the sampler. The reconstructed output can be determined by using Equation (4.4.9) to obtain

$$x_a(t) = \sum_{n=-\infty}^{\infty} x_a(nT)\frac{\sin[\pi(t - nT)/T]}{\pi(t - nT)/T} \tag{7.5.20}$$

Since the ideal low-pass filter is not causal and hence not physically realizable, in practice we cannot exactly recover $x_a(t)$ from its sample values. Thus, any practical reconstruction filter can only give an approximation to the analog signal. Indeed, as can be seen from Equation. (7.5.20), in order to reconstruct $x_a(t)$ exactly, we need all sample values $x_a(nT)$ for n in the range $(-\infty, \infty)$. However, any realizable filter can use only past samples to reconstruct $x_a(t)$. Among such realizable filters are the *hold circuits*, which are based on approximating $x_a(t)$ in the range $nT \leq t < (n + 1)T$ in a series as

$$\hat{x}_a(t) = x_a(nT) + x_a'(nT)(t - nT) + \frac{1}{2!}x_a''(nT)(t - nT)^2 + \tag{7.5.21}$$

The derivatives are approximated in terms of past sampled values; for example $x_a'(nT) = [x_a(nT) - x_a((n - 1)T)]/T$.

The most widely used of these filters is the zero-order hold, which can be easily implemented. The zero-order hold corresponds to retaining only the first term on the right-hand side in Eq. (7.5.21). That is, the output of the hold is given by

$$\hat{x}_a(t) = x_a(nT) \qquad nT \leq t < (n + 1)T \tag{7.5.22}$$

In other words, the zero-order hold provides a staircase approximation to the analog signal, as shown in Fig. 7.5.3.

Let $g_{h0}(t)$ denote the impulse response of the zero-order hold, obtained by applying a unit impulse $\delta(n)$ to the circuit Since all values of the input are zero except at $n - 0$, it follows that

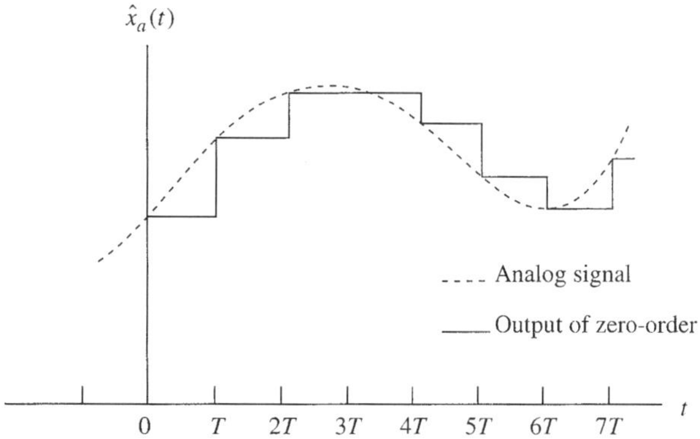

Figure 7.5.3 Recontruction of sampled signal using zero-order hold.

$$g_{h0}(t) = \begin{cases} 1, & 0 \le t < T \\ 0, & \text{otherwise} \end{cases} \tag{7.5.23}$$

with corresponding transfer function

$$G_{h0}(S) = \frac{1 - e^{-sT}}{s} \tag{7.5.24}$$

In order to compare the zero-order hold with the ideal reconstruction filter, let us replace s by $j\omega$ in Eq. (7.5.22) to get

$$\begin{aligned} G_{h0}(\omega) &= \frac{1 - e^{-j\omega T}}{j\omega T} = \frac{2e^{-j(\omega T/2)}}{\omega}\left[\frac{e^{j(\omega T/2)} - e^{-j(\omega T/2)}}{2j}\right] \\ &= T\frac{\sin(\pi\omega/\omega_s)}{\pi\omega/\omega_s} e^{-j(\pi\omega/\omega_s)} \end{aligned} \tag{7.5.25}$$

where we have used $T = 2\pi/\omega_s$.

Figure 7.5.4 shows the magnitude and phase spectra of the zero-order hold as a function of ω. The figure also shows the magnitude and phase spectra of the ideal reconstruction filter matched to ω_s. The presence of the *side lobes* in $G_{h0}(\omega)$ introduces distortion in the reconstructed signal, even when there is no aliasing distortion during sampling. Since the energy in the side lobes is much less in the case of higher order hold circuits, the reconstructed signal obtained with these filters is much closer to the original analog signal.

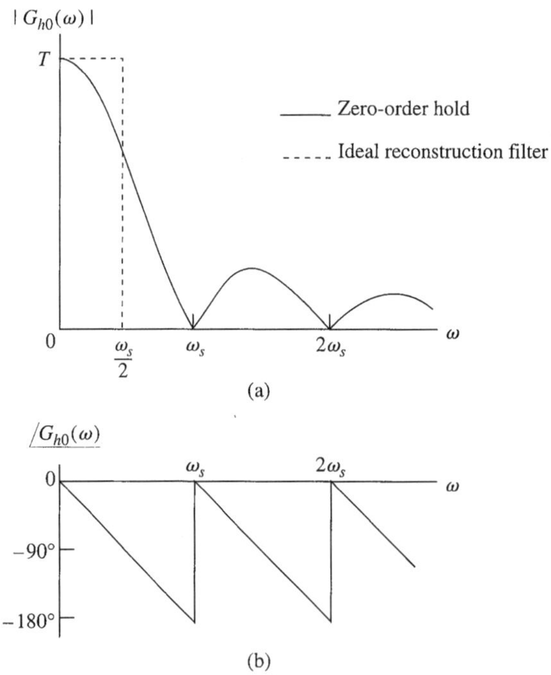

Figure 7.5.4 (a) Magnitude spectrum and (b) phase spectrum of zero-order hold.

An alternative scheme that is also easy to implement obtains the reconstructed signal $\hat{x}_a(t)$ in the interval $[(n-1)T, nT]$ as the straight line joining the values $x_a[(n-1)T]$ and $x_a(nT)$. This interpolator is called a linear interpolator and is described by the input-output relation

$$\hat{x}_a(t) = x_a(nT)\left[1 + \frac{t-nT}{T}\right] - x_a[(n-1)T]\left[\frac{t-nT}{T}\right], (n-1)T \le t \le nT \qquad (7.5.26)$$

It can be easily verified that the impulse response of the linear interpolator is

$$g_l(t) = \begin{cases} 1 - \dfrac{|t|}{T}, & |t| \le T \\ 0, & \text{otherwise} \end{cases} \qquad (7.5.27)$$

with corresponding frequency function

$$G_l(\omega) = T\left[\frac{\sin(\omega T/2)^2}{\omega T/2}\right] \qquad (7.5.28)$$

Note that this interpolator is noncausal. Nonetheless, it applies in areas such as the processing of still image frames, in which interpolation is done in the spatial domain.

7.5.2 Sampling-Rate Conversion

We will conclude our consideration of sampled analog symbols with a brief discussion of changing the sampling rate of a sampled signal. In many applications, we may have to change the sampling rate of a signal as the signal undergoes successive stages of processing by digital filters. For example, if we process the sampled signal by a low-pass digital filter, the bandwidth of the filter output will be less than that of the input. Thus, it is not necessary to retain the same sampling rate at the output. As another example, in many telecommunications systems, the signals involved are of different types with different bandwidths. These signals will therefore have to be processed at different rates.

One method of changing the sampling rate is to pass the sampled signal through a reconstruction filter and resample the resulting signal. Here, we will explore an alternative, which is to change the effective sampling rate in the digital domain. The two basic operations necessary for accomplishing such a conversion are *decimation* (or *downsampling*) and *interpolation* (or *upsampling*).

Suppose we have an analog signal band limited to a frequency ω_0, which has been sampled at a rate T to get the discrete-time signal $x(n)$, with $x(n) = x_a(nT)$. Decimation involves reducing the sampling frequency so that the new sampling rate is $T' = MT$. The new sampling frequency will thus be $\omega_s' = \omega_s/M$. We will restrict ourselves to integer values of M, so that decimation is equivalent to retaining one of every M samples of the sampled signal $x(n)$. The decimated signal is given by

$$x_d(n) = x(Mn) = x_a(nMT) \qquad (7.5.29)$$

Since the effective sampling rate is now $T' = MT$, for no aliasing in the sampled signal, we must have

$$T' \leq \frac{\pi}{\omega_0}$$

or equivalently,

$$MT \leq \frac{\pi}{\omega_0} \tag{7.5.30}$$

For a fixed T, Equation (7.5.30) provides an upper limit on the maximum value that M can take.

If there is no aliasing in the decimated signal, we can use Equation (7.5.18) to write

$$\begin{aligned} X_d(\Omega) &= \frac{1}{T'} X_a\left(\frac{\Omega}{T'}\right), \qquad -\pi \leq \Omega \leq \pi \\ &= \frac{1}{MT} X_a\left(\frac{1}{M}\frac{\Omega}{T}\right), \qquad -\pi \leq \Omega \leq \pi \end{aligned}$$

Since $X(\Omega)$, the discrete-time Fourier transform of the analog signal sampled at the rate T, is equal to

$$X(\Omega) = \frac{1}{T} X_a\left(\frac{\Omega}{T}\right), \; -\pi \leq \Omega \leq \pi$$

it follows that

$$X_d(\Omega) = \frac{1}{M} X\left(\frac{\Omega}{M}\right) \tag{7.5.31}$$

That is, $X_d(\Omega)$ is equal to $X(\Omega)$ amplitude scaled by a factor $1/M$ and frequency scaled by the same factor. This is illustrated in Figure 7.5.5 for the case where $T = 0.4\pi/\omega_0$ and $T' = 2T$.

Increasing the effective sampling rate of an analog signal implies that, given a signal $x(n)$ obtained by sampling an analog signal $x_a(t)$ at a rate T, we want to determine a signal $x_i(n)$ that corresponds to sampling $x_a(t)$ at a rate $T'' = T/L$, where $L > 1$. That is,

$$x_i(n) = x(n/L) = x_a(nT/L) \tag{7.5.32}$$

This process is known as interpolation, since, for each n, it involves reconstructing the missing samples $x(n + mL)$, $m = 1, 2, \ldots, L - 1$. We will again restrict ourselves to integer values of L.

The spectrum of the interpolated signal can be found by replacing T with T/L in Equation (7.5.18). Thus, in the range $-\pi \leq \Omega \leq \pi$,

$$\begin{aligned} X_i(\Omega) &= \frac{L}{T} X_a\left(L\frac{\Omega}{T}\right) \\ &= \begin{cases} LX(L\Omega), & |\Omega| \leq \dfrac{\omega_0 T}{L} \\ 0, & \dfrac{\omega_0 T}{L} < |\Omega| < \pi \end{cases} \end{aligned} \tag{7.5.33}$$

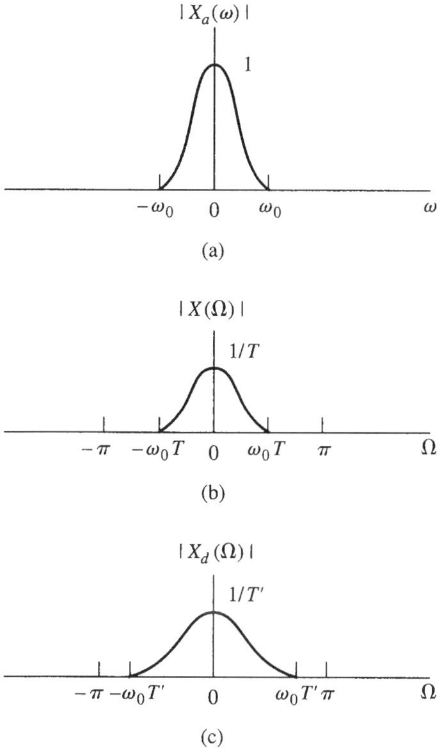

Figure 7.5.5 Illustration of decimation. (a) Spectrum of analog signal. (b) Spectrum of $x(n)$ with sampling rate T. (c) Spectrum of decimated signal corresponding to rate $T' = MT$. Figures correspond to $T = 0.4\pi/\omega_0$ and $M = 2$.

As a first step in determining $x_i(n)$ from $x(n)$, let us replace the missing samples by zeros to form the signal

$$x_l(n) = \begin{cases} x(n/L), & n = 0, \pm L, \pm 2L, \ldots \\ 0, & \text{otherwise} \end{cases} \tag{7.5.34}$$

Then

$$\begin{aligned} X_l(\Omega) &= \sum_{n=-\infty}^{\infty} x_l(n) e^{-j\Omega n} \\ &= \sum_{n=-\infty}^{\infty} x(n/L) e^{-j\Omega n} \\ &= \sum_{k=-\infty}^{\infty} x(k) e^{-j\Omega k L} = X(L\Omega) \end{aligned} \tag{7.5.35}$$

so that $X_l(\Omega)$ is a frequency-scaled version of $X(\Omega)$. The relation between these various spectra is shown in Figure 7.5.6, for the case when $T = 0.4\pi/\omega_0$ and $L = 2$.

From the figure, it is clear that if we pass $x_l(n)$ through a low-pass digital filter with gain L and cutoff frequency $\omega_0 T/L$, the output will correspond to $x_i(n)$. Interpolation by a factor L therefore consists of interspersing $L - 1$ zeros between samples and then low-pass filtering the resulting signal.

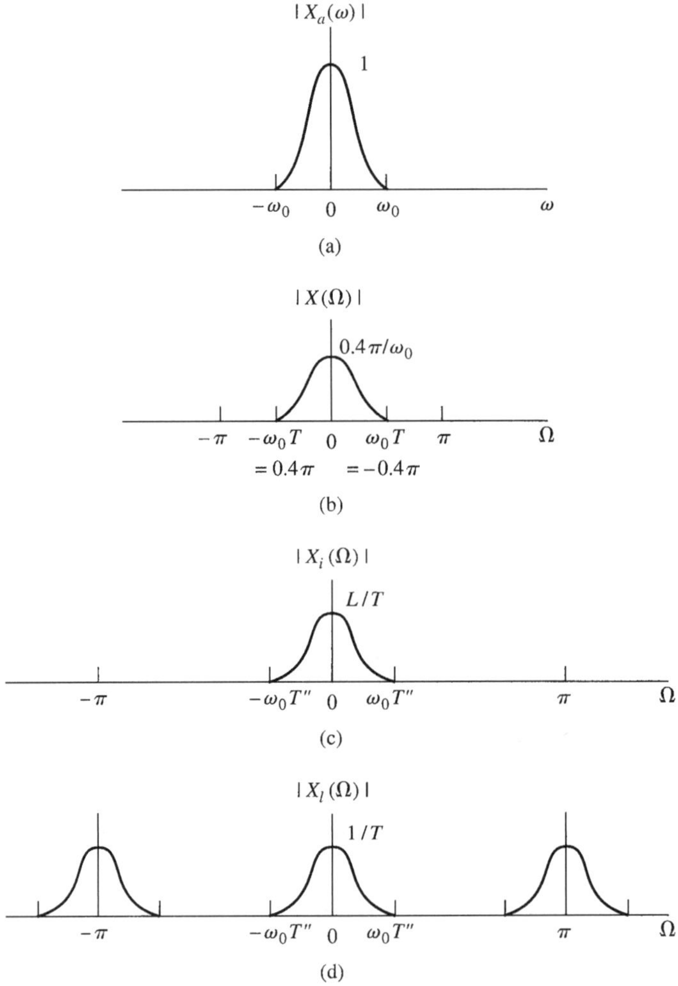

Figure 7.5.6 Illustration of interpolation. (a) Spectrum of analog signal. (b) Spectrum of $x(n)$ with sampling rate T. (c) Spectrum of interpolated signal corresponding to rate $T'' = T/L$. (d) Spectrum of signal $x_l(n)$. Figures correspond to $T = 0.4\pi/\omega_0$ and $L = 2$.

Example 7.5.2

Consider the signal of Example 7.5.1, which was band limited to $\omega_0 = 10{,}000\pi$ rad/sec, so that $T_{\max} = 100\mu s$. Suppose $x_a(t)$ is sampled at $T = 25\mu s$ to obtain the signal $x(n)$ with spectrum $X(\Omega)$ as shown in Figure 7.5.7 (b). If we want to decimate $x(n)$ by a factor M without introducing aliasing, it is clear from Equation (7.5.30) that $M \leq (\pi/\omega_0 T) = 4$.

Suppose we decimate $x(n)$ by $M = 3$, so that the effective sampling rate is $T' = 75\mu s$. It follows from Equation (7.5.31) and Figure 7.5.5(c) that the spectrum of the decimated

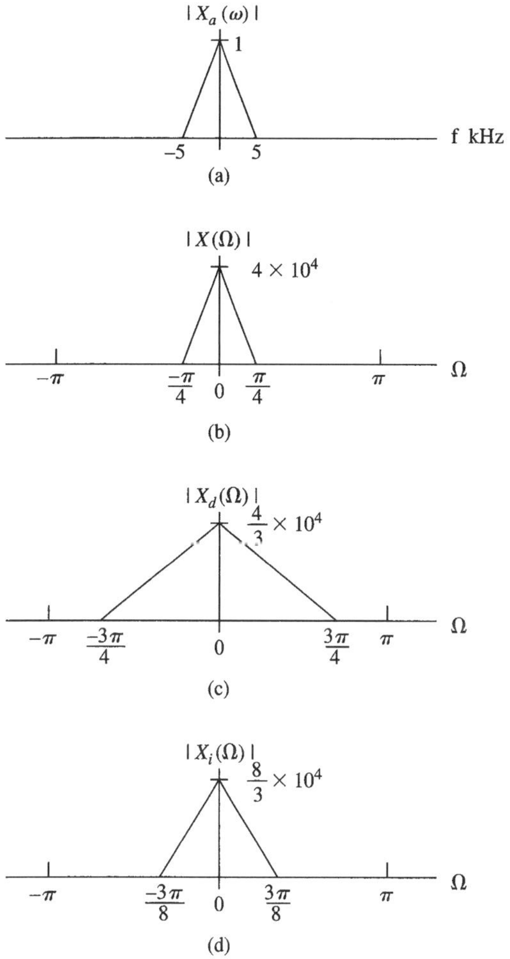

Figure 7.5.7 Spectra for Example 7.5.2. (a) Analog spectrum. (b) Spectrum of sampled signal. (c) Spectrum of decimated signal. (d) Spectrum of interpolated signal after decimation.

signal, $X_d(\Omega)$, is found by amplitude and frequency scaling $X(\Omega)$ by the factor 1/3. The resulting spectrum is shown in Figure 7.5.7(c).

Let us now interpolate the decimated signal $x_d(n)$ by $L = 2$ to form the interpolated signal $x_i(n)$. It follows from Equation (7.5.33) that

$$X_i(\Omega) = \begin{cases} 2X_d(2\Omega), & |\Omega| \le \dfrac{\omega_0 T'}{L} = \dfrac{3\pi}{8} \\ 0, & \dfrac{3\pi}{8} < |\Omega| < \pi \end{cases}$$

Figure 7.5.7 (d) shows the spectrum of the interpolated signal. From our earlier discussion, it follows that interpolation is achieved by interspersing a zero between each two samples of $x_d(n)$ and low-pass filtering the result with a filter with gain 2 and cutoff frequency $(3\pi/8)$ rad/sec.

Note that the combination of decimation and interpolation gives us an effective sampling rate of $T'' = MT/L = 37.5\mu s$. In general, by suitably choosing M and L, we can change the sampling rate by any rational multiple of it.

7.5.3 A/D and D/A Conversion

The application of digital signal processing techniques in areas such as communication, speech, and image processing, to cite only a few, has been significantly facilitated by the rapid development of new technologies and some important theoretical contributions. In such applications, the underlying analog signals must be converted into a sequence of samples before they can be processed. After processing, the discrete-time signals must be converted back into analog form. The term "digital signal processing" usually implies that after time sampling, the amplitude of the signal is quantized into a finite number of levels and converted into binary form suitable for processing using for example, a digital computer. The process of converting an analog signal to a binary representation is referred to as *analog-to-digital* (A/D) conversion, and the process of converting the binary representation back into an analog signal is called *digital-to-analog* (D/A) conversion. Figure 7.5.8 shows a functional block diagram for processing an analog signal using digital signal-processing techniques.

As we have seen, the sampling operation converts the analog signal into a discrete-time signal whose amplitude can take on a continuum of values; that is, the amplitude is represented by an *infinite-precision* number. The quantization operation converts the amplitude into a *finite-precision* number. The binary coder converts this finite precision number into a string of ones and zeros. While each of the operations depicted in the figure can introduce errors in the representation of the analog signal in digital form, in most applications the encoding and decoding processes can be carried out without any significant error. We will, therefore, not consider these processes any further. We have already discussed the sampling operation, associated errors, and methods for reducing these errors. We have also considered various schemes for the reconstruction of sampled signals. In this section, we will briefly look at the quantization of the amplitude of a discrete-time signal.

The quantizer is essentially a device that assigns one of a finite set of values to a signal which can take on a continuum of values over its range. Let $[x_l, x_h]$ denote the range of values, D, of the signal $x_a(t)$. We divide the range into intervals $[x_{i-1}, x_i]$, $i = 1, 2, \dots N$, with $x_1 = x_l$ and $x_N = x_h$. We then assign a value y_i, $i = 1, 2 \dots,$

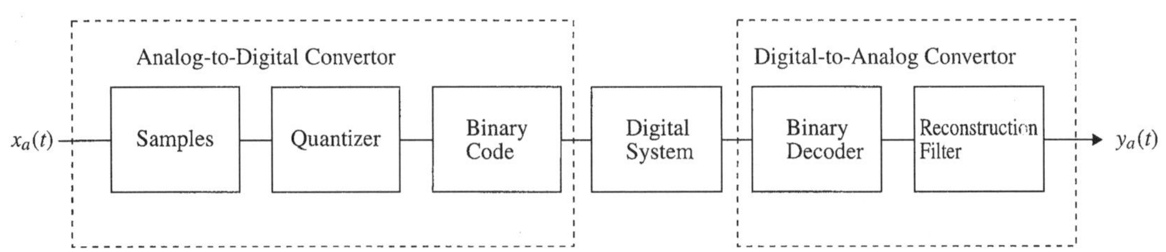

Figure 7.5.8 Functional block diagram of the A/D and D/A processes.

N to the signal whenever $x_{i-1} \leq x_a(t) < x_i$. Thus, N represents the number of levels of the quantizer. The x_i are known as the decision levels, and the y_i are known as the reconstruction levels. Even though the dynamic range of the input signal is usually not known exactly, and the values x_l and x_h are educated guesses, it may be expected that values of $x_a(t)$ outside this range occur not too frequently. All values $x_a(t) < x_l$ are assigned to x_l, while all values $x_a(t) > x_h$ are assigned to x_h.

For best performance, the decision and reconstruction levels must be chosen to match the characteristics of the input signals. This is, in general, a fairly complex procedure. However, optimal quantizers have been designed for certain classes of signals. *Uniform quantizers* are often used in practice because they are easy to implement. In such quantizers, the differences $x_i - x_{i-1}$ and $y_i - y_{i-1}$ are chosen to be the same value—say, Δ—which is referred to as the step size. The step size is related to the dynamic range D and the number of levels N as

$$\Delta = \frac{D}{N} \tag{7.5.36}$$

Figures 7.5.9 (a) and (b) show two variations of the uniform quantizer—namely, the *midriser* and the *midtread*. The difference between the two is that the output in the midriser quantizer is not assigned a value of zero. The midtread quantizer is useful in situations where the signal level is very close to zero for significant lengths of time—for example, the level of the error signal in a control system.

Since there are eight and seven output levels, respectively, for the quantizers shown in Figures 7.5.9(a) and (b), if we use a fixed-length code word, each output value can be represented by a three-bit code word, with one code word left over for the midtread quantizer. In what follows, we will restrict our discussion to the midriser quantizer. In that case, for a quantizer with N levels, each output level can be represented by a code word of length

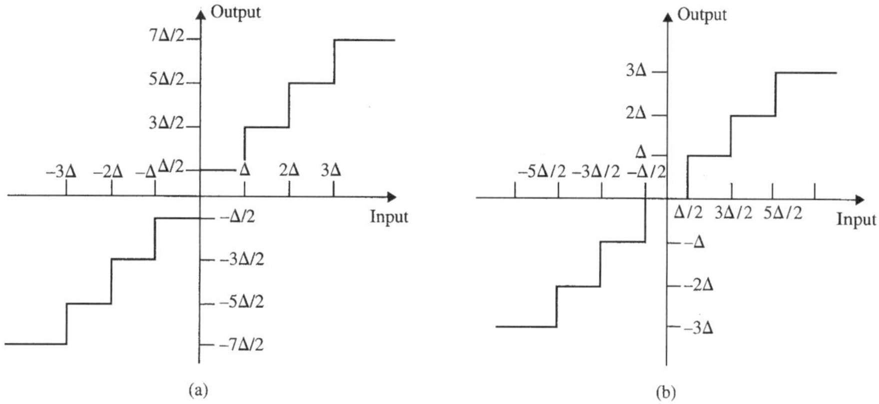

Figure 7.5.9 Input/output relation for uniform quantizer. (a) Midriser quantizer. (b) Midtread quantizer.

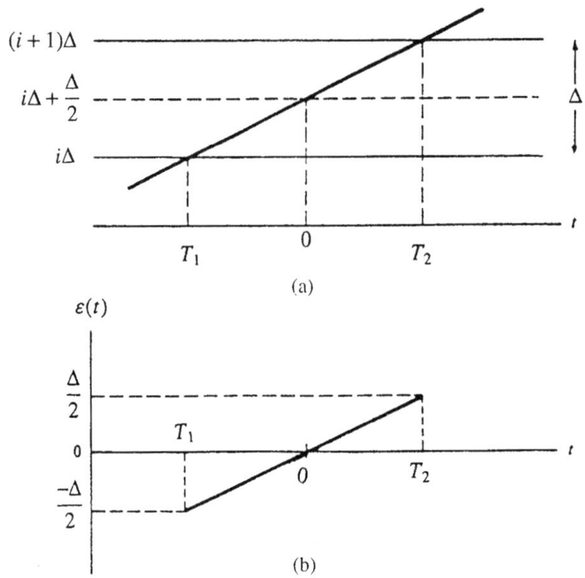

Figure 7.5.10 Quantization error. (a) Quantizer input. (b) Error.

$$B = \log_2 N = \log_2 \frac{D}{\Delta} \tag{7.5.37}$$

The proper analysis of the errors introduced by quantization requires the use of techniques that are outside the scope of this book. However, we can get a fairly good understanding of these errors by assuming that the input to the quantizer is a signal which increases linearly with time at a rate S units/s. Then the input assumes values in any specific range of the quantizer—say, $[i\Delta, (i+1)\Delta]$—for a duration $[T_1, T_2]$, with $T_2 - T_1 = \Delta/S$ as shown in Figure 7.5.10. The quantizer input over this time can be easily verified to be

$$x_a(t) = \frac{\Delta}{T_2 - T_1}(t - T_1) + i\Delta$$

while the output is

$$x_Q(t) = i\Delta + \frac{\Delta}{2}$$

The quantization error, $\varepsilon(t)$, is defined as the difference between the input and the output. We have

$$\begin{aligned}\varepsilon(t) = x_a(t) - x_Q(t) &= \frac{\Delta}{T_2 - T_1}(t - T_1) - \frac{\Delta}{2} \\ &= \frac{\Delta}{T_2 - T_1}\left[t - \frac{T_2 + T_1}{2}\right]\end{aligned} \tag{7.5.38}$$

It is clear that $\varepsilon(t)$ also increases linearly from $-\Delta/2$ to $\Delta/2$ during the interval $[T_1, T_2]$. The mean-square value of the error signal is therefore given by (see Problem 7.27)

$$E = \frac{1}{T_2 - T_1}\int_{T_1}^{T_2} \varepsilon^2(t)\,dt = \frac{\Delta^2}{12} \tag{7.5.39}$$
$$= \frac{D^2}{12}2^{-2B}$$

where the last step follows from Equation (7.5.37). E is usually referred to as the quantization noise power.

It can be shown that if the number of quantizer levels, N, is very large, Equation (7.5.39) still provides a very good approximation to the mean-square value of the quantization error for a wide variety of input signals.

In conclusion, we note that a quantitative measure of the quality of a quantizer is the signal-to-noise ratio (SNR), which is defined as the ratio of the quantizer input signal power P_x to the quantizer noise power E. From Equation (7.5.39), we can write

$$\text{SNR} = \frac{P_x}{E} = 12P_xD^{-2}2^{2B} \tag{7.5.40}$$

In decibels,

$$(\text{SNR})_{\text{dB}} = 10\log_{10}\text{SNR}$$
$$= 10\log_{10}(12) + 10\log_{10}P_x - 20\log_{10}D + 20B\log_{10}(2) \tag{7.5.41}$$

That is,

$$(\text{SNR})_{\text{dB}} = 10.79 + 10\log_{10}P_x + 6.02B - 20\log_{10}D \tag{7.5.42}$$

As can be seen from the last equation, increasing the code-word length by one bit results in an approximately 6-dB improvement in the quantizer SNR. The equation also shows that the assumed dynamic range of the quantizer must be matched to the input signal. The choice of a very large value for D reduces the SNR.

Example 7.5.3

Let the input to the quantizer be the signal

$$x_a(t) = A\sin\omega_0 t$$

The dynamic range of this signal is $2A$, and the signal power is $P_x = A^2/2$. The use of Equation (7.5.41) gives the SNR for this input as

$$(\text{SNR})_{\text{dB}} = 20\log_{10}(1.5) + 6.02B = 1.76 + 6.02B$$

Note that in this case D was exactly equal to the dynamic range of the input signal. The SNR is independent of the amplitude A of the signal.

7.6 SUMMARY

- A periodic discrete-time signal $x(n)$ with period N can be represented by the discrete-time Fourier series (DTFS)

$$x(n) = \sum_{k=0}^{N-1} a_k \exp\left[j\frac{2\pi}{N}kn\right]$$

- The DTFS coefficients a_k are given by

$$a_k = \frac{1}{N}\sum_{n=0}^{N-1} x(n)\exp\left[-j\frac{2\pi}{N}kn\right]$$

- The coefficients a_k are periodic with period N, so that $a_k = a_k + N$.
- The DTFS is a finite sum over only N terms. It provides an *exact* alternative representation of the time signal, and issues such as convergence or the Gibbs phenomenon do not arise.
- If a_k are the DTFS coefficients of the signal $x(n)$, then the coefficients of $x(n - m)$ are equal to $a_k \exp[-j(2\pi/N)km]$.
- If the periodic sequence $x(n)$ with DTFS coefficients a_k is input into an LTI system with impulse response $h(n)$, the DTFS coefficients b_k of the output $y(n)$ are given by

$$b_k = a_k H\left(\frac{2\pi}{N}k\right)$$

where

$$H(\Omega) = \sum_{n=-\infty}^{\infty} h(n)\exp[-j\Omega n]$$

- The discrete-time Fourier transform (DTFT) of an aperiodic sequence $x(n)$ is given by

$$X(\Omega) = \sum_{n=-\infty}^{\infty} x(n)\exp[-j\Omega n]$$

- The inverse relationship is given by

$$x(n) = \frac{1}{2\pi}\int_0^{2\pi} X(\Omega)\exp[j\Omega n]\,d\Omega$$

- The DTFT variable Ω has units of radians.
- The DTFT is periodic in Ω with period 2π, so that $X(\Omega) = X(\Omega + 2\pi)$.
- Other properties of the DTFT are similar to those of the continuous-time Fourier transform. In particular, if $y(n)$ is the convolution of $x(n)$ and $h(n)$, then,

$$Y(\Omega) = H(\Omega)X(\Omega)$$

- When the analog signal $x_a(t)$ is sampled, the resulting signal can be considered to be either a CT signal $x_s(t)$ with Fourier transform $X_s(\omega)$ or a DT sequence $x(n)$ with DTFT $X(\Omega)$. The relation between the two is

$$X_s(\omega) = X(\Omega)\big|_{\Omega=\omega T}$$

- The preceding equation can be used to derive the impulse modulation model for sampling:

$$x_s(t) = x_a(t)\sum_{n=-\infty}^{\infty}\delta(t - nT)$$

- The transform $X_s(\omega)$ is related to $X_a(\omega)$ by

$$X_s(\omega) = \frac{1}{T}\sum_{r=-\infty}^{\infty} X_a(\omega + r\omega_s)$$

- We can use the last equation to derive the *sampling theorem*, which gives the minimum rate at which an analog signal must be sampled to permit error-free reconstruction. If the signal has a bandwidth of ω_c, then $T < 2\pi/\omega_c$.
- The ideal reconstruction filter is an ideal low-pass filter. Hold circuits are practical reconstruction filters that approximate the analog signal.
- The zero-order hold provides a staircase approximation to the analog signal.
- Changing the sampling rate involves decimation and interpolation. Decimation by a factor M implies retaining only one of every M samples. Interpolation by a factor L requires reconstruction of $L - 1$ missing samples between every two samples of the original sampled signal.
- By using a suitable combination of M and L, the sampling rate can be changed by any rational factor.
- The process of representing an analog signal by a string of binary numbers is known as analog-to-digital (AD) conversion. Conceptually, the process consists of sampling, amplitude quantization, and conversion to a binary code.
- The process of digital-to-analog (D/A) conversion consists of decoding the digital sequence and passing it through a reconstruction filter.
- A quantizer outputs one of a finite set of values corresponding to the amplitude of the input signal.
- Uniform quantizers are widely used in practice. In such quantizers, the output values differ by a constant value. The SNR of a uniform quantizer increases by approximately 6 dB per bit.

7.7 CHECKLIST OF IMPORTANT TERMS

Aliasing
Convergence of DTFS
Decimation
Discrete-time Fourier series
Discrete-time Fourier transform
DTFS coefficients
Impulse-modulation model
Interpolation
Inverse DTFT
Periodicity of DTFS coefficients
Periodicity of DTFT
Sampling of analog signals
Sampling theorem
Zero-order hold

7.8 PROBLEMS

7.1. Determine the Fourier-series representation for each of the following discrete-time signals. Plot the magnitude and phase of the Fourier coefficients a_k.

(a) $x(n) = \cos\frac{3\pi n}{4}$

(b) $x(n) = 4\cos\frac{2\pi n}{3}\sin\frac{\pi n}{2}$

(c) $x(n)$ is the periodic extension of the sequence $\{1, -1, 0, 1, -1\}$.

(d) $x(n)$ is periodic with period 8, and

$$x(n) = \begin{cases} 1, & 0 \le n \le 3 \\ 0, & 4 \le n \le 7 \end{cases}$$

(e) $x(n)$ is periodic with period 6, and

$$x(n) = \begin{cases} n, & 0 \le n \le 3 \\ 0, & 4 \le n \le 5 \end{cases}$$

(f) $x(n) = \sum_{k=-\infty}^{\infty} (-1)^k\delta(n-k) + \cos\frac{2\pi n}{3}$

7.2. Given a periodic sequence $x(n)$ with the following Fourier-series coefficients, determine the sequence:

(a) $a_k = 1 + \frac{1}{2}\cos\frac{\pi k}{2} + 2\cos\frac{\pi k}{4}, \quad 0 \le k \le 8$

(b) $a_k = \begin{cases} 1, & 0 \le k \le 3 \\ 0, & 4 \le k \le 7 \end{cases}$

(c) $a_k = \exp[-j\pi k/4], \quad 0 \le k \le 7$

(d) $a_k = \{1, 0, -1, 0, 1\}$

7.3. Let a_k represent the Fourier series coefficients of the periodic sequence $x(n)$ with period N. Find the Fourier-series coefficients of each of the following signals in terms of a_k:

(a) $x(n - n_0)$

(b) $x(-n)$

(c) $(-1)^n x(n)$

(d)

$$y(n) = \begin{cases} x(n), & n \text{ even} \\ 0, & n \text{ odd} \end{cases}$$

(*Hint*: $y(n)$ can be written as $\frac{1}{2}[x(n) + (-1)^n x(n)]$.)

(e)

$$y(n) = \begin{cases} x(n), & n \text{ odd} \\ 0, & n \text{ even} \end{cases}$$

(f) $y(n) = x_e(n)$

(g) $y(n) = x_o(n)$

7.4. Show that for a real periodic sequence $x(n)$, $a_k = a_{N-k}^*$.

7.5. Find the Fourier-series representation of the output $y(n)$ when each of the periodic signals in Problem 7.1 is input into a system with impulse response $h(n) = \left(\frac{1}{2}\right)^n u(n)$.

7.6. Repeat Problem 7.5 if $h(n) = \left(\frac{1}{2}\right)^{|n|}$

7.7. Let $x(n)$, $h(n)$, and $y(n)$ be periodic sequences with the same period N, and let a_k, b_k, and c_k be the respective Fourier-series coefficients.

(a) Let $y(n) = x(n)h(n)$. Show that

$$c_k = \sum_{\langle N \rangle} a_m b_{k-m} = \sum_{\langle N \rangle} a_{k-m} b_m$$
$$= a_k \circledast b_k$$

(b) Let $y(n) = x(n) \circledast h(n)$. Show that

$$c_k = N a_k b_k$$

7.8. Consider the periodic extensions of the following pairs of signals:

(a) $x(n) = \{1, 0, 1, 0, 0, 1\}$

$h(n) = \{1, -1, 1, 0, -1, 1\}$

(b) $x(n) = \cos \dfrac{\pi n}{3}$

$h(n) = \{1, -1, 1, 1, -1, 1\}$

(c) $x(n) = 2\cos \dfrac{\pi n}{2}$

$h(n) = \left\{1, \dfrac{-1}{2}, \dfrac{1}{4}, \dfrac{-1}{8}\right\}$

(d) $x(n) = 1, \quad 0 \le n \le 7$

$$h(n) = \begin{cases} n + 1, & 0 \le n \le 3 \\ -n + 8, & 4 \le n \le 7 \end{cases}$$

Let $y(n) = h(n)x(n)$. Use the results of Problem 7.7 to find the Fourier-series coefficients for $y(n)$.

7.9. Repeat Problem 7.8 if $y(n) = h(n) \circledast x(n)$

7.10. Show that

$$\frac{1}{2\pi} \int_0^{2\pi} \exp[j\Omega(n - k)]\, d\Omega = \delta(n - k)$$

7.11. By successively differentiating Equation (7.3.2) with respect to Ω, show that

$$\mathcal{F}\{n^p x(n)\} = j^p \frac{d^p X(\Omega)}{d\Omega}$$

7.12. Use the properties of the discrete-time transform to determine $X(\Omega)$ for the following sequences:

(a) $x(n) = \begin{cases} 1, & 0 \le n \le n_0 \\ 0, & \text{otherwise} \end{cases}$

(b) $x(n) = n\left(\dfrac{1}{3}\right)^{|n|}$

(c) $x(n) = a^n \cos \Omega_0 n\, u(n), \quad |a| < 1$

(d) $x(n) = \exp[j3n]$

(e) $x(n) = \exp\left[j\frac{\pi}{8}n\right]$

(f) $x(n) = \frac{1}{2}\sin 3n + 4\cos\frac{\pi}{8}n$

(g) $x(n) = \alpha^n[u(n) - u(n - n_0)]$

(h) $x(n) = \frac{\sin(\pi n/3)}{\pi n}$

(i) $x(n) = \frac{\sin(\pi n/3)\sin(\pi n/2)}{\pi^2 n^2}$

(j) $x(n) = \frac{\sin(\pi n/3)\sin(\pi n/2)}{\pi n}$

(k) $x(n) = (n+1)a^n u(n), \quad |a| < 1$

7.13. Find the discrete-time sequence $x(n)$ with transforms in the range $0 \leq \Omega < 2\pi$ as follows:

(a) $X(\Omega) = -j\pi\delta\left(\Omega - \frac{\pi}{3}\right) + \pi\delta\left(\Omega - \frac{2\pi}{3}\right) + \pi\delta\left(\Omega - \frac{4\pi}{3}\right) + j\pi\delta\left(\Omega - \frac{5\pi}{3}\right)$

(b) $X(\Omega) = 4\sin 5\Omega + 2\cos\Omega$

(c) $X(\Omega) = \frac{4}{[\exp(-j\Omega) - 2]^2}$

(d) $X(\Omega) = \frac{1 - \frac{5}{6}\exp[-j\Omega]}{1 + \frac{1}{12}\exp[-j\Omega] - \frac{1}{12}\exp[-j2\Omega]}$

7.14. Show that

$$\sum_{n=-\infty}^{\infty} |x(n)|^2 = \frac{1}{2\pi}\int_{-\pi}^{\pi} |X(\Omega)|^2 d\Omega$$

7.15. The following signals are input into a system with impulse response $h(n) = \left(\frac{1}{2}\right)^n u(n)$. Use Fourier transforms to find the output $y(n)$ in each case.

(a) $x(n) = \left(\frac{1}{3}\right)^n\left(\cos\frac{\pi n}{2}\right)u(n)$

(b) $x(n) = \left(\frac{1}{3}\right)^n \sin\left(\frac{\pi n}{2}\right)u(n)$

(c) $x(n) = \left(\frac{1}{3}\right)^{|n|}$

(d) $x(n) = n\left(\frac{1}{3}\right)^{|n|} u(n)$

7.16. Repeat Problem 7.15 if $h(n) = \delta(n-1) + \left(\frac{1}{4}\right)^n u(n)$.

7.17. For the LTI system with impulse response

$$h(n) = \frac{\sin(\pi n/2)}{\pi n}$$

find the output if the input $x(n)$ is as follows:

(a) A periodic square wave with period 6 such that

$$x(n) = \begin{cases} 1, & 0 \leq n \leq 3 \\ 0, & 4 \leq n \leq 5 \end{cases}$$

(b) $x(n) = \sum_{k=-\infty}^{\infty} [\delta(n - 2k) - \delta(n - 1 - 2k)]$

7.18. Repeat Problem 7.17 if

$$h(n) = 2\frac{\sin(\pi n/4)}{\pi n}$$

7.19. (a) Use the time-shift property of the Fourier transform to find $H(\Omega)$ for the systems in Problem 6.18.

(b) Find $h(n)$ for these systems by inverse tranforming $H(\Omega)$.

7.20. The frequency response of a discrete-time system is given by

$$H(\Omega) = \frac{\frac{1}{2} + \frac{1}{12}\exp[-j\Omega]}{1 + \frac{5}{6}\exp[-j\Omega] + \frac{1}{6}\exp[-j2\Omega]}$$

(a) Find the impulse response of the system.

(b) Find the difference equation representation of the system.

(c) Find the response of the system if the input is the signal $\left(\frac{1}{2}\right)^n u(n)$.

7.21. A discrete-time system has a frequency response

$$H(\Omega) = \frac{\alpha + \exp[-j\Omega]}{1 + \beta\exp[-j\Omega]}, \quad |\beta| < 1$$

Assume that β is fixed. Find α such that $H(\Omega)$ is an all-pass function—that is, $|H(j\Omega)|$ is a constant for all Ω. (Do not assume that β is real.)

7.22. (a) Consider the causal system with frequency response

$$H(\Omega) = \frac{1 + a\exp[-j\Omega] + b\exp[-j2\Omega]}{b + a\exp[-j\Omega] + \exp[-j2\Omega]}$$

Show that this is an all-pass function if a and b are real.

(b) Let $H(\Omega) = N(\Omega)/D(\Omega)$, where $N(\Omega)$ and $D(\Omega)$ are polynomials in $\exp[-j\Omega]$. Can you generalize your result in part (a) to find the relation between $N(\Omega)$ and $D(\Omega)$ so that $H(\Omega)$ is an all-pass function?

7.23. An analog signal $x_a(t) = 5\cos(200\pi t - 30°)$ is sampled at a frequency f_s in Hz.

(a) Plot the Fourier spectrum of the sampled signal if f_s is (i) 150 Hz (ii) 250 Hz.

(b) Explain whether $x_a(t)$ can be recovered from the samples, and if so, how.

7.24. Derive Equation (7.5.27) for the impulse response of the linear interpolator of Equation (7.5.26), and show that the corresponding frequency function is as given in Equation (7.5.28).

7.25. A low-pass signal with a bandwidth of 1.5 kHz is sampled at a rate of 10,000 samples/s.

(a) We want to decimate the sampled signal by a factor M. How large can M be without introducing aliasing distortion in the decimated signal?

(b) Explain how you can change the sampling rate from 10,000 samples/s to 4000 samples/s.

7.26. An analog signal with spectrum

$$X_a(\omega) = \begin{cases} 1 - \dfrac{|\omega|}{1000}, & |\omega| \le 1000 \\ 0, & \text{otherwise} \end{cases}$$

is sampled at a frequency $\omega_s = 10{,}000$ rad/s.

(a) Sketch the spectrum of the sampled signal.

(b) If it is desired to decimate the signal by a factor M, what is the largest value of M that can be used without introducing aliasing distortion?

(c) Sketch the spectrum of the decimated signal if $M = 4$.

(d) The decimated signal in (c) is to be processed by an interpolator to obtain a sampling frequency of 7500 rad/s. Sketch the spectrum of the interpolated signal.

7.27. Verify for the uniform quantizer that the mean-square value of the error, E, is equal to $\Delta^2/12$, where Δ is the step size.

7.28. A uniform quantizer is to be designed for a signal with amplitude assumed to lie in the range ± 20.

(a) Find the number of quantizer levels needed if the quantizer SNR must be at least 5 dB. Assume that the signal power is 10.

(b) If the dynamic range of the signal is $[-10, 10]$, what is the resulting SNR?

7.29. Repeat Problem 7.27 if the quantizer SNR is to be at least 10 db.

Chapter 8

The Z-Transform

8.1 INTRODUCTION

In this chapter, we study the Z-transform, which is the discrete-time counterpart of the Laplace transform that we studied in Chapter 5. Just as the Laplace transform provides us a frequency-domain technique for analyzing signals for which the Fourier transform does not exist, the Z-transform enables us to analyze certain discrete-time signals that do not have a discrete-time Fourier transform. As might be expected, the properties of the Z-transform closely resemble those of the Laplace transform, so that the results are similar to those of Chapter 5. However, as with Fourier transforms of continuous and discrete-time signals, there are certain differences.

The relationship between the Laplace transform and the Z-transform can be established by considering the sequence of samples obtained by sampling an analog signal $x_a(t)$. In our discussion of sampled signals in Chapter 7, we saw that the output of the sampler could be considered to be either the continuous-time signal

$$x_s(t) = \sum_{n=-\infty}^{\infty} x_a(nT)\delta(t - nT) \tag{8.1.1}$$

or the discrete-time signal

$$x(n) = x_a(nT) \tag{8.1.2}$$

Thus, the Laplace transform of $x_s(t)$ is

$$\begin{aligned} X_s(s) &= \int_{-\infty}^{\infty} \sum_{n=-\infty}^{\infty} x_a(nT)\exp[-st]\,dt \\ &= \sum_{n=-\infty}^{\infty} x_a(nT)\exp[-nTs]\,dt \end{aligned} \tag{8.1.3}$$

where the last step follows from the sifting property of the δ function. If we make the substitution $z = \exp[Ts]$, then

$$X_s(S)\big|_{z=\exp[sT]} = \sum_{n=-\infty}^{\infty} x_a(nT)z^{-n} \tag{8.1.4}$$

The summation on the right side of Equation (8.1.4) is usually written as $X(z)$ and defines the Z-transform of the discrete-time signal $x(n)$.

We have, in fact, already encountered the Z-transform in Section 7.1, where we discussed the response of a linear, discrete-time, time-invariant system to exponential inputs. There we saw that if the input to the system was $x(n) = z^n$, the output was

$$y(n) = H(z)z^n \tag{8.1.5}$$

where $H(z)$ was defined in terms of the impulse response $h(n)$ of the system as

$$H(z) = \sum_{n=-\infty}^{\infty} h(n)z^{-n} \tag{8.1.6}$$

Equation (8.1.6) thus defines the Z-transform of the sequence $h(n)$. We will formalize this definition in the next section and subsequently investigate the properties and look at the applications of the Z-transform.

8.2 THE Z-TRANSFORM

The Z-transform of a discrete-time sequence $x(n)$ is defined as

$$X(z) = \sum_{n=-\infty}^{\infty} x(n)z^{-n} \tag{8.2.1}$$

where z is a complex variable. For convenience, we sometimes denote the Z-transform as $Z[x(n)]$. For causal sequences, the Z-transform becomes

$$X(z) = \sum_{n=0}^{\infty} x(n)z^{-n} \tag{8.2.2}$$

To distinguish between the two definitions, as with the Laplace transform, the transform in Equation (8.2.1) is usually referred to as the bilateral transform, and the transform in Equation (8.2.2) is referred to as the unilateral transform.

Example 8.2.1

Consider the unit-sample sequence

$$x(n) = \begin{cases} 1, & n = 0 \\ 0, & n \neq 0 \end{cases} \tag{8.2.3}$$

The use of Equation (8.2.1) yields

$$X(z) = 1 \cdot z^0 = 1 \tag{8.2.4}$$

Example 8.2.2

Let $x(n)$ be the sequence obtained by sampling the continuous-time function

$$x(t) = \exp[-at]u(t) \tag{8.2.5}$$

every T seconds. Then

$$x(n) = \exp[-anT]u(n) \tag{8.2.6}$$

so that, from Equation (8.2.2), we have

$$X(z) = \sum_{n=0}^{\infty} \exp[-anT]z^{-n} = \sum_{n=0}^{\infty} [\exp[-aT]z^{-1}]^n$$

Using Equation (6.3.7), we can write this in closed form as

$$X(z) = \frac{1}{1 - \exp[-aT]z^{-1}} = \frac{z}{z - \exp[-aT]} \tag{8.2.7}$$

Example 8.2.3

Consider the two sequences

$$x(n) = \begin{cases} \left(\frac{1}{2}\right)^n, & n \geq 0 \\ 0, & n < 0 \end{cases} \tag{8.2.8}$$

and

$$y(n) = \begin{cases} -\left(\frac{1}{2}\right)^n, & n < 0 \\ 0, & n \geq 0 \end{cases} \tag{8.2.9}$$

Using the definition of the Z-transform, we can write

$$X(z) = \sum_{n=0}^{\infty} \left(\frac{1}{2}\right)^n z^{-n} = \sum_{n=0}^{\infty} \left(\frac{1}{2}z^{-1}\right)^n \tag{8.2.10}$$

We can obtain a closed-form expression for $X(z)$ by again using Equation (6.3.7), so that

$$X(z) = \frac{1}{1 - \frac{1}{2}z^{-1}} = \frac{z}{z - \frac{1}{2}} \tag{8.2.11}$$

Similarly, we get

$$Y(z) = \sum_{n=-\infty}^{-1} -\left(\frac{1}{2}\right)^n z^{-n} = -\sum_{n=-\infty}^{-1} \left(\frac{1}{2}z^{-1}\right)^n = -\sum_{m=1}^{\infty} (2z)^m \tag{8.2.12}$$

which yields the closed form

$$Y(z) = -\frac{2z}{1 - 2z} = \frac{z}{z - \frac{1}{2}} \tag{8.2.13}$$

As can be seen, the expressions for the two transforms, $X(z)$ and $Y(z)$, are identical. Seemingly, the two totally different sequences $x(n)$ and $y(n)$ have the same Z-transform. The difference, of course, as with the Laplace transform, is in the two different regions of convergence for $X(z)$ and $Y(z)$, where the region of convergence is those values of z for which the power series in Equation (8.2.1) or (8.2.2) exists—that is, has a finite value. Since Equation (6.3.7) is a geometric series, the sum can be put in closed form only when the summand has a magnitude less than unity. Thus, the expression for $X(z)$ given in Equation (8.2.11) is valid (that is, $X(z)$ exists) only if

$$\left|\tfrac{1}{2}z^{-1}\right| < 1 \quad \text{or} \quad |z| > \tfrac{1}{2} \tag{8.2.14}$$

Similarly, from Equation (8.2.13), $Y(z)$ exists if

$$|2z| < 1 \quad \text{or} \quad |z| < \tfrac{1}{2} \tag{8.2.15}$$

Equations (8.2.14) and (8.2.15) define the regions of convergence for $X(z)$ and $Y(z)$, respectively. These regions are plotted in the complex z-plane in Figure 8.2.1.

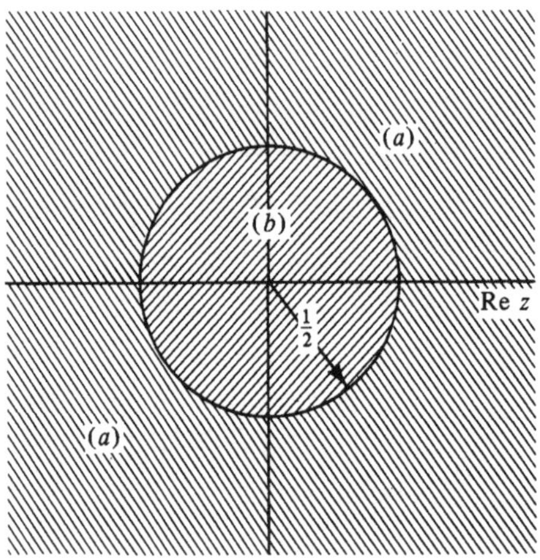

Figure 8.2.1 Regions of convergence (ROCs) of the Z-transforms for Example 8.2.3. (a) ROC for $X(z)$. (b) ROC for $Y(z)$.

From Example 8.2.3, it follows that, in order to uniquely relate a Z-transform to a time function, we must specify the region of convergence of the Z-transform.

8.3 CONVERGENCE OF THE Z-TRANSFORM

Consider a sequence $x(n)$ with Z-transform

$$X(z) = \sum_{n=-\infty}^{\infty} x(n) z^{-n} \tag{8.3.1}$$

We want to determine the values of z for which $X(z)$ exists. In order to do so, we represent z in polar form as $z = r\exp[j\theta]$ and write

$$X(z) = \sum_{n=-\infty}^{\infty} x(n)(r\exp[j\theta])^{-n}$$

$$= \sum_{n=-\infty}^{\infty} x(n)r^{-n}\exp[-jn\theta] \tag{8.3.2}$$

Let $x_+(n)$ and $x_-(n)$ denote the causal and anticausal parts of $x(n)$, respectively. That is,

$$x_+(n) = x(n)u(n)$$
$$x_-(n) = x(n)u(-n-1) \tag{8.3.3}$$

We substitute Equation (8.3.3) into Equation (8.3.2) to get

$$X(z) = \sum_{n=-\infty}^{-1} x_-(n)r^{-n}\exp[-jn\theta] + \sum_{n=0}^{\infty} x_+(n)r^{-n}\exp[-jn\theta]$$

$$= \sum_{m=1}^{\infty} x_-(-m)r^{m}\exp[jm\theta] + \sum_{n=0}^{\infty} x_+(n)r^{-n}\exp[-jn\theta]$$

$$\le \sum_{m=1}^{\infty} |x_-(-m)|r^{m} + \sum_{n=0}^{\infty} |x_+(n)|r^{-n} \tag{8.3.4}$$

For $X(z)$ to exist, each of the two terms on the right-hand side of Equation (8.3.4) must be finite. Suppose there exist constants M, N, R_-, and R_+ such that

$$|x_-(n)| < MR_-^n \quad \text{for } n < 0, \qquad |x_+(n)| < NR_+^n \quad \text{for } n \ge 0 \tag{8.3.5}$$

We can substitute these bounds in Equation (8.3.4) to obtain

$$X(z) \le M\sum_{m=1}^{\infty} R_-^{-m}r^{m} + N\sum_{n=0}^{\infty} R_+^{n}r^{-n} \tag{8.3.6}$$

Clearly, the first sum in Equation (8.3.6) is finite if $r/R_- < 1$, and the second sum is finite if $R_+/r < 1$. We can combine the two relations to determine the region of convergence for $X(z)$ as

$$R_+ < r < R_-$$

Figure 8.3.1 shows the region of convergence in the z plane as the annular region between the circles of radii R_- and R_+. The part of the transform corresponding to the causal sequence $x_+(n)$ converges in the region for which $r > R_+$ or, equivalently, $|z| > R_+$. That is, the region of convergence is outside the circle with radius R_+. Similarly, the transform corresponding to the anticausal sequence $x_-(n)$ converges for $r < R_-$ or, equivalently, $|z| < R_-$, so that the region of convergence is inside the circle of radius R_-. $X(z)$ does not exist if $R_- < R_+$.

We recall from our discussion of the Fourier transform of discrete-time signals in Chapter 7 that the frequency variable Ω takes on values in $[0, 2\pi]$. For a fixed value of r, it follows from a comparison of Equation (8.3.2) with Equation (7.3.12) that $X(z)$ can be interpreted as the discrete-time Fourier transform of the signal $x(n)r^{-n}$. This corresponds to evaluating $X(z)$ along the circle of radius r in the z plane. If we set $r = 1$, that is, for values of z along the circle with unity radius, $X(z)$ reduces

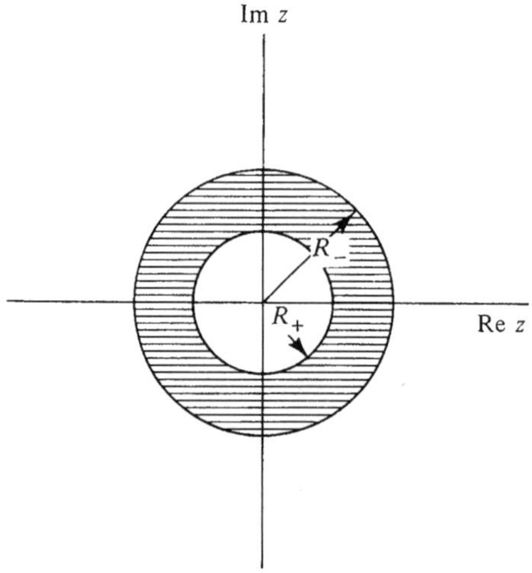

Figure 8.3.1 Region of convergence for a general noncausal sequence.

to the discrete-time Fourier transform of $x(n)$, assuming that the transform exists. That is, for sequences which possses both a discrete-time Fourier transform and a Z-transform, we have

$$X(\Omega) = X(z)\big|_{z=\exp[-j\Omega]} \tag{8.3.7}$$

The circle with radius unity is referred to as the unit circle.

In general, if $x(n)$ is the sum of several sequences, $X(z)$ exists only if there is a set of values of z for which the transforms of each of the sequences forming the sum converge. The region of convergence is thus the intersection of the individual regions of convergence. If there is no common region of convergence, then $X(z)$ does not exist.

Example 8.3.1

Consider the function

$$x(n) = \left(\frac{1}{3}\right)^n u(n)$$

Clearly, $R_+ = 1/3$ and $R_- = \infty$, so that the region of convergence is

$$|z| > \frac{1}{3}$$

The Z-transform of $x(n)$ is

$$X(z) = \frac{1}{1 - \frac{1}{3}z^{-1}} = \frac{z}{z - \frac{1}{3}}$$

which has a pole at $z = 1/3$. The region of convergence is thus outside the circle enclosing the pole of $X(z)$. Now let us consider the function

$$x(n) = \left(\frac{1}{2}\right)^n u(n) + \left(\frac{1}{3}\right)^n u(n)$$

which is the sum of the two functions

$$x_1(n) = \left(\frac{1}{2}\right)^n u(n) \qquad \text{and} \qquad x_2(n) = \left(\frac{1}{3}\right)^n u(n)$$

From the preceding example, the region of convergence for $X_1(z)$ is

$$|z| > \frac{1}{2}$$

whereas for $X_2(z)$ it is

$$|z| > \frac{1}{3}$$

Thus, the region of convergence for $X(z)$ is the intersection of these two regions and is

$$|z| > \text{Max}\left(\frac{1}{2}, \frac{1}{3}\right) = \frac{1}{2}$$

It can easily be verified that

$$X(z) = \frac{z}{z - \frac{1}{2}} + \frac{z}{z - \frac{1}{3}} = \frac{2z^2 - \frac{5}{6}z}{(z - \frac{1}{2})(z - \frac{1}{3})}$$

Hence, the region of convergence is outside the circle that includes both poles of $X(z)$.

The foregoing example shows that for a causal sequence, the region of convergence is outside of a circle which is such that all the poles of the transform $X(z)$ are within this circle. We may similarly conclude that for an anticausal function, the region of convergence is inside a circle such that all the poles are external to the circle.

If the region of convergence is an annular region, then the poles of $X(z)$ outside this annulus correspond to the anticausal part of the function, while the poles inside the annulus correspond to the causal part.

Example 8.3.2

The function

$$x(n) = \begin{cases} 3^n, & n < 0 \\ \left(\frac{1}{3}\right)^n, & n = 0, 2, 4, \text{etc.} \\ \left(\frac{1}{2}\right)^n, & n = 1, 3, 5, \text{etc.} \end{cases}$$

has the Z-transform

$$X(z) = \sum_{n=-\infty}^{-1} 3^n z^{-n} + \sum_{\substack{n=0,\\ n \text{ even}}}^{\infty} \left(\frac{1}{3}\right)^n z^{-n} + \sum_{\substack{n=0,\\ n \text{ odd}}}^{\infty} \left(\frac{1}{2}\right)^n z^{-n}$$

Let $n = -m$ in the first sum, $n = 2m$ in the second sum, and $n = 2m + 1$ in the third sum. Then

$$X(z) = \sum_{m=1}^{\infty}\left(\frac{1}{3}z\right)^m + \sum_{m=0}^{\infty}(\tfrac{1}{9}z^{-2})^m + \frac{z^{-1}}{2}\sum_{m=0}^{\infty}\left(\frac{1}{4}z^{-2}\right)^m$$
$$= \frac{\frac{1}{3}z}{1-\frac{1}{3}z} + \frac{1}{1-\frac{1}{9}z^{-2}} + \frac{\frac{1}{2}z^{-1}}{1-\frac{1}{4}z^{-2}}$$
$$= -\frac{z}{z-3} + \frac{z^2}{z^2-\frac{1}{9}} + \frac{2z}{z^2-\frac{1}{4}}$$

As can be seen, $X(z)$ has poles at $z = 3$, 1/3, −1/3, 1/2, and −1/2. The pole at 3 corresponds to the anticausal part, and the others are causal poles. Figure 8.3.2 shows the locations of these poles in the z plane, as well as the region of convergence, which, according to our previous discussion, is $1/2 < |z| < 3$.

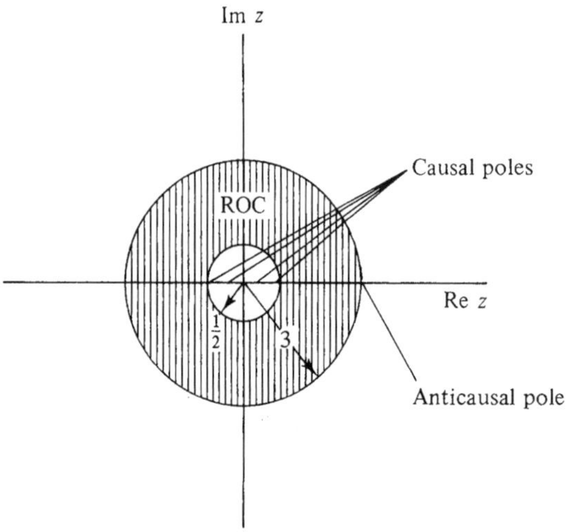

Figure 8.3.2 Pole locations and region of convergence for Example 8.3.2.

Example 8.3.3

Let $x(n)$ be a finite sequence that is zero for $n < n_0$ and $n > n_f$. Then

$$X(z) = x(n_0)z^{-n_0} + x(n_0+1)z^{-(n_0+1)} + \cdots + x(n_f)z^{-n_f}$$

Since $X(z)$ is a polynomial in z (or z^{-1}), $X(z)$ converges for all finite values of z, except $z = 0$ for $n_f > 0$. The poles of $X(z)$ are at infinity if $n_0 < 0$ and at the origin if $n_f > 0$.

From the previous example, it can be seen that if we form a sequence $y(n)$ by adding a finite-length sequence to a sequence $x(n)$, the region of convergence of $Y(z)$ is the same as that of $X(z)$, except possibly for $z = 0$.

Example 8.3.4

Consider the *right-sided* sequence

$$y(n) = 3\left(\frac{1}{2}\right)^{(n+5)} u(n+5)$$

By writing $y(n)$ as the sum of the finite sequence $3(1/2)^n[u(n+5) - u(n)]$ and the sequence $x(n) = 3/32(1/2)^n u(n)$, it becomes clear that the ROC of $Y(z)$ is the same as that of $X(z)$, namely, $|z| > 1/2$.

Similarly, the sequence

$$y(n) = -\left(\frac{1}{2}\right)^n u(-n+5)$$

can be considered to be the sum of the sequence $x(n) = -32(1/2)^n u(-n-1)$ and the finite sequence $-32(1/2)^n[u(n) - u(n-6)]$. It follows that $Y(z)$ converges for $0 < |z| < 1/2$.

In the rest of this chapter, we restrict ourselves to causal signals and systems, for which we will be concerned only with the unilateral transform. In the next section, we discuss some of the relevant properties of the unilateral Z-transform. Many of these properties carry over to the bilateral transform.

8.4 PROPERTIES OF THE Z-TRANSFORM

Recall the definition of the Z-transform of a causal sequence $x(n)$:

$$X(z) = \sum_{n=0}^{\infty} x(n)z^{-n} \tag{8.4.1}$$

We can directly use Equation (8.4.1) to derive the Z-transforms of common discrete-time signals, as the following example shows.

Example 8.4.1

(a) For the δ function, we saw that the Z-transform is

$$Z[\delta(n)] = 1 \cdot z^0 = 1 \tag{8.4.2}$$

(b) Let

$$x(n) = \alpha^n u(n)$$

Then

$$X(z) = \sum_{n=0}^{\infty} \alpha^n z^{-n} = \frac{1}{1 - \alpha z^{-1}} = \frac{z}{z - \alpha}, \quad |z| > |\alpha| \tag{8.4.3}$$

By letting $\alpha = 1$, we obtain the transform of the unit-step function:

$$Z[u(n)] = \frac{z}{z-1}, \quad |z| > 1 \tag{8.4.4}$$

(c) Let

$$x(n) = \cos \Omega_0 n u(n) \tag{8.4.5}$$

By writing $x(n)$ as

$$x(n) = \frac{1}{2}[\exp[j\Omega_0 n] + \exp[-j\Omega_0 n]]u(n)$$

and using the result of (b), it follows that

$$X(z) = \frac{1}{2}\frac{1}{z - \exp[j\Omega_0]} + \frac{1}{2}\frac{z}{z - \exp[-j\Omega_0]}$$

$$= \frac{z(z - \cos\Omega_0)}{z^2 - 2z\cos\Omega_0 + 1} \tag{8.4.6}$$

Similarly, the Z-transform of the sequence

$$x(n) = \sin\Omega_0 n u(n) \tag{8.4.7}$$

is

$$X(z) = \frac{z\sin\Omega_0}{z^2 - 2z\cos\Omega_0 + 1} \tag{8.4.8}$$

Let $x(n)$ be noncausal, with $x_+(n)$ and $x_-(n)$ denoting its causal and anticausal parts, respectively, as in Equation (8.3.3). Then

$$X(z) = X_+(z) + X_-(z) \tag{8.4.9}$$

Now,

$$X_-(z) = \sum_{n=-\infty}^{-1} x_-(n)z^{-n} \tag{8.4.10}$$

By making the change of variable $m = -n$ and noting that $x_-(0) = 0$, we can write

$$X_-(z) = \sum_{m=0}^{\infty} x_-(-m)z^{m} \tag{8.4.11}$$

Let us denote $x_-(-m)$ by $x_1(m)$. It then follows that

$$X_-(z) = X_1(z^{-1}) \tag{8.4.12}$$

where $X_1(z)$ denotes the Z-transform of the *causal* sequence $x_-(-n)$

Example 8.4.2

Let

$$x(n) = \left(\frac{1}{2}\right)^{|n|}$$

Then

$$x_+(n) = \left(\frac{1}{2}\right)^{n}, \quad n \geq 0$$

$$x_-(n) = \left(\frac{1}{2}\right)^{-n}, \quad n < 0$$

and

$$x_1(n) = x_-(-n) = \left(\frac{1}{2}\right)^n \quad n > 0$$

$$= \left(\frac{1}{2}\right)^n u(n) - \delta(n)$$

From Example 8.4.1, we can write

$$X_+(z) = \frac{z}{z - \frac{1}{2}}, \quad |z| > \tfrac{1}{2}$$

and

$$X_1(z) = \frac{z}{z - \frac{1}{2}} - 1 = \frac{\frac{1}{2}}{z - \frac{1}{2}}, \quad |z| > \tfrac{1}{2}$$

so that

$$X_-(z) = -\frac{z}{z - 2}, \quad |z| < 2$$

and

$$X(z) = \frac{z}{z - \frac{1}{2}} - \frac{z}{z - 2}, \quad \tfrac{1}{2} < |z| < 2$$

Thus, a table of transforms of causal time functions can be used to find the Z-transforms of noncausal functions. Table 8.1 lists the transform pairs derived in Example 8.4.1, as well as a few others. The additional pairs can be derived directly by using Equation (8.4.1) or by using the properties of the Z-transform. We discuss a few of these properties next. Since they are similar to those of the other transforms we have discussed so far, we state many of the more familiar properties and do not derive them in detail.

8.4.1 Linearity

If $x_1(n)$ and $x_2(n)$ are two sequences with transforms $X_1(z)$ and $X_2(z)$, respectively, then

$$Z[a_1 x_1(n) + a_2 x_2(n)] = a_1 X_1(z) + a_2 X_2(z) \tag{8.4.13}$$

where a_1 and a_2 are arbitrary constants.

8.4.2 Time Shifting

Let $x(n)$ be a causal sequence and let $X(z)$ denote its transform. Then, for any integer $n_0 > 0$,

$$\begin{aligned} Z[x(n+n_0)] &= \sum_{n=0}^{\infty} x(n+n_0)z^{-n} \\ &= \sum_{m=n_0}^{\infty} x(m)z^{-(m-n_0)} \\ &= z^{n_0}\left[\sum_{m=0}^{\infty} x(m)z^{-m} - \sum_{m=0}^{n_0-1} x(m)z^{-m}\right] \\ &= z^{n_0}\left[X(z) - \sum_{m=0}^{n_0-1} x(m)z^{-m}\right] \end{aligned} \tag{8.4.14}$$

Similarly,

$$\begin{aligned} Z[x(n-n_0)] &= \sum_{n=0}^{\infty} x(n-n_0)z^{-n} \\ &= \sum_{m=-n_0}^{\infty} x(m)z^{-(m+n_0)} \\ &= z^{-n_0}\left[\sum_{m=0}^{\infty} x(m)z^{-m} + \sum_{m=-n_0}^{-1} x(m)z^{-m}\right] \\ &= z^{-n_0}\left[X(z) + \sum_{m=-n_0}^{-1} x(m)z^{-m}\right] \end{aligned} \tag{8.4.15}$$

Example 8.4.3

Consider the difference equation

$$y(n) - \frac{1}{2}y(n-1) = \delta(n)$$

with the initial condition

$$y(-1) = 3$$

In order to find $y(n)$ for $n \geq 0$, we use Equation (8.4.15) and take transforms on both sides of the difference equation, gettting

$$Y(z) - \frac{1}{2}z^{-1}[Y(z) + y(-1)z] = 1$$

We now substitute the initial condition and rearrange terms, so that

$$Y(z) = \frac{\frac{5}{2}}{1 - \frac{1}{2}z^{-1}} = \frac{5}{2}\frac{z}{z - \frac{1}{2}}$$

It follows from Example 8.4.1(b) that

$$y(n) = \frac{5}{2}\left(\frac{1}{2}\right)^n, \quad n \geq 0$$

Example 8.4.4

Solve the difference equation

$$y(n + 2) - y(n + 1) + \frac{2}{9}y(n) = x(n)$$

for $y(n)$, $n \geq 0$, if $x(n) = u(n)$, $y(1) = 1$, and $y(0) = 1$.

Using Equation (8.4.14), we have

$$z^2[Y(z) - y(0) - y(1)z^{-1}] - z[Y(z) - y(0)] + \frac{2}{9}Y(z) = X(z)$$

Substituting $X(z) = z/(z - 1)$ and using the given initial conditions, we get

$$\left(z^2 - z + \frac{2}{9}\right)Y(z) = \frac{z}{z - 1} + z^2 = z\frac{z^2 - z + 1}{z - 1}$$

Writing $Y(z)$ as

$$Y(z) = z\frac{z^2 - z + 1}{(z - 1)(z - \frac{1}{3})(z - \frac{2}{3})}$$

and expanding the fractional term in partial fractions yields

$$\begin{aligned} Y(z) &= z\left[\frac{\frac{9}{2}}{z - 1} + \frac{\frac{7}{2}}{z - \frac{1}{3}} - \frac{7}{z - \frac{2}{3}}\right] \\ &= \frac{\frac{9}{2}z}{z - 1} + \frac{\frac{7}{2}z}{z - \frac{1}{3}} - \frac{7z}{z - \frac{2}{3}} \end{aligned}$$

From Example 8.4.1(b), it follows that

$$y(n) = \frac{9}{2}u(n) + \frac{7}{2}\left(\frac{1}{3}\right)^n u(n) - 7\left(\frac{2}{3}\right)^n u(n)$$

8.4.3 Frequency Scaling

The Z-transform of a sequence $a^n x(n)$ is

$$\begin{aligned} Z[a^n x(n)] &= \sum_{n=0}^{\infty} a^n x(n) z^{-n} = \sum_{n=0}^{\infty} x(n)(a^{-1}z)^{-n} \\ &= X(a^{-1}z) \end{aligned} \tag{8.4.16}$$

Example 8.4.5

We can use the scaling property to derive the transform of the signal

$$y(n) = (a^n \cos \Omega_0 n)u(n)$$

from the transform of

$$x(n) = (\cos \Omega_0 n)u(n)$$

which, from Equation (8.4.6), is

$$X(z) = \frac{z(z - \cos \Omega_0)}{z^2 - 2z \cos \Omega_0 + 1}$$

Thus,

$$Y(z) = \frac{a^{-1}z(a^{-1}z - \cos \Omega_0)}{a^{-2}z^2 - 2a^{-1}z \cos \Omega_0 + 1}$$
$$= \frac{z(z - a \cos \Omega_0)}{z^2 - 2a \cos \Omega_0 z + a^2}$$

Similarly, the transform of

$$y(n) = a^n (\sin \Omega_0 n)u(n)$$

is, from Equation (8.4.8),

$$Y(z) = \frac{az \sin \Omega_0}{z^2 - 2a \cos \Omega_0 z + a^2}$$

8.4.4 Differentiation with Respect to z

If we differentiate both sides of Equation (8.4.1) with respect to z, we obtain

$$\frac{dX(z)}{dz} = \sum_{n=0}^{\infty} (-n)x(n)z^{-n-1}$$
$$= -z^{-1} \sum_{n=0}^{\infty} nx(n)z^{-n}$$

from which it follows that

$$Z[nx(n)] = -z \frac{d}{dz} X(z) \tag{8.4.17}$$

By successively differentiating with respect to z, this result can be generalized as

$$Z[n^k x(n)] = \left(-z \frac{d}{dz}\right)^k X(z) \tag{8.4.18}$$

Example 8.4.6

Let us find the transform of the function

$$y(n) = n(n + 1)u(n)$$

From Equation (8.4.17), we have

$$Z[nu(n)] = -z \frac{d}{dz} Z[u(n)] = -z \frac{d}{dz} \frac{z}{z - 1} = \frac{z}{(z - 1)^2}$$

and

$$Z[n^2u(n)] = \left(-z\frac{d}{dz}\right)^2 Z[u(n)] = -z\frac{d}{dz}\left\{-z\frac{d}{dz}Z[u(n)]\right\}$$

$$= -z\frac{d}{dz}\frac{z}{(z-1)^2} = \frac{z(z+1)}{(z-1)^3}$$

so that

$$Y(z) = \frac{z(z+1)}{(z-1)^3} + \frac{z}{(z-1)^2} = \frac{2z^2}{(z-1)^3}$$

8.4.5 Initial Value

For a causal sequence $x(n)$, we can write Equation (8.4.1) explicitly as

$$X(z) = x(0) + x(1)z^{-1} + x(2)z^{-2} + \cdots + x(n)z^{-n} + \cdots \tag{8.4.19}$$

It can be seen that as $z \to \infty$, the term $z^{-n} \to 0$ for each $n > 0$, so that

$$\lim_{z\to\infty} X(z) = x(0) \tag{8.4.20}$$

Example 8.4.7

We will determine the initial value $x(0)$ for the signal with transform

$$X(z) = \frac{z^3 - \frac{3}{4}z^2 + 2z - \frac{5}{4}}{(z-1)(z-\frac{1}{3})(z^2 - \frac{1}{2}z + 1)}$$

Use the initial value theorem gives

$$x(0) = \lim_{z\to\infty} X(z) = 1$$

The initial value theorem is a convenient tool for checking if the Z-transform of a given signal is in error. Partial fraction expansion of $X(z)$ gives

$$X(z) = \frac{z}{z-1} + \frac{z}{z-\frac{1}{3}} - \frac{z-\frac{1}{4}}{z^2-\frac{1}{2}z+1}$$

so that

$$x(n) = u(n) + \left(\frac{1}{3}\right)^n u(n) - \left(\frac{1}{2}\right)^n \cos\left(\frac{\pi}{3}n\right)$$

The initial value is $x(0) = 1$ which agrees with the result above.

8.4.6 Final Value

From the time-shift theorem, we have

$$Z[x(n) - x(n-1)] = (1 - z^{-1})X(z) \tag{8.4.21}$$

The left-hand side of Equation (8.4.21) can be written as

$$\sum_{n=0}^{\infty}[x(n) - x(n-1)]z^{-n} = \lim_{N\to\infty}\sum_{n=0}^{N}[x(n) - x(n-1)]z^{-n}$$

If we now let $z \to 1$, Equation (8.4.21) can be written as

$$\lim_{z\to 1}(1 - z^{-1})X(z) = \lim_{N\to\infty}\sum_{n=0}^{N}[x(n) - x(n-1)]$$

$$= \lim_{N\to\infty} x(N) = x(\infty) \tag{8.4.22}$$

assuming $x(\infty)$ exists.

Example 8.4.8

By applying the final value theorem, we can find the final value of the signal of Example 8.4.7 as

$$x(\infty) = \lim_{z\to 1}\frac{z-1}{z}X(z) = \lim_{z\to 1}\left[\frac{z^3 - \frac{3}{4}z^2 + 2z - \frac{5}{4}}{z(z-\frac{1}{3})(z^2 - \frac{1}{2}z + 1)}\right]$$

so that

$$x(\infty) = 1$$

which again agrees with the final value of $x(n)$ given in the previous example.

Example 8.4.9

Let us consider the signal $x(n) = 2^n u(n)$ with Z-transform given by

$$X(z) = \frac{z}{z-2}$$

Application of the final value theorem yields

$$x(\infty) = \lim_{z\to 1}\frac{z-1}{z}\frac{z}{z-2} = 1$$

Clearly this result is incorrect since $x(n)$ grows without bound and hence has no final value.

This example shows that the final value theorem must be used with care. As noted earlier it gives the correct result only if the final value exists.

8.4.7 Convolution

If $y(n)$ is the convolution of two sequences $x(n)$ and $h(n)$, then, in a manner analogous to our derivation of the convolution property for the discrete-time Fourier transform, we can show that

$$Y(z) = H(z)X(z) \tag{8.4.23}$$

Recall that

$$Y(z) = \sum_{n=0}^{\infty} y(n) z^{-n}$$

so that $y(n)$ is the coefficient of the nth term in the power-series expansion of $Y(z)$. It follows that when we multiply two power series or polynomials $X(z)$ and $H(z)$, the coefficients of the resulting polynomial are the convolutions of the coefficients in $x(n)$ and $h(n)$.

Example 8.4.10

We want to use the Z-transform to find the convolution of the following two sequences, which were considered in Example 6.3.4:

$$h(n) = \{1, 2, 0, -1, 1\} \qquad \text{and} \qquad x(n) = \{1, 3, -1, -2\}$$

The respective transforms are

$$H(z) = 1 + 2z^{-1} - z^{-3} + z^{-4}$$

and

$$H(z) = 1 + 3z^{-1} - z^{-2} - 2z^{-3}$$

so that

$$Y(z) = 1 + 5z^{-1} + 5z^{-2} - 5z^{-3} - 6z^{-4} + 4z^{-5} + z^{-6} - 2z^{-7}$$

It follows that the resulting sequence is

$$y(n) = \{1, 5, 5, -5, -6, 4, 1, -2\}$$

This is the same answer that was obtained in Example 6.3.4.

The Z-transform properties discussed in this section are summarized in Table 8-1. Table 8-2, which is a table of Z-transform pairs of causal time functions, gives, in addition to the transforms of discrete-time sequences, the transforms of several sampled-time functions. These transforms can be obtained by fairly obvious modifications of the derivations discussed in this section and are left as exercises for the reader.

TABLE 8-1
Z-Transform Properties

1. Linearity	$a_1x_1(n) + a_2x_2(n)$	$a_1X_1(z) + a_2X_2(z)$	(8.4.13)
2. Time shift	$x(n + n_0)$	$z^{n_0}\left[X(z) - \sum_{m=0}^{n_0-1} x(m)z^{-m}\right]$	(8.4.14)
	$x(n - n_0)$	$z^{-n_0}\left[X(z) + \sum_{m=-n_0}^{-1} x(m)z^{-m}\right]$	(8.4.15)
3. Frequency scaling	$a^nx(n)$	$X(a^{-1}z)$	(8.4.16)
4. Multiplication by n	$nx(n)$	$-z\frac{d}{dz}X(z)$	(8.4.17)
	$n^kx(n)$	$\left(-z\frac{d}{dz}\right)^k X(z)$	(8.4.18)
5. Convolution	$x_1(n) * x_2(n)$	$X_1(z)X_2(z)$	(8.4.23)

8.5 THE INVERSE Z-TRANSFORM

There are several methods for finding a sequence $x(n)$, given its Z-transform $X(z)$. The most direct is by using the *inversion integral*;

$$x(n) = \frac{1}{2\pi j}\oint_\Gamma X(z)z^{n-1}dz \tag{8.5.1}$$

where $\oint_\Gamma$ represents integration along the closed contour Γ in the counterclockwise direction in the z plane. The contour must be chosen to lie in the region of convergence of $X(z)$.

Equation (8.5.1) can be derived from Equation (8.4.1) by multiplying both sides by z^{k-1} and integrating over Γ so that

$$\frac{1}{2\pi j}\oint_\Gamma X(z)z^{k-1}dz = \frac{1}{2\pi j}\oint_\Gamma \sum_{n=0}^{\infty} x(n)z^{k-n-1}dz$$

By the Cauchy integral theorem,

$$\oint_\Gamma z^{k-n-1}dz = \begin{cases} 2\pi j, & k = n \\ 0, & k \neq n \end{cases}$$

so that

$$\oint_\Gamma X(z)z^{k-1}dz = 2\pi jx(k)$$

from which it follows that

$$x(k) = \frac{1}{2\pi j}\oint_\Gamma X(z)z^{k-1}dz$$

TABLE 8-2
***Z*-Transform Pairs**

$x(n)$ for $n \geq 0$	$X(z)$	Radius of convergence $\lvert z \rvert > R$
1. $\delta(n)$	1	0
2. $\delta(n-m)$	z^{-m}	0
3. $u(n)$	$\dfrac{z}{z-1}$	1
4. n	$\dfrac{z}{(z-1)^2}$	1
5. n^2	$\dfrac{z(z+1)}{(z-1)^3}$	1
6. a^n	$\dfrac{z}{z-a}$	$\lvert a \rvert$
7. na^n	$\dfrac{az}{(z-a)^2}$	$\lvert a \rvert$
8. $(n+1)a^n$	$\dfrac{z^2}{(z-a)^2}$	$\lvert a \rvert$
9. $\dfrac{(n+1)(n+2)\cdots(n+m)a^n}{m!}$	$\dfrac{z^{m+1}}{(z-a)^{m+1}}$	$\lvert a \rvert$
10. $\cos \Omega_0 n$	$\dfrac{z(z-\cos\Omega_0)}{z^2-2z\cos\Omega_0+1}$	1
11. $\sin \Omega_0 n$	$\dfrac{z\sin\Omega_0}{z^2-2z\cos\Omega_0+1}$	1
12. $a^n \cos \Omega_0 n$	$\dfrac{z(z-a\cos\Omega_0)}{z^2-2za\cos\Omega_0+a^2}$	$\lvert a \rvert$
13. $a^n \sin \Omega_0 n$	$\dfrac{za\sin\Omega_0}{z^2-2za\cos\Omega_0+a^2}$	$\lvert a \rvert$
14. $\exp[-anT]$	$\dfrac{z}{z-\exp[-aT]}$	$\lvert \exp[-aT] \rvert$
15. nT	$\dfrac{Tz}{(z-1)^2}$	1
16. $nT\exp[-anT]$	$\dfrac{Tz\exp[-aT]}{[z-\exp[-aT]]^2}$	$\lvert \exp[-aT] \rvert$
17. $\cos n\omega_0 T$	$\dfrac{z(z-\cos\omega_0 T)}{z^2-2z\cos\omega_0 T+1}$	1
18. $\sin n\omega_0 T$	$\dfrac{z\sin\omega_0 T}{z^2-2z\cos\omega_0 T+1}$	1
19. $\exp[-anT]\cos n\omega_0 T$	$\dfrac{z[z-\exp[-aT]\cos\omega_0 T]}{z^2-2z\exp[-aT]\cos\omega_0 T+\exp[-2aT]}$	$\lvert \exp[-aT] \rvert$
20. $\exp[-anT]\sin n\omega_0 T$	$\dfrac{z[z-\exp[-aT]\sin\omega_0 T]}{z^2-2z\exp[-aT]\cos\omega_0 T+\exp[-2aT]}$	$\lvert \exp[-aT] \rvert$

We can evaluate the integral on the right side of Equation (8.5.1) by using the residue theorem. However, in many cases, this is not necessary, and we can obtain the inverse transform by using other methods.

We assume that $X(z)$ is a rational function in z of the form

$$X(z) = \frac{b_0 + b_1 z + \cdots + b_M z^M}{a_0 + a_1 z + \cdots + a_N z^N}, \qquad M \le N \tag{8.5.2}$$

with region of convergence outside all the poles of $X(z)$.

8.5.1 Inversion by a Power-Series Expansion

If we express $X(z)$ in a power series in z^{-1}, $x(n)$ can easily be determined by identifying it with the coefficient of z^{-n} in the power-series expansion. The power series can be obtained by arranging the numerator and denominator of $X(z)$ in descending powers of z and then dividing the numerator by the denominator using long division.

Example 8.5.1

Determine the inverse Z-transform of the function

$$X(z) = \frac{z}{z - 0.1}, \qquad |z| > 0.1$$

Since we want a power-series expansion in powers of z^{-1}, we divide the numerator by the denominator to obtain

$$\begin{array}{r|l}
 & 1 + 0.1z^{-1} + (0.1)^2 z^{-2} + (0.1)^3 z^{-3} + \cdots \\ \hline
z - 0.1 & z \\
 & \underline{z - 0.1} \\
 & \quad 0.1 \\
 & \quad \underline{0.1 \quad - (0.1)^2 z^{-1}} \\
 & \qquad\qquad (0.1)^2 z^{-1} \\
 & \qquad\qquad \underline{(0.1)^2 z^{-1} - (0.1)^2 z^{-2}} \\
 & \qquad\qquad\qquad\qquad (0.1)^3 z^{-2}
\end{array}$$

We can write, therefore,

$$X(z) = 1 + 0.1z^{-1} + (0.1)^2 z^{-1} + (0.1)^3 z^{-3} + \cdots$$

so that

$$x(0) = 1, \quad x(1) = 0.1, \quad x(2) = (0.1)^2, \quad x(3) = (0.1)^3, \quad \text{etc.}$$

It can easily be seen that this corresponds to the sequence

$$x(n) = (0.1)^n u(n)$$

Although we were able to identify the general expression for $x(n)$ in the last example, in most cases it is not easy to identify the general term from the first few sample values. However, in those cases where we are interested in only a few sample values of

$x(n)$, this technique can readily be applied. For example, if $x(n)$ in the last example represented a system impulse response, then, since $x(n)$ decreases very rapidly to zero, we can for all practical purposes evaluate just the first few values of $x(n)$ and assume that the rest are zero. The resulting error in our analysis of the system should prove to be negligible in most cases.

It is clear from our definition of the Z-transform that the series expansion of the transform of a causal sequence can have only negative powers of z. A consequence of this result is that, if $x(n)$ is causal, the degree of the denominator polynomial in the expression for $X(z)$ in Equation (8.5.2) must be greater than or equal to the degree of the numerator polynomial. That is, $N \geq M$.

Example 8.5.2

We want to find the inverse transform of

$$X(z) = \frac{z^3 - z^2 + z - \frac{1}{16}}{z^3 - \frac{5}{4}z^2 + \frac{1}{2}z - \frac{1}{16}}, \qquad |z| > \tfrac{1}{2}$$

Carrying out the long division yields the series expansion

$$X(z) = 1 + \frac{1}{4}z^{-1} + \frac{13}{16}z^{-2} + \frac{57}{64}z^{-3} + \cdots$$

from which it follows that

$$x(0) = 1, \quad x(1) = \frac{1}{4}, \quad x(2) = \frac{13}{16}, \quad x(3) = \frac{57}{64}, \quad \text{etc.}$$

In this example, it is not easy to determine the general expression for $x(n)$, which, as we see in the next section, is

$$x(n) = \delta(n) - 9\left(\frac{1}{2}\right)^n u(n) + 5n\left(\frac{1}{2}\right)^n u(n) + 9\left(\frac{1}{4}\right)^n u(n)$$

8.5.2 Inversion by Partial-Fraction Expansion

For rational functions, we can obtain a partial-fraction expansion of $X(z)$ over its poles just as in the case of the Laplace transform. We can then, in view of the uniqueness of the Z-transform, use the table of Z-transform pairs to identify the sequences corresponding to the terms in the partial-fraction expansion.

Example 8.5.3

Consider $X(z)$ of Example 8.5.2:

$$X(z) = \frac{z^3 - z^2 + z - \frac{1}{16}}{z^3 - \frac{5}{4}z^2 + \frac{1}{2}z - \frac{1}{16}}, \ |z| > \tfrac{1}{2}$$

In order to obtain the partial-fraction expansion, we first write $X(z)$ as the sum of a constant and a term in which the degree of the numerator is less than that of the denominator:

$$X(z) = 1 + \frac{\frac{1}{4}z^2 + \frac{1}{2}z}{z^3 - \frac{5}{4}z^2 + \frac{1}{2}z - \frac{1}{16}}$$

In factored form, this becomes

$$X(z) = 1 + \frac{\frac{1}{4}z(z+2)}{(z-\frac{1}{2})^2(z-\frac{1}{4})}$$

We can make a partial-fraction expansion of the second term and try to identify terms from Table 8-2. However, the entries in the table have a factor z in the numerator. We therefore write $X(z)$ as

$$X(z) = 1 + z\frac{\frac{1}{4}(z+2)}{(z-\frac{1}{2})^2(z-\frac{1}{4})}$$

If we now make a partial-fraction expansion of the fractional term, we obtain

$$X(z) = 1 + z\left(\frac{-9}{z-\frac{1}{2}} + \frac{\frac{5}{2}}{(z-\frac{1}{2})^2} + \frac{9}{z-\frac{1}{4}}\right)$$

$$= 1 - 9\frac{z}{z-\frac{1}{2}} + 5\frac{z/2}{(z-\frac{1}{2})^2} + 9\frac{z}{z-\frac{1}{4}}$$

From Table 8-2, we can now write

$$x(n) = \delta(n) - 9\left(\frac{1}{2}\right)^n u(n) + 5n\left(\frac{1}{2}\right)^n u(n) + 9\left(\frac{1}{4}\right)^n u(n)$$

Example 8.5.4

Solve the difference equation

$$y(n) - \frac{3}{4}y(n-1) + \frac{1}{8}y(n-2) = 2\sin\frac{n\pi}{2}$$

with initial conditions

$$y(-1) = 2 \qquad \text{and} \qquad y(-2) = 4$$

This is the problem considered in Example 6.5.4. To solve it, we first find $Y(z)$ by transforming the difference equation, with the use of Equation (8.4.15):

$$Y(z) - \frac{3}{4}z^{-1}[Y(z) + 2z] + \frac{1}{8}z^{-2}[Y(z) + 4z^2 + 2z] = \frac{2z}{z^2+1}$$

Collecting terms in $Y(z)$ gives

$$\left(1 - \frac{3}{4}z^{-1} + \frac{1}{8}z^{-2}\right)Y(z) = 1 - \frac{1}{4}z^{-1} + \frac{2z}{z^2+1}$$

from which it follows that

$$Y(z) = \frac{z^2 - \frac{1}{4}z}{z^2 - \frac{3}{4}z + \frac{1}{8}} + \frac{2z^3}{(z^2+1)(z^2 - \frac{3}{4} + \frac{1}{8})}, \qquad |z| > 1$$

Carrying out a partial-fraction expansion of the terms on the right side along the lines of the previous example yields

$$Y(z) = \frac{z}{z-\frac{1}{2}} + \frac{8}{5}\frac{z}{z-\frac{1}{2}} - \frac{8}{17}\frac{z}{z-\frac{1}{4}} + \frac{\frac{112}{85}z - \frac{96}{85}z^2}{z^2+1}$$

$$= \frac{13}{5}\frac{z}{z-\frac{1}{2}} - \frac{8}{17}\frac{z}{z-\frac{1}{4}} + \frac{112}{85}\frac{z}{z^2+1} - \frac{96}{85}\frac{z^2}{z^2+1}$$

The first two terms on the right side correspond to the homogeneous solution, and the last two terms correspond to the particular solution. From Table 8-2, it follows that

$$y(n) = \frac{13}{5}\left(\frac{1}{2}\right)^n - \frac{8}{17}\left(\frac{1}{4}\right)^n + \frac{112}{85}\sin\frac{n\pi}{2} - \frac{96}{85}\cos\frac{n\pi}{2}, \qquad n \geq 0$$

which agrees with our earlier solution.

Example 8.5.5

Let us find the inverse transform of the function

$$X(z) = \frac{1}{(z-\frac{1}{2})(z-\frac{1}{4})}, \qquad |z| > \frac{1}{2}$$

Direct partial-fraction expansion yields

$$X(z) = \frac{4}{z-\frac{1}{2}} - \frac{4}{z-\frac{1}{4}}$$

which can be written as

$$X(z) = z^{-1}\frac{4z}{z-\frac{1}{2}} - 4z^{-1}\frac{4z}{z-\frac{1}{4}}$$

We can now use the table of transforms and the time-shift theorem, Equation (8.4.15), to write the inverse transform as

$$x(n) = 4\left(\frac{1}{2}\right)^{n-1}u(n-1) - 4\left(\frac{1}{4}\right)^{n-1}u(n-1)$$

Alternatively, we can write

$$X(z) = \frac{z}{z(z-\frac{1}{2})(z-\frac{1}{4})}$$

and expand in partial fractions along the lines of the previous example to get

$$X(z) = z\left(\frac{8}{z} + \frac{8}{z-\frac{1}{2}} - \frac{16}{z-\frac{1}{4}}\right) = 8 + \frac{8z}{z-\frac{1}{2}} - \frac{16z}{z-\frac{1}{4}}$$

We can directly use Table 8-2 to write $x(n)$ as

$$x(n) = 8\,\delta(n) + 8\left(\frac{1}{2}\right)^n u(n) - 16\left(\frac{1}{4}\right)^n u(n)$$

To verify that this is the same solution as obtained previously, we note that for $n = 0$, we have

$$x(0) = 8 + 8 - 16 = 0$$

For $n \geq 1$, we get

$$x(n) = 8\left(\frac{1}{2}\right)^n - 16\left(\frac{1}{2}\right)^n = 4\left(\frac{1}{2}\right)^{n-1} - 4\left(\frac{1}{4}\right)^{n-1}$$

Thus, either method gives the same answer.

8.6 Z-TRANSFER FUNCTIONS OF CAUSAL DISCRETE-TIME SYSTEMS

We saw that, given a causal system with impulse response $h(n)$, output corresponding to any input $x(n)$ is given by the convolution sum:

$$y(n) = \sum_{k=0}^{\infty} h(k)x(n-k) \tag{8.6.1}$$

In terms of the respective Z-transforms, the output can be written as

$$Y(z) = H(z)X(z) \tag{8.6.2}$$

where

$$H(z) = Z[h(n)] = \frac{Y(z)}{X(z)} \tag{8.6.3}$$

represents the system transfer function.

As can be seen from Equation (8.6.2), the Z-transform of a causal function contains only negative powers of z. Consequently, when the transfer function $H(z)$ of a causal system is expressed as the ratio of two polynomials in z, the degree of the denominator polynomial must be at least as large as the degree of the numerator polynomial. That is, if

$$H(z) = \frac{\beta_M z^M + \beta_{M-1} z^{M-1} + \cdots + \beta_1 z + \beta_0}{\alpha_N z^N + \alpha_{N-1} z^{N-1} + \cdots + \alpha_1 z + \alpha_0} \tag{8.6.4}$$

then $N \geq M$ if the system is causal. On the other hand, if we write $H(z)$ as the ratio of two polynomials in z^{-1}, i.e.,

$$H(z) = \frac{b_0 + b_1 z^{-1} + \cdots + b_{M-1} z^{-M+1} + b_M z^{-M}}{a_0 + a_1 z^{-1} + \cdots + \alpha_{N-1} z^{-N+1} + \alpha_N z^{-N}} \tag{8.6.5}$$

then if the system is causal, $b_0 \neq 0$.

Given a system described by the difference equation

$$\sum_{k=0}^{N} a_k y(n-k) = \sum_{k=0}^{M} b_k x(n-k) \tag{8.6.6}$$

we can find the transfer function of the system by taking the Z-transform on both sides of the equation. We note that in finding the impulse response of a system, and consequently, in finding the transfer function, the system must be initially relaxed. Thus, if we assume zero initial conditions, we can use the shift theorem to get

$$\left[\sum_{k=0}^{M} b_k z^{-k}\right] Y(z) = \left[\sum_{k=0}^{N} a_k z^{-k}\right] X(z) \tag{8.6.7}$$

so that

$$H(z) = \frac{\sum_{k=0}^{M} b_k z^{-k}}{\sum_{k=0}^{N} a_k z^{-k}} \tag{8.6.8}$$

The corresponding impulse response can be found as

$$h(n) = Z^{-1}[H(z)] \tag{8.6.9}$$

It is clear that the poles of the system transfer function are the same as the characteristic values of the corresponding difference equation. From our discussion of stability in Chapter 6, it follows that for the system to be stable, the poles must lie within the unit circle in the z plane. Consequently, for a stable, causal function, the ROC includes the unit circle.

We illustrate these results by the following examples.

Example 8.6.1

Let the step response of a linear, time-invariant, causal system be

$$y(n) = \frac{6}{5}u(n) - \frac{1}{3}\left(\frac{1}{2}\right)^n u(n) + \frac{2}{15}\left(-\frac{1}{4}\right)^n u(n)$$

To find the transfer function $H(z)$ of this system, we note that

$$\begin{aligned} Y(z) &= \frac{6}{5}\frac{z}{(z-1)} - \frac{1}{3}\frac{z}{(z-\frac{1}{2})} + \frac{2}{15}\frac{z}{(z+\frac{1}{4})} \\ &= \frac{z^3 - \frac{1}{4}z^2}{(z-1)(z-\frac{1}{2})(z+\frac{1}{4})} \end{aligned}$$

Since

$$X(z) = \frac{z}{z-1}$$

it follows that

$$\begin{aligned} H(z) &= \frac{Y(z)}{X(z)} = \frac{z^2 - \frac{1}{4}z}{(z-\frac{1}{2})(z+\frac{1}{4})} \\ &= \frac{2}{3}\frac{z}{z+\frac{1}{4}} + \frac{1}{3}\frac{z}{z-\frac{1}{2}} \end{aligned} \tag{8.6.10}$$

Thus, the impulse response of the system is

$$h(n) = \frac{2}{3}\left(-\frac{1}{4}\right)^n u(n) + \frac{1}{3}\left(\frac{1}{2}\right)^n u(n)$$

Since both poles of the system are within the unit circle, the system is stable.

We can find the difference-equation representation of the system by rewriting Equation (8.6.10) as

$$\frac{Y(z)}{X(z)} = \frac{1 - \frac{1}{4}z^{-1}}{(1 - \frac{1}{2}z^{-1})(1 + \frac{1}{4}z^{-1})} = \frac{1 - \frac{1}{4}z^{-1}}{1 - \frac{1}{4}z^{-1} - \frac{1}{8}z^{-2}}$$

Cross multiplying yields

$$\left[1 - \frac{1}{4}z^{-1} - \frac{1}{8}z^{-2}\right]Y(z) = \left[1 - \frac{1}{4}z^{-1}\right]X(z)$$

Taking the inverse transformation of both sides of this equation, we obtain

$$y(n) - \frac{1}{4}y(n-1) - \frac{1}{8}y(n-2) = x(n) - \frac{1}{4}x(n-1)$$

Example 8.6.2

Consider the system described by the difference equation

$$y(n) - 2y(n-1) + 2y(n-2) = x(n) + \tfrac{1}{2}x(n-1)$$

We can find the transfer function of the system by Z-transforming both sides of this equation. With all initial conditions assumed to be zero, the use of Equation (8.4.15) gives

$$Y(z) - 2z^{-1}Y(z) + z^{-2}Y(z) = X(z) + \tfrac{1}{2}z^{-1}X(z)$$

so that

$$H(z) = \frac{Y(z)}{X(z)} = \frac{1 + \frac{1}{2}z^{-1}}{1 - 2z^{-1} + 2z^{-2}} = \frac{z^2 + \frac{1}{2}z}{z^2 - 2z + 2}$$

The zeros of this system are at $z = 0$ and $z = -(1/2)$, while the poles are at $z = 1 \pm j1$. Since the poles are outside the unit circle, the system is unstable. Figure 8.6.1 shows the location of the poles and zeros of $H(z)$ in the z plane. The graph is called a pole-zero plot.

The impulse response of the system found by writing $H(z)$ as

$$H(z) = \frac{z(z-1)}{z^2 - 2z + 2} + \frac{3}{2}\frac{\frac{1}{2}z}{z^2 - 2z + 2}$$

and using Table 8-2 is

$$h(n) = (\sqrt{2})^n \cos\left(\frac{\pi}{4}n\right)u(n) + \frac{3}{2}(\sqrt{2})^n \sin\left(\frac{\pi}{4}n\right)u(n)$$

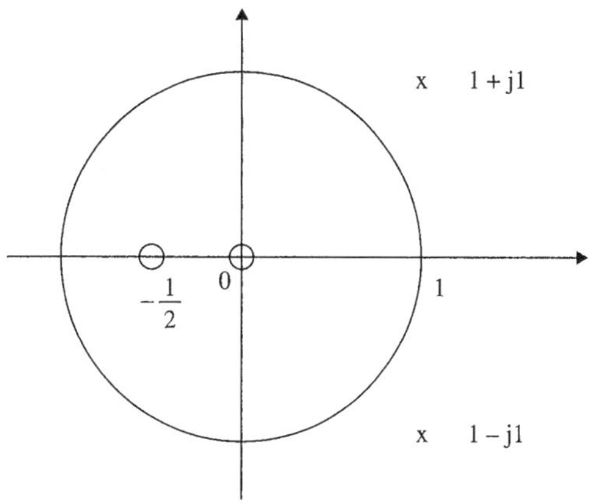

Figure 8.6.1 Pole-zero plot for Example 8.6.2.

Example 8.6.3

Consider the system shown in Figure 8.6.2, in which

$$H(z) = \frac{0.8Kz}{(z-0.8)(z-0.5)}$$

where K is a constant gain.

The transfer function of the system can be derived by noting that the output of the summer can be written as

$$E(z) = X(z) - Y(z)$$

so that the system output is

$$\begin{aligned} Y(z) &= H(z)E(z) \\ &= [X(z) - Y(z)]H(z) \end{aligned}$$

Substituting for $H(z)$ and simplifying yields

$$Y(z) = \frac{H(z)}{1 + H(z)}X(z) = \frac{0.8Kz}{(z-0.8)(z-0.5) + 0.8Kz}X(z)$$

The transfer function for the feedback system is therefore

$$T(z) = \frac{Y(z)}{X(z)} = \frac{0.8Kz}{z^2 + (0.8K - 1.3)z + 0.04}$$

The poles of the system can be determined as the roots of the equation

$$z^2 + (0.8K - 1.3)z + 0.04 = 0$$

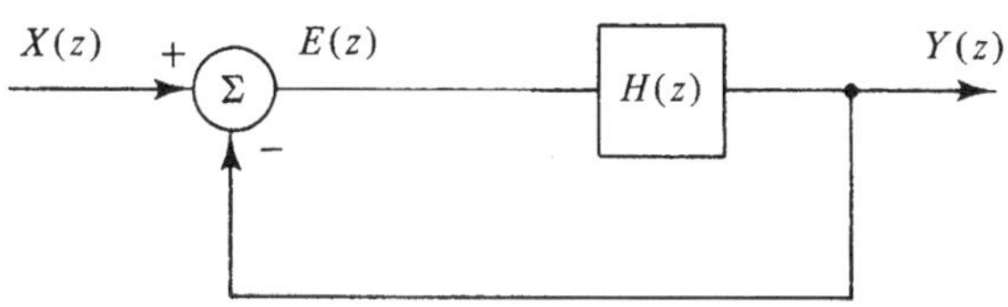

Figure 8.6.2 Block diagram of control system of Example 8.6.1.

For $K = 1$, the two roots are

$$z_1 = 0.1 \qquad \text{and} \qquad z_2 = 0.4$$

Since both roots are inside the unit circle, the system is stable. With $K = 4$, however, the roots are

$$z_1 = 0.0213 \qquad \text{and} \qquad z_2 = 1.87875$$

Since one of the roots is now outside the unit circle, the system is unstable.

8.7 Z-TRANSFORM ANALYSIS OF STATE-VARIABLE SYSTEMS

As we have seen in many of our discussions, the use of frequency-domain techniques considerably simplifies the analysis of linear, time-invariant systems. In this section, we consider the Z-transform analysis of discrete-time systems that are represented by a set of state equations of the form

$$\mathbf{v}(n+1) = \mathbf{A}\mathbf{v}(n) + \mathbf{b}x(n), \quad \mathbf{v}(0) = \mathbf{v}_0 \qquad (8.7.1)$$
$$y(n) = \mathbf{c}\mathbf{v}(n) + dx(n)$$

As we will see, the use of Z-transforms is useful both in deriving state-variable representations from the transfer function of the system and in obtaining the solution to the state equations.

In Chapter 6, starting from the difference-equation representation, we derived two alternative state-space representations. Here, we start with the transfer-function representation and derive two more representations, namely, the parallel and cascade forms. In order to show how this can be done, let us consider a simple first-order system described by the state-variable equations

$$v(n+1) = av(n) + bx(n) \qquad (8.7.2)$$
$$y(n) = cv(n) + dx(n)$$

From these equations it follows that

$$V(z) = \frac{b}{z-a}X(z)$$

Thus, the system can be represented by the block diagram of Figure 8.7.1. Note that as far as the relation between $Y(z)$ and $X(z)$ is concerned, the gains b and c at the input and output can be arbitrary as long as their product is equal to bc.

We use this block diagram and the corresponding equation, Equation (8.7.2), to obtain the state-variable representation for a general system by writing $H(z)$ as a combination of such blocks and associating a state variable with the output of each block. As in continuous-time systems, if we use a partial-fraction expansion over the poles of $H(z)$, we get the parallel form of the state equations, whereas if we represent $H(z)$ as a cascade of such blocks, we get the cascade representation. To obtain the two forms

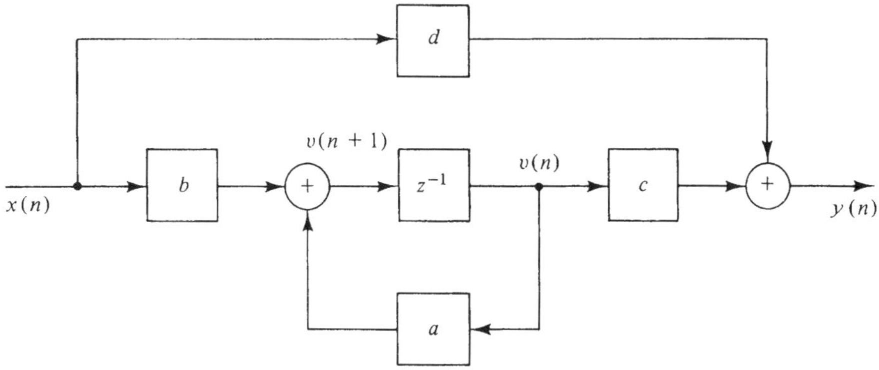

Figure 8.7.1 Block diagram of a first-order state-space system.

discussed in Chapter 6, we represent the system as a cascade of two blocks, with one block consisting of all the poles and the other block all the zeros. If the poles are in the first block and the zeros in the second block, we get the second canonical form. The first canonical form also can be derived, by putting the zeros in the first block and the poles in the second block. However, this derivation is not very straightforward, since it involves manipulating the first block to eliminate terms involving positive powers of z.

Example 8.7.1

Consider the system with transfer function

$$H(z) = \frac{3z + \frac{3}{4}}{z^2 + \frac{1}{4}z - \frac{1}{8}} = \frac{3z + \frac{3}{4}}{(z + \frac{1}{2})(z - \frac{1}{4})}$$

Expanding $H(z)$ by partial fractions, we can write

$$H(z) = \frac{1}{z + \frac{1}{2}} + \frac{2}{z - \frac{1}{4}}$$

with the corresponding block-diagram representation shown in Figure 8.7.2(a). By using the state variables identified in the figure, we obtain the following set of equations:

$$\left(z + \frac{1}{2}\right)V_1(z) = X(z)$$

$$\left(z - \frac{1}{4}\right)V_2(z) = 2X(z)$$

$$Y(z) = V_1(z) + 2V_2(z)$$

The corresponding equations in the time domain are

$$v_1(n + 1) = -\frac{1}{2}v_1(n) + x(n)$$

$$v_2(n + 1) = \frac{1}{4}v_2(n) + 2x(n)$$

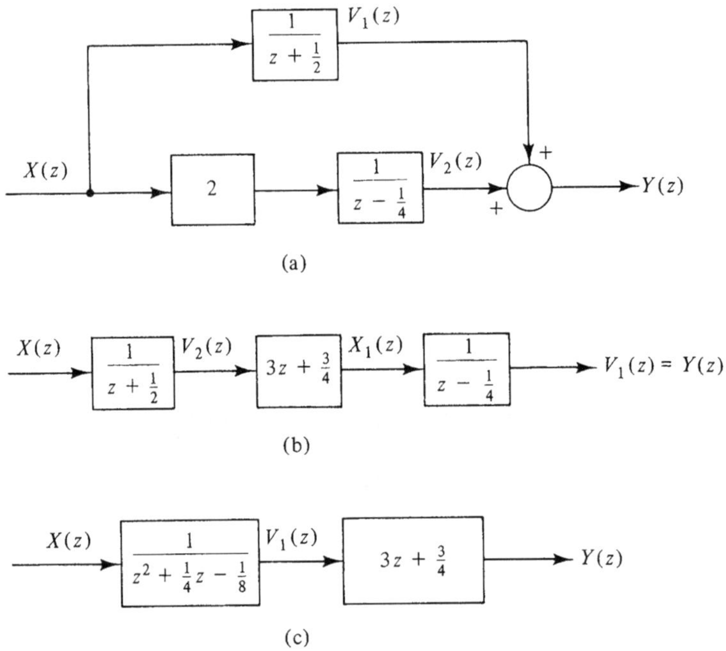

Figure 8.7.2 Block-diagram representations for Example 8.7.1.

$$y(n) = v_1(n) + 2v_2(n)$$

If we use the block-diagram representation of Fig. 8.7.2(b), with the states as shown, we have

$$\left(z - \frac{1}{4}\right)V_1(z) = X_1(z)$$

$$X_1(z) = \left(3z + \frac{3}{4}\right)V_2(z)$$

$$\left(z + \frac{1}{2}\right)V_2(z) = X(z)$$

$$Y(z) = V_1(z)$$

which, in the time domain, are equivalent to

$$v_1(n+1) = \frac{1}{4}v_1(n) + x_1(n)$$

$$x_1(n) = 3v_2(n+1) + \frac{3}{4}v_2(n)$$

$$v_2(n+1) = -\frac{1}{2}v_2(n) + x(n)$$

$$y(n) = v_1(n)$$

Eliminating $x_1(n)$ and rearranging the equations yields

$$v_1(n+1) = \frac{1}{4}v_1(n) - \frac{3}{4}v_2(n) + 3x(n)$$

$$v_2(n+1) = -\frac{1}{2}v_2(n) + x(n)$$

$$y(n) = v_1(n)$$

To get the second canonical form, we use the block diagram of Figure 8.7.2(c) to get

$$\left(z^2 + \frac{1}{4}z - \frac{1}{8}\right)V_1(z) = X(z)$$

$$Y(z) = \left(3z + \frac{3}{4}\right)V_1(z)$$

By defining

$$zV_1(z) = V_2(z)$$

we can write

$$zV_2(z) + \frac{1}{4}V_2(z) - \frac{1}{8}V_1(z) = X(z)$$

$$Y(z) = \frac{3}{4}V_1(z) + 3V_2(z)$$

Thus, the corresponding state equations are

$$v_1(n+1) = v_2(n)$$

$$v_2(n+1) = \frac{1}{8}v_1(n) - \frac{1}{4}v_2(n) + x(n)$$

$$y(n) = \frac{3}{4}v_1(n) + 3v_2(n)$$

As in the continuous-time case, in order to avoid working with complex numbers, for systems with complex conjugate poles or zeros, we can combine conjugate pairs. The representation for the resulting second-order term can then be obtained in either the first or the second canonical form. As an example, for the second-order system described by

$$Y(z) = \frac{b_0 + b_1 z^{-1} + b_2 z^{-2}}{1 + a_1 z^{-1} + a_2 z^{-2}} X(z) \tag{8.7.3}$$

we can obtain the simulation diagram by writing

$$Y(z) = (b_0 + b_1 z^{-1} + b_2 z^{-2})V(z) \tag{8.7.4a}$$

where

$$V(z) = \frac{1}{1 + a_1 z^{-1} + a_2 z^{-2}} X(z)$$

or equivalently,

$$V(z) = -a_1 z^{-1}V(z) - a_2 z^{-2}V(z) + X(z) \tag{8.7.4b}$$

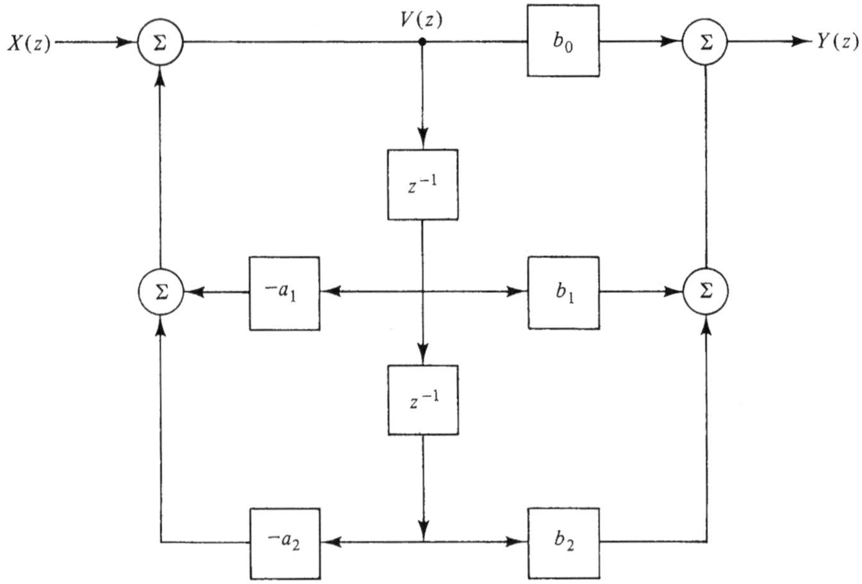

Figure 8.7.3 Simulation diagram for a second-order discrete-time system.

We generate $V(z)$ as the sum of $X(z)$, $-a_1z^{-1}V(z)$, and $-a_2z^{-2}V(z)$ and form $Y(z)$ as the sum of $b_0V(z)$ and $b_2z^{-2}V(z)$ to get the simulation diagram shown in Figure 8.7.3.

Example 8.7.2

Consider the system with transfer function

$$H(z) = \frac{1 + 2.5z^{-1} + z^{-2}}{(1 + 0.5z^{-1} + 0.8z^{-2})(1 + 0.3z^{-1})}$$

By treating this as the cascade combination of the two systems

$$H_1(z) = \frac{1 + 0.5z^{-1}}{1 + 0.5z^{-1} + 0.8z^{-2}}, \quad H_2(z) = \frac{1 + 2z^{-1}}{1 + 0.3z^{-1}}$$

we can draw the simulation diagram using Figure 8.7.3, as shown in Figure 8.7.4.

Using the outputs of the delays as state variables, we get the following equations:

$$\hat{V}(z) = zV_1(z) = -0.3V_1(z) + X_1(z)$$

$$X_1(z) = V(z) + 0.5V_2(z)$$

$$zV_2(z) = V(z) = -0.5V_2(z) - 0.8V_3(z) + X(z)$$

$$zV_3(z) = V_2(z)$$

$$Y(z) = \hat{V}(z) + 2V_1(z)$$

Eliminating $V(z)$ and $\hat{V}(z)$ and writing the equivalent time-domain equations yields

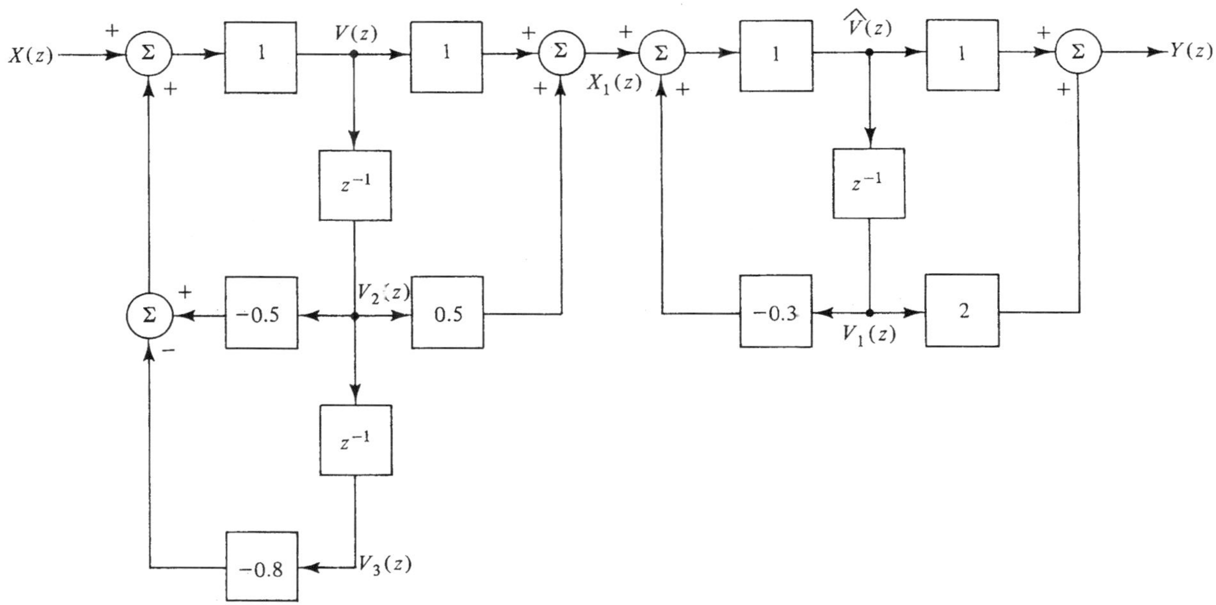

Figure 8.7.4 Simulation diagram for Example 8.7.2.

$$v_1(n+1) = -0.3v_1(n) - 0.8v_3(n) + x(n)$$
$$v_2(n+1) = -0.5v_2(n) - 0.8v_3(n) + x(n)$$
$$v_3(n+1) = v_2(n)$$
$$y(n) = 1.7v_1(n) - 0.8v_3(n) + x(n)$$

Clearly, by using different combinations of first- and second-order sections, we can obtain several different realizations of a given transfer function.

We now consider the frequency-domain solution of the state equations of Equation (8.7.1), which we repeat for convenience.

$$\mathbf{v}(n+1) = \mathbf{A}\mathbf{v}(n) + \mathbf{b}x(n), \quad \mathbf{v}(0) = \mathbf{v}_0 \tag{8.7.5a}$$

$$y(n) = \mathbf{c}\mathbf{v}(n) + dx(n) \tag{8.7.5b}$$

Z-transforming both sides of Equation (8.7.5a) yields

$$z[\mathbf{V}(z) - \mathbf{v}_0] = \mathbf{A}\mathbf{V}(z) + \mathbf{b}X(z) \tag{8.7.6}$$

Solving for $\mathbf{V}(z)$, we get

$$\mathbf{V}(z) = z(z\mathbf{I} - \mathbf{A})^{-1}\mathbf{v}_0 + (z\mathbf{I} - \mathbf{A})^{-1}\mathbf{b}X(z) \tag{8.7.7}$$

It follows from Equation (8.7.5b) that

$$Y(z) = \mathbf{c}z(z\mathbf{I} - \mathbf{A})^{-1}\mathbf{v}_0 + \mathbf{c}(z\mathbf{I} - \mathbf{A})^{-1}\mathbf{b}X(z) + dX(z) \tag{8.7.8}$$

We can determine the transfer function of this system by setting $\mathbf{v}(0) = 0$ to get

$$Y(z) = [\mathbf{c}(z\mathbf{I} - \mathbf{A})^{-1}\mathbf{b} + d]X(z) \tag{8.7.9}$$

It follows that

$$H(z) = \frac{Y(z)}{X(z)} = \mathbf{c}(z\mathbf{I} - \mathbf{A})^{-1}\mathbf{b} + d \tag{8.7.10}$$

Recall from Equation (6.7.16) that the time-domain solution of the state equations is

$$\mathbf{v}(n) = \mathbf{\Phi}(n)\mathbf{v}_0 + \sum_{j=0}^{n-1} \mathbf{\Phi}(n-1-j)\mathbf{b}x(j) \tag{8.7.11}$$

Z-transforming both sides of Equation (8.7.11) yields

$$\mathbf{V}(z) = \mathbf{\Phi}(z)\mathbf{v}_0 + z^{-1}\mathbf{\Phi}(z)\mathbf{b}X(z) \tag{8.7.12}$$

Comparing Equations (8.7.12) and (8.7.7), we obtain

$$\mathbf{\Phi}(z) = z(z\mathbf{I} - \mathbf{A})^{-1} \tag{8.7.13}$$

or equivalently,

$$\mathbf{\Phi}(n) = \mathbf{A}^n = Z^{-1}[z(z\mathbf{I} - \mathbf{A})^{-1}] \tag{8.7.14}$$

Equation (8.7.14) gives us an alternative method of determining $\mathbf{A}^n$.

Example 8.7.3

Consider the system

$$v_1(n+1) = v_2(n)$$

$$v_2(n+1) = \frac{1}{8}v_1(n) - \frac{1}{4}v_2(n) + x(n)$$

$$y(n) = v_1(n)$$

which was discussed in Examples 6.7.2, 6.7.3, and 6.7.4. We find the unit-step response of this system for the case when $\mathbf{v}(0) = [1 \quad -1]^T$. Since

$$A = \begin{bmatrix} 0 & 1 \\ \frac{1}{8} & -\frac{1}{4} \end{bmatrix}$$

it follows that

$$(z\mathbf{I} - \mathbf{A})^{-1} = \begin{bmatrix} z & -1 \\ -\frac{1}{8} & z+\frac{1}{4} \end{bmatrix}^{-1} = \frac{1}{z^2 + \frac{1}{4}z - \frac{1}{8}} \begin{bmatrix} z+\frac{1}{4} & 1 \\ \frac{1}{8} & z \end{bmatrix}$$

so we can write

$$\mathbf{\Phi}(z) = z(z\mathbf{I} - \mathbf{A})^{-1} = z \begin{bmatrix} \dfrac{\frac{2}{3}}{z-\frac{1}{4}} + \dfrac{\frac{1}{3}}{z+\frac{1}{2}} & \dfrac{\frac{4}{3}}{z-\frac{1}{4}} - \dfrac{\frac{4}{3}}{z+\frac{1}{2}} \\ \dfrac{\frac{1}{6}}{z-\frac{1}{4}} - \dfrac{\frac{1}{6}}{z+\frac{1}{2}} & \dfrac{\frac{1}{3}}{z-\frac{1}{4}} + \dfrac{\frac{2}{3}}{z+\frac{1}{2}} \end{bmatrix}$$

We therefore have

$$\mathbf{A}^n = \mathbf{\Phi}(n) = \begin{bmatrix} \frac{2}{3}\left(\frac{1}{4}\right)^n + \frac{1}{3}\left(-\frac{1}{2}\right)^n & \frac{4}{3}\left(\frac{1}{4}\right)^n - \frac{4}{3}\left(-\frac{1}{2}\right)^n \\ \frac{1}{6}\left(\frac{1}{4}\right)^n - \frac{1}{6}\left(-\frac{1}{2}\right)^n & \frac{1}{3}\left(\frac{1}{4}\right)^n + \frac{2}{3}\left(-\frac{1}{2}\right)^n \end{bmatrix}$$

which is the same result that was obtained in Example 6.7.2. From Equation (8.7.7), for the given initial condition,

$$\mathbf{V}(z) = (z\mathbf{I} - A)^{-1}\begin{bmatrix} 1 \\ -1 \end{bmatrix} + (z\mathbf{I} - A)^{-1}\begin{bmatrix} 0 \\ 1 \end{bmatrix}\frac{z}{z-1}$$

Multiplying terms and expanding in partial fractions yields

$$\mathbf{V}(z) = \begin{bmatrix} \dfrac{\frac{8}{9}z}{z-1} + \dfrac{\frac{23}{9}z}{z+\frac{1}{2}} - \dfrac{\frac{22}{9}z}{z-\frac{1}{4}} \\ \dfrac{\frac{8}{9}z}{z-1} - \dfrac{\frac{23}{18}z}{z+\frac{1}{2}} - \dfrac{\frac{11}{18}z}{z-\frac{1}{4}} \end{bmatrix}$$

so that

$$\mathbf{v}(n) = \begin{bmatrix} v_1(n) \\ v_2(n) \end{bmatrix} = \begin{bmatrix} \frac{8}{9} + \frac{23}{9}\left(-\frac{1}{2}\right)^n - \frac{22}{9}\left(\frac{1}{4}\right)^n \\ \frac{8}{9} - \frac{23}{18}\left(-\frac{1}{2}\right)^n - \frac{11}{18}\left(\frac{1}{4}\right)^n \end{bmatrix}$$

To find the output, we note that

$$Y(z) = [1 \quad 0]\begin{bmatrix} V_1(z) \\ V_2(z) \end{bmatrix} = V_1(z)$$

and

$$y(n) = v_1(n)$$

These are exactly the same as the results in Example 6.7.3. Finally, we have, from Equation (8.7.10),

$$\begin{aligned} H(z) &= [1 \quad 0]\frac{1}{(z+\frac{1}{2})(z-\frac{1}{4})}\begin{bmatrix} z+\frac{1}{4} & 1 \\ \frac{1}{8} & z \end{bmatrix}\begin{bmatrix} 0 \\ 1 \end{bmatrix} \\ &= \frac{1}{(z+\frac{1}{2})(z-\frac{1}{4})} \\ &= \frac{\frac{4}{3}}{z-\frac{1}{4}} - \frac{\frac{4}{3}}{z+\frac{1}{2}} \end{aligned}$$

so that

$$h(n) = \frac{4}{3}\left(\frac{1}{4}\right)^{n-1} - \frac{4}{3}\left(-\frac{1}{2}\right)^{n-1}, \quad n \geq 1$$

Since $h(0) = 0$, we can write the last equation as

$$h(n) = \frac{4}{3}\left(\frac{1}{4}\right)^{n} - \frac{4}{3}\left(-\frac{1}{2}\right)^{n}, \quad n \geq 0$$

which is the result obtained in Example 6.7.4.

8.8 RELATION BETWEEN THE Z-TRANSFORM AND THE LAPLACE TRANSFORM

The relation between the Laplace transform and the Z-transform of the sequence of samples obtained by sampling analog signal $x_a(t)$ can easily be developed from our discussion of sampled signals in Chapter 7. There we saw that the output of the sampler could be considered to be either the continuous-time signal

$$x_s(t) = \sum_{n=-\infty}^{\infty} x_a(nT)\delta(t - nT) \tag{8.8.1}$$

or the discrete-time signal

$$x(n) = x_a(nT) \tag{8.8.2}$$

The Laplace transformation of Equation (8.8.1) yields

$$X_s(s) = \sum_{n=-\infty}^{\infty} x_a(nT)\exp[-nTs] \tag{8.8.3}$$

If we make the substitution $z = \exp[Ts]$, then

$$X_s(s)\big|_{z=\exp[Ts]} = \sum_{n=-\infty}^{\infty} x_a(nT)z^{-n} \tag{8.8.4}$$

We recognize that the right-hand side of Equation (8.8.4) is the Z-transform, $X(z)$, of the sequence $x(n)$. Thus, the Z-transform can be viewed as the Laplace transform of the sampled function $x_s(t)$ with the change of variable

$$z = \exp[Ts] \tag{8.8.5}$$

Equation (8.8.5) defines a mapping of the s plane to the z plane. To determine the nature of this mapping, let $s = \sigma + j\omega$, so that

$$z = \exp[\sigma T]\exp[j\omega T]$$

Since $|z| = \exp[\sigma T]$, it is clear that if $\sigma < 0$, $|z| < 1$. Thus, any point in the left half of the s plane is mapped into a point inside the unit circle in the z plane. Similarly, since, for $\sigma > 0$, we have $|z| > 1$, a point in the right half of the s plane is mapped into a point outside the unit circle in the z plane. For $\sigma = 0$, $|z| = 1$, so that the $j\omega$-axis of the s plane is mapped into the unit circle in the z plane. The origin in the s plane corresponds to the point $z = 1$.

Finally, let s_k denote a set of points that are spaced vertically apart from any point s_0 by multiples of the sampling frequency $\omega_s = 2\pi/T$. That is,

$$s_k = s_0 + jk\omega_s, \qquad k = 0, \pm 1, \pm 2, \cdots$$

Then we have

$$d_k = \exp[Ts_k] = e^{T(s_0 + jk\omega_s)} = \exp[Ts_0] = z_0$$

since $\exp[jk\omega_s T] = \exp[j2k\pi]$. That is, the points s_k all map into the same point $z_0 = \exp[Ts_0]$ in the z plane. We can thus divide the s plane into horizontal strips, each of width ω_s. Each of these strips is then mapped onto the entire z plane. For convenience, we choose the strips to be symmetric about the horizontal axis. This is summarized in Figure 8.8.1, which shows the mapping of the s plane into the z plane.

We have already seen that $X_s(\omega)$ is periodic with period ω_s. Equivalently, $X(\Omega)$ is periodic with period 2π. This is easily seen to be a consequence of the result that the process of sampling essentially divides the s plane into a set of identical horizontal strips of width ω_s. The fact that the mapping from this plane to the z plane is not unique (the same point in the z plane corresponds to several points in the s plane) is a consequence of the fact that we can associate any one of several analog signals with a given set of sample values.

8.9 SUMMARY

- The Z-transform is the discrete-time counterpart of the Laplace transform.
- The bilateral Z-transform of the discrete-time sequence $x(n)$ is defined as

$$X(z) = \sum_{n=-\infty}^{\infty} x(n)z^{-n}$$

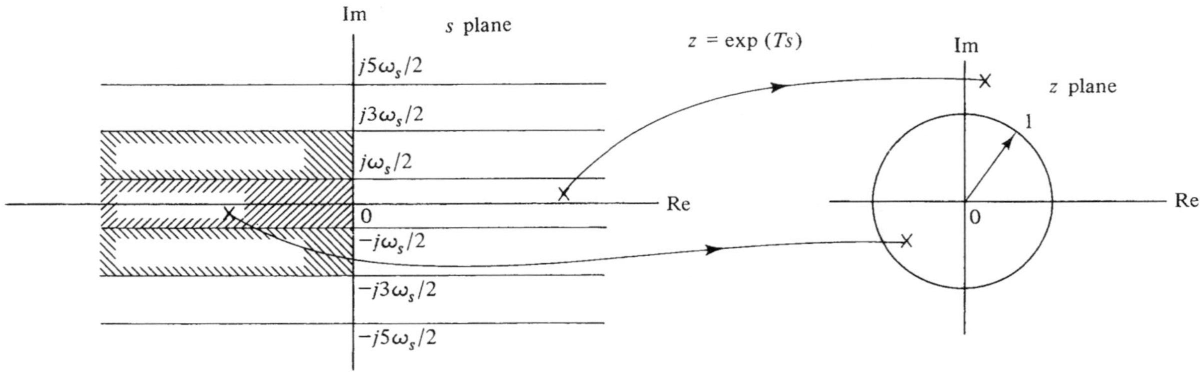

Figure 8.8.1 Mapping of the s plane onto the z plane by the transformation $z = \exp[Ts]$.

- The unilateral Z-transform of a causal signal $x(n)$ is defined as

$$X(z) = \sum_{n=0}^{\infty} x(n) z^{-n}$$

- The region of convergence (ROC) of the Z-transform consists of those values of z for which the sum converges.
- For causal sequences, the ROC in the z plane lies outside a circle containing all the poles of $X(z)$. For anticausal signals, the ROC is inside the circle such that all poles of $X(z)$ are external to this circle. If $x(n)$ consists of both a causal and an anticausal part, then the ROC is an annular region, such that the poles outside this region correspond to the anticausal part of $x(n)$, and the poles inside the annulus correspond to the causal part.
- The Z-transform of an anticausal sequence $x_{-}(n)$ can be determined from a table of unilateral transforms as

$$X_{-}(z) = Z\{x_{-}(-n)\}$$

- Expanding $X(z)$ in partial fractions and identifying the inverse of each term from a table of Z-transforms is the most convenient method for determining $x(n)$. If only the first few terms of the sequence are of interest, $x(n)$ can be obtained by expanding $X(z)$ in a power series in z^{-1} by a process of long division.
- The properties of the Z-transform are similar to those of the Laplace transform. Among the applications of the Z-transform are the solution of difference equations and the evaluation of the convolution of two discrete sequences.
- The time-shift property of the Z-transform can be used to solve difference equations.
- If $y(n)$ represents the convolution of two discrete sequences $x(n)$ and $h(n)$, then

$$Y(z) = H(z)X(z)$$

- The transfer function $H(z)$ of a system with input $x(n)$, impulse response $h(n)$, and output $y(n)$ is

$$H(z) = Z\{h(n)\} = \frac{Y(z)}{X(z)}$$

- Simulation diagrams for discrete-time systems in the z domain can be used to obtain state-variable representations. The solutions to the state equations in the z domain are given by

$$\mathbf{V}(z) = z(z\mathbf{I} - \mathbf{A})^{-1}\mathbf{v}_0 + (z\mathbf{I} - \mathbf{A})^{-1}\mathbf{b}X(z)$$

$$Y(z) = \mathbf{c}\mathbf{V}(z) + dX(z)$$

- The transfer function is given by

$$H(z) = \mathbf{c}(z\mathbf{I} - \mathbf{A})^{-1}\mathbf{b} + d$$

- The state-transition matrix can be obtained as

$$\mathbf{\Phi}(n) = \mathbf{A}^n = Z^{-1}[z(z\mathbf{I} - \mathbf{A})^{-1}]$$

- The relation between the Laplace transform and the Z-transform of the sampled analog signal $x_a(t)$ is

$$X(z)\big|_{z=\exp[Ts]} = X_s(s)$$

- The transformation $z = \exp[Ts]$ represents a mapping from the s plane to the z plane in which the left half of the s plane is mapped inside the unit circle in the z plane, the $j\omega$-axis is mapped into the unit circle, and the right half of the s plane is mapped outside the unit circle. The mapping efffectively divides the s plane into horizontal strips of width ω_s, each of which is mapped into the entire z plane.

8.10 CHECKLIST OF IMPORTANT TERMS

Bilateral Z-transform
Mapping of the s plane into the z plane
Partial-fraction expansion
Power-series expansion
Region of convergence
Simulation diagrams
Solution of difference equations
State-transition matrix
State-variable representations
Transfer function
Unilateral Z-transform

8.11 PROBLEMS

8.1. Determine the Z-transforms and the regions of convergence for the following sequences:

(a) $x(n) = (-3)^n u(-n-1)$

(b) $x(n) = \begin{cases} 1, & -5 \le n \le 5 \\ 0, & \text{otherwise} \end{cases}$

(c) $x(n) = \begin{cases} \left(\dfrac{1}{3}\right)^n, & n \ge 0 \\ 3^n, & n < 0 \end{cases}$

(d) $x(n) = 2\delta(n) - 2^n u(n)$

8.2. The Z-transform of a sequence $x(n)$ is

$$X(z) = \frac{z^3 + 4z^2 - \frac{11}{6}}{z^3 + \frac{7}{6}z^2 - \frac{3}{2}z + \frac{1}{3}}$$

(a) Plot the locations of the poles and zeros of $X(z)$.

(b) Identify the causal pole(s) if the ROC is (i) $|z| < \frac{1}{3}$, (ii) $|z| > 2$

(c) Find $x(n)$ in both cases.

8.3. Use the definition and the properties of the Z-transform to find $X(z)$ for the following causal sequences:

(a) $x(n) = n\, a^n \sin \Omega_0 n$

(b) $x(n) = n^2 \cos \Omega_0 n$

(c) $x(n) = n\left(\dfrac{1}{2}\right)^n + (n-1)\left(\dfrac{1}{3}\right)^n$

(d) $x(n) = \delta(n-2) + nu(n)$

(e) $x(n) = 2\exp[-n]\sin\left(\frac{2}{5}\pi n\right)$

8.4. Determine the Z-transform of the sequences that result when the following causal continuous-time signals are sampled uniformly every T seconds:

(a) $x(t) = t\cos 1000\pi t$

(b) $x(t) = t\exp[-3(t-1)]$

8.5. Find the inverse of each of the following Z-transforms by means of (i) power series expansion and (ii) partial-fraction expansion. Use a mathematical software package to verify your partial-fraction expansions. Assume that all sequences are causal.

(a)

$$X(z) = \frac{1 + \frac{1}{2}z^{-1}}{1 - \frac{3}{4}z^{-1} + \frac{1}{8}z^{-2}}$$

(b)

$$X(z) = \frac{(z + \frac{1}{2})(z + \frac{1}{4})}{(z - \frac{3}{8})(z - \frac{1}{4})}$$

(c)

$$X(z) = \frac{1}{(z - \frac{1}{2})^3}$$

(d)

$$\frac{z(z+2)}{z^2 + 4z + 3}$$

8.6. Find the inverse transform of

$$X(z) = \log(1 - \tfrac{1}{3}z^{-1})$$

by the following methods:

(a) Use the series expansion

$$\log(1-a) = -\sum_{i=1}^{\infty}\frac{a^i}{i}, \qquad |a| < 1$$

(b) Differentiate $X(z)$ and use the properties of the Z-transform.

8.7. Use Z-transforms to find the convolution of the following causal sequences:

(a)

$$h(n) = \left(\frac{1}{2}\right)^n, \qquad x(n) = \begin{cases} 1, & 0 \le n \le 10 \\ 0, & \text{otherwise} \end{cases}$$

(b)

$$h(n) = \left(\frac{1}{3}\right)^n, \qquad x(n) = \begin{cases} 2, & 0 \le n \le 5 \\ 1, & 6 \le n \le 9 \end{cases}$$

(c)

$$h(n) = \{1, -1, 2, -1, 1\}, \qquad x(n) = \{1, 0, -2, 3\}$$

8.8. Find the step response of the system with transfer function

$$H(z) = \frac{z - 2}{z^2 + \frac{1}{6}z - \frac{1}{6}}$$

8.9. **(a)** Solve the following difference equations using Z-transforms:

(i) $y(n) - y(n-1) + y(n-2) = x(n)$

$y(-1) = 1, \quad y(-2) = 0, \quad x(n) = (\frac{1}{2})^n u(n)$

(ii) $y(n) - \frac{1}{6}y(n-1) - \frac{1}{6}y(n-2) = x(n) - \frac{1}{2}x(n-1)$

$y(-1) = 0, \quad y(-2) = 0, \quad x(n) = (\frac{1}{3})^n u(n)$

(b) Verify your result using any mathematical software package.

8.10. Solve the difference equations of Problem 6.17 using Z-transforms.

8.11. **(a)** Find the transfer functions of the systems in Problem 6.17, and plot the pole-zero locations in the z plane.

(b) What is the corresponding impulse response?

8.12. **(a)** When input $x(n) = u(n) + (-\frac{1}{2})^n u(n)$ is applied to a linear, causal, time-invariant system, the output is

$$y(n) = 6\left(-\frac{1}{4}\right)^n u(n) - 6\left(-\frac{1}{3}\right)^n u(n)$$

Find the transfer function of the system.

(b) What is the difference-equation representation of the system?

8.13. Find the transfer function of the system

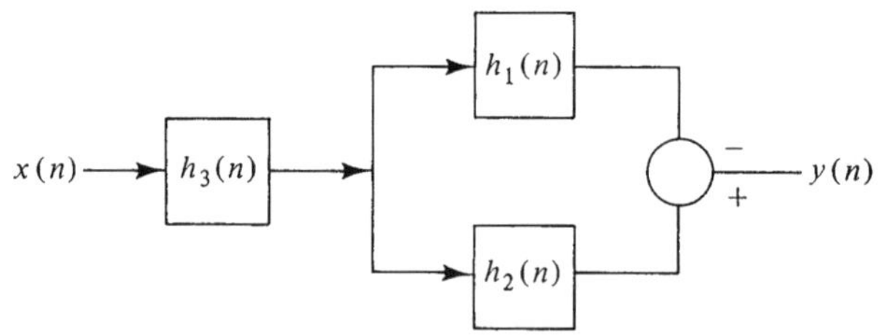

if

$$h_1(n) = (n-1)u(n)$$
$$h_2(n) = \delta(n) + nu(n-1) + \delta(n-2)$$
$$h_3(n) = \left(\frac{1}{2}\right)^n u(n)$$

8.14. **(a)** Show that a simplified criterion for the polynomial $F(z) = z^2 + a_1 z + a_2$ to have all its poles within the unit circle in the z plane is

$$|F(0)| < 1, \quad F(-1) > 0, \quad F(1) > 0$$

(b) Use this criterion to find the values of K for which the system of Example 8.6.3 is stable.

8.15. The transfer function of a linear, causal, time-invariant system is

$$H(z) = \frac{Kz(1-\alpha)}{z^2 + (K - 1 - \alpha)z + (1-K)\alpha}$$

where K and α are constant. Find the range of values of K and α for which the system is stable, and plot this region in the K-α plane.

8.16. Obtain a realization of the following transfer function as a combination of first- and second-order sections in (a) cascade and (b) parallel.

$$H(z) = \frac{(1 + 0.5z^{-1} + 0.06z^{-2})(1 + 1.7z^{-1} + 0.72z^{-2})}{(1 + 0.4z^{-1} + 0.8z^{-2})(1 - 0.25z^{-1} - 0.125z^{-2})}$$

8.17. (a) Find the state-transition matrix for the systems of Problem 6.28, using the Z-transform.

(b) Use the frequency-domain technique to find the unit-step response of these systems, assuming that $\mathbf{v}(0) = \mathbf{0}$.

(c) Find the transfer function from the state representation.

(d) Verify your result using any mathemathical software package.

8.18. Repeat Problem 8.17 for the state equations obtained in Problem 6.26.

8.19. A low-pass analog signal $x_a(t)$ with a bandwidth of 3 kHz is sampled at an appropriate rate to yield the discrete-time sequence $x(nT)$.

(a) What sampling rate should be chosen to avoid aliasing?

(b) For the sampling rate chosen in part (a), determine the primary and secondary strips in the s plane.

8.20. The signal $x_a(t) = 10 \cos 600\pi t \sin 2400\pi t$ is sampled at rates of (i) 800 Hz, (ii) 1600 Hz, and (iii) 3200 Hz.

(a) Plot the frequencies present in $x_a(t)$ as poles at the appropriate locations in the s plane.

(b) Determine the frequencies present in the sampled signal for each sampling rate. On the s plane, indicate the primary and secondary strips for each case, and plot the frequencies in the sampled signal as poles at appropriate locations.

(c) From your plots in part (b), can you determine which sampling rate will enable an error-free reconstruction of the analog signal?

(d) Verify your answer in part (c) by plotting the spectra of the analog and sampled signals.

8.21. In the text, we saw that the zero-order represents an easily implementable approximation to the ideal reconstruction filter. However, the zero-order hold gives only a staircase approximation to the analog signal. In the first-order hold, the output in the interval $nT \le t < (n + 1)T$ is given by

$$y(t) = x_a(nT) + \frac{t - nT}{T}[x_a(nT) - x_a(nT - T)]$$

Find the transfer function $G_{h1}(s)$ of the first order hold, and compare its frequency response with that of the ideal reconstruction filter matched to the rate T.

8.22. As we saw in Chapter 4, filters are used to modify the frequency content of signals in an appropriate manner. A technique for designing digital filters is based on transforming an analog filter into an equivalent digital filter. In order to do so, we have to obtain a relation between the Laplace and Z-transform variables. In Section 8.8, we discussed one such relation based on equating the sample values of an analog signal with a discrete-time signal. The relation obtained was

$$z = \exp[Ts]$$

We can obtain other such relations by using different equivalences. For example, by equating the s-domain transfer function of the derivative operator and the Z-domain transfer function of its backward-difference approximation, we can write

$$s = \frac{1 - z^{-1}}{T}$$

or equivalently,

$$z = \frac{1}{1 - sT}$$

Similarly, equating the integral operator with the trapezoidal approximation (see Problem 6.16) yields

$$s = \frac{2}{T}\frac{1 - z^{-1}}{1 + z^{-1}}$$

or

$$z = \frac{1 + (T/2)s}{1 - (T/2)s}$$

(a) Derive the two alternative relations between the s and z planes just given.

(b) Discuss the mapping of the s plane into the z plane using the two relations.

Chapter 9

The Discrete Fourier Transform

9.1 INTRODUCTION

From our discussions so far, we see that transform techniques play a very useful role in the analysis of linear, time-invariant systems. Among the many applications of these techniques are the spectral analysis of signals, the solution of differential or difference equations, and the analysis of systems in terms of a frequency response or transfer function. With the tremendous increase in the use of digital hardware in recent years, interest has centered upon transforms that are especially suited for machine computation. In this chapter we study one such transform, namely, the discrete Fourier transform (DFT), which can be viewed as a logical extension of the Fourier transforms discussed earlier.

In order to motivate our definition of the DFT, let us assume that we are interested in finding the Fourier transform of an analog signal $x_a(t)$ using a digital computer. Since such a computer can store and manipulate only a finite set of numbers, it is necessary to represent $x_a(t)$ by a finite set of values. The first step in doing so is to sample the signal to obtain a discrete sequence $x_a(n)$. Because the analog signal may not be time limited, the next step is to obtain a finite set of samples of the discrete sequence by means of truncation. Without loss of generality, we can assume that these samples are defined for n in the range $[0, N-1]$. Let us denote this finite sequence by $x(n)$, which we can consider to be the product of the infinite sequence $x_a(n)$ and the *window* function

$$w(n) = \begin{cases} 1, & 0 \le n \le N-1 \\ 0, & \text{otherwise} \end{cases} \tag{9.1.1}$$

so that

$$x(n) = x_a(n)w(n) \tag{9.1.2}$$

Since we now have a discrete sequence, we can take the discrete-time Fourier transform of the sequence as

$$X(\Omega) = \sum_{n=0}^{N-1} x(n) \exp[-j\Omega n] \tag{9.1.3}$$

This is still not in a form suitable for machine computation, since Ω is a continuous variable taking values in $[0, 2\pi]$. The final step, therefore, is to evaluate $X(\Omega)$ at only a finite number of values Ω_k by a process of sampling uniformly in the range $[0, 2\pi]$. We obtain

$$X(\Omega_k) = \sum_{n=0}^{N-1} x(n) \exp[-j\Omega_k n], \qquad k = 0, 1, \dots, M-1 \tag{9.1.4}$$

where

$$\Omega_k = \frac{2\pi}{M} k \tag{9.1.5}$$

The number of frequency samples, M, can be any value. However, we choose it to be the same as the number of time samples, N. With this modification, and writing $X(\Omega_k)$ as $X(k)$, we finally have

$$X(k) = \sum_{n=0}^{N-1} x(n) \exp\left[-j\frac{2\pi}{N} nk\right] \tag{9.1.6}$$

An assumption that is implicit in our derivations is that $x(n)$ can take any value in the range $(-\infty, \infty)$—that is, that $x(n)$ can be represented to infinite precision. However, the computer can use only a finite word-length representation. Thus, we *quantize* the dynamic range of the signal to a finite number of levels. In many applications, the error that arises in representing an infinite-precision number by a finite word can be made small, in comparison to the errors introduced by sampling, by a suitable choice of quantization levels. We therefore assume that $x(n)$ can assume any value in $(-\infty, \infty)$.

Although Equation (9.1.6) can be considered to be an approximation to the continuous-time Fourier transform of the signal $x_a(t)$, it defines the discrete Fourier transform of the N-point sequence $x(n)$. We will investigate the nature of this approximation in Section 9.6, where we consider the spectral estimation of analog signals using the DFT. However, as we will see in subsequent sections, although the DFT is similar to the discrete-time Fourier transform that we studied in Chapter 7, some of its properties are quite different.

One of the reasons for the widespread use of the DFT and other discrete transforms is the existence of algorithms for their fast and efficient computation on a computer. For the DFT, these algorithms collectively go under the name of fast Fourier transform (FFT) algorithms. We discuss two popular versions of the FFT in Section 9.5.

9.2 THE DISCRETE FOURIER TRANSFORM AND ITS INVERSE

Let $x(n), n = 0, 1, 2, \ldots, N - 1$, be an N-point sequence. We define the discrete Fourier transform of $x(n)$ as

$$X(k) = \sum_{n=0}^{N-1} x(n) \exp\left[-j\frac{2\pi}{N}nk\right] \tag{9.2.1}$$

The inverse discrete Fourier-transform (IDFT) relation is given by

$$x(n) = \frac{1}{N}\sum_{k=0}^{N-1} X(k) \exp\left[j\frac{2\pi}{N}nk\right] \tag{9.2.2}$$

To derive this relation, we replace n by p in the right side of Equation (9.2.1) and multiply by $\exp[j2\pi nk/N]$ to get

$$X(k) \exp\left[j\frac{2\pi}{N}nk\right] = \sum_{p=0}^{N-1} x(p) \exp\left[j\frac{2\pi}{N}k(n-p)\right] \tag{9.2.3}$$

If we now sum over k in the range $[0, N - 1]$, we obtain

$$\sum_{k=0}^{N-1} X(k) \exp\left[j\frac{2\pi}{N}nk\right] = \sum_{p=0}^{N-1}\sum_{k=0}^{N-1} x(p) \exp\left[j\frac{2\pi}{N}k(n-p)\right] \tag{9.2.4}$$

In Equation (7.2.12) we saw that

$$\sum_{k=0}^{N-1} \exp\left[j\frac{2\pi}{N}k(n-p)\right] = \begin{cases} 0, & n \neq p \\ N, & n = p \end{cases}$$

so that the right-hand side of Equation (9.2.4) evaluates to $Nx(n)$, and Equation (9.2.2) follows.

We saw that $X(\Omega)$ is periodic in Ω with period 2π; thus, $X(\Omega_k) = X(\Omega_k + 2\pi)$. This can be written as

$$X(k) = X(\Omega_k) = X(\Omega_k + 2\pi) = X\left(\frac{2\pi}{N}(k + N)\right) = X(k + N) \tag{9.2.5}$$

That is, $X(k)$ is periodic with period N.

We now show that $x(n)$, as determined from Equation (9.2.2), is also periodic with period N. From that equation, we have

$$\begin{aligned} x(n + N) &= \sum_{k=0}^{N-1} X(k) \exp\left[j\frac{2\pi}{N}(n + N)k\right] \\ &= \sum_{k=0}^{N-1} X(k) \exp\left[j\frac{2\pi}{N}nk\right] \\ &= x(n) \end{aligned} \tag{9.2.6}$$

That is, the IDFT operation yields a periodic sequence, of which only the first N values, corresponding to one period, are evaluated. Hence, in all operations involving the DFT and the IDFT, we are effectively replacing the finite sequence $x(n)$ by its periodic extension. We can therefore expect that there is a connection between the Fourier-series expansion of periodic discrete-time sequences that we discussed in Chapter 7 and the DFT. In fact, a comparison of Equations (9.2.1) and (9.2.2) with Equations (7.2.15) and (7.2.16) shows that the DFT $X(k)$ of finite sequence $x(n)$ can be interpreted as the coefficient a_k in the Fourier series representation of its periodic extension $x_p(n)$, multiplied by the period N. (The two can be made identical by including the factor $1/N$ with the DFT rather than with the IDFT.)

9.3 PROPERTIES OF THE DFT

We now consider some of the more important properties of the DFT. As might be expected, they closely parallel properties of the discrete-time Fourier transform. In considering the properties of the DFT, it is helpful to remember that we are in essence replacing an N-point sequence by its periodic extension. Thus, operations such as time shifting must be considered to be operations on a periodic sequence. However, we are interested only in the range $[0, N-1]$, so that the shift can be interpreted as a circular shift, as explained in Section 6.4.

Since the DFT is evaluated at frequencies in the range $[0, 2\pi]$, which are spaced apart by $2\pi/N$, in considering the DFT of two signals simultaneously, the frequencies corresponding to the DFT must be the same for any operation to be meaningful. This means that the length of the sequences considered must be the same. If this is not the case, it is usual to augment the signals by an appropriate number of zeros, so that all the signals considered are of the same length. (Since it is assumed that the signals are of finite length, adding zeros does not change the essential nature of the signal.)

9.3.1 Linearity

Let $X_1(k)$ and $X_2(k)$ be the DFTs of the two sequences $x_1(n)$ and $x_2(n)$. Then

$$\text{DFT}[a_1x_1(n) + a_2x_2(n)] = a_1X_1(k) + a_2X_2(k) \tag{9.3.1}$$

for any constants a_1 and a_2.

9.3.2 Time Shifting

For any real integer n_0,

$$\begin{aligned}\text{DFT}[x(n+n_0)] &= \sum_{n=0}^{N-1} x(n+n_0)\exp\left[-j\frac{2\pi}{N}kn\right] \\ &= \sum_{\langle N\rangle} x(m)\exp\left[-j\frac{2\pi}{N}k(m-n_0)\right] \\ &= \exp\left[j\frac{2\pi}{N}kn_0\right]X(k)\end{aligned} \tag{9.3.2}$$

where, as explained before, the shift is a circular shift.

9.3.3 Alternative Inversion Formula

By writing the IDFT formula, Equation (9.2.2), as

$$x(n) = \frac{1}{N}\left[\sum_{k=0}^{N-1} X^*(k)\exp\left[-j\frac{2\pi}{N}nk\right]\right]^*$$

$$= \frac{1}{N}\{\text{DFT}[X^*(k)]\}^* \tag{9.3.3}$$

we can interpret $x(n)$ as the complex conjugate of the DFT of $X^*(k)$ multiplied by $1/N$. Thus, the same algorithm used to calculate the DFT can be used to evaluate the IDFT.

9.3.4 Time Convolution

We saw in our earlier discussions of different transforms that the inverse transform of the product of two transforms corresponds to a convolution of the corresponding time functions. With this in view, let us determine IDFT of the function $Y(k) = H(k)X(k)$. We have

$$y(n) = \text{IDFT}[Y(k)]$$

$$= \frac{1}{N}\sum_{k=0}^{N-1} Y(k)\exp\left[j\frac{2\pi}{N}nk\right]$$

$$= \frac{1}{N}\sum_{k=0}^{N-1} H(k)X(k)\exp\left[j\frac{2\pi}{N}nk\right]$$

Using the definition of $H(k)$, we get

$$y(n) = \frac{1}{N}\sum_{k=0}^{N-1}\left(\sum_{m=0}^{N-1} h(m)\exp\left[-j\frac{2\pi}{N}mk\right]\right)X(k)\exp\left[j\frac{2\pi}{N}nk\right]$$

Interchanging the order of summation and using Equation (9.2.2), we obtain

$$y(n) = \sum_{m=0}^{N-1} h(m)x(n-m) \tag{9.3.4}$$

A comparison with Equation (6.4.1) shows that the right-hand side of Equation (9.3.4) corresponds to the periodic convolution of the two sequences $x(n)$ and $h(n)$.

Example 9.3.1

Consider the periodic convolution $y(n)$ of the two sequences

$$h(n) = \{1, 3, -1, -2\} \qquad \text{and} \qquad x(n) = \{1, 2, 0, -1\}$$

Here $N = 4$, so that $\exp[j(2\pi/N)] = j$. By using Equation (9.2.1), we can calculate the DFTs of the two sequences as

$$H(0) = h(0) + h(1) + h(2) + h(3) = 1$$

$$H(1) = h(0) + h(1)\exp\left[-j\frac{\pi}{2}\right] + h(2)\exp[-j\pi] + h(3)\exp\left[-j\frac{3\pi}{2}\right] = 2 - j5$$

$$H(2) = h(0) + h(1)\exp[-j\pi] + h(2)\exp[-j2\pi] + h(3)\exp[-j3\pi] = -1$$

$$H(3) = h(0) + h(1)\exp\left[-j\frac{3\pi}{2}\right] + h(2)\exp[-j3\pi] + h(3)\exp\left[-j\frac{9\pi}{2}\right] = 2 + j5$$

and

$$X(0) = x(0) + x(1) + x(2) + x(3) = 2$$

$$X(1) = x(0) + x(1)\exp\left[-j\frac{\pi}{2}\right] + x(2)\exp[-j\pi] + x(3)\exp\left[-j\frac{3\pi}{2}\right] = 1 - 3j$$

$$X(2) = x(0) + x(1)\exp[-j\pi] + x(2)\exp[-j2\pi] + x(3)\exp[-j3\pi] = 0$$

$$X(3) = x(0) + x(1)\exp\left[-j\frac{3\pi}{2}\right] + x(2)\exp[-j3\pi] + x(3)\exp\left[-j\frac{9\pi}{2}\right] = 1 + 3j$$

so that

$$\begin{aligned} Y(0) &= H(0)X(0) = 2 \\ Y(1) &= H(1)X(1) = -13 - j11 \\ Y(2) &= H(2)X(2) = 0 \\ Y(3) &= H(3)X(3) = -13 + j11 \end{aligned}$$

We can now use Equation (9.2.2) to find $y(n)$ as

$$y(0) = \tfrac{1}{4}[Y(0) + Y(1) + Y(2) + Y(3)] = -6$$

$$y(1) = \tfrac{1}{4}\left[Y(0) + Y(1)\exp\left[j\frac{\pi}{2}\right] + Y(2)\exp[j\pi] + Y(3)\exp\left[j\frac{3\pi}{2}\right]\right] = 6$$

$$y(2) = \tfrac{1}{4}(Y(0) + Y(1)\exp[j\pi] + Y(2)\exp[j2\pi] + Y(3)\exp[j3\pi]) = 7$$

$$y(3) = \tfrac{1}{4}\left(Y(0) + Y(1)\exp\left[j\frac{3\pi}{2}\right] + Y(2)\exp[j3\pi] + Y(3)\exp\left[j\frac{9\pi}{2}\right]\right) = -5$$

which is the same answer as was obtained in Example 6.4.1.

9.3.5 Relation to the Discrete-Time Fourier and Z-Transforms

From Equation (9.2.1), we note that the DFT of an N-point sequence $x(n)$ can be written as

$$\begin{aligned} X(k) &= \sum_{n=0}^{N-1} x(n)\exp[-j\Omega n]\big|_{\Omega=\frac{2\pi}{N}k} \\ &= X(\Omega)\big|_{\Omega=\frac{2\pi}{N}k} \end{aligned} \tag{9.3.5}$$

That is, the DFT of the sequence $x(n)$ is its discrete-time Fourier transform $X(\Omega)$ evaluated at N equally spaced points in the range $[0, 2\pi)$.

For a sequence for which both the discrete-time Fourier transform and the Z-transform exist, it follows from Equation (8.3.7) that

$$X(k) = X(z)\big|_{z=\exp[j(2\pi/N)k]} \tag{9.3.6}$$

so that the DFT is the Z-transform evaluated at N equally spaced points along the unit circle in the z plane.

9.3.6 Matrix Interpretation of the DFT

We can express the DFT relation of Equation (9.2.1) compactly as a matrix operation on the *data vector* $\mathbf{x} = [x(0)x(1)\cdots x(N-1)]^T$. For convenience, let us denote $\exp[-j2\pi/N]$ by W_N. We can then write

$$X(k) = \sum_{n=0}^{N-1} x(n)W_N^{kn} \qquad k = 0, 1, \ldots, N-1 \tag{9.3.7}$$

Let $\mathbf{W}$ be the matrix whose (k, n)th element $[\mathbf{W}]_{kn}$ is equal to W_N^{kn}. That is,

$$\mathbf{W} = \begin{bmatrix} W_N^0 & W_N^0 & W_N^0 & \cdot & W_N^0 \\ W_N^0 & W_N^1 & W_N^2 & \cdot & W_N^{N-1} \\ \cdot & \cdot & \cdot & \cdot & \cdot \\ \cdot & \cdot & \cdot & \cdot & \cdot \\ W_N^0 & W_N^{N-1} & W_N^{2(N-1)} & \cdot & W_N^{(N-1)(N-1)} \end{bmatrix} \tag{9.3.8}$$

It then follows that the *transform vector* $\mathbf{X} = [X(0)X(1)\cdots X(N-1)]^T$ can be obtained as

$$\mathbf{X} = \mathbf{W}\,\mathbf{x} \tag{9.3.9}$$

The matrix $\mathbf{W}$ is usually referred to as the DFT matrix. Clearly, $[\mathbf{W}]_{kn} = [\mathbf{W}]_{nk}$, so that $\mathbf{W}$ is symmetric ($\mathbf{W} = \mathbf{W}^{\mathrm{T}}$).

From Equation (9.2.2), we can write

$$x(n) = \frac{1}{N}\sum_{k=0}^{N-1} x(n)W_N^{-nk} \tag{9.3.10}$$

Since $W_N^{-1} = W_N^*$, where $*$ represents the complex conjugate, it follows that the IDFT relation can be written in matrix form as

$$\mathbf{x} = \frac{1}{N}\mathbf{W}^*\mathbf{X} \tag{9.3.11}$$

Solving for $\mathbf{x}$ from Equation (9.3.9) gives

$$\mathbf{x} = \mathbf{W}^{-1}\mathbf{X} \tag{9.3.12}$$

It therefore follows that

$$\mathbf{W}^{-1} = \frac{1}{N}\mathbf{W}^* \tag{9.3.13}$$

or equivalently,

$$\mathbf{W}^*\mathbf{W} = N\mathbf{I}_N \tag{9.3.14}$$

where $\mathbf{I}_N$ is the identity matrix of dimension $N \times N$. Since $\mathbf{W}$ is a symmetric matrix, we can write Equation (9.3.14) as

$$\mathbf{W}^{*T}\mathbf{W} = N\mathbf{I}_N \tag{9.3.15}$$

In general, a matrix $\mathbf{A}$ that satisfies $\mathbf{A}^{*T}\mathbf{A} = \mathbf{I}$ is called a *unitary matrix*. A real matrix $\mathbf{A}$ that satisfies $\mathbf{A}^T\mathbf{A} = \mathbf{I}$ is said to be an *orthogonal matrix*. The matrix $\mathbf{W}$, as defined in Equation (9.3.8), is not strictly a unitary matrix, as can be seen from Equation (9.3.15). However, it was pointed out in Section 9.2 that the factor $1/N$ could be used with either the DFT or the IDFT relation. Thus, if we associate a factor $1/\sqrt{N}$ with both the DFT and IDFT relations and let $W_N = 1/\sqrt{N}\exp[-j2\pi/N]$ in the definition of the matrix $\mathbf{W}$ in Equation (9.3.8), it follows that

$$\mathbf{X} = \mathbf{W}\,\mathbf{x} \quad \text{and} \quad \mathbf{x} = \mathbf{W}^*\mathbf{X} \tag{9.3.16}$$

with $\mathbf{W}$ being a unitary matrix. The DFT is therefore a *unitary transform*; often, however, it is simply referred to as an orthogonal transform.

Other useful orthogonal transforms can be defined by replacing the DFT matrix in Equation (9.3.8) by other unitary or orthogonal matrices. Examples are the Walsh-Hadamard transform and the discrete cosine transform, which have applications in areas such as speech and image processing. As with the DFT, the utility of these transforms arises from the existence of fast and efficient algorithms for their computation.

9.4 LINEAR CONVOLUTION USING THE DFT

We have seen that one the primary uses of transforms is in the analysis of linear time-invariant systems. For a linear, discrete-time-system with impulse response $h(n)$, the output due to any input $x(n)$ is given by the *linear* convolution of the two sequences. However, the product $H(k)X(k)$ of the two DFTs corresponds to a *periodic* convolution of $h(n)$ and $x(n)$. A question that naturally arises is whether the DFT can be used to perform a linear convolution. In order to answer this question, let us assume that $h(n)$ and $x(n)$ are of length M and N, respectively, so that $h(n)$ is zero outside the range $[0, M-1]$, and $x(n)$ is zero outside the range $[0, N-1]$. We also assume that $M < \mathrm{N}$.

Recall that in Chapter 6 we defined the periodic convolution of two finite-length sequences of equal length as the convolution of their periodic extensions. For the two sequences $h(n)$ and $x(n)$ considered here, we can zero-pad both to the length $K \geq \text{Max}(M, N)$, to form the augmented sequences $h_a(n)$ and $x_a(n)$, respectively. We can now define the K-point periodic convolution, $y_p(n)$, of the sequences as the convolution of their periodic extensions. We note that while $y_p(n)$ is a K-point sequence, the linear convolution of $h(n)$ and $x(n)$, $y_l(n)$, has length $L = M + N - 1$.

As Example 6.4.2 shows, for $K \leq L$, $y_p(n)$ corresponds to the sequence obtained by adding in, or *time-aliasing*, the last $L - K$ values of $y_l(n)$ to the first $L - K$ points. Thus, the first $L - K$ points of $y_p(n)$ will not correspond to $y_l(n)$, while the remaining $2K - L$

points will be the same in both sequences. Clearly, if we choose $K = L$, $y_p(n)$ and $y_l(n)$ will be identical.

Most available routines for the efficient computation of the DFT assume that the length of the sequence is a power of 2. In that case, K is chosen as the smallest power of 2 that is larger than L. When $K > L$, the first L points of $y_p(n)$ will be identical to $y_l(n)$, while the remaining $K - L$ values will be zero.

We will now show that the K-point periodic convolution of $h(n)$ and $x(n)$ is identical to the linear convolution of the two functions if $K = L$. We note that

$$y_l(n) = \sum_{m=-\infty}^{\infty} h(m)x(n-m) \tag{9.4.1}$$

Now, $h(m)$ is zero for $m \in [0, M-1]$, and $x(n-m)$ is zero for $(n-m) \in [0, N-1]$, so that we have the following.

$0 \le n \le M-1$:

$$\begin{aligned} y_l(n) &= \sum_{m=0}^{n} h(m)x(n-m) \\ &= h(0)x(n) + h(1)x(n-1) + \cdots + h(n)x(0) \end{aligned}$$

$M \le n \le N-1$:

$$\begin{aligned} y_l(n) &= \sum_{m=n-M+1}^{n} h(m)x(n-m) \\ &= h(n-M+1)x(M-1) + h(n-M+2)x(M-2) \\ &\quad + \cdots + h(n)x(0) \end{aligned}$$

$N \le n \le M+N-2$:

$$\begin{aligned} y_l(n) &= \sum_{m=n-M+1}^{N-1} h(m)x(n-m) \\ &= h(n-M+1)x(M-1) \\ &\quad + \cdots + h(N+1)x(n-N+1) \end{aligned} \tag{9.4.2}$$

On the other hand,

$$y_p(n) = \sum_{m=0}^{k-1} h_a(m)x_a(n-m) \tag{9.4.3}$$

Since the sequence $x_a(n-m)$ is obtained by circularly shifting $x_a(n)$, it follows that

$$x_a(n-m) = \begin{cases} x_a(n-m), & m \le n \\ x_a(n-m+K), & n+1 \le m \le K-1 \end{cases} \tag{9.4.4}$$

so that

$$\begin{aligned} y_p(n) = h_a(0)x_a(n) + \cdots + h_a(n)x_a(0) + h_a(n+1)x_a(K-1) \\ + h_a(n+2)x_a(K-2) + \cdots + h_a(K-1)x_a(n+1) \end{aligned} \tag{9.4.5}$$

Now, if we use the fact that

$$h_a(n) = \begin{cases} h(n), & 0 \le n \le M-1 \\ 0, & \text{otherwise} \end{cases}$$

$$x_a(n) = \begin{cases} x(n), & 0 \le n \le N-1 \\ 0, & \text{otherwise} \end{cases} \tag{9.4.6}$$

we can easily verify that $y_p(n)$ is exactly the same as $y_l(n)$ for $0 \le n \le N + M - 2$ and is zero for $N + M - 1 \le n \le K - 1$.

In sum, in order to use the DFT to perform the linear convolution of the M-point sequence $h(n)$ and the N-point sequence $x(n)$, we augment both sequences with zeros to form the K-point sequences $h_a(n)$ and $x_a(n)$, with $K \ge M + N - 1$. We determine the product of the corresponding DFTs, $H_a(k)$ and $X_a(k)$. Then

$$y_l(n) = \text{IDFT}[H_a(k)X_a(k)] \tag{9.4.7}$$

9.5 FAST FOURIER TRANSFORMS

As indicated earlier, one of the main reasons for the popularity of the DFT is the existence of efficient algorithms for its computation. Recall that the DFT of the sequence $x(n)$ is given by

$$X(k) = \sum_{n=0}^{N-1} x(n)\exp\left[-j\frac{2\pi}{N}nk\right], \qquad k = 0, 1, 2, \ldots, N-1 \tag{9.5.1}$$

For convenience, let us denote $\exp[-j2\pi/N]$ by W_N, so that

$$X(k) = \sum_{n=0}^{N-1} x(n)W_N^{kn}, \qquad k = 0, 1, 2, \ldots, N-1 \tag{9.5.2}$$

which can be explicitly written as

$$X(k) = x(0)W_N^0 + x(1)W_N^k + x(2)W_N^{2k} + \cdots + x(N-1)W_N^{(N-1)k} \tag{9.5.3}$$

It follows from Equation (9.5.3) that the determination of each $X(k)$ requires N complex multiplications and N complex additions, where we have also included trivial multiplications by ± 1 or $\pm j$ in the count, in order to have a simple method for comparing the computational complexity of different algorithms. Since we have to evaluate $X(k)$ for $k = 0, 1, \cdots, N-1$, it follows that the direct determination of the DFT requires N^2 complex multiplications and N^2 complex additions, which can become prohibitive for large values of N. Thus, procedures that reduce the computational burden are of considerable interest. These procedures are known as fast Fourier-transform (FFT) algorithms. The basic idea in all of the approaches is to divide the given sequence into subsequences of smaller length. We then combine these smaller DFTs suitably to obtain the DFT of the original sequence.

In this section, we derive two versions of the FFT algorithm, assuming that the data length N is a power of 2, so that N is of the form $N = 2^P$, where P is a positive integer.

That is, $N = 2, 4, 8, 16, 32$, etc. Accordingly, the algorithms are referred to as radix-2 algorithms.

9.5.1 The Decimation-in-Time Algorithm

In the decimation-in-time (DIT) algorithm, we divide $x(n)$ into two subsequences, each of length $N/2$, by grouping the even-indexed samples and the odd-indexed samples together. We can then write $X(k)$ in Equation (9.5.3) as

$$X(k) = \sum_{n \text{ even}} x(n) W_N^{nk} + \sum_{n \text{ odd}} x(n) W_N^{nk} \tag{9.5.4}$$

Letting $n = 2r$ in the first sum and $n = 2r + 1$ in the second sum, we can write

$$\begin{aligned} X(k) &= \sum_{r=0}^{N/2-1} x(2r) W_N^{2rk} + \sum_{r=0}^{N/2-1} x(2r+1) W_N^{(2r+1)k} \\ &= \sum_{r=0}^{N/2-1} g(r) W_N^{2rk} + W_N^k \sum_{r=0}^{N/2-1} h(r) W_N^{2rk} \end{aligned} \tag{9.5.5}$$

where $g(r) = x(2r)$ and $h(r) = x(2r + 1)$.

Note that for an $N/2$-point sequence $y(n)$, the DFT is given by

$$\begin{aligned} Y(k) &= \sum_{n=0}^{N/2-1} y(n) W_{N/2}^{nk} \\ &= \sum_{n=0}^{N/2-1} y(n) W_N^{2nk} \end{aligned} \tag{9.5.6}$$

where the last step follows from the relation

$$W_{N/2} = \exp\left[-j\frac{2\pi}{N/2}\right] = \left(\exp\left[-j\frac{2\pi}{N}\right]\right)^2 = W_N^2 \tag{9.5.7}$$

Thus, Equation (9.5.5) can be written as

$$X(k) = G(k) + W_N^k H(k), \qquad k = 0, 1, \ldots, N - 1 \tag{9.5.8}$$

where $G(k)$ and $H(k)$ denote the $N/2$-point DFTs of the sequences $g(r)$ and $h(r)$, respectively. Since $G(k)$ and $H(k)$ are periodic with period $N/2$, we can write Equation (9.5.8) as

$$\begin{aligned} X(k) &= G(k) + W_N^k H(k), \qquad k = 0, 1, \ldots, \frac{N}{2} - 1 \\ X\left(k + \frac{N}{2}\right) &= G(k) + W_N^{k+N/2} H(k) \end{aligned} \tag{9.5.9}$$

The steps involved in determining $X(k)$ can be illustrated by drawing a *signal-flow graph* corresponding to Equation (9.5.9). In the graph, we associate a *node* with each signal. The interrelations between the nodes are indicated by drawing appropriate lines (branches) with arrows pointing in the direction of the signal flow. Hence, each branch has an input signal and an output signal. We associate a weight with each branch that determines the transmittance between the input and output signals. When not indicated on the graph, the transmittance of any branch is assumed to be 1. The signal at any node is the sum of the outputs of all the branches entering the node. These concepts are illustrated in Figure 9.5.1, which shows the signal-flow graph for the computations involved in Equation (9.5.9) for a particular value of k. Figure 9.5.2 shows the signal-flow graph for computing $X(k)$ for an eight-point sequence. As can be seen from the graph, to determine $X(k)$, we first compute the two four-point DFTs $G(k)$ and $H(k)$ of the sequences $g(r) = \{x(0), x(2), x(4), x(6)\}$ and $h(r) = \{x(1), x(3), x(5), x(7)\}$ and combine them appropriately.

We can determine the number of computations required to find $X(k)$ using this procedure. Each of the two DFTs requires $(N/2)^2$ complex multiplications and $(N/2)^2$

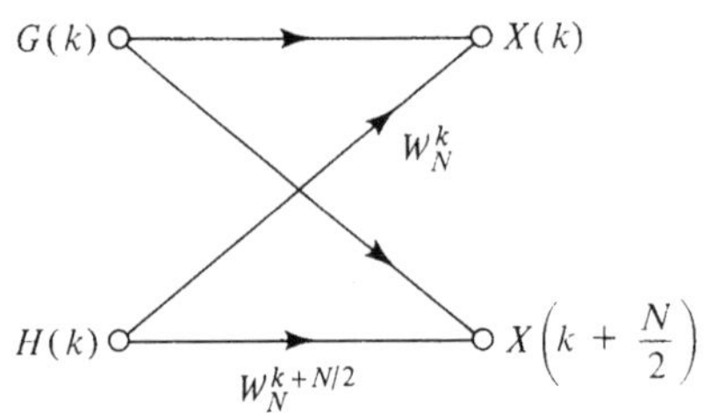

Figure 9.5.1 Signal-flow graph for Equation (9.5.9)

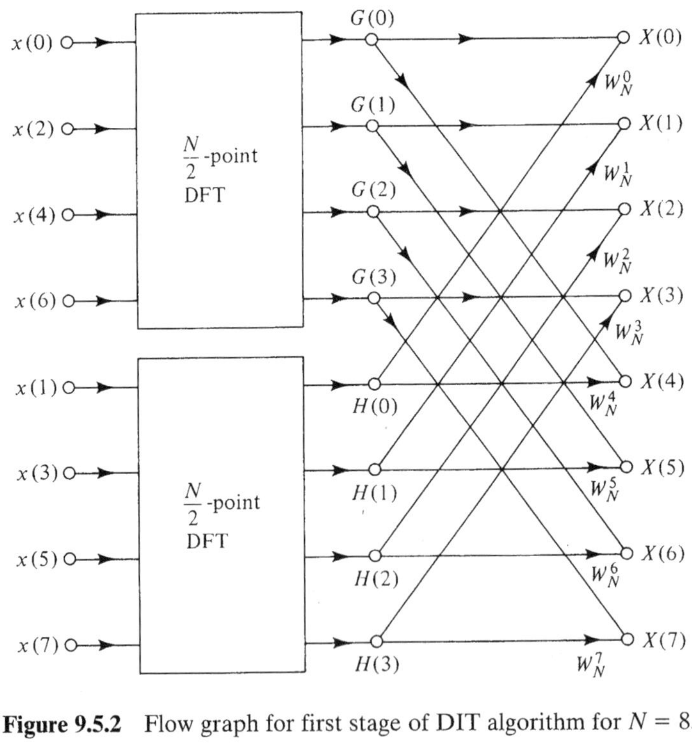

Figure 9.5.2 Flow graph for first stage of DIT algorithm for $N = 8$.

complex additions. Combining the two DFTs requires N complex multiplications and N complex additions. Thus, the computation of $X(k)$ using Equation (9.5.8) requires $N + N^2/2$ complex additions and multiplications, compared to N^2 complex multiplications and additions for direct computation.

Since $N/2$ is also even, we can consider using the same procedure for determining the $N/2$-point DFTs $G(k)$ and $H(k)$ by first determining the $N/4$-point DFTs of appropriately chosen sequences and combining them. For $N = 8$, this involves dividing the sequence $g(r)$ into the two sequences $\{x(0), x(4)\}$ and $\{x(2), x(6)\}$ and the sequence $h(r)$ into $\{x(1), x(5)\}$ and $\{x(3), x(7)\}$. The resulting computations for finding $G(k)$ and $H(k)$ are illustrated in Figure 9.5.3.

Clearly, this procedure can be continued by further subdividing the subsequences until we get a set of two-point sequences. Figure 9.5.4 illustrates the computation of the DFT of a two-point sequence $y(n) = \{y(0), y(1)\}$. The complete flow graph for the computation of an eight-point DFT is shown in Figure 9.5.5.

A careful examination of the flow graph in the latter figure leads to several observations. First, the number of stages in the graph is 3, which equals $\log_2 8$. In general,

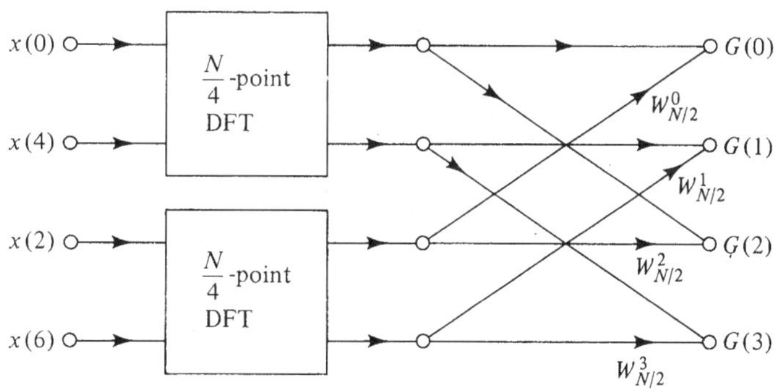

(a)

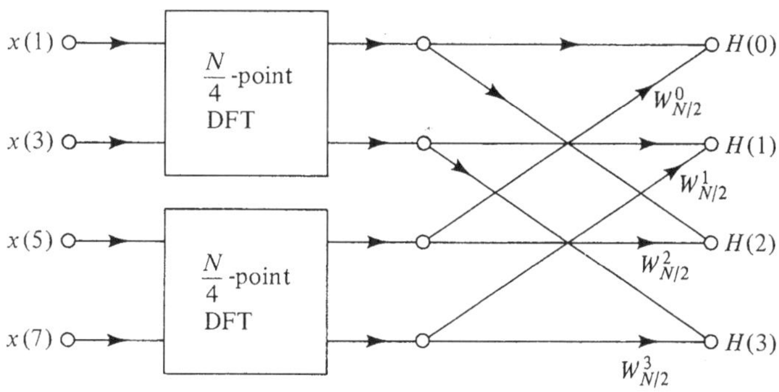

(b)

Figure 9.5.3 Flow graph for computation of four-point DFT.

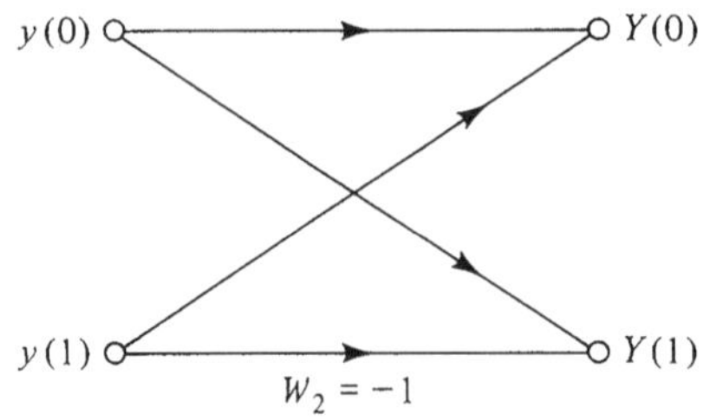

Figure 9.5.4 Flow graph for computation of two-point DFT.

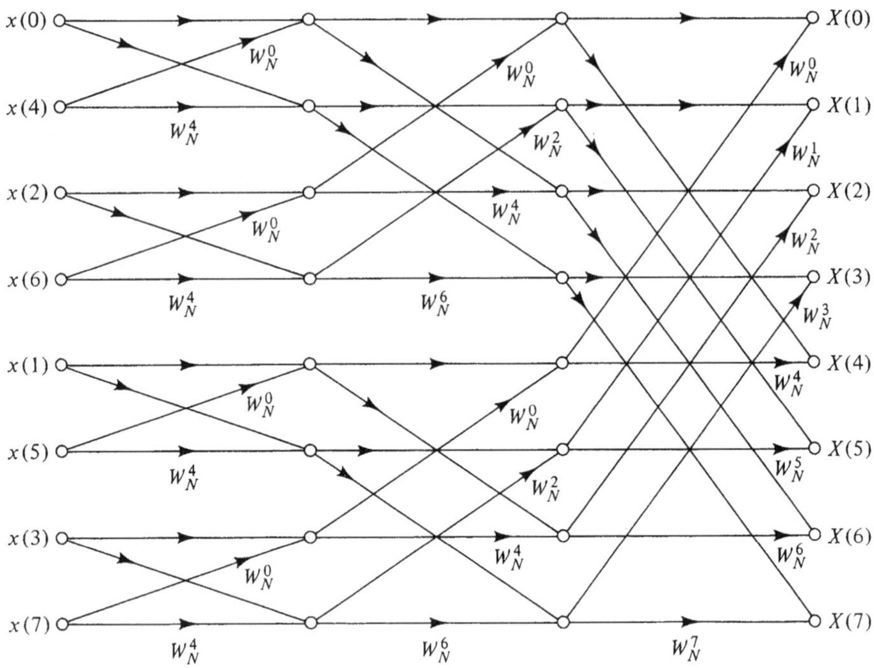

Figure 9.5.5 Complete flow graph for computation of the DFT for $N = 8$.

the number of stages is equal to $\log_2 N$. Second, each stage of the computation requires eight complex multiplications and additions. For general N, we require N complex multiplications and additions, leading to a total of $N \log_2 N$ operations.

The ordering of the input to the flow graph, which is 0, 4, 2, 6, 1, 5, 3, 7, is determined by *bit reversing* the natural numbers 0, 1, 2, 3, 4, 5, 6, 7. To obtain the bit-reversed order, we reverse the bits in the binary representation of the numbers in their natural order and obtain their decimal equivalents, as illustrated in Table 9-1.

Finally, the procedure permits *in-place computation*; that is, the results of the computations at any stage can be stored in the same locations as those of the input to that stage. To illustrate this, let us consider the computation of $X(0)$ and $X(4)$. Both of these computations require the quantities $G(0)$ and $H(0)$ as inputs. Since $G(0)$ and $H(0)$ are not required for determining any other value of $X(k)$, once $X(0)$ and $X(4)$ have been determined, they can be stored in the same locations as $G(0)$ and $H(0)$. Sim-

TABLE 9-1
Bit-reversed order for $N = 8$

Decimal Number	Binary Representation	Bit-Reversed Representation	Decimal Equivalent
0	000	000	0
1	001	100	4
2	010	010	2
3	011	110	6
4	100	001	1
5	101	101	5
6	110	011	3
7	111	111	7

ilarly, the locations of $G(1)$ and $H(1)$ can be used to store $X(1)$ and $X(5)$, and so on. Thus, only $2N$ storage locations are needed to complete the computations.

9.5.2 The Decimation-in-Frequency Algorithm

The decimation-in-frequency (DIF) algorithm is obtained essentially by dividing the output sequence $X(k)$, rather than input sequence $x(n)$, into smaller subsequences. To derive this algorithm, we group the first $N/2$ points and the last $N/2$ points of the sequence $x(n)$ together and write

$$\begin{aligned} X(k) &= \sum_{n=0}^{(N/2)-1} x(n)W_N^{nk} + \sum_{n=N/2}^{N-1} x(n)W_N^{nk} \\ &= \sum_{n=0}^{N/2-1} x(n)W_N^{nk} + W_N^{N/2k}\sum_{n=0}^{N/2-1} x\left(n+\frac{N}{2}\right)W_N^{nk} \end{aligned} \tag{9.5.10}$$

A comparison with Equation (9.5.6) shows that even though the two sums in the right side of Equation (9.5.10) are taken over $N/2$ values of n, they do not represent DFTs. We can combine the two terms in Equation (9.5.10) by noting that $W_N^{Nk/2} = (-1)^k$, to get

$$X(k) = \sum_{n=0}^{(N/2)-1}\left[x(n) + (-1)^k x\left(n+\frac{N}{2}\right)\right]W_N^{nk} \tag{9.5.11}$$

Let

$$g(n) = x(n) + x\left(n+\frac{N}{2}\right)$$

and

$$h(n) = \left[x(n) - x\left(n+\frac{N}{2}\right)\right]W_N^{n} \tag{9.5.12}$$

where $0 \le n \le (N/2) - 1$.

For k even, we can set $k = 2r$ and write Equation (9.5.11) as

$$X(2r) = \sum_{n=0}^{(N/2)-1} g(n) W_N^{2rn} = \sum_{n=0}^{(N/2)-1} g(n) W_{N/2}^{rn} \tag{9.5.13}$$

Similarly, setting $k = 2r + 1$ gives the expression for odd values of k:

$$X(2r+1) = \sum_{n=0}^{(N/2)-1} h(n) W_N^{2rn} = \sum_{n=0}^{(N/2)-1} h(n) W_{N/2}^{rn} \tag{9.5.14}$$

Equations (9.5.13) and (9.5.14) represent the $(N/2)$-point DFTs of the sequences $G(k)$ and $H(k)$, respectively. Thus, the computation of $X(k)$ involves first forming the sequences $g(n)$ and $h(n)$ and then computing their DFTs to obtain the even and odd values of $X(k)$. This is illustrated in Figure 9.5.6 for the case where $N = 8$. From the figure, we see that $G(0) = X(0)$, $G(1) = X(2)$, $G(2) = X(4)$, $G(3) = X(6)$, $H(0) = X(1)$, $H(1) = X(3)$, $H(2) = X(5)$, and $H(3) = X(7)$.

We can proceed to determine the two $(N/2)$-point DFTs $G(k)$ and $H(k)$ by computing the even and odd values separately using a similar procedure. That is, we form the sequences

$$g_1(n) = g(n) + g\left(n + \frac{N}{4}\right)$$

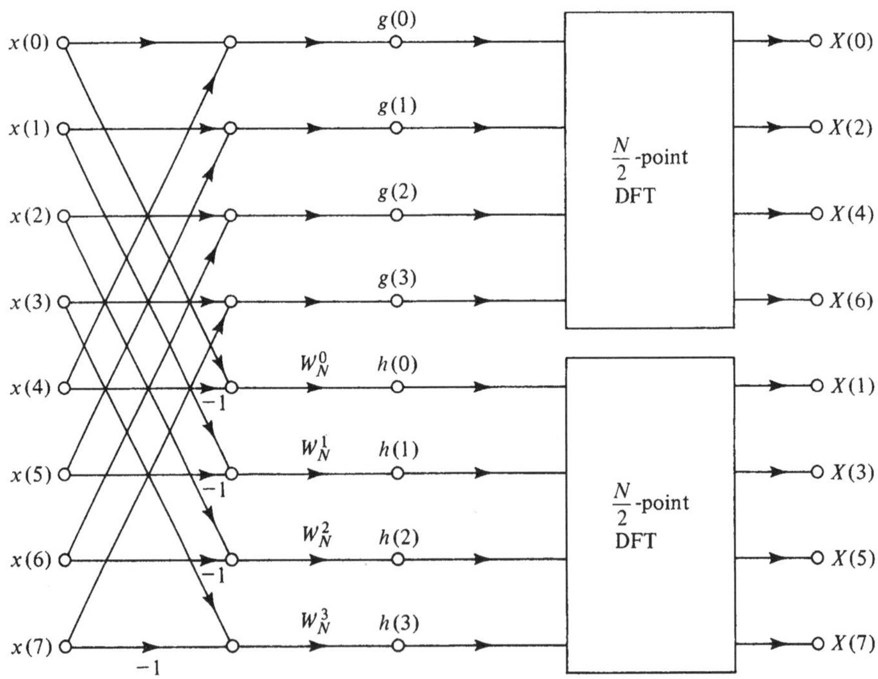

Figure 9.5.6 First stage of DIF graph.

$$g_2(n) = \left[g(n) - g\left(n + \frac{N}{4}\right)\right] W_{N/2}^n \tag{9.5.15}$$

and

$$h_1(n) = h(n) + h\left(n + \frac{N}{4}\right)$$

$$h_2(n) = \left[h(n) - h\left(n + \frac{N}{4}\right)\right] W_{N/2}^n \tag{9.5.16}$$

Then the $(N/4)$-point DFTs, $G_1(k)$, $G_2(k)$ and $H_1(k)$, $H_2(k)$, correspond to the even and odd values of $G(k)$ and $H(k)$, respectively, as shown in Figure 9.5.7 for $N = 8$.

We can continue this procedure until we have a set of two-point sequences, which, as can be seen from Figure 9.5.4, are implemented by adding and subtracting the input values. Figure 9.5.8 shows the complete flow graph for the computation of an eight-

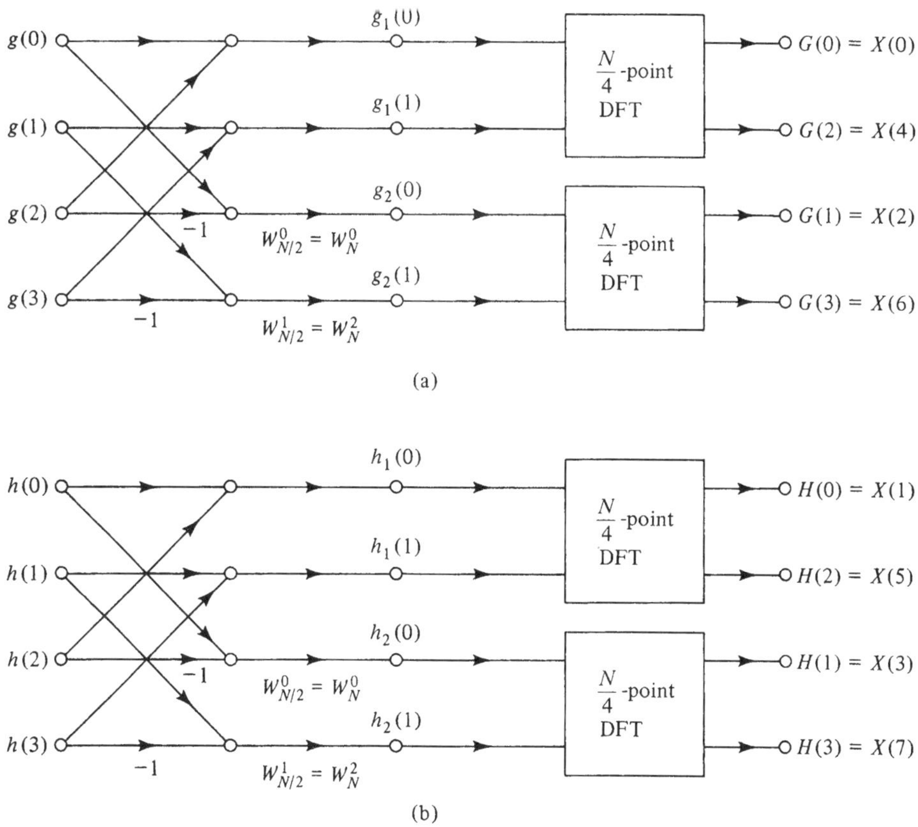

Figure 9.5.7 Flow graph for the $N/4$-point DFTs of $G(k)$ and $H(k)$, $N = 8$.

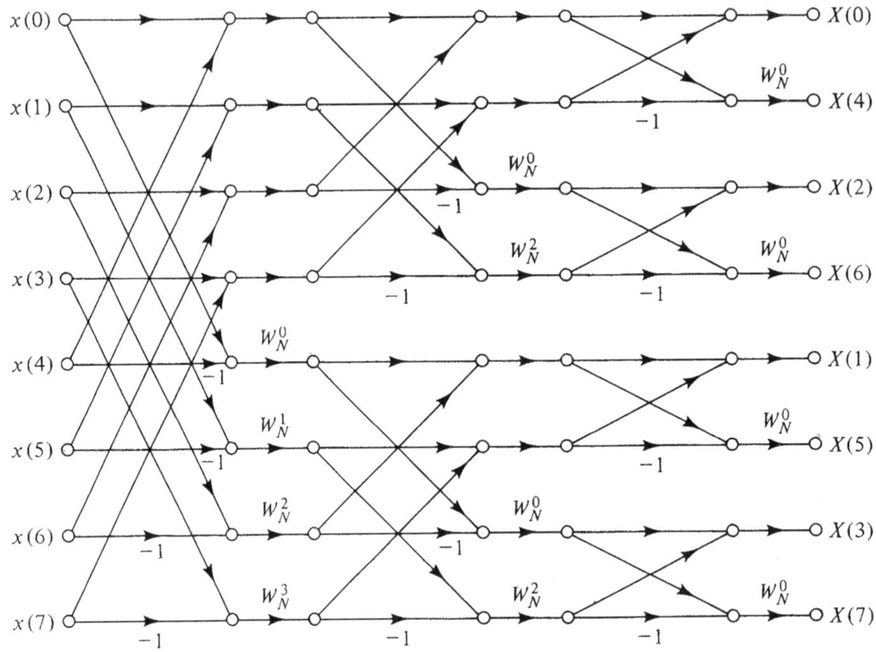

Figure 9.5.8 Complete flow graph for DIF algorithm, $N = 8$.

point DFT. As can be seen from the figure, the input in this case is in its natural order, and the output is in bit-reversed order. However, the other observations made in reference to the DIT algorithm, such as the number of computations, and the in-place nature of the computations apply to the DIF algorithm also. We can modify the signal-flow graph of Figure 9.5.7 to get a DIF algorithm in which the input is in scrambled (bit-reversed) order and the output is in the natural order. We can also obtain a DIT algorithm for which the input is in the natural order. In both cases, we can modify the graphs to give an algorithm in which both the input and the ouput are in their natural order. However, in this case, the in-place property of the algorithm will no longer hold. Finally, as noted earlier (see Equation (9.3.3)), the FFT algorithm can be used to find the IDFT in an efficient manner.

9.6 SPECTRAL ESTIMATION OF ANALOG SIGNALS USING THE DFT

The DFT represents a transformation of a finite-length discrete-time signal $x(n)$ into the frequency domain and is similar to the other frequency-domain transforms that we have discussed, with some significant differences. As we have seen, however, for analog signals $x_a(t)$, we can consider the DFT as an approximation to the continuous-time Fourier transform $X_a(\omega)$. It is therefore of interest to study how closely the DFT approximates the true spectrum of the signal.

As noted earlier, the first step in obtaining the DFT of signal $x_a(t)$ is to convert it into a discrete-time signal $x_s(t)$ by sampling at a uniform rate. The process of sampling, as we saw, can be modeled by multiplying the signal $x_a(t)$ by the impulse train

$$p_T(t) = \sum_{n=-\infty}^{\infty} \delta(t - nT)$$

so that we have

$$x_s(t) = x_a(t)p_T(t) \tag{9.6.1}$$

The corresponding Fourier transform is obtained from Equation (7.5.12):

$$X_s(\omega) = \frac{1}{T}\sum_{m=-\infty}^{\infty} X_a(\omega + m\omega_S) \tag{9.6.2}$$

These steps and the others involved in obtaining the DFT of the signal $x_a(t)$ are illustrated in Figure 9.6.1. The figures on the left correspond to the time functions, and the figures on the right correspond to their Fourier transforms. Figure 9.6.1(a) shows a typical analog signal that is multiplied by the impulse sequence shown in Figure 9.6.1(b) to yield the sampled signal of Figure 9.6.1(c). The Fourier transform of the impulse sequence $p_T(t)$, also shown in Figure 9.6.1(b), is a sequence of impulses of strength $1/T$ in the frequency domain, with spacing ω_s. The spectrum of the sampled signal is the convolution of the transform-domain functions in Figures 9.6.1(a) and 9.6.1(b) and is thus an aliased version of the spectrum of the analog signal, as shown in Figure 9.6.1(c). Thus, the spectrum of the sampled signal is a periodic repetition, with period ω_s, of the spectrum of the analog signal $x_a(t)$.

If the signal $x_a(t)$ is band-limited, we can avoid aliasing errors by sampling at a rate that is above the Nyquist rate. If the signal is not band limited, aliasing effects cannot be avoided. They can, however, be minimized by choosing the sampling rate to be the maximum feasible. In many applications, it is usual to low-pass filter the analog signal prior to sampling in order to minimize aliasing errors.

The second step in the procedure is to truncate the sampled signal by multiplying by the window function $w(t)$. The length of the data window T_0 is related to the number of data points N and sampling interval T by

$$T_0 = NT \tag{9.6.3}$$

Figure 9.6.1(d) shows the rectangular window function

$$w_R(t) = \begin{cases} 1, & -\dfrac{T}{2} \le t < T_0 - \dfrac{T}{2} \\ 0, & \text{otherwise} \end{cases} \tag{9.6.4}$$

The shift of $T/2$ from the origin is introduced in order to avoid having data samples at points of discontinuity of the window function. The Fourier transform is

$$W_R(\omega) = T_0\frac{\sin \omega T_0/2}{\omega T_0/2}\exp\left[-\frac{j\omega(T_0 - T)}{2}\right] \tag{9.6.5}$$

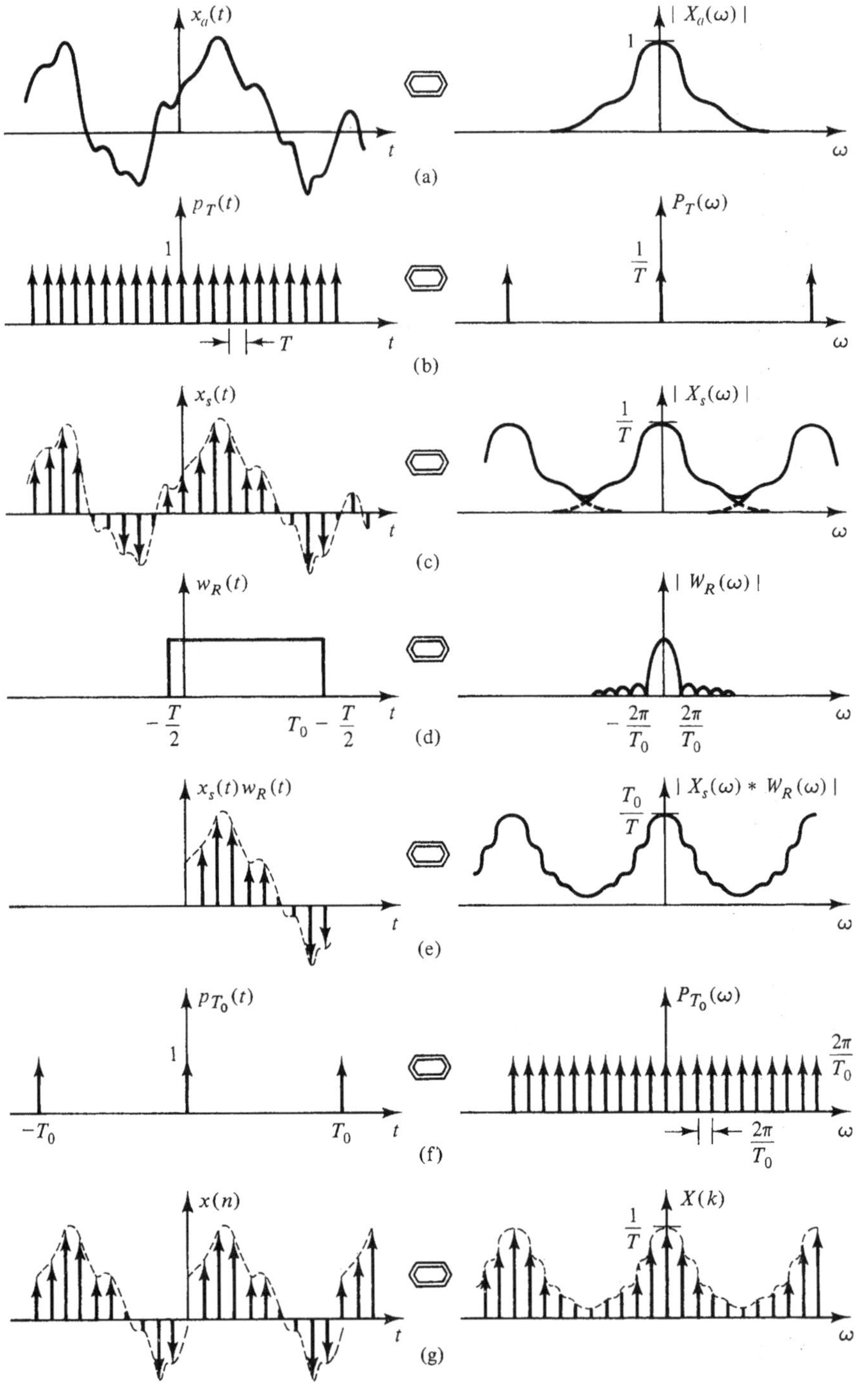

Figure 9.6.1 Discrete Fourier transform of an analog signal. (Adapted with permission from E. Oran Brigham, *The Fourier Transform*, Prentice-Hall, 1987.)

and Figure 9.6.1(e) shows the truncated sampled function. The corresponding Fourier transform is obtained as the convolution of the two transforms $X_s(\omega)$ and $X_R(\omega)$. The effect of this convolution is to introduce a ripple into the spectrum.

The final step is to sample the spectrum at equally spaced points in the frequency domain. Since the number of frequency points in the range $0 \leq \omega < \omega_s$ is equal to the number of data points N, the spacing between frequency samples is ω_s/N, or equivalently, $2\pi/T_0$, as can be seen by using Equation (9.6.3). Just as we assumed that the sampled signal in the time domain could be modeled as the modulation (multiplication) of the analog signal $x_a(t)$ by the impulse train $p_T(t)$, the sampling operation in the frequency domain can be modeled as the multiplication of the transform $X_s(\omega) * W_R(\omega)$ by the impulse train in the frequency domain:

$$p_{T_0}(\omega) = \frac{2\pi}{T_0} \sum_{m=-\infty}^{\infty} \delta\left(\omega - m\frac{2\pi}{T_0}\right) \tag{9.6.6}$$

Note that the inverse transform of $p_{T_0}(\omega)$ is also an impulse train, as shown in Figure 9.6.1(f):

$$p_{T_0}(t) = \sum_{m=-\infty}^{\infty} \delta(t - mT_0) \tag{9.6.7}$$

Since multiplication in the frequency domain corresponds to convolution in the time domain, the sampling operation in the frequency domain yields the convolution of the signal $x_s(t)w_R(t)$ and the impulse train $p_{T_0}(t)$. The result, as shown in Figure 9.6.1(g), is the periodic extension of the signal $x_s(t)w_R(t)$, with period T_0. This result also follows from the symmetry between time-domain and frequency-domain operations, from which it can be expected that sampling in the frequency domain causes aliasing in the time domain. This is a restatement of our earlier conclusion that in operations involving the DFT, the original sequence is replaced by its periodic extension.

As can be seen from Figure 9.6.1, for a general analog signal $x_a(t)$, the spectrum as obtained by the DFT is somewhat different from the true spectrum $X_s(\omega)$. There are two principal sources of error introduced in the process of determining the DFT of $x_a(t)$. The first, of course, is the aliasing error introduced by sampling. As discussed earlier, we can reduce aliasing errors either by increasing the sampling rate or by prefiltering the signal to eliminate its high-frequency components.

The second source of error is the windowing operation, which is equivalent to convolving the spectrum of the sampled signal with the Fourier transform of the window signal. Unfortunately, this introduces ripples into the spectrum, due to the convolution operation causing the signal component in $x_s(t)$ at any frequency to be spread over, or to *leak* into, other frequencies. For there to be no leakage, the Fourier transform of the window function must be a delta function. This corresponds to a window function that is constant for all time, which implies no windowing. Thus, windowing necessarily causes leakage. We can seek to minimize leakage by choosing a window function whose Fourier transform is as close to a delta function as possible. The rectangular window function is not generally used, since it does not approximate a delta function very well.

For the rectangular window defined as

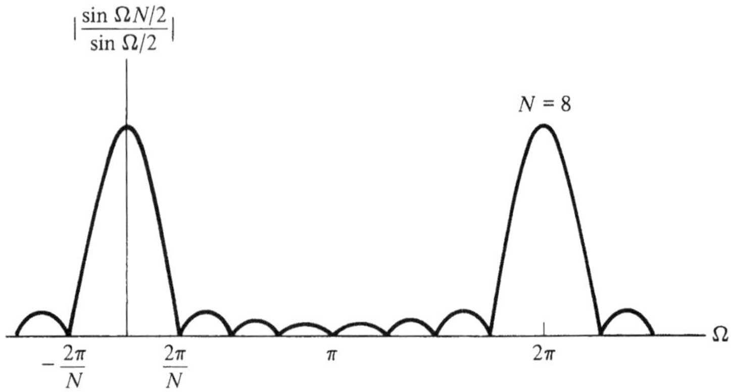

Figure 9.6.2 Magnitude spectrum of a rectangular window.

$$w_R(n) = \begin{cases} 1, & 0 \le n \le N-1 \\ 0, & \text{otherwise} \end{cases} \tag{9.6.8}$$

the frequency response is

$$W_R(\Omega) = \exp\left[-j\Omega\,\frac{(N-1)}{2}\right]\frac{\sin \Omega N/2}{\sin \Omega/2} \tag{9.6.9}$$

Figure 9.6.2 shows $|W_R(\Omega)|$, which consists of a main lobe extending from $\Omega = -2\pi/N$ to $2\pi/N$ and a set of side lobes. The area under the side lobes, which is a significant percentage of the area under the main lobe, contributes to the smearing of the DFT spectrum.

It can be shown that window functions which taper smoothly to zero at both ends give much better results. For these windows, the area under the side lobes is a much smaller percentage of the area under the main lobes. An example is the Hamming window, defined as

$$w_H(n) = 0.54 - 0.46 \cos \frac{2\pi n}{N-1}, \qquad 0 \le n \le N-1 \tag{9.6.10}$$

Figure 9.6.3(a) compares the rectangular and Hamming windows. Figures 9.6.3(b) and 9.6.3(c) show the magnitude spectra of the rectangular and Hamming windows, respectively. These are conventionally plotted in units of decibels (dB). As can be seen from the figure, whereas the rectangular window has a narrower main lobe than the Hamming window, the attenuation of the side lobes is much higher with the Hamming window.

A factor that has to be considered is the *frequency resolution*, which refers to the spacing between samples in the frequency domain. If the frequency resolution is too low, the frequency samples may be too far apart, and we may miss critical information in the spectrum. For example, we may assume that a single peak exists at a frequency where there actually are two closely spaced peaks in the spectrum. The frequency resolution is

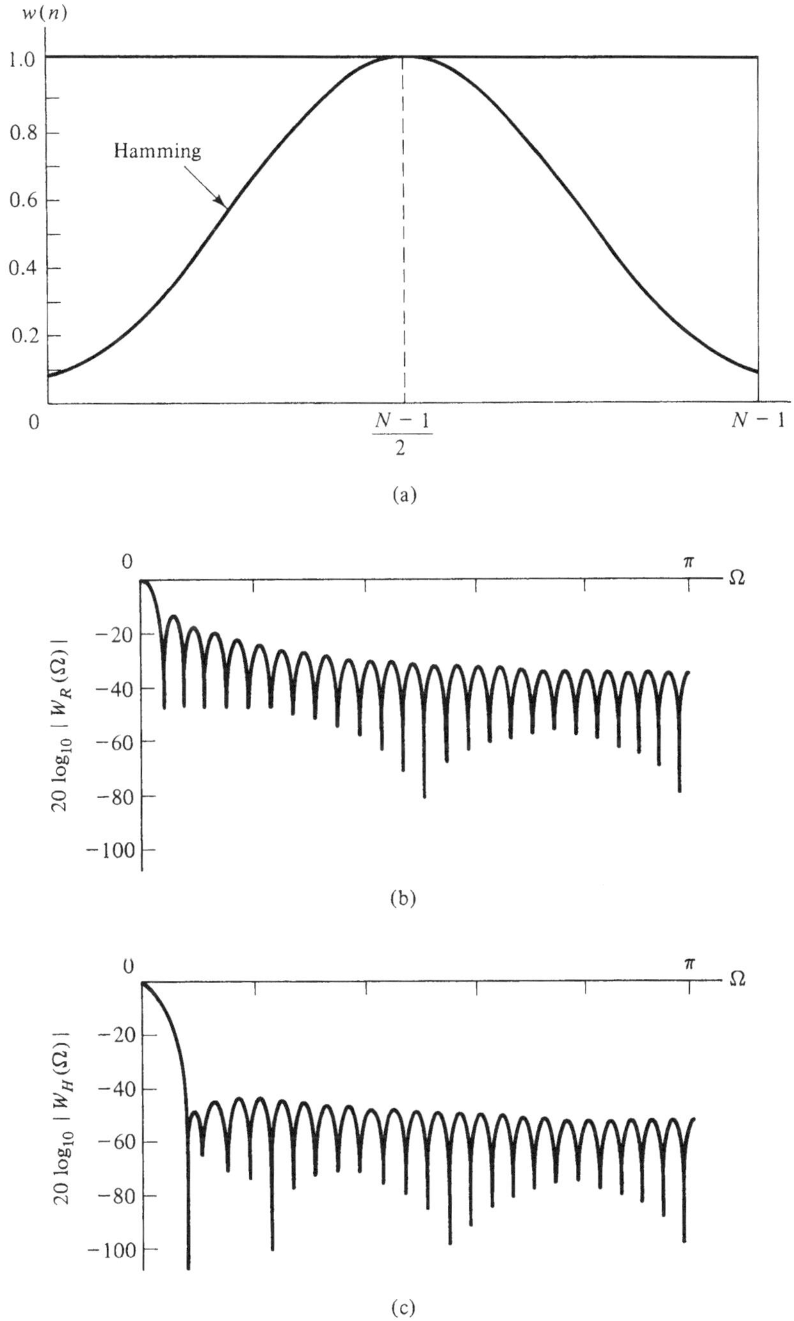

Figure 9.6.3 Comparison of rectangular and Hamming windows. (a) Time functions. (b) Spectrum of rectangular window. (c) Spectrum of Hamming window.

$$\Delta\omega = \frac{\omega_s}{N} = \frac{2\pi}{NT} = \frac{2\pi}{T_0} \tag{9.6.11}$$

where T_0 refers to the length of the data window. It is clear from Equation (9.6.11) that, to improve the frequency resolution, we have to use a longer data record. If the record length is fixed and we need a higher resolution in the spectrum, we can consider padding the data sequence with zeros, thereby increasing the number of samples from N to some new value $N_0 > N$. This is equivalent to using a window of longer duration $T_1 > T_0$ on the modified signal now defined as

$$x(t) = \begin{cases} x_a(t), & 0 \le t \le T_0 \\ 0, & T_0 < t \le T_1 \end{cases} \tag{9.6.12}$$

Example 9.6.1

Suppose we want to use the DFT to find the spectrum of an analog signal that has been prefiltered by passing it through a low-pass filter with a cutoff of 10 kHz. The desired frequency resolution is less than 0.1 Hz.

The sampling theorem gives the minimum sampling frequency for this signal as $f_s = 20$ kHz, so that

$$T \le 0.05 \text{ ms}$$

The duration of the data window can be determined from the desired frequency resolution Δf as

$$T_0 = \frac{1}{\Delta f} \ge 10 \text{ s}$$

from which it follows that

$$N = \frac{T_0}{T} \ge 2 \times 10^5$$

Assuming that we want to use a radix-2 FFT routine, we choose N to be 262,144 ($= 2^{18}$), which is the smallest power of 2 satisfying the constraint on N. If we choose $f_s = 20$ kHz, T_0 must be chosen to be 13.1072 s.

Example 9.6.2

In this example, we illustrate the use of the DFT in finding the Fourier spectrum of analog signals. Let us consider the signal

$$x_a(t) = \cos 400\pi t$$

Since the signal consists of a single frequency, its continuous-time Fourier transform is a pair of δ functions occurring at ± 200 Hz.

Figure 9.6.4 shows the magnitude of the DFT spectrum $X(k)$ of the signal for data lengths of 32, 64, and 128 samples obtained by using a rectangular window. The signal was sampled at a rate of 2 kHz, which is considerably higher than the Nyquist rate of 400 Hz. As can be seen from the figure, the DFT spectrum exhibits two peaks in each case. If we let k_p denote the location of the first peak, the second peak occurs at $N - k_p$ in all cases. This is to be expected, since $X(-k) = X(N - k)$. The analog frequencies corresponding to the two peaks can be determined to be $f_0 = \pm k_p T/N$.

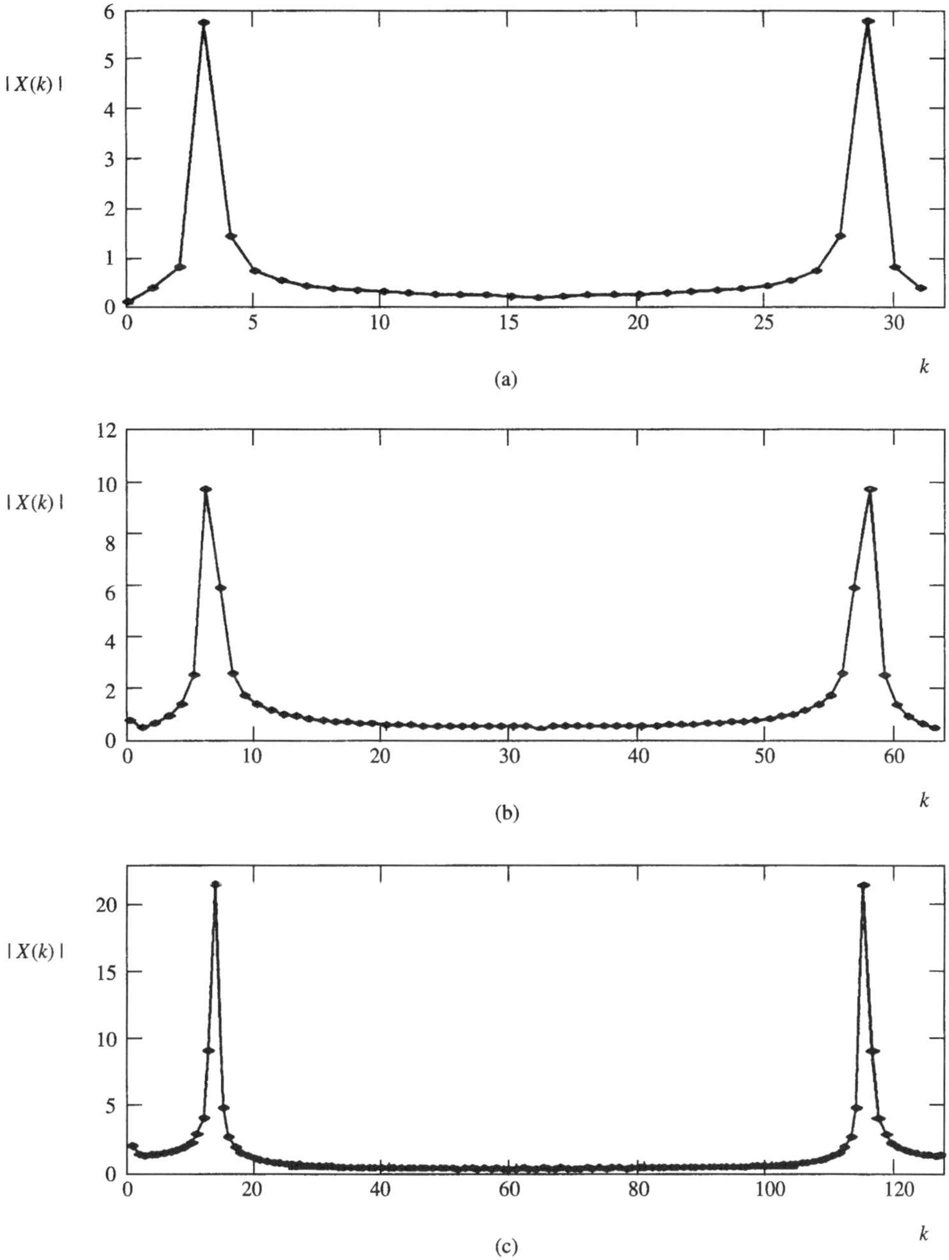

Figure 9.6.4 DFT spectrum of analog signal $x_a(t)$ using rectangular window. (a) $N = 32$. (b)$N = 64$. (c) $N = 128$.

Figure 9.6.5 shows the results of using a Hamming window on the sampled signal for data lengths of 32, 64, and 128 samples. The DFT spectrum again exhibits two peaks at the same locations as before.

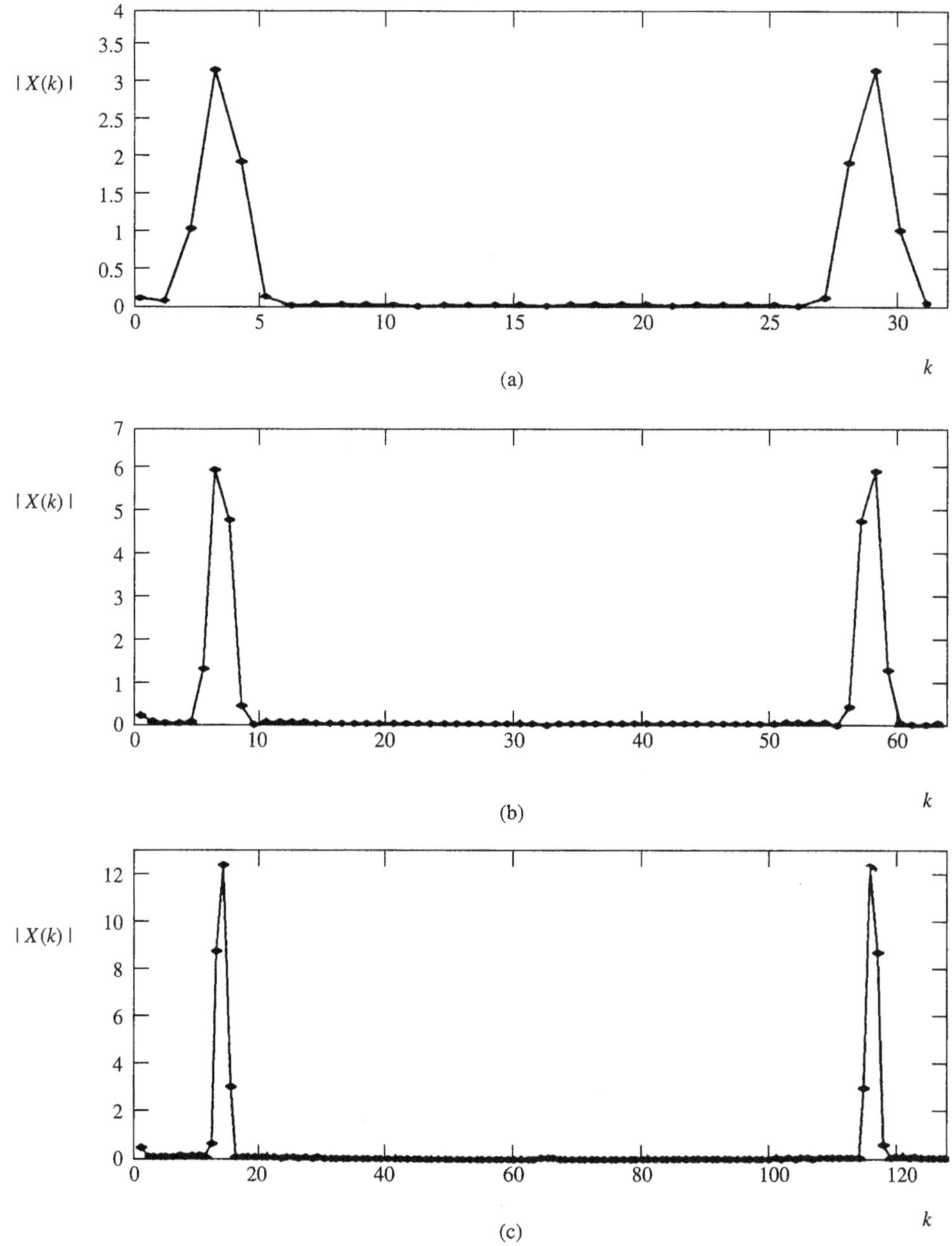

Figure 9.6.5 DFT spectrum of analog signal $x_a(t)$ using Hamming window. (a) $N = 32$. (b)$N = 64$. (c) $N = 128$.

With both the rectangular and Hamming windows, the first peak occurs at $k_p = 3, 6$, and 13 for $N = 32, 64$, and 128 samples, respectively. These correspond to analog frequencies of 187.5 Hz, 187.5 Hz, and 190.0625 Hz. Thus, as the number of data samples increases, the peak moves closer to the actual analog frequency. Note that the peaks become sharper as N (and hence the resolution in the digital frequency domain) increases.

The figures also show that the spectrum obtained using the Hamming window is somewhat smoother than that resulting from the rectangular window.

Suppose we add another frequency to our analog signal, so that the signal is now

$$x_b(t) = \cos 400\pi t + \cos 440\pi t$$

To resolve the two sinusoids in the signal, the frequency resolution Δf must be less than 20 Hz. The duration of the data window, T_0, must therefore be chosen to be greater than 1/20 s. If the sampling rate is 2 kHz, the number N of discrete-time samples needed to resolve the two frequencies must be chosen to be larger than 100.

Figure 9.6.6 shows the DFT spectrum of the signal for data lengths of 64, 128, and 256 samples using a rectangular window, while Figure 9.6.7 shows the corresponding results obtained using a Hamming window. With both windows, the 64-point DFT is unable to resolve the two peaks. For a window length of 128, there are two large values of $|X(k)|$ at values of $k = 13$ and 14. The corresponding frequencies in the analog domain are equal to 203.125 Hz and 218.75 Hz, respectively. Thus, even though the two frequencies do not appear as peaks in the DFT spectrum, it is nevertheless possible to identify them.

For $N = 256$, there are two clearly identifiable peaks in the spectrum at $k = 26$ and 28. These again correspond to 203.125 Hz and 218.75 Hz in the analog frequency domain.

9.7 SUMMARY

- The discrete Fourier transform (DFT) of the finite-length sequence $x(n)$ of length N is defined as

$$X(k) = \sum_{n=0}^{N-1} x(n) W_N^{nk}$$

where

$$W_N = \exp\left[-j\frac{2\pi}{N}\right]$$

- The inverse discrete Fourier transform (IDFT) is defined by

$$x(n) = \frac{1}{N}\sum_{n=0}^{N-1} X(k) W_N^{-nk}$$

- The DFT of an N-point sequence is related to its Z-transform as

$$X(k) = X(z)\big|_{z = W_N^k}$$

- The sequence $X(k)$, $k = 0, 1, 2, \ldots, \mathrm{N} - 1$, is periodic with period N. The sequence $x(n)$ obtained by determining the IDFT of $X(k)$ is also periodic with period N.

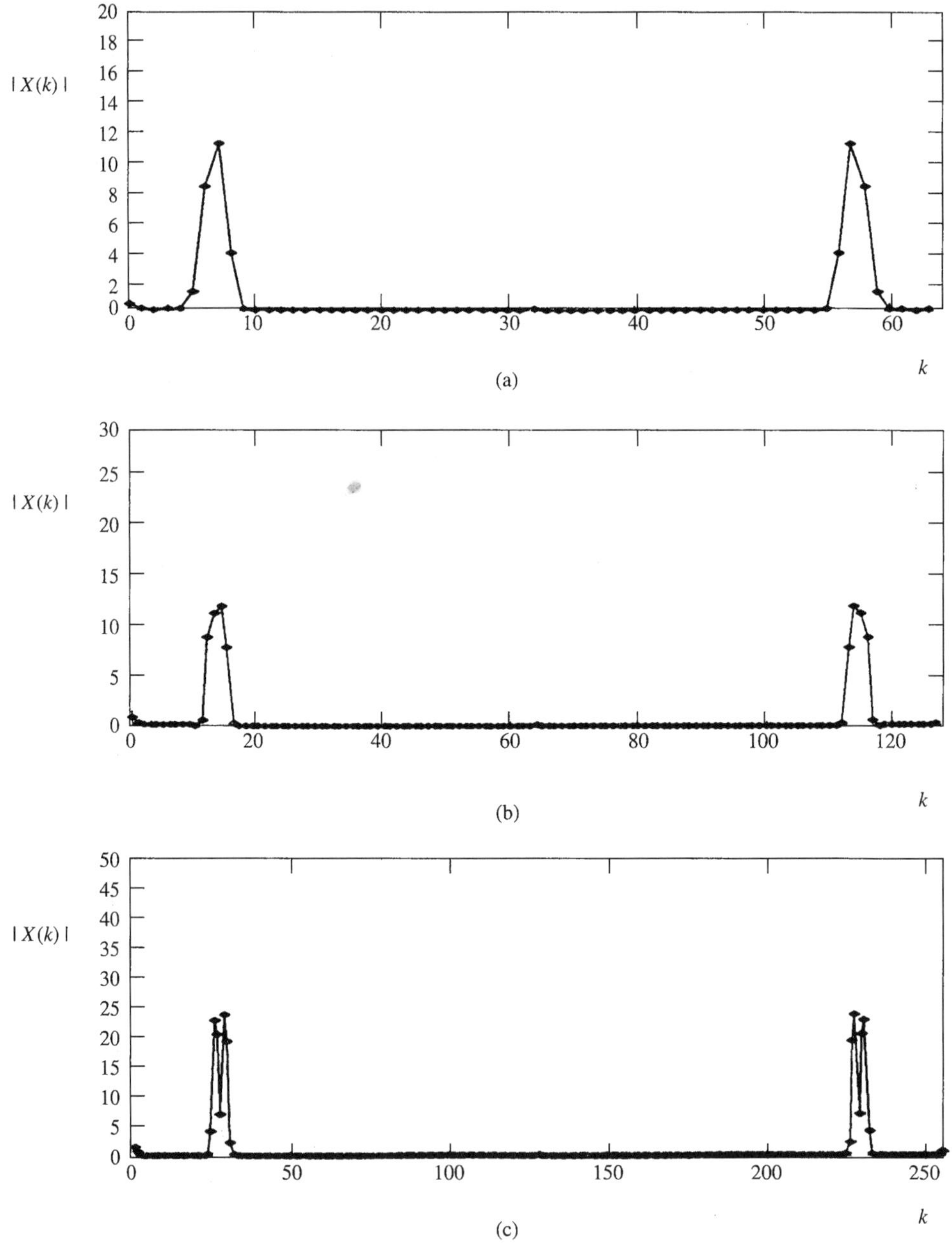

Figure 9.6.6 DFT spectrum of analog signal $x_b(t)$ using rectangular window. (a) $N = 64$. (b) $N = 128$. (c) $N = 256$.

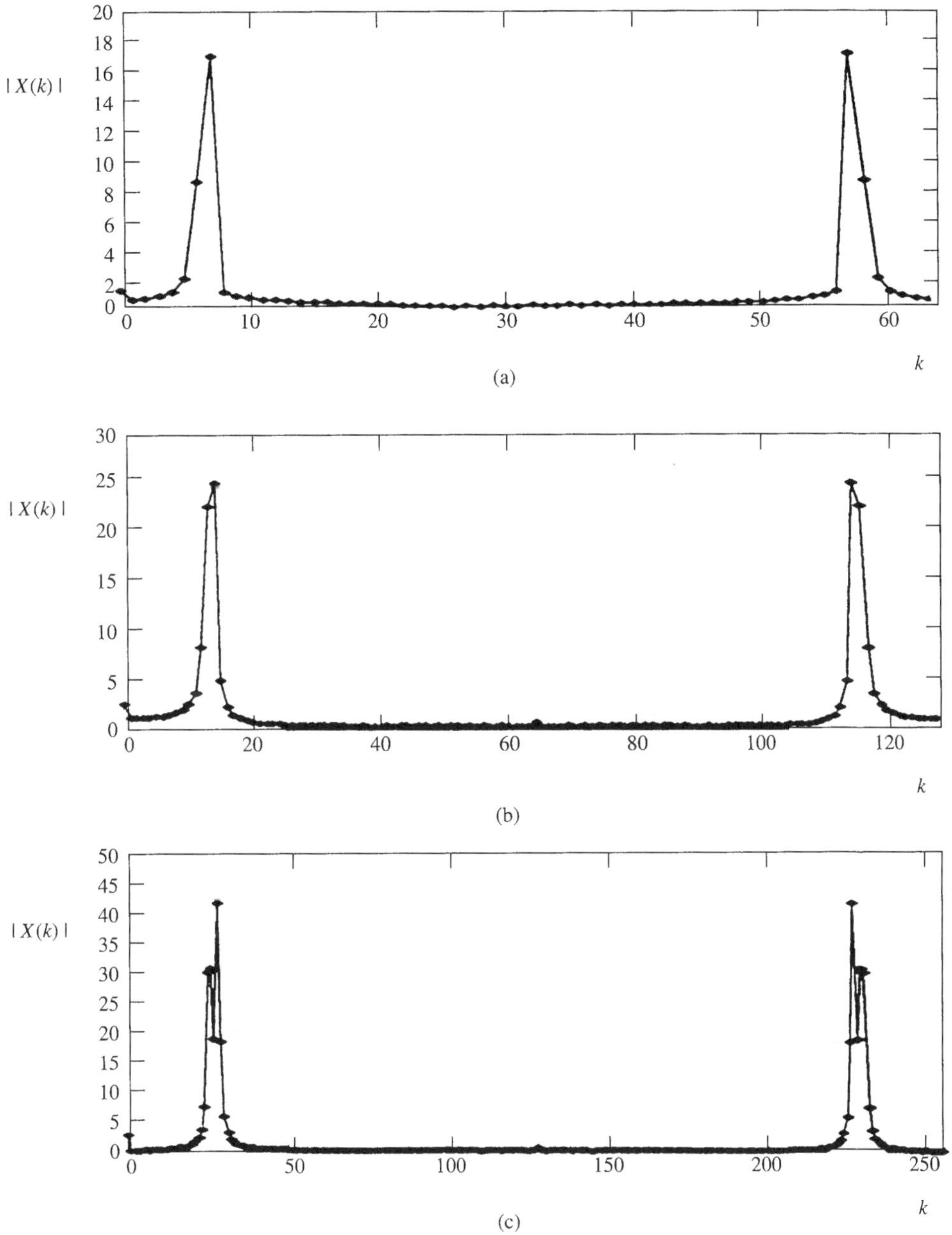

Figure 9.6.7 DFT spectrum of analog signal $x_b(t)$ using Hamming window. (a) $N = 64$. (b)$N = 128$. (c) $N = 256$.

- In all operations involving the DFT and the IDFT, the sequence $x(n)$ is effectively replaced by its periodic extension $x_p(n)$.
- $X(k)$ is equal to Na_k, where a_k is the coefficient of the discrete-time Fourier-series representation of $x_p(n)$.
- The properties of the DFT are similar to those of the other Fourier transforms, with some significant differences. In particular, the DFT performs cyclic or periodic convolution instead of the linear convolution needed for the analysis of LTI systems.
- To perform a linear convolution of an N-point sequence with an M-point sequence, the sequences must be padded with zeros so that both are of length $N + M - 1$.
- Algorithms for efficient and fast machine computation of the DFT are known as fast Fourier-transform (FFT) algorithms.
- For sequences whose length is an integer power of 2, the most commonly used FFT algorithms are the decimation-in-time (DIT) and decimation-in-frequency (DIF) algorithms.
- For in-place computation using either the DIT or the DIF algorithm, either the input or the output must be in bit-reversed order.
- The DFT provides a convenient method for the approximate determination of the spectra of analog signals. Care must be taken, however, to minimize errors caused by sampling and windowing the analog signal to obtain a finite-length discrete-time sequence.
- Aliasing errors can be reduced by choosing a higher sampling rate or by prefiltering the analog signal. Windowing errors can be reduced by choosing a window function that tapers smoothly to zero at both ends.
- The spectral resolution in the analog domain is directly proportional to the data length.

9.8 CHECKLIST OF IMPORTANT TERMS

Aliasing
Analog spectrum
Bit-reversed order
Decimation-in-frequency algorithm
Decimation-in-time algorithm
Discrete Fourier transform (DFT)
Error reduction
Fast Fourier transform (FFT)
In-place computation
Inverse discrete Fourier transform (IDFT)
Linear convolution
Periodic convolution
Periodicity of DFT and IDFT
Prefiltering
Spectral resolution
Windowing
Zero padding

9.9 PROBLEMS

9.1. Compute the DFT of the following N-point sequences:

(a) $x(n) = \begin{cases} 1, & n = n_0, \quad 0 < n_0 < N - 1 \\ 0, & \text{otherwise} \end{cases}$

(b) $x(n) = (-1)^n$

(c) $x(n) = \begin{cases} 1, & n \text{ even} \\ 0, & \text{otherwise} \end{cases}$

9.2. Show that if $x(n)$ is a real sequence, $X(N - k) = X^*(k)$.

9.3. Let $x(n)$ be an N-point sequence with DFT $X(k)$. Find the DFT of the following sequences in term of $X(k)$:

(a) $y_1(n) = \begin{cases} x\left(\dfrac{n}{2}\right), & n \text{ even} \\ 0, & n \text{ odd} \end{cases}$

(b) $y_2(n) = x(N - n - 1), \qquad 0 \le n \le N - 1$

(c) $y_3(n) = x(2n), \qquad 0 \le n \le N - 1$

(d) $y_4(n) = \begin{cases} x(n), & 0 \le n \le N - 1 \\ 0, & N \le n \le 2N - 1 \end{cases}$

9.4. Let $x(n)$ be a real eight-point-sequence, and let

$$y(n) = \begin{cases} x(n), & 0 \le n \le 7 \\ x(n - 8), & 8 \le n \le 15 \end{cases}$$

Find $Y(k)$, given that

$$X(0) = 1;\ X(1) = 1 + 2j;\ X(2) = 1 - j1;\ X(3) = 1 + j1;\ \text{and } X(4) = 2.$$

9.5. **(a)** Use the DFT to find the periodic convolution of the following sequences:

(i) $x(n) = \{1, -1, -1, 1, -1, 1\}$ and $h(n) = \{1, 2, 3, 3, 2, 1\}$

(ii) $x(n) = \{1, -2, -1, 1\}$ and $h(n) = \{1, 0, 0, 1\}$

(b) Verify your results using any mathematical software package.

9.6. Repeat Problem 9.5 for the linear convolution of the sequences in the problem.

9.7. Let $X(\Omega)$ denote the Fourier transform of the sequence $x(n) = (1/3)^n u(n)$, and let $y(n)$ denote an eight-point sequence such that its DFT, (k), corresponds to eight equally spaced samples of $X(\Omega)$. That is,

$$Y(k) = X\left(\frac{2\pi}{8} k\right) \qquad k = 0, 1, \dots, 7$$

What is $y(n)$?

9.8. Derive Parseval's relation for the DFT:

$$\sum_{n=0}^{N-1} |x(n)|^2 = \frac{1}{N} \sum_{k=0}^{N-1} |X(k)|^2$$

9.9. Suppose we want to evaluate the discrete-time Fourier transform of an N-point sequence $x(n)$ at M equally spaced points in the range $[0, 2\pi]$. Explain how we can use the DFT to do this if (a) $M > N$ and (b) $M < N$.

9.10. Let $x(n)$ be an N-point sequence. It is desired to find 128 equally spaced samples of the spectrum $X(\Omega)$ in the range $7\pi/16 \le \Omega \le 15\pi/16$, using a radix-2 FFT algorithm. Describe a procedure for doing so if (i) $N = 1000$, (ii) $N = 120$.

9.11. Suppose we want to evaluate the DFT of an N-point sequence $x(n)$ using a hardware processor that can only do M-point FFTs, where M is an integer multiple of N. Assuming that additional facilities for storage, addition, or multiplication are available, show how this can be done.

9.12. Given a six-point sequence $x(n)$, we can seek to find its DFT by subdividing it into three two-point DFTs that can then be combined to give $X(k)$. Draw a signal-flow graph to evaluate $X(k)$ using this procedure.

9.13. Draw a signal-flow graph for computing a nine-point DFT as the sum of three three-point DFTs.

9.14. Analog data that has been prefiltered to 20 kHz must be spectrum analyzed to a resolution of less than 0.25 Hz using a radix-2 algorithm. Determine the necessary data length T_0.

9.15. For the analog signal in Problem 9.14, what is the frequency resolution if the signal is sampled at 40 kHz to obtain 4096 samples?

9.16. The analog signal $x_a(t)$ of duration 24 s is sampled at the rate of $42\frac{2}{3}$ Hz and the DFT of the resulting samples taken.

(a) What is the frequency resolution in the analog domain?

(b) What is the digital frequency spacing for the DFT taken?

(c) What is the highest analog frequency that does not cause aliasing?

9.17. The following represent the DFT values $X(k)$ of an analog signal $x_a(t)$ that has been sampled to yield 16 samples:

$$X(0) = 2, X(3) = 4 - j4, X(5) = -2, X(8) = -\tfrac{1}{2}, X(11) = -2, X(13) = 4 + j4$$

All other values are zero.

(a) Find the corresponding $x(n)$.

(b) What is the digital frequency resolution?

(c) Assuming that the sampling interval is 0.25 s, find the analog frequency resolution. What is the duration T_0 of the analog signal?

(d) For the sampling rate in part (c), what is the highest analog frequency that can be present in $x_a(t)$ without causing aliasing?

(e) Find T_0 to give an analog frequency resolution that is twice that in part (c).

9.18. Given two real N-point sequences $f(n)$ and $g(n)$, we can find their DFTs simultaneously by computing a single N-point DFT of the complex sequence

$$x(n) = f(n) + jg(n)$$

We show how to do this in the following:

(a) Let $h(n)$ be any real N-point sequence. Show that

$$\text{Re}\{H(k)\} = H_e(k) = \frac{H(k) + H(N-k)}{2}$$

$$\text{Im}\{H(k)\} = H_o(k) = \frac{H(k) - H(N-k)}{2}$$

(b) Let $h(n)$ be purely imaginary. Show that

$$\text{Re}\{H(k)\} = H_o(k)$$
$$\text{Im}\{H(k)\} = H_e(k)$$

(c) Use your results in Parts (a) and (b) to show that

$$F(k) = X_{\text{Re}}(k) + jX_{\text{Io}}(k)$$
$$G(k) = X_{\text{Ie}}(k) - jX_{\text{Ro}}(k)$$

where $X_{\text{Re}}(k)$ and $X_{\text{Ro}}(k)$ represent the even and odd parts of $X_R(k)$, the real part of $X(k)$, and $X_{\text{Ie}}(k)$ and $X_{\text{Io}}(k)$ represent the even and odd parts of $X_I(k)$ the imaginary part of $X(k)$.

9.19. **(a)** The signal $x_a(t) = 4\cos(2\pi t/3)$ is sampled at discrete instants T to generate 32 points of the sequence $x(n)$. Find the DFT of the sequence if $T = 15/16$, and plot the magnitude and phase of the sequence. Use a rectangular window in finding the DFT.

(b) Determine the Fourier transform of $x_a(t)$, and compare its magnitude and phase with the results of Part (a).

(c) Repeat Parts (a) and (b) if $T = 0.1$ s.

9.20. Repeat problem 9.19 with a Hamming window. Comment on your results.

9.21. We want to determine the Fourier transform of the amplitude-modulated signal $x_a(t) = 10\cos(2000\pi t)\cos(100\pi t)$ using the DFT. Choose an appropriate duration T_0 over which the signal must be observed in order to clearly distinguish all the frequencies in $x_a(t)$. Assume a sampling interval of $T = 0.4$ ms.

(a) Use a rectangular window, and find the DFT of the sampled signal for $N = 128$, $N = 256$, and $N = 512$ samples.

(b) Determine the Fourier transform of $x_a(t)$, and compare its magnitude and phase with the results of Part (a).

9.22. Repeat Problem 9.21 with a Hamming window. Comment on your results.

Chapter 10

Design of Analog and Digital Filters

10.1 INTRODUCTION

Earlier we saw that when we apply an input to a system, it is modified or transformed at the output. Typically, we would like to design the system such that it modifies the input in a specified manner. When the system is designed to remove certain unwanted components of the input signal, it is usually referred to as a filter. When the unwanted components are described in terms of their frequency content, the filters, as discussed in Chapter 4, are said to be frequency selective. Although many applications require only simple filters that can be designed using a brute-force method, the design of more complicated filters requires the use of sophisticated techniques. In this chapter, we consider some techniques for the design of both continuous-time and discrete-time frequency-selective filters.

As noted in Chapter 4, an ideal frequency-selective filter passes certain frequencies without any change and completely stops the other frequencies. The range of frequencies that are passed without attenuation is the passband of the filter, and the range of frequencies that are not passed constitutes the stop band. Thus, for ideal continuous-time filters, the magnitude transfer function of the filter is given by $|H(\omega)| = 1$ in the passband and $|H(\omega)| = 0$ in the stop band. Frequency-selective filters are classified as low-pass, high-pass, band-pass, or band-stop filters, depending on the band of frequencies that either are passed through without attenuation or are completely stopped. Figure 10.1.1 shows the characteristics of these filters.

Similar definitions carry over to discrete-time filters, with the distinction that the frequency range of interest in this case is $0 \leq \Omega < 2\pi$, since $H(\Omega)$ is now a periodic function with period 2π. Figure 10.1.2 shows the discrete-time counterparts of the filters shown in Fig. 10.1.1.

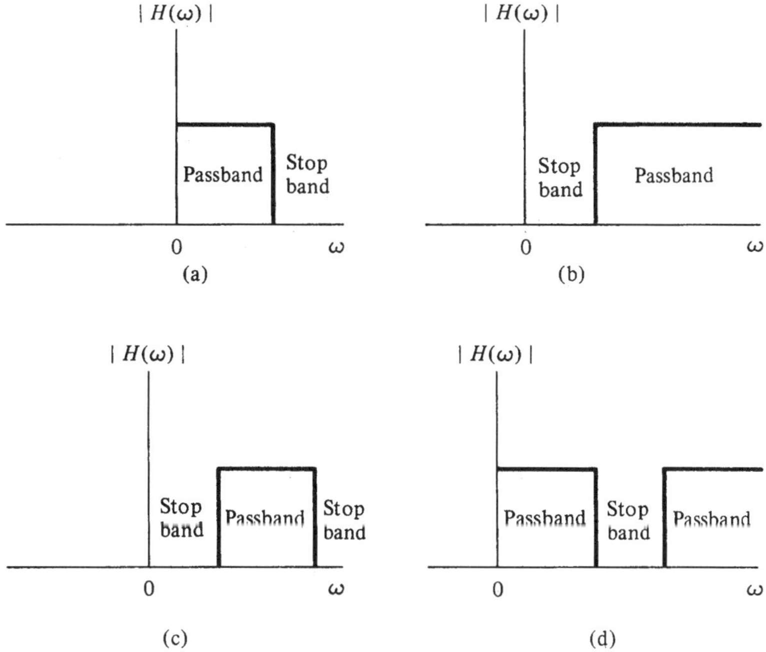

Figure 10.1.1 Ideal continuous-time frequency-selective filters.

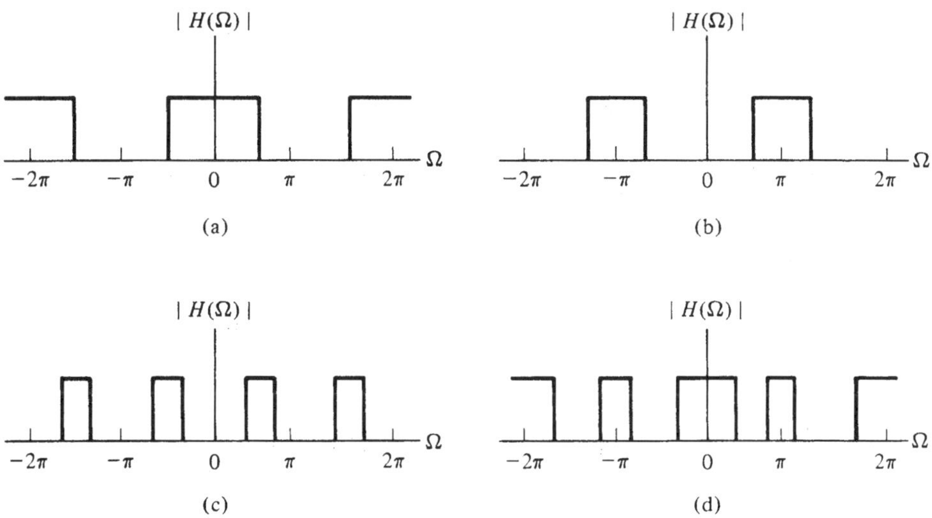

Figure 10.1.2 Ideal discrete-time frequency-selective filters.

In practice, we cannot obtain filter characteristics with abrupt transitions between passbands and stop bands, as shown in Figures 10.1.1 and 10.1.2. This can easily be seen by considering the impulse response of the ideal low-pass filter, which is noncausal and hence not physically realizable. To obtain practical filters, we therefore have to relax

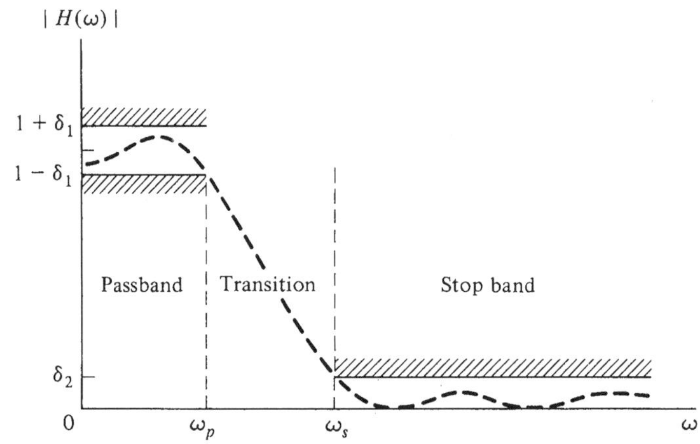

Figure 10.1.3 Specification for practical low-pass filter.

our requirements on $|H(\omega)|$ (or $|H(\Omega)|$) in the passbands and stop bands, by permitting deviations from the ideal response, as well as specifying a transition band between the passbands and stop bands. Thus, for a continuous-time low-pass filter, the specifications can be of the form

$$\begin{aligned} 1 - \delta_1 \leq |H(\omega)| \leq 1 + \delta_1, \qquad & |\omega| \leq \omega_p \\ |H(\omega)| \leq \delta_2, \qquad & |\omega_s| \leq \omega \end{aligned} \tag{10.1.1}$$

where ω_p and ω_s are the passband and stop band cutoff frequencies, respectively. The range of frequencies between ω_p and ω_s is the transition band, depicted in Figure 10.1.3.

Often, the filter is specified to have a peak gain of unity. The corresponding specifications for the filter frequency response can be easily determined from Figure 10.1.3 by amplitude scaling by a factor of $1/(1 + \delta_1)$. Specifications for discrete-time filters are given in a similar manner as

$$\begin{aligned} \big||H(\Omega)| - 1\big| \leq \delta_1, \qquad & |\Omega| \leq \Omega_p \\ |H(\Omega)| \leq \delta_2, \qquad & \Omega_s \leq |\Omega| \leq \pi \end{aligned} \tag{10.1.2}$$

Given a set of specifications, filter design consists of obtaining an analytical approximation to the desired filter characteristics in the form of a filter transfer function $H(s)$ for continuous-time systems and $H(z)$ for discrete-time systems. Once the transfer function has been determined, we can obtain a realization of the filter, as discussed in earlier chapters. We consider the design of two standard analog filters in Section 10.3 and examine digital filter design in Section 10.4.

In our discussion of filter design, we confine ourselves to low-pass filters, since, as is shown in the next section, a low-pass filter can be converted to one of the other types of filters by using appropriate frequency transformations. Thus, given a specification for any other type of filter, we can convert this specification into an equivalent one for a low-pass filter, obtain the corresponding transfer function $H(s)$ (or $H(z)$), and convert the transfer function back into the desired range.

10.2 FREQUENCY TRANSFORMATIONS

As indicated before, frequency transformations are useful for converting a frequency-selective filter from one type to another. For example, suppose we are given a continuous-time low-pass filter transfer function $H(s)$ with a normalized cutoff frequency of unity. We now verify that the transformation which converts it into a low-pass filter with a cutoff frequency ω_c is

$$s^{\#} = s\omega_c \tag{10.2.1}$$

where $s^{\#}$ represents the transformed frequency variable. Since

$$\omega^{\#} = \omega\omega_c \tag{10.2.2}$$

it is clear that the frequency range $0 \leq |\omega| \leq 1$ is mapped into the range $0 \leq |\omega^{\#}| \leq \omega_c$ Thus, $H(s^{\#})$ represents a low-pass filter with a cutoff frequency of ω_c.

More generally, the transformation

$$s^{\#} = s\frac{\omega_c^{\#}}{\omega_c} \tag{10.2.3}$$

transforms a low-pass filter with a cutoff frequency ω_c to a low-pass filter with a cutoff frequency of $\omega_c^{\#}$. Similarly, the transformation

$$s^{\#} = \frac{\omega_c}{s} \tag{10.2.4}$$

transforms a normalized low-pass filter to a high-pass filter with a cutoff frequency of ω_c. This can be easily verified by noting that in this case we have

$$\omega^{\#} = -\frac{\omega_c}{\omega} \tag{10.2.5}$$

so that the point $|\omega| = 1$ corresponds to the point $|\omega^{\#}| = \omega_c$. Also, the range $|\omega| \leq 1$ is mapped onto the ranges defined by $\omega_c \leq |\omega^{\#}| \leq \infty$.

Next we consider the transformation of the normalized low-pass filter to a band-pass filter with lower and upper cutoff frequencies given by ω_{c_1} and ω_{c_2}, respectively. The required transformation is given in terms of the bandwidth of the filter,

$$\mathrm{BW} = \omega_{c_2} - \omega_{c_1} \tag{10.2.6}$$

and the frequency,

$$\omega_0 \sqrt{\omega_{c_2}\omega_{c_1}} \tag{10.2.7}$$

as

$$s = \frac{\omega_0}{\mathrm{BW}}\left(\frac{s^{\#}}{\omega_0} + \frac{\omega_0}{s^{\#}}\right) \tag{10.2.8}$$

This transformation maps $\omega = 0$ into the points $\omega^{\#} = \pm\omega_0$ and the segment $|\omega| \leq 1$ to the segments $\omega_{c_2} \geq |\omega^{\#}| \geq \omega_{c_1}$.

Finally, the band-stop filter is obtained through the transformation

$$s = \frac{\text{BW}}{\omega_0\left(\dfrac{s^{\#}}{\omega_0} + \dfrac{\omega_0}{s^{\#}}\right)} \tag{10.2.9}$$

where BW and ω_0 are defined similarly to the way they were in the case of the band-pass filter. The transformations are summarized in Table 10-1.

TABLE 10-1
Frequency transformations from low-pass analog filter response.

Filter Type	Transformation	
Low Pass	$\dfrac{s^{\#}}{\omega_c}$	
High Pass	$\dfrac{\omega_c}{s^{\#}}$	
Band Pass	$\dfrac{\omega_0}{\text{BW}}\left(\dfrac{s^{\#}}{\omega_0} + \dfrac{\omega_0}{s^{\#}}\right)$,	$\omega_0 = \sqrt{\omega_{c_2}\omega_{c_1}}$
Band Stop	$\dfrac{\text{BW}}{\omega_0\left(\dfrac{s^{\#}}{\omega_0} + \dfrac{\omega_0}{s^{\#}}\right)}$,	$\text{BW} = \omega_{c_2} - \omega_{c_1}$

We now consider similar transformations for the discrete-time problem. Thus, suppose that we are given a discrete-time low-pass filter with a cutoff frequency Ω_c, and we want to obtain a low-pass filter with cutoff frequency $\Omega_c^{\#}$. The required transformation is

$$z^{\#} = \frac{z - \alpha}{1 - \alpha z} \tag{10.2.10}$$

More conventionally, this is written as

$$(z^{\#})^{-1} = \frac{z^{-1} - \alpha}{1 - \alpha z^{-1}} \tag{10.2.11}$$

By setting $z = \exp[j\Omega]$ in the right side of Equation (10.2.11), it follows that

$$z^{\#} = \exp\left[j\tan^{-1}\frac{(1 - \alpha^2)\sin\Omega}{2\alpha + (1 + \alpha^2)\cos\Omega}\right] \tag{10.2.12}$$

Thus, the transformation maps the unit circle in the z plane into the unit circle in the $z^{\#}$ plane. The required value of α can be determined by setting $z^{\#} = \exp[j\Omega_c^{\#}]$ and $\Omega = \Omega_c$ in Equation (10.2.11), yielding

$$\alpha = \frac{\sin[(\Omega_c - \Omega_c^{\#})/2]}{\sin[(\Omega_c + \Omega_c^{\#})/2]} \tag{10.2.13}$$

TABLE 10-2
Frequency transformations from low-pass digital filter response.

Filter Type	Transformation	Associated Formulas
Low Pass	$(z^{\#})^{-1} = \dfrac{z^{-1} - \alpha}{1 - \alpha z^{-1}}$	$\alpha = \dfrac{\sin \dfrac{\Omega_c - \Omega_c^{\#}}{2}}{\sin \dfrac{\Omega_c + \Omega_c^{\#}}{2}}$ $\Omega_c^{\#}$ = desired cutoff frequency
High Pass	$-\dfrac{z^{-1} + \alpha}{1 + \alpha z^{-1}}$	$\alpha = -\dfrac{\cos \dfrac{\Omega_c^{\#} - \Omega_c}{2}}{\cos \dfrac{\Omega_c^{\#} + \Omega_c}{2}}$
Band Pass	$\dfrac{z^{-2} - \dfrac{2\alpha k}{k+1} z^{-1} + \dfrac{k-1}{k+1}}{\dfrac{k-1}{k+1} z^{-2} - \dfrac{2\alpha k}{k+1} + 1}$	$\alpha = \dfrac{\cos \dfrac{\Omega_{c_2}^{\#} + \Omega_{c_1}^{\#}}{2}}{\cos \dfrac{\Omega_{c_2}^{\#} \quad \Omega_{c_1}^{\#}}{2}}$ $k = \cot \dfrac{\Omega_{c_2}^{\#} - \Omega_{c_1}^{\#}}{2} \tan \dfrac{\Omega_c}{2}$ $\Omega_{c_1}^{\#}, \Omega_{c_2}^{\#}$ = desired lower and upper cutoff frequencies, respectively
Band Stop	$\dfrac{z^{-2} - \dfrac{2\alpha}{1+k} z^{-1} + \dfrac{1-k}{1+k}}{\dfrac{1-k}{1+k} z^{-2} - \dfrac{2\alpha}{1+k} z^{-1} + 1}$	$\alpha = \dfrac{\cos \dfrac{\Omega_{c_2}^{\#} + \Omega_{c_1}^{\#}}{2}}{\cos \dfrac{\Omega_{c_2}^{\#} - \Omega_{c_1}^{\#}}{2}}$ $k = \tan \dfrac{\Omega_{c_2}^{\#} - \Omega_{c_1}^{\#}}{2} \tan \dfrac{\Omega_c}{2}$

Transformations for converting a low-pass filter into a high-pass, band-pass, or band-stop filter can be similarly defined and are summarized in Table 10-2.

10.3 DESIGN OF ANALOG FILTERS

The design of practical filters starts with a prescribed set of specifications, such as those given in Equation (10.1.1) or depicted in Figure 10.1.2. Whereas procedures are available for the design of several different analog filters, we consider the design of two standard filters, namely, the Butterworth and Chebyshev filters. The Butterworth filter provides an approximation to a low-pass characteristic that approaches zero smoothly. The Chebyshev filter provides an approximation that oscillates in the passband, but monotonically decreases in the transition and stop bands.

10.3.1 The Butterworth Filter

The Butterworth filter is characterized by the magnitude function

$$|H(\omega)|^2 = \frac{1}{1 + (\omega)^{2N}} \tag{10.3.1}$$

where N denotes the order of the filter. It is clear from this equation that the magnitude is a monotonically decreasing function of ω, with its maximum value of unity occurring at $\omega = 0$. For $\omega = 1$, the magnitude is equal to $1/\sqrt{2}$, for all values of N. Thus, the normalized Butterworth filter has a 3-dB cutoff frequency of unity.

Figure 10.3.1 shows a plot of the magnitude characteristic of this filter as a function of ω for various values of N. The parameter N determines how closely the Butterworth characteristic approximates the ideal filter. Clearly, the approximation improves as N is increased.

The Butterworth approximation is called a maximally flat approximation, since, for a given N, the maximal number of derivatives of the magnitude function is zero at the origin. In fact, the first $2N - 1$ derivatives of $|H(\omega)|$ are zero at $\omega = 0$, as we can see by expanding $|H(\omega)|$ in a power series about $\omega = 0$:

$$|H(\omega)|^2 = 1 - \tfrac{1}{2}\omega^{2N} + \tfrac{3}{8}\omega^{4N} - \cdots \tag{10.3.2}$$

To obtain the filter transfer function $H(s)$, we use

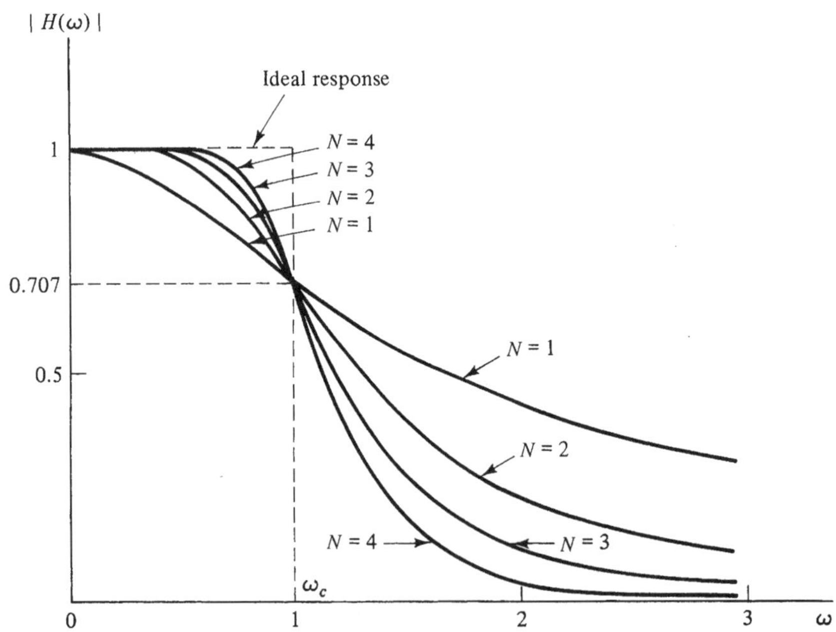

Figure 10.3.1 Magnitude plot of normalized Butterworth filter.

$$H(s)H(-s)\big|_{s=j\omega} = |H(\omega)|^2 \tag{10.3.3}$$

$$= \frac{1}{1+\left[\dfrac{(j\omega)^2}{j^2}\right]^N}$$

so that

$$H(s)H(-s) = \frac{1}{1+\left(\dfrac{s}{j}\right)^{2N}} \tag{10.3.4}$$

From Equation (10.3.4), it is clear that the poles of $H(s)$ are given by the roots of the equation

$$\left(\frac{s}{j}\right)^{2N} = -1 \tag{10.3.5}$$

$$= \exp[j(2k-1)\pi], \qquad k = 0, 1, 2, \dots, 2N-1$$

It follows that the roots are given by

$$s_k = \exp[j(2k+N-1)\pi/2N] \qquad k = 0, 1, 2, \dots, 2N-1 \tag{10.3.6}$$

By substituting $s_k = \sigma_k + j\omega_k$, we can write the real and imaginary parts as

$$\sigma_k = \cos\left(\frac{2k+N-1}{2N}\pi\right)$$

$$= \sin\left(\frac{2k-1}{N}\frac{\pi}{2}\right)$$

$$\omega_k = \sin\left(\frac{2k+N-1}{2N}\pi\right) \tag{10.3.7}$$

$$= \cos\left(\frac{2k-1}{N}\frac{\pi}{2}\right)$$

As can be seen from Equation (10.3.6), Equation (10.3.5) has $2N$ roots spaced uniformly around the unit circle at intervals of $\pi/2N$ radians. Since $2k-1$ cannot be even, it is clear that there are no roots on the $j\omega$ axis, so that there are exactly N roots each in the left and right half planes. Now, the poles and zeros of $H(s)$ are the mirror images of the poles and zeros of $H(-s)$. Thus, in order to get a stable transfer function, we simply associate the roots in the left half plane with $H(s)$.

As an example, for $N = 3$, from Equation (10.3.6), the roots are located at

$$s_0 = \exp\left[j\frac{\pi}{3}\right], \qquad s_1 = \exp\left[j\frac{2\pi}{3}\right], \qquad s_2 = \exp[j\pi],$$

$$s_3 = \exp\left[j\frac{4\pi}{3}\right], \qquad s_4 = \exp\left[j\frac{5\pi}{3}\right], \qquad s_5 = 1$$

as shown in Figure 10.3.2.

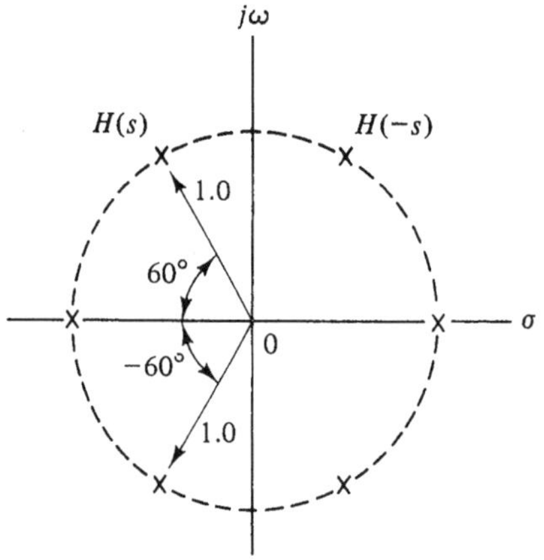

Figure 10.3.2 Roots of the Butterworth polynomial for $N = 3$.

To get a stable transfer function, we choose as the poles of $H(s)$ the left-half plane roots, so that

$$H(s) = \frac{1}{[s - \exp[j2\pi/3]][s - \exp[j\pi]][s - \exp[j4\pi/3]]} \tag{10.3.8}$$

The denominator can be expanded to yield

$$H(s) = \frac{1}{(s^2 + s + 1)(s + 1)} \tag{10.3.9}$$

Table 10-3 lists the denominator of the Butterworth transfer function in factored form for values of N ranging from $N = 1$ to $N = 8$. When these factors are multiplied, the result is a polynomial of the form

$$S(s) = a_N s^N + a_{N-1} s^{N-1} + \cdots + a_1 s + 1 \tag{10.3.10}$$

These coefficients are listed in Table 10-4 for $N = 1$ to $N = 8$.

To obtain a filter with 3-dB cutoff at ω_c, we replace s in $H(s)$ by s/ω_c. The corresponding magnitude characteristic is

TABLE 10-3
Butterworth polynomials (factored form)

n	
1	$s + 1$
2	$s^2 + \sqrt{2}s + 1$
3	$(s^2 + s + 1)(s + 1)$
4	$(s^2 + 0.7653s + 1)(s^2 + 1.84776s + 1)$
5	$(s + 1)(s^2 + 0.6180s + 1)(s^2 + 1.6180s + 1)$
6	$(s^2 + 0.5176s + 1)(s^2 + \sqrt{2}s + 1)(s^2 + 1.9318s + 1)$
7	$(s + 1)(s^2 + 0.4450s + 1)(s^2 + 1.2456s + 1)(s^2 + 1.8022s + 1)$
8	$(s^2 + 0.3986s + 1)(s^2 + 1.1110s + 1)(s^2 + 1.6630s + 1)(s^2 + 1.9622s + 1)$

TABLE 10-4
Butterworth polynomials

n	a_1	a_2	a_3	a_4	a_5	a_6	a_7	a_8
1	1							
2	$\sqrt{2}$	1						
3	2	2	1					
4	2.613	3.414	2.613	1				
5	3.236	5.236	5.236	3.236	1			
6	3.864	7.464	9.141	7.464	3.864	1		
7	4.494	10.103	14.606	14.606	10.103	4.494	1	
8	5.126	13.128	21.828	25.691	21.848	13.138	5.126	1

$$|H(\omega)|^2 = \frac{1}{1 + (\omega/\omega_c)^{2N}} \tag{10.3.11}$$

Let us now consider the design of a low-pass Butterworth filter that satisfies the following specifications:

$$\begin{aligned} |H(\omega)| &\geq 1 - \delta_1, & |\omega| &\leq \omega_p \\ &\leq \delta_2, & |\omega| &> \omega_s \end{aligned} \tag{10.3.12}$$

Since the Butterworth filter is defined by the parameters N and ω_c, we need two equations to determine these quantities. From the monotonic nature of the magnitude response, it is clear that the specifications are satisfied if we choose

$$|H(\omega_p)| = 1 - \delta_1 \tag{10.3.13}$$

and

$$|H(\omega_s)| = \delta_2 \tag{10.3.14}$$

Substituting these relations into Equation (10.3.11) yields

$$\left(\frac{\omega_p}{\omega_c}\right)^{2N} = \left(\frac{1}{1 - \delta_1}\right)^2 - 1$$

and

$$\left(\frac{\omega_s}{\omega_c}\right)^{2N} = \frac{1}{\delta_2^2} - 1$$

Eliminating ω_c from these two equations and solving for N results in

$$N = \frac{1}{2}\left[\frac{\log \dfrac{\delta_1(2 - \delta_1)\delta_2^2}{(1 - \delta_1)^2(1 - \delta_2^2)}}{\log \dfrac{\omega_p}{\omega_s}}\right] \tag{10.3.15}$$

Since N must be an integer, we round up the value of N obtained from Equation (10.3.15) to the nearest integer. This value of N can now be used in either Equation (10.3.13) or Equation (10.3.14) to determine ω_c. If ω_c is determined from Equation (10.3.13), the passband specifications are met exactly, whereas the stopband specifications are exceeded. But if we use Equation (10.3.14) to determine ω_c, the reverse is true. The steps in finding $H(s)$ are summarized as follows:

1. Determine N from Equation (10.3.15), using the values of δ_1, δ_2, ω_p, and ω_s, and round–up to the nearest integer.
2. Determine ω_c, using either Equation (10.3.13) or Equation (10.3.14).
3. For the value of N calculated in Step 1, determine the denominator polynomial of the normalized Butterworth filter, using either Table 10-3 or Table 10-4 (for values of $N \leqslant 8$) or using Equation (10.3.8), and form $H(s)$.
4. Find the unnormalized transfer function by replacing s in $H(s)$ found in Step 3 by s/ω_c. The filter so obtained will have a dc gain of unity. If some other dc gain is desired, $H(s)$ must be multiplied by the desired gain.

Example 10.3.1

We will design Butterworth filter to have an attenuation of no more than 1 dB for $|\omega| \le 2000$ rad/s and at least 15 dB for $|\omega| \ge 5000$ rad/s. From the specifications

$$20 \log_{10}(1 - \delta_1) = -1 \qquad \text{and} \qquad 20 \log_{10}\delta_2 = -15$$

so that $\delta_1 = 0.1087$ and $\delta_2 = 0.1778$. Substituting these values into Equation (10.3.15) yields a value of 2.6045 for N. Thus we choose N to be 3 and obtain the normalized filter from Table 10-3 as

$$H(s) = \frac{1}{s^3 + 2s^2 + 2s + 1}$$

Use of Equation (10.3.14) yields $\omega_c = 2826.8$ rads/s.

The unnormalized filter is therefore equal to

$$\begin{aligned} H(s) &= \frac{1}{(s/2826.8)^3 + 2(s/2826.8)^2 + 2(s/2826.8) + 1} \\ &= \frac{(2826.8)^3}{s^3 + 2(2826.8)s^2 + 2(2826.8)^2 s + (2826.8)^3} \end{aligned}$$

Figure 10.3.3 shows a plot of the magnitude of the filter as a function of ω. As can be seen from the plot, the filter meets the spot-band specifications, and the passband specifications are exceeded.

10.3.2 The Chebyshev Filter

The Butterworth filter provides a good approximation to the ideal low-pass characteristic for values of ω near zero, but has a low falloff rate in the transition band. We now consider the Chebyshev filter, which has ripples in the passband, but has a sharper cutoff in the transition band. Thus, for filters of the same order, the Chebyshev filter has

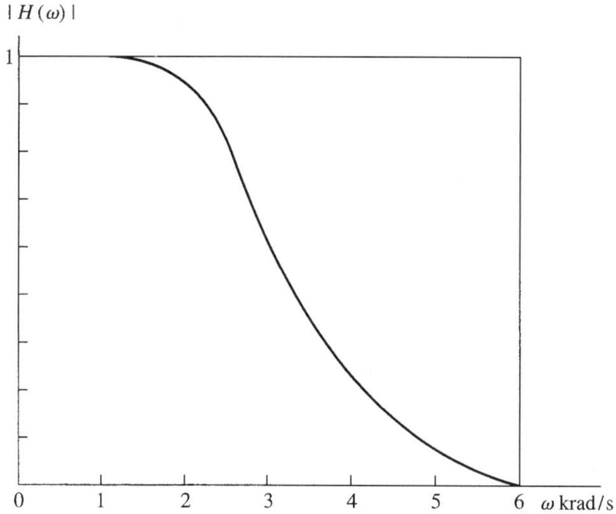

Figure 10.3.3 Magnitude function of the Butterworth filter of Example 10.3.1.

a smaller transition band than the Butterworth filter. Since the derivation of the Chebyshev approximation is quite complicated, we do not give the details here, but only present the steps needed to determine $H(s)$ from the specifications.

The Chebyshev filter is based on Chebyshev cosine polynomials, defined as

$$\begin{aligned} C_N(\omega) &= \cos(N\cos^{-1}\omega), \qquad |\omega| \le 1 \\ &= \cosh(N\cosh^{-1}\omega), \qquad |\omega| > 1 \end{aligned} \tag{10.3.16}$$

Chebyshev polynomials are also defined by the recursion formula

$$C_N(\omega) = 2\omega C_{N-1}(\omega) - C_{N-2}(\omega) \tag{10.3.17}$$

with $C_0(\omega) = 1$ and $C_1(\omega) = \omega$.

The Chebyshev low-pass characteristic of order N is defined in terms of $C_N(\omega)$ as

$$|H(\omega)|^2 = \frac{1}{1 + \varepsilon^2 C_N^2(\omega)} \tag{10.3.18}$$

To determine the behavior of this characteristic, we note that for any N, the zeros of $C_N(\omega)$ are located in the interval $|\omega| \le 1$. Further, for $|\omega| \le 1$, $|C_N(\omega)| \le 1$, and for $|\omega| > 1$, $|C_N(\omega)|$ increases rapidly as $|\omega|$ becomes large. It follows that in the interval $|\omega| \le 1$, $|H(\omega)|^2$ oscillates about unity such that the maximum value is 1 and the minimum is $1/(1 + \varepsilon^2)$. As $|\omega|$ increases, $|H(\omega)|^2$ approaches zero rapidly, thus providing an approximation to the ideal low-pass characteristic.

The magnitude characteristic corresponding to the Chebyshev filter is shown in Figure 10.3.4. As can be seen from the figure, $|H(\omega)|$ ripples between 1 and $1/\sqrt{1 + \varepsilon^2}$. Since $C_N^2(1) = 1$ for all N, it follows that for $\omega = 1$,

$$|H(1)| = \frac{1}{\sqrt{1 + \varepsilon^2}} \tag{10.3.19}$$

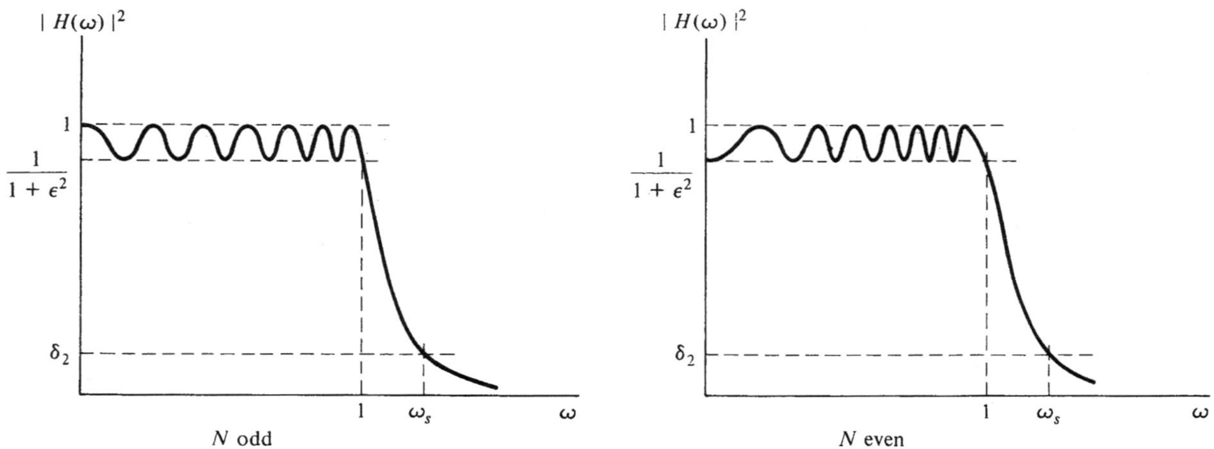

Figure 10.3.4 Magnitude characteristic of the Chebyshev approximation.

For large values of ω—that is, values in the stop band—we can approximate $|H(\omega)|$ as

$$|H(\omega)| = \frac{1}{\varepsilon C_N(\omega)} \tag{10.3.20}$$

The dB attenuation (or loss) from the value at $\omega = 0$ can thus be written as

$$\begin{aligned} \text{loss} &= -20 \log_{10}|H(\omega)| \\ &= 20 \log \varepsilon + 20 \log C_N(\omega) \end{aligned} \tag{10.3.21}$$

For large ω, $C_N(\omega)$ can be approximated by $2^{N-1}\omega^N$, so that we have

$$\text{loss} = 20 \log \varepsilon + 6(N-1) + 20N \log \omega \tag{10.3.22}$$

Equations (10.3.19) and (10.3.22) can be used to determine the two parameters N and ε required for the Chebyshev filter. The parameter ε is determined by using the passband specifications in Equation (10.3.19). This value is then used in Equation (10.3.22), along with the stop-band specifications, to determine N. In order to find $H(s)$, we introduce the parameter

$$\beta = \frac{1}{N} \sinh^{-1} \frac{1}{\varepsilon} \tag{10.3.23}$$

The poles of $H(s)$, $s_k = \sigma_k + j\omega_k$, $k = 0, 1, \ldots, N-1$, are given by

$$\begin{aligned} \sigma_k &= \sin\left(\frac{2k-1}{N}\right)\frac{\pi}{2} \sinh \beta \\ \omega_k &= \cos\left(\frac{2k-1}{N}\right)\frac{\pi}{2} \cosh \beta \end{aligned} \tag{10.3.24}$$

It follows that the poles are located on an ellipse in the s plane given by

$$\frac{\sigma_k^2}{\sinh^2 \beta} + \frac{\omega_k^2}{\cosh^2 \beta} = 1 \tag{10.3.25}$$

The major semiaxis of the ellipse is on the $j\omega$ axis, the minor semiaxis is on the σ axis, and the foci are at $\omega = \pm 1$, as shown in Figure 10.3.5. The 3-dB cutoff frequency occurs at the point where the ellipse intersects the $j\omega$ axis—that is, at $\omega = \cosh \beta$.

It is clear from Equation (10.3.24) that the Chebyshev poles are related to the Butterworth poles of the same order. The relation between these poles is shown in Figure 10.3.6 for $N = 3$ and can be used to determine the locations of the Chebyshev poles geometrically. The corresponding $H(s)$ is obtained from the left-half plane poles.

Example 10.3.2

We consider the design of a Chebyshev filter to have an attenuation of no more than 1 dB for $|\omega| \leq 1000$ rads/s and at least 10 dB for $|\omega| \geq 5000$ rads/s.

We will first normalize ω_p to 1, so that $\omega_s = 5$. From the passband specifications, we have, from Equation (10.3.19)

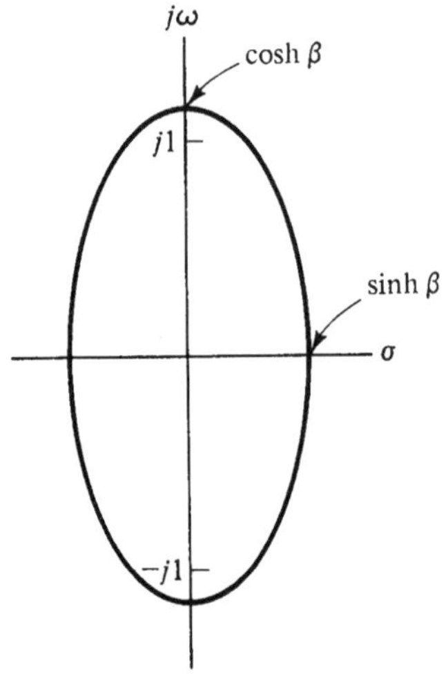

Figure 10.3.5 Poles of the Chebyshev filter.

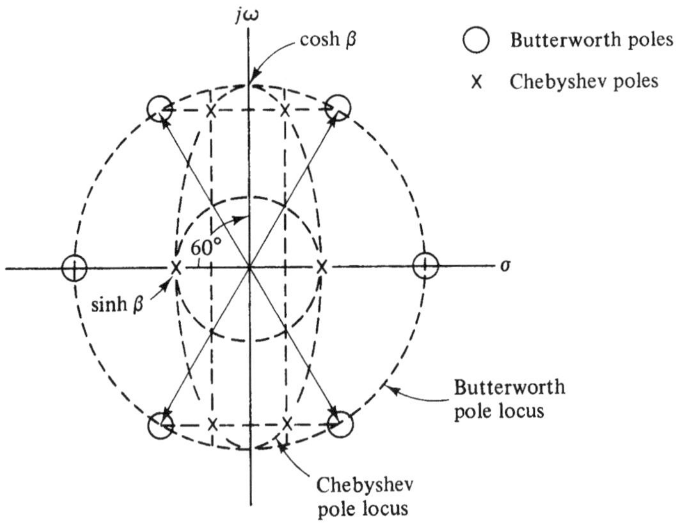

Figure 10.3.6 Relation between the Chebyshev and Butterworth poles for $N = 3$.

$$20 \log_{10} \frac{1}{\sqrt{1 + \varepsilon^2}} = -1$$

It follows that $\varepsilon = 0.509$. From Equation 10.3.22,

$$10 = 20 \log_{10} 0.509 + 6(N - 1) + 20N \log_{10} 5$$

so that $N = 1.0943$. Thus, we use a value of $N = 2$. The parameter β can be determined from Equation (10.3.23) to be equal to 0.714. To find the Chebyshev poles, we determine the poles for the corresponding Butterworth filter of the same order and multiply the real parts by $\sinh \beta$ and the imaginary parts by $\cosh \beta$. From Table 10-3, the poles of the normalized Butterworth filter are given by

$$\frac{s}{\omega_p} = -\frac{1}{\sqrt{2}} \pm j\frac{1}{\sqrt{2}}$$

where $\omega_p = 1000$. The Chebyshev poles are, then,

$$s = -\frac{1000}{\sqrt{2}}(\sinh 0.714) \pm j\frac{1000}{\sqrt{2}}(\cosh 0.714)$$

$$= -545.31 \pm j892.92$$

Hence,

$$H(s) = \frac{1}{(s + 545.31)^2 + (892.92)^2}$$

The corresponding filter with a dc gain of unity is given by

$$H(s) = \frac{(545.31)^2 + (892.92)^2}{(s + 545.31)^2 + (892.92)^2}$$

The magnitude characteristic for this filter is shown in Figure 10.3.7.

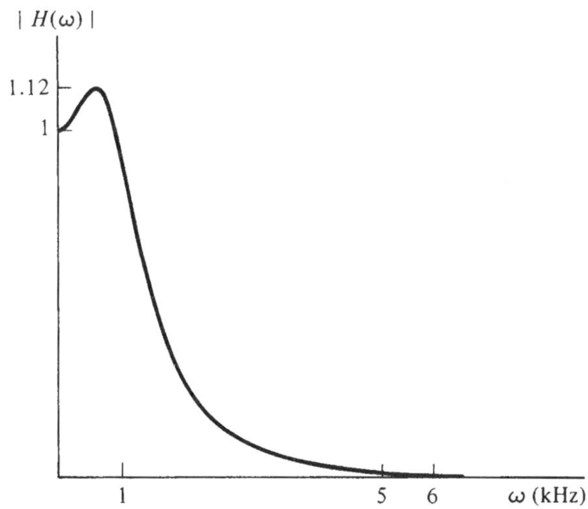

Figure 10.3.7 Magnitude characteristic of the Chebyshev filter of Example 10.3.2.

An approximation to the ideal low-pass characteristic, which, for a given order of filter has an even smaller transition band than the Chebyshev filter, can be obtained in terms of Jacobi elliptic sine functions. The resulting filter is called an elliptic filter. The design of this filter is somewhat complicated and is not discussed here. We note, however, that the magnitude characteristic of the elliptic filter has ripples in both the passband and the stop band. Figure 10.3.8 shows a typical elliptic-filter characteristic.

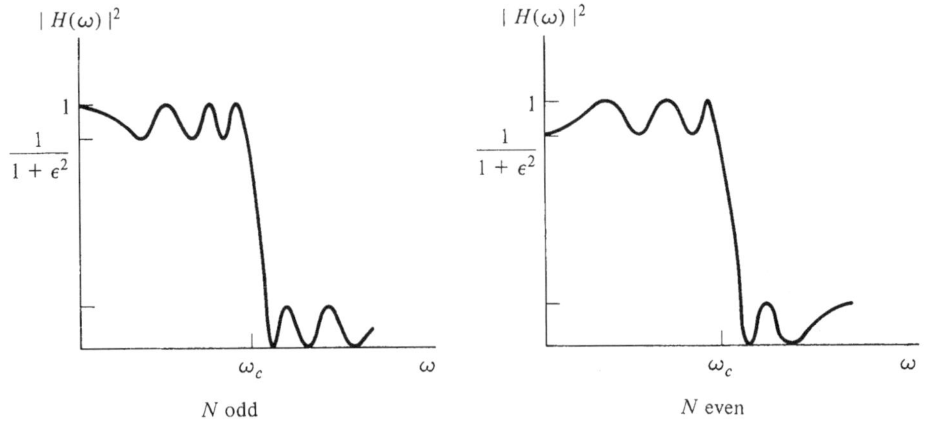

Figure 10.3.8 Magnitude characteristic of an elliptic filter.

10.4 DIGITAL FILTERS

In recent years, digital filters have supplanted analog filters in many applications because of their higher reliability, flexibility, and superior performance. The digital filter is designed to alter the spectral characteristics of a discrete-time input signal in a specified manner, in much the same way as the analog filter does for continuous-time signals. The specifications for the digital filter are given in terms of the discrete-time Fourier-transform variable Ω, and the design procedure consists of determining the discrete-time transfer function $H(z)$ that meets these specifications. We refer to $H(z)$ as the digital filter.

In certain applications in which a continuous-time signal is to be filtered, the analog filter is implemented as a digital filter for the reasons given. Such an implementation involves an analog-to digital conversion of the continuous-time signal, to obtain a digital signal that is filtered using a digital filter. The output of the digital filter is then converted back into a continuous-time signal by a digital-to-analog converter. In obtaining this equivalent digital realization of an analog filter, the specifications for the analog filter, which are in terms of the continuous-time Fourier-transform variable ω, must be transformed into an equivalent set of specifications in terms of the variable Ω.

As we saw earlier, digital systems (and, hence, digital filters) can be either FIR or IIR filters. The FIR digital filter, of course, has no counterpart in the analog domain. However, as we saw in previous sections, there are several well-established techniques for designing IIR filters. It would appear reasonable, therefore, to try and use these techniques for the design of IIR digital filters. In the next section, we discuss two commonly used methods for designing IIR digital filters based on analog-filter design techniques. For reasons discussed in the previous section, we confine our discussions to the design of low-pass filters. The procedure essentially involves converting the given digital-filter specifications to equivalent analog specifications, designing an analog filter that meets these specifications, and finally, converting the analog-filter transfer function $H_a(s)$ into an equivalent discrete-time transfer function $H(z)$.

10.4.1 Design of IIR Digital Filters Using Impulse Invariance

A fairly straightforward method for establishing an equivalence between a discrete-time system and a corresponding analog system is to require that the responses of the two systems to a test input match in a certain sense. To obtain a meaningful match, we assume that the output $y_a(t)$ of the continuous-time system is sampled at an appropriate rate T. We can then require that the sampled output $y_a(nT)$ be equal to the output $y(n)$ of the discrete-time system. If we now choose the test input as a unit impulse, we require that the impulse responses of the two systems be the same at the sampling instants, so that

$$h_a(nT) = h(n) \tag{10.4.1}$$

The technique is thus referred to as impulse-invariant design.

It follows from Equation (10.4.1) and our discussions in Section 7.5 that the relation between the digital frequency Ω and the analog frequency ω under this equivalence is given by Equation (7.5.10),

$$\omega = \frac{\Omega}{T} \tag{10.4.2}$$

Equation (10.4.2) can be used to convert the digital-filter specifications to equivalent analog-filter specifications. Once the analog filter $H_a(s)$ is determined, we can obtain the digital filter $H(z)$ by finding the sampled impulse response $h_a(nT)$ and taking its Z-transform. In most cases, we can go directly from $H_a(s)$ to $H(z)$ by expanding $H_a(s)$ in partial fractions and determining the corresponding Z-transform of each term from a table of transforms, as shown in Table 10-5. The steps can be summarized as follows:

1. From the specified passband and stop-band cutoff frequencies, Ω_p and Ω_s respectively, determine the equivalent analog frequencies, ω_p and ω_s.
2. Determine the analog transfer function $H_a(s)$, using the techniques of Section 10.3.
3. Expand $H_a(s)$ in partial fractions, and determine the Z-transform of each term from a table of transforms. Combine the terms to obtain $H(z)$.

While the impulse-invariant technique is fairly straightforward to use, it suffers from one disadvantage, namely, that we are in essence obtaining a discrete-time system from a continuous-time system by the process of sampling. We recall that sampling introduces aliasing and that the frequency response corresponding to the sequence $h_a(nT)$ is obtained from Equation (7.5.9) as

$$H(\Omega) = \frac{1}{T}\sum_{k=-\infty}^{\infty} H_a\left(\Omega + \frac{2\pi}{T}k\right) \tag{10.4.3}$$

so that

$$H(\Omega) = \frac{1}{T}H_a(\Omega) \tag{10.4.4}$$

only if

TABLE 10-5
Laplace transforms and their Z-transform equivalents

Laplace Transform, $H(s)$	Z-Transform, $H(z)$
$\frac{1}{s^2}$	$\frac{Tz}{(z-1)^2}$
$\frac{2}{s^3}$	$\frac{T^2 z(z+1)}{(z-1)^3}$
$\frac{1}{s+a}$	$\frac{z}{z-\exp[-aT]}$
$\frac{1}{(s+a)^2}$	$\frac{Tz\exp[-aT]}{(z-\exp[-aT])^2}$
$\frac{1}{(s+a)(s+b)}$	$\frac{1}{(b-a)}\left(\frac{z}{z-\exp[-aT]}-\frac{z}{z-\exp[-bT]}\right)$
$\frac{a}{s^2(s+a)}$	$\frac{Tz}{(z-1)^2}-\frac{(1-\exp[-aT])z}{a(z-1)(z-\exp[-aT])}$
$\frac{1}{(s+a)^2}$	$\frac{Tz\exp[-aT]}{(z-\exp[-aT])^2}$
$\frac{a^2}{s(s+a)^2}$	$\frac{z}{z-1}-\frac{z}{z-\exp[-aT]}-\frac{aT\exp[-aT]z}{(z-\exp[-aT])^2}$
$\frac{\omega_0}{s^2+\omega_0^2}$	$\frac{z\sin\omega_0 T}{z^2-2z\cos\omega_0 T+1}$
$\frac{s}{s^2+\omega_0^2}$	$\frac{z(z-\cos\omega_0 T)}{z^2-2z\cos\omega_0 T+1}$
$\frac{\omega_0}{(s+a)^2+\omega_0^2}$	$\frac{z\exp[-aT]\sin\omega_0 T}{z^2-2z\exp[-aT]\cos\omega_0 T+\exp[-2aT]}$
$\frac{s+a}{(s+a)^2+\omega_0^2}$	$\frac{z^2-z\exp[-aT]\cos\omega_0 T}{z^2-2z\exp[-aT]\cos\omega_0 T+\exp[-2aT]}$

$$H_a(\omega) = 0, \qquad |\omega| \geq \frac{\pi}{T} \tag{10.4.5}$$

which is not the case with practical low-pass filters. Thus, the resulting digital filter does not exactly meet the original design specifications.

It may appear that one way to reduce aliasing effects is to decrease the sampling interval T. However, since the analog passband cutoff frequency is given by $\omega_p = \Omega_p/T$, decreasing T has the effect of increasing ω_p, thereby increasing aliasing. It follows, therefore, that the choice of T has no effect on the performance of the digital filter and can be chosen to be unity.

For implementing an analog filter as a digital filter, we can follow exactly the same procedure as before, except that Step 1 is not required, since the specifications are now given directly in the analog domain. From Equation (10.4.3), when the analog filter is sufficiently band limited, the corresponding digital filter has a gain of $1/T$, which can become extremely high for low values of T. Generally, therefore, the resulting transfer function $H(z)$ is multiplied by T. The choice of T is usually determined by hardware considerations. We illustrate the procedure by the following example.

Example 10.4.1

Find the digital equivalent of the analog Butterworth filter derived in Example 10.3.1 using the impulse-invariant method.

From Example 10.3.1, with $\omega_c = 2826.8$, the filter transfer function is

$$H(s) = \frac{(2826.8)^3}{s^3 + 2(2826.8)s^2 + 2(2826.8)^2 s + (2826.8)^3}$$

$$= \frac{2826.8}{s + 2836.8} - \frac{2826.8(s + 1413.4)}{(s + 1413.4)^2 + (2448.1)^2} + \frac{0.5(2826.8)^2}{(s + 1413.4)^2 + (2448.1)^2}$$

We can determine the equivalent Z-transfer function from Table 10.5 as

$$H(z) = 2826.8\left[\frac{z}{z - e^{-1413.4T}} - \frac{z^2 - ze^{-1413.4T}[\cos(2448.1T) + \sin(2448.1T)]}{z^2 - 2ze^{-1413.4T}\cos(2448.1T) + e^{-2826.8T}}\right]$$

If the sampling interval T is assumed to be 1 ms, we get

$$H(z) = 2826.8\left[\frac{z}{z - 0.2433} - \frac{z^2 - 0.0973z}{z^2 - 0.3742z + 0.0592}\right]$$

which can be amplitude normalized as desired.

Example 10.4.2

Let us consider the design of a Butterworth low-pass digital filter that meets the following specifications. The passband magnitude should be constant to within 2 dB for frequencies below 0.2π radians, and the stop-band magnitude in the range $0.4\pi < \Omega < \pi$ should be less than 10 dB. Assume that the magnitude at $\Omega = 0$ is normalized to unity.

With $\Omega_p = 0.2\pi$ and $\Omega_s = 0.4\pi$, since the Butterworth filter has a monotonic magnitude characteristic, it is clear that to meet the specifications, we must have

$$20\log_{10}|H(0.2\pi)| = -2, \qquad \text{or} \qquad |H(0.2\pi)|^2 = 10^{-0.2}$$

and

$$20\log_{10}|H(0.4\pi)| = -10, \qquad \text{or} \qquad |H(0.4\pi)|^2 = 10^{-1}$$

For the impulse-invariant design technique, we obtain the equivalent analog domain specifications by setting $\omega = \Omega T$, with $T = 1$, so that

$$|H_a(0.2\pi)|^2 = 10^{-0.2}$$
$$|H_a(0.4\pi)|^2 = 10^{-1}$$

For the Butterworth filter,

$$|H_a(j\omega)|^2 = \frac{1}{1 + (\omega/\omega_c)^{2N}}$$

where ω_c and N must be determined from the specifications. This yields the two equations

$$1 + \left(\frac{0.2\pi}{\omega_c}\right)^{2N} = 10^{0.2}$$

$$1 + \left(\frac{0.4\pi}{\omega_c}\right)^{2N} = 10$$

Solving for N gives $N = 1.9718$, so that we choose $N = 2$. With this value of N, we can solve for ω_c from either of the last two equations. If we use the first equation, we just meet the passband specifications, but more than meet the stop-band specifications, whereas if we use the second equation, the reverse is true. Assuming we use the first equation, we get

$$\omega_c = 0.7185 \text{ rad/s}$$

The corresponding Butterworth filter is given by

$$H_a(s) = \frac{1}{(s/\omega_c)^2 + \sqrt{2}(s/\omega_c) + 1} = \frac{0.5162}{s^2 + 1.016s + 0.5162}$$

with impulse response

$$h_a(t) = 1.01 \exp[-0.508t] \sin 0.508t\, u(t)$$

The impulse response of the digital filter obtained by sampling $h_a(t)$ with $T = 1$ is

$$h(n) = 1.01 \exp[-0.508n] \sin 0.508\, n\, u(n)$$

By taking the corresponding Z-transform and normalizing so that the magnitude at $\Omega = 0$ is unity, we obtain

$$H(z) = \frac{0.5854z}{z^2 - 1.051z + 0.362}$$

Figure 10.4.1 shows a plot of $|H(\Omega)|$ for Ω in the range $[0, \pi/2]$. For this particular example, the analog filter is sufficiently band limited, so that the effects of aliasing are not noticeable. This is not true in general, however. One possibility in such a case is to choose a higher value of N than is obtained from the specifications.

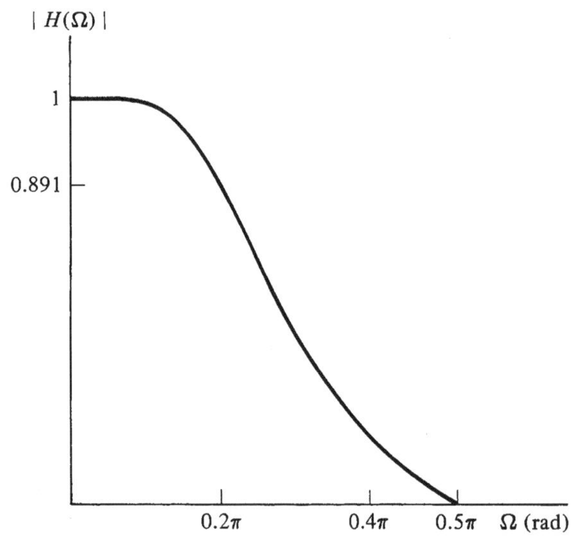

Figure 10.4.1 Magnitude function of the Butterworth design of Example 10.4.2.

10.4.2 IIR Design Using the Bilinear Transformation

As stated earlier, digital-filter design based on analog filters involves converting discrete-domain specifications into the analog domain. The impulse-invariant design does this by using the transformation

$$\omega = \Omega/T$$

or equivalently,

$$z = \exp[Ts]$$

We saw, however, that because of the nature of this mapping, as discussed in Section 8.8, the impulse-invariant design leads to aliasing problems. One approach to overcoming aliasing is to use a transformation that maps the Z-domain onto a domain that is similar to the s domain, in that the unit circle in the z plane maps into the vertical axis in the new domain, the interior of the unit circle maps onto the open left half plane, and the exterior of the circle maps onto the open right half plane. We can then treat this new plane as if it were the analog domain and use standard techniques for obtaining the equivalent analog filter. The specific transformation that we use is

$$s = \frac{2}{T}\frac{1 - z^{-1}}{1 + z^{-1}} \tag{10.4.6}$$

or equivalently,

$$z = \frac{1 + (T/2)s}{1 - (T/2)s} \tag{10.4.7}$$

where T is a parameter that can be chosen to be any convenient value. It can easily be verified by setting $z = r^{-1}\exp[j\Omega]$ that this transformation, which is referred to as the bilinear transformation, does indeed satisfy the three requirements that we mentioned earlier. We have

$$s = \sigma + j\omega = \frac{2}{T}\frac{1 - r\exp[-j\Omega]}{1 + r\exp[-j\Omega]}$$

$$= \frac{2}{T}\frac{1 - r^2}{1 + r^2 + 2r\cos\Omega} + j\frac{2}{T}\frac{2r\sin\Omega}{1 + r^2 + 2r\cos\Omega}$$

For $r < 1$, clearly, $\sigma > 0$, and for $r > 1$, we have $\sigma < 0$. For $r = 1$, s is purely imaginary, with

$$\omega = \frac{2}{T}\frac{\sin\Omega}{1 + \cos\Omega} = \frac{2}{T}\tan\frac{\Omega}{2}$$

This relationship is plotted in Figure 10.4.2.

The procedure for obtaining the digital filter $H(z)$ can be summarized as follows:

1. From the given digital-filter specifications, find the corresponding analog-filter specifications by using the relation

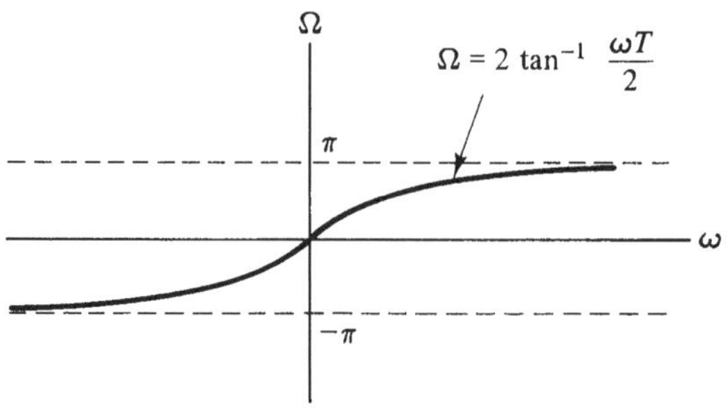

Figure 10.4.2 Relation between Ω and ω under the bilinear transformation.

$$\omega = \frac{2}{T}\tan\frac{\Omega}{2} \tag{10.4.8}$$

where T can be chosen arbitrarily, e.g., $T = 2$.

2. Find the corresponding analog-filter function $H_a(s)$. Then find the equivalent digital filter as

$$H(z) = H_a(s)\big|_{s=\frac{2}{T}\frac{1-z^{-1}}{1+z^{-1}}} \tag{10.4.9}$$

The following example illustrates the use of the bilinear transform in digital filter design.

Example 10.4.3

We consider the problem of Example 10.4.2, but will now obtain a Butterworth design using the bilinear transform method. With $T = 2$, we determine the corresponding pass-band and stop-band cutoff frequencies in the analog domain as

$$\omega_p = \tan\frac{0.2\pi}{2} = 0.3249$$

$$\omega_s = \tan\frac{0.4\pi}{2} = 0.7265$$

To meet the specifications, we now set

$$1 + \left(\frac{0.3249}{\omega_c}\right)^{2N} = 10^{-0.2}$$

$$1 + \left(\frac{0.7265}{\omega_c}\right)^{2N} = 10^{-1}$$

and solve for N to get $N = 1.695$. Choosing $N = 2$ and determining ω_c as before gives

$$\omega_c = 0.4195$$

The corresponding analog filter is

$$H_a(s) = \frac{0.176}{s^2 + 0.593s + 0.176}$$

We can now obtain the digital filter $H(z)$ with gain at $\Omega = 0$ normalized to be unity as

$$H(z) = H_a(s)\big|_{s=\frac{1+z^{-1}}{1-z^{-1}}} = \frac{0.1355(z+1)^2}{z^2 - 2.174z + 1.716}$$

Figure 10.4.3 shows the magnitude characteristic of the digital filter for Ω in the range $[0, \pi/2]$.

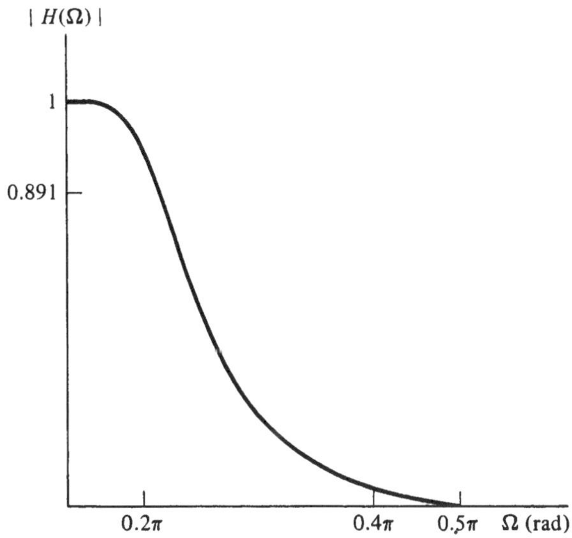

Figure 10.4.3 Magnitude characteristic of the filter for Example 10.4.3 using the bilinear method.

10.4.3 FIR Filter Design

In our earlier discussions, we noted that it is desirable that a filter have a linear phase characteristic. Although an IIR digital filter does not in general have a linear phase, we can obtain such a characteristic with a FIR digital filter. In this section, we consider a technique for the design of FIR digital filters.

We first establish that a FIR digital filter of length N has a linear phase characteristic, provided that its impulse response satisfies the symmetry condition

$$h(n) = h(N - 1 - n) \tag{10.4.10}$$

This can be easily verified by determining $H(\Omega)$. We consider the case of N even and N odd separately. For N even, we write

$$\begin{aligned} H(\Omega) &= \sum_{n=0}^{N-1} h(n)\exp[-j\Omega n] \\ &= \sum_{n=0}^{N/2-1} h(n)\exp[-j\Omega n] + \sum_{n=N/2}^{N-1} h(n)\exp[-j\Omega n] \end{aligned}$$

Now we replace n by $N - n - 1$ in the second term in the last equation and use Equation (10.4.10) to get

$$H(\Omega) = \sum_{n=0}^{(N/2)-1} h(n)\exp[-j\Omega n] + \sum_{n=0}^{(N/2)-1} h(n)\exp[-j\Omega(N-1-n)]$$

which can be written as

$$H(\Omega) = \left\{\sum_{n=0}^{N/2-1} 2h(n)\cos\left[\Omega\left(n - \frac{N-1}{2}\right)\right]\right\}\exp\left[-j\Omega\left(\frac{N-1}{2}\right)\right]$$

Similarly, for N odd, we can show that

$$H(\Omega) = \left\{h\left(\frac{N-1}{2}\right) + \sum_{n=0}^{(N-3)/2} 2h(n)\cos\left[\Omega\left(n - \frac{N-1}{2}\right)\right]\right\}\exp\left[-j\Omega\left(\frac{N-1}{2}\right)\right]$$

In both these cases, the term in braces is real, so that the phase of $H(\Omega)$ is given by the complex exponential. It follows that the system has a linear phase shift, with a corresponding delay of $(N-1)/2$ samples.

Given a desired frequency response $H_d(\Omega)$, such as an ideal low-pass characteristic, which is symmetric about the origin, the corresponding impulse response $h_d(n)$ is symmetric about the point $n = 0$, but, in general, is of infinite duration. The most direct way of obtaining an equivalent FIR filter of length N is to just truncate this infinite sequence. The truncation operation, as in our earlier discussion of the DFT in Chapter 9, can be considered to result from multiplying the infinite sequence by a window sequence $w(n)$. If $h_d(n)$ is symmetric about $n = 0$, we get a linear phase filter that is, however, noncausal. We can get a causal impulse response by shifting the truncated sequence to the right by $(N-1)/2$ samples. The desired digital filter $H(z)$ is then determined as the Z-transform of this truncated, shifted sequence. We summarize these steps as follows:

1. From the desired frequency-response characteristic $H_d(\Omega)$, find the corresponding impulse response $h_d(n)$.
2. Multiply $h_d(n)$ by the window function $w(n)$.
3. Find the impulse response of the digital filter as

$$h(n) = h_d[n - (N-1)/2]w(n)$$

and determine $H(z)$. Alternatively, we can find the Z-transform $H'(z)$ of the sequence $h_d(n)w(n)$ and find $H(z)$ as

$$H(z) = z^{-(N-1)/2}H'(z)$$

As noted before, we encountered the truncation operation in Step 2 in our earlier discussion of the DFT in Chapter 9. There it was pointed out that truncation causes the frequency response of the filter to be smeared.

In general, windows with wide main lobes cause more spreading than those with narrow main lobes. Figure 10.4.4 shows the effect of using the rectangular window on the ideal low-pass filter characteristic. As can be seen, the transition width of the resulting filter is approximately equal to the main lobe width of the window function and is, hence, inversely proportional to the window length N. The choice of N, therefore, involves a compromise between transition width and filter length.

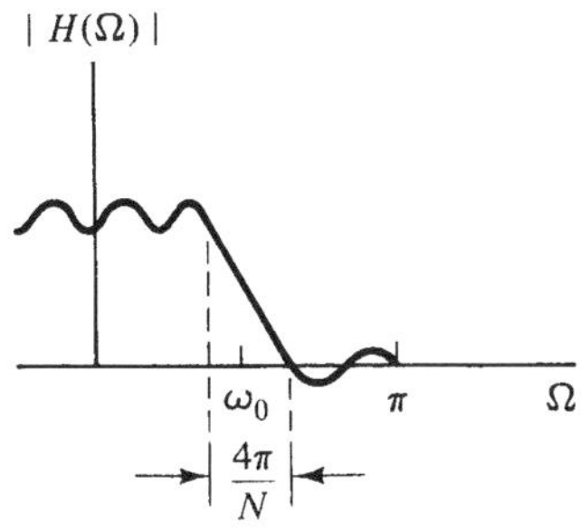

Figure 10.4.4 Frequency response obtained by using rectangular window on ideal filter response.

The following are some commonly used window functions.

Rectangular:

$$w_R(n) = \begin{cases} 1, & 0 \le n \le N-1 \\ 0, & \text{elsewhere} \end{cases} \tag{10.4.11a}$$

Bartlett:

$$w_B(n) = \begin{cases} \dfrac{2n}{N-1}, & 0 \le n \le \dfrac{N-1}{2} \\ 2 - \dfrac{2n}{N-1}, & \dfrac{N-1}{2} \le n \le N-1 \\ 0, & \text{elsewhere} \end{cases} \tag{10.4.11b}$$

Hanning:

$$w_{\text{Han}}(n) = \begin{cases} \frac{1}{2}\left(1 - \cos\dfrac{2\pi n}{N-1}\right), & 0 \le n \le N-1 \\ 0, & \text{elsewhere} \end{cases} \tag{10.4.11c}$$

Hamming:

$$w_{\text{Ham}}(n) = \begin{cases} 0.54 - 0.46\cos\dfrac{2\pi n}{N-1}, & 0 \le n \le N-1 \\ 0, & \text{elsewhere} \end{cases} \tag{10.4.11d}$$

Blackman:

$$w_{\text{Bl}}(n) = \begin{cases} 0.42 - 0.5\cos\dfrac{2\pi n}{N-1} + 0.08\cos\dfrac{4\pi n}{N-1}, & 0 \le n \le N-1 \\ 0, & \text{elsewhere} \end{cases} \tag{10.4.11e}$$

Kaiser:

$$w_K(n) = \begin{cases} \dfrac{I_0\left(\alpha\left[\left(\dfrac{N-1}{2}\right)^2 - \left(n - \dfrac{N-1}{2}\right)^2\right]^{1/2}\right)}{I_0\left[\alpha\left(\dfrac{N-1}{2}\right)\right]}, & 0 \le n \le N-1 \\ 0, & \text{elsewhere} \end{cases} \qquad (10.4.11f)$$

where $I_0(x)$ is the modified zero-oder Bessel function of the first kind given by $I_0(x) = \int_0^{2\pi} \exp[x\cos\theta]\,d\theta/2\pi$ and α is a parameter that effects the relative widths of the main and side lobes. When α is zero, we get the rectangular window, and for $\alpha = 5.414$, we get the Hamming window. In general, as α becomes larger, the main lobe becomes wider and the side lobes smaller. Of the windows described previously, the most commonly used is the Hamming window, and the most versatile is the Kaiser window.

Example 10.4.4

Let us consider the design of a nine-point FIR digital filter to approximate an ideal low-pass digital filter with a cutoff frequency $\Omega_c = 0.2\pi$. The impulse response of the desired filter is

$$h_d(n) = \frac{1}{2\pi}\int_{-.2\pi}^{.2\pi} \exp[j\Omega n]\,d\Omega = \frac{\sin 0.2\pi n}{\pi n}$$

For a rectangular window of length 9, the corresponding impulse response is obtained by evaluating $h_d(n)$ for $-4 \le n \le 4$. We obtain

$$h_d(n) = \left\{\frac{0.147}{\pi}, \frac{0.317}{\pi}, \frac{0.475}{\pi}, \frac{0.588}{\pi}, \underset{\uparrow}{1}, \frac{0.588}{\pi}, \frac{0.475}{\pi}, \frac{0.317}{\pi}, \frac{0.147}{\pi}\right\}$$

The filter function is

$$H'(z) = \frac{0.147}{\pi}z^4 + \frac{0.317}{\pi}z^3 + \frac{0.475}{\pi}z^2 + \frac{0.588}{\pi}z + 1$$
$$+ \frac{0.588}{\pi}z^{-1} + \frac{0.475}{\pi}z^{-2} + \frac{0.317}{\pi}z^{-3} + \frac{0.147}{\pi}z^{-4}$$

so that

$$H(z) = z^{-4}H'(z) = \frac{0.147}{\pi}(1 + z^{-8}) + \frac{0.317}{\pi}(z^{-1} + z^{-7})$$
$$+ \frac{0.475}{\pi}(z^{-2} + z^{-6}) + \frac{0.588}{\pi}(z^{-3} + z^{-5}) + z^{-4}$$

For $N = 9$, the Hamming window defined in Equation (10.4.11d) is given by the sequence

$$w(n) = \{0.081, 0.215, 0.541, 0.865, \underset{\uparrow}{1}, 0.865, 0.541, 0.215, 0.081\}$$

Hence, we have

$$h_d(n)W(n) = \left\{\frac{0.012}{\pi}, \frac{0.068}{\pi}, \frac{0.257}{\pi}, \frac{0.508}{\pi}, \underset{\uparrow}{1}, \frac{0.508}{\pi}, \frac{0.257}{\pi}, \frac{0.068}{\pi}, \frac{0.012}{\pi}\right\}$$

The filter function is

$$H'(z) = \frac{0.012}{\pi}z^4 + \frac{0.0068}{\pi}z^3 + \frac{0.257}{\pi}z^2 + \frac{0.508}{\pi}z + 1$$

$$+ \frac{0.508}{\pi}z^{-1} + \frac{0.257}{\pi}z^{-2} + \frac{0.068}{\pi}z^{-3} + \frac{0.012}{\pi}z^{-4}$$

Finally,

$$H(z) = z^{-4}H'(z) = \frac{0.012}{\pi}(1 + z^{-8}) + \frac{0.068}{\pi}(z^{-1} + z^{-7}) + \frac{0.257}{\pi}(z^{-2} + z^{-6})$$

$$+ \frac{0.508}{\pi}(z^{-3} + z^{-5}) + z^{-4}$$

The frequency responses of the filters obtained using both the rectangular and Hamming windows are shown in Figure 10.4.5, with the gain at $\Omega = 0$ normalized to unity. As can be seen from the figure, the response corresponding to the Hamming window is smoother than the one for the rectangular window.

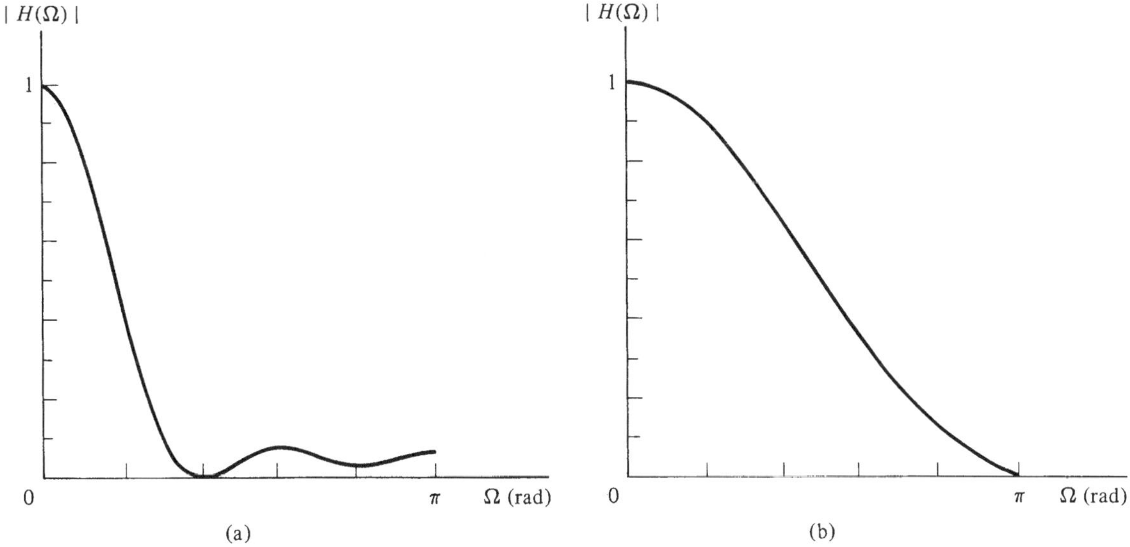

Figure 10.4.5 Response of the FIR digital filter of Example 10.4.4. (a) Rectangular window. (b) Hamming window.

Example 10.4.5

FIR digital filters can be used to approximate filters such as the ideal differentiator or the Hilbert transformer, which cannot be implemented in the analog domain. In the analog domain, the ideal differentiator is described by the frequency response

$$H(\omega) = j\omega$$

while the Hilbert transformer is described by the frequency response

$$H(\omega) = -j\,\mathrm{sgn}\,(\omega)$$

To design a discrete-time implementation of such filters, we start by specifying the desired response in the frequency domain as

$$H_d(\omega) = H(\omega), \qquad -\frac{\omega_s}{2} \le \omega \le \frac{\omega_s}{2}$$

where $\omega_s = 2\pi/T$ for some choice of T. Equivalently,

$$H_d(\Omega) = H(\omega T), \qquad -\pi \le \Omega \le \pi$$

where $\Omega = \omega T$. Since $H_d(\Omega)$ is periodic in Ω with period 2π, we can expand it in a Fourier series as

$$H_d(\Omega) = \sum_{n=-\infty}^{\infty} h_d(n)e^{-jn\Omega}$$

where the coefficients $h_d(n)$ are the corresponding impulse response samples, given by

$$h_d(n) = \frac{1}{2\pi}\int_{-\pi}^{\pi} H_d(\Omega)e^{jn\Omega}\,d\Omega$$

As we have seen earlier, if the desired frequency function $H_d(\Omega)$is purely real, the impulse response is even and symmetric; that is, $h_d(n) = h_d(-n)$. On the other hand, if the frequency response is purely imaginary, the impulse response is odd and symmetric, so that $h_d(n) = -h_d(-n)$.

We can now design a FIR digital filter by following the procedure given earlier. We will illustrate this for the case of the Hilbert transformer. This transformer is used to generate signals that are in phase quadrature to an input sinusoidal signal (or, more generally, an input narrow-band waveform). That is, if the input to a Hilbert transformer is the signal $x_a(t) = \cos\omega_0 t$, the output is $y_a(t) = \sin\omega_0 t$. The Hilbert transformer is used in communication systems in various modulation schemes.

The impulse response for the Hilbert transformer is obtained as

$$h_d(n) = \frac{1}{2\pi}\int_{-\pi}^{\pi} -j\,\mathrm{sgn}\,(\Omega)e^{jn\Omega}\,d\Omega$$

$$= \begin{cases} 0, & n \text{ even} \\ \dfrac{2}{n\pi} & n \text{ odd} \end{cases}$$

For a rectangular window of length 15, we obtain

$$h_d(n) = \left\{-\frac{2}{7\pi}, 0, -\frac{2}{5\pi}, 0, -\frac{2}{3\pi}, 0, -\frac{2}{\pi}, 0, \frac{2}{\pi}, 0, \frac{2}{3\pi}, 0, \frac{2}{5\pi}, 0, \frac{2}{7\pi}\right\}$$

which can be realized with a delay of seven samples by the transfer fuction

$$H(z) = \frac{2}{\pi}\left[-\frac{1}{7} - \frac{1}{5}z^{-2} - \frac{1}{3}z^{-4} - z^{-6} + z^{-8} + \frac{1}{3}z^{-10} + \frac{1}{5}z^{-12} + \frac{1}{7}z^{-14}\right]$$

The frequency response $H(\Omega)$ of this filter is shown in Figure 10.4.6. As can be seen, the response exhibits considerable ripple. As discussed previously, the ripples can be reduced by using window functions other than the rectangular. Also shown in the figure is the response corresponding to the Hamming window.

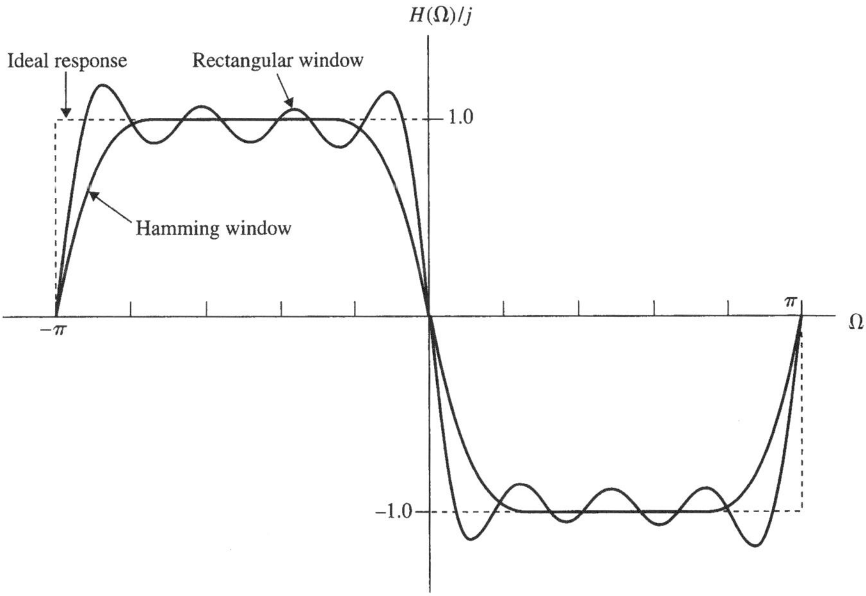

Figure 10.4.6 Frequency response of Hilbert transformer.

10.4.4 Computer-Aided Design of Digital Filters

In recent years, the use of computer-aided techniques for the design of digital filters has become widespread, and several software packages are available for such design. Techniques have been developed for both FIR and IIR filters that, in general, involve the minimization of a suitably chosen cost function. Given a desired frequency-response characteristic $H_d(\Omega)$, a filter of either the FIR or IIR type and of fixed order is selected. We express the frequency response of this filter, $H(\Omega)$, in terms of the vector **a** of filter coefficients. The difference between the two responses, which represents the deviation from the desired response, is a function of **a**. We associate a cost function with this difference and seek the set of filter coefficients **a** that minimizes this cost function. A typical cost function is of the form

$$J(\mathbf{a}) = \int_{-\pi}^{\pi} W(\Omega)|H_d(\Omega) - H(\Omega)|^2 d\Omega \tag{10.4.12}$$

where $W(\Omega)$ is a nonnegative weighting function that reflects the significance attached to the deviation from the desired response in a particular range of frequencies. $W(\Omega)$ is chosen to be relatively large over that range of frequencies considered to be important.

Quite often, instead of minimizing the deviation at all frequencies, as in Equation (10.4.12), we can choose to do so only at a finite number of frequencies. The cost function then becomes

$$J(\mathbf{a}) = \sum_{i=1}^{M} W(\Omega_i)|H_d(\Omega_i) - H(\Omega_i)|^2 \tag{10.4.13}$$

where Ω_i, $1 \le i \le M$, are a set of frequency samples over the range of interest. Typically, the minimization problem is quite complex, and the resulting equations cannot be solved analytically. An iterative search procedure is usually employed to determine the optimum set of filter coefficients. We start with an arbitrary initial choice for the filter coefficients and successively adjust them such that the resulting cost function is reduced at each step. The procedure stops when a further adjustment of the coefficients does not result in a reduction in the cost function. Several standard algorithms and software packages are available for determining the optimum filter coefficients.

A popular technique for the design of FIR filters is based on the fact that the frequency response of a linear-phase FIR filter can be expressed as a trignometric polynomial similar to the Chebyshev polynomial. The filter coefficients are chosen to minimize the maximum deviation from the desired response. Again, computer programs are available to determine the optimum filter coefficients.

10.5 SUMMARY

- Frequency-selective filters are classified as low-pass, high-pass, band-pass, or band-stop filters.
- The passband of a filter is the range of frequencies that are passed without attenuation. The stop band is the range of frequencies that are completely attenuated.
- Filter specifications usually specify the permissible deviation from the ideal characteristic in both passband and stop band, as well as specifying a transition band between the two.
- Filter design consists of obtaining an analytical approximation to the desired filter characteristic in the form of a filter transfer function, given as $H(s)$ for analog filters and $H(z)$ for digital filters.
- Frequency transformations can be used for converting one type of filter to another.
- Two popular low-pass analog filters are the Butterworth and Chebyshev filters. The Butterworth filter has a monotonically decreasing characteristic that goes to zero smoothly. The Chebyshev filter has a ripple in the passband, but is monotonically decreasing in the transition and stop bands.

- A given set of specifications can be met by a Chebyshev filter of lower order than a Butterworth filter.
- The poles of the Butterworth filter are spaced uniformly around the unit circle in the s plane. The poles of the Chebyshev filter are located in an ellipse on the s plane and can be obtained geometrically from the Butterworth poles.
- Digital filters can be either IIR or FIR.
- Digital IIR filters can be obtained from equivalent analog designs by using either the impulse-invariant technique or the bilinear transformation.
- Digital filters designed using impulse invariance exhibit distortion due to aliasing. No aliasing distortion arises from the use of the bilinear transformation method.
- Digital FIR filters are often chosen to have a linear phase characteristic. One method of obtaining an FIR filter is to determine the impulse response $h_d(n)$ corresponding to the desired filter characteristic $H_d(\Omega)$ and to truncate the resulting sequence by multiplying it by an appropriate window function.
- For a given filter length, the transition band depends on the window function.

10.6 CHECKLIST OF IMPORTANT TERMS

Aliasing error
Analog filter
Band-pass filter
Band-stop filter
Bilinear transformation
Butterworth filter
Chebyshev filter
Digital filter
Filter specifications
FIR filter
Frequency-selective filter
Frequency transformations
High-pass filter
Impulse invariance
IIR filter
Linear phase characteristic
Low-pass filter
Passband
Transition band
Window function

10.7 PROBLEMS

10.1. Design an analog low-pass Butterworth filter to meet the following specifications: the attenuation to be less than 1.5 dB up to 1 kHz and to be at least-15 dB for frequencies greater than 4 kHz.

10.2. Use the frequency transformations of Section 10.2 to obtain an analog Butterworth filter with an attenuation of less than 1.5 dB for frequencies up to 3 kHz, from your design in Problem 10.1.

10.3. Design a Butterworth band-pass filter to meet the following specifications:

$$\omega_{c_1} = \text{lower cutoff frequency} = 200 \text{ Hz}$$

$$\omega_{c_2} = \text{upper cutoff frequency} = 300 \text{ Hz}$$

The attenuation in the passband is to be less than 1 dB. The attenuation in the stop band is to be at lcast 10 dB.

10.4. A Chebyshev low-pass filter is to be designed to have a passband ripple $\leqslant 2$ dB and a cutoff frequency of 1500 Hz. The attenuation for frequencies greater than 5000 Hz must be at least 20 dB. Find ε, N, and $H(s)$.

10.5. Consider the third-order Butterworth and Chebyshev filters with the 3-dB cutoff frequency normalized to 1 in both cases. Compare and comment on the corresponding characteristics in both passbands and stop bands.

10.6. In Problem 10.5, what order of Butterworth filter compares to the Chebyshev filter of order 3?

10.7. Design a Chebyshev filter to meet the specifications of Problem 10.1. Compare the frequency response of the resulting filter to that of the Butterworth filter of Problem 10.1.

10.8. Obtain the digital equivalent of the low-pass filter of Problem 10.1 using the impulse-invariant method. Assume a sampling frequency of (a) 6 kHz, (b) 10 kHz.

10.9. Plot the frequency responses of the digital filters of Problem 10.8. Comment on your results.

10.10. The bilinear transform technique enables us to design IIR digital filters using standard analog designs. However, if we want to replace an analog filter by an equivalent A/D digital filter-D/A combination, we have to prewarp the given cutoff frequencies before designing the analog filter. Thus, if we want to replace an analog Butterworth filter by a digital filter, we first design the analog filter by replacing the passband and stop-band cutoff frequencies, ω_p and ω_s, respectively, by

$$w_p^d = \frac{2}{T}\tan\frac{\omega_p T}{2}$$

$$w_s^d = \frac{2}{T}\tan\frac{\omega_s T}{2}$$

The equivalent digital filter is then obtained from the analog design by using Equation (10.4.9). Use this method to obtain a digital filter to replace the analog filter in Problem 10.1. Assume that the sampling frequency is 3 kHz.

10.11. Repeat Problem 10.10 for the bandpass filter of Problem 10.4.

10.12. **(a)** Show that the frequency response $H(\Omega)$ of a filter is (i) purely real if the impulse response $h(n)$ is even and symmetric (i.e., $h(n) = -h(-n)$) and (ii) purely imaginary if $h(n)$ is odd and symmetric (i.e., $h(n) = -h(-n)$).

(b) Use your result in Part (a) to determine the phase of an N-point FIR filter if (i) $h(n) = h(N - 1 - n)$ and (ii) $h(n) = -h(N - 1 - n)$.

10.13. **(a)** The ideal differentatior has frequency response

$$H_d(\Omega) = j\Omega \quad 0 \leqslant |\Omega| \leqslant \pi$$

Show that the Fourier series coefficient for $H_d(\Omega)$ are

$$h_d(n) = \frac{(-1)^n}{n}$$

(b) Hence, design a 10-point differentiator using both rectangular and Hanning windows.

10.14. **(a)** Design an 11-tap FIR filter ($N = 12$) to approximate the ideal low-pass characteristic with cutoff $\pi/6$ radians.

(b) Plot the frequency response of the filter you designed in Part (a).

(c) Use the Hanning window to modify the results of Part (a). Plot the frequency response of the resulting filter, and comment on it.

Appendix A

Complex Numbers

Many engineering problems can be treated and solved by methods of complex analysis. Roughly speaking, these problems can be subdivided into two large classes. The first class consists of elementary problems for which the knowledge of complex numbers and calculus is sufficient. Applications of this class of problems are in differential equations, electric circuits, and the analysis of signals and systems. The second class of problems requires detailed knowledge of the theory of complex analytic functions. Problems in areas such as electrostatics, electromagnetics, and heat transfer belong to this category.

In this appendix, we concern ourselves with problems of the first class. Problems of the second class are beyond the scope of the text.

A.1 DEFINITION

A complex number $z = x + jy$, where $j = \sqrt{-1}$, consists of two parts, a real part x and an imaginary part y.[1] This form of representation for complex numbers is called the rectangular or Cartesian form, since z can be represented in rectangular coordinates by the point (x, y), as shown in Figure A.1.

The horizontal x axis is called the real axis, and the vertical y axis is called the imaginary axis. The x-y plane in which the complex numbers are represented in this way is called the complex plane. Two complex numbers are equal if their real parts are equal and their imaginary parts are equal.

The complex number z can also be written in polar form. The polar coordinates r and θ are related to the Cartesian coordinates x and y by

[1] Mathematicians use i to represent $\sqrt{-1}$, but electrical engineers use j for this purpose because i is usually used to represent current in electric circuits.

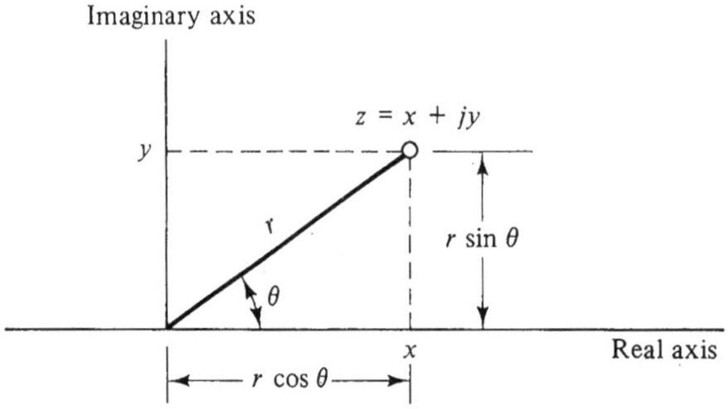

Figure A.1 The complex number z in the complex plane.

$$x = r\cos\theta \qquad \text{and} \qquad y = r\sin\theta \tag{A.1}$$

Hence, a complex number $z = x + jy$ can be written as

$$z = r\cos\theta + jr\sin\theta \tag{A.2}$$

This is known as the polar form, or trigonometric form, of a complex number. By using Euler's identity,

$$\exp[j\theta] = \cos\theta + j\sin\theta$$

we can express the complex z number in Equation (A.2) in exponential form as

$$z = r\exp[j\theta] \tag{A.3}$$

where r, the magnitude of z, is denoted by $|z|$. From Figure A.1,

$$|z| = r = \sqrt{x^2 + y^2} \tag{A.4}$$

$$\theta = \arctan\frac{y}{x} = \arcsin\frac{y}{r} = \arccos\frac{x}{r} \tag{A.5}$$

The angle θ is called the argument of z, denoted $\measuredangle z$, and measured in radians. The argument is defined only for nonzero complex numbers and is determined only up to integer multiples of 2π. The value of θ that lies in the interval

$$-\pi < \theta \le \pi$$

is called the principal value of the argument of z. Geometrically, $|z|$ is the length of the vector from the origin to the point z in the complex plane, and θ is the directed angle from the positive x axis to z.

Example A.1

For the complex number $z = 1 + j\sqrt{3}$,

$$r = \sqrt{1^2 + (\sqrt{3})^2} = 2 \qquad \text{and} \qquad \measuredangle z = \arctan\sqrt{3} = \frac{\pi}{3} + 2n\pi$$

The principal value of $\measuredangle z$ is $\pi/3$, and therefore,

$$z = 2(\cos\pi/3 + j\sin\pi/3)$$

The complex conjugate of z is defined as

$$z^* = x - jy \tag{A.6}$$

Since

$$z + z^* = 2x \quad \text{and} \quad z - z^* = 2jy \tag{A.7}$$

it follows that

$$\text{Re}\{z\} = x = \frac{1}{2}(z + z^*), \quad \text{and} \quad \text{Im}\{z\} = y = \frac{1}{2j}(z - z^*) \tag{A.8}$$

Note that if $z = z^*$, then the number is real, and if $z = -z^*$, then the number is purely imaginary.

A.2 ARITHMETIC OPERATIONS

A.2.1 Addition and Subtraction

The sum and difference of two complex numbers are respectively defined by

$$z_1 + z_2 = (x_1 + x_2) + j(y_1 + y_2) \tag{A.9}$$

and

$$z_1 - z_2 = (x_1 - x_2) + j(y_1 - y_2) \tag{A.10}$$

These are demonstrated geometrically in Figure A.2 and can be interpreted in accordance with the "parallelogram law" by which forces are added in mechanics.

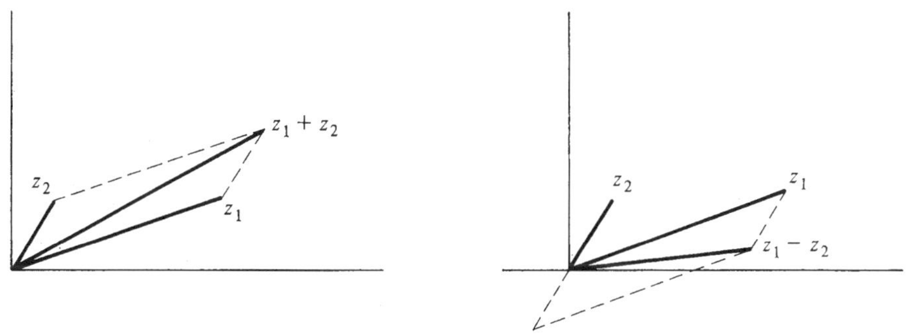

Figure A.2 Addition and subtraction of complex numbers.

A.2.2 Multiplication

The product $z_1 z_2$ is given by

$$\begin{aligned} z_1 z_2 &= (x_1 + jy_1)(x_2 + jy_2) \\ &= (x_1 x_2 - y_1 y_2) + j(x_1 y_2 + x_2 y_1) \end{aligned} \tag{A.11}$$

In polar form, this becomes

$$z_1 z_2 = r_1 \exp[j\theta_1] r_2 \exp[j\theta_2] = r_1 r_2 \exp[j(\theta_1 + \theta_2)] \tag{A.12}$$

That is, the magnitude of the product of two complex numbers is the product of the magnitudes of the two numbers, and the angle of the product is the sum of the two angles.

A.2.3 Division

Division is defined as the inverse of multiplication. The quotient z_1/z_2 is obtained by multiplying both the numerator and denominator by the conjugate of z_2:

$$\begin{aligned}\frac{z_1}{z_2} &= \frac{(x_1 + jy_1)}{(x_2 + jy_2)} \\ &= \frac{(x_1 + jy_1)(x_2 - jy_2)}{x_2^2 + y_2^2} \\ &= \frac{x_1x_2 + y_1y_2}{x_2^2 + y_2^2} + j\frac{x_2y_1 - x_1y_2}{x_2^2 + y_2^2}\end{aligned} \tag{A.13}$$

Division is performed easily in polar form as follows:

$$\begin{aligned}\frac{z_1}{z_2} &= \frac{r_1 \exp[j\theta_1]}{r_2 \exp[j\theta_2]} \\ &= \frac{r_1}{r_2} \exp[j(\theta_1 - \theta_2)]\end{aligned} \tag{A.14}$$

That is, the magnitude of the quotient is the quotient of the magnitudes, and the angle of the quotient is the difference of the angle of the numerator and the angle of the denominator.

For any complex numbers z_1, z_2, and z_3, we have the following:

- Commutative laws:

$$\begin{cases} z_1 + z_2 = z_2 + z_1 \\ z_1 z_2 \quad\;\; = z_2 z_1 \end{cases} \tag{A.15}$$

- Associative laws:

$$\begin{cases} (z_1 + z_2) + z_3 = z_1 + (z_2 + z_3) \\ z_1(z_2 z_3) \qquad\;\; = (z_1 z_2) z_3 \end{cases} \tag{A.16}$$

- Distributive law:

$$z_1(z_2 + z_3) = z_1 z_2 + z_1 z_3 \tag{A.17}$$

A.3 POWERS AND ROOTS OF COMPLEX NUMBERS

The nth power of the complex number

$$z = r\exp[j\theta]$$

is

$$z^n = r^n\exp[jn\theta] = r^n(\cos n\theta + j\sin n\theta)$$

from which we obtain the so-called formula of De Moivre:[2]

$$(\cos\theta + j\sin\theta)^n = (\cos n\theta + j\sin n\theta) \tag{A.18}$$

For example,

$$(1 + j1)^5 = (\sqrt{2}\exp[j\pi/4])^5 = 4\sqrt{2}\exp[j5\pi/4] = -4 - j4$$

The nth root of a complex number z is the number w such that $w^n = z$ Thus, to find the nth root of z, we must solve the equation

$$w^n - |z|\exp[j\theta] = 0$$

which is of degree n and, therefore, has n roots. These roots are given by

$$\begin{aligned} w_1 &= |z|^{1/n}\exp\left[j\frac{\theta}{n}\right] \\ w_2 &= |z|^{1/n}\exp\left[j\frac{\theta + 2\pi}{n}\right] \\ w_3 &= |z|^{1/n}\exp\left[j\frac{\theta + 4\pi}{n}\right] \\ &\vdots \\ w_n &= |z|^{1/n}\exp\left[j\frac{\theta + 2(n-1)\pi}{n}\right] \end{aligned} \tag{A.19}$$

For example, the five roots of $32\exp[j\pi]$ are

$$w_1 = 2\exp\left[j\frac{\pi}{5}\right], \qquad w_2 = 2\exp\left[j\frac{3\pi}{5}\right], \qquad w_3 = 2\exp[j\pi],$$

$$w_4 = 2\exp\left[j\frac{7\pi}{5}\right], \qquad w_5 = 2\exp\left[j\frac{9\pi}{5}\right]$$

Notice that the roots of a complex number lie on a circle in the complex-number plane. The radius of the circle is $|z|^{1/n}$. The roots are uniformly distributed around the circle,

[2] Abraham De Moivre (1667-1754) is a French mathematician who introduced imaginary quantities in trigonometry and contributed to the theory of mathematical probability.

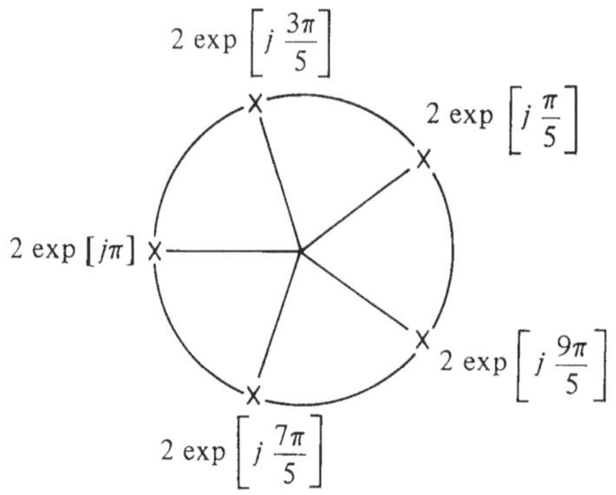

Figure A.3 Roots of 32 exp$[j\pi]$.

and the angle between adjacent roots is $2\pi/n$ radians. The five roots of 32 exp$[j\pi]$ are shown in Figure A.3.

A.4 INEQUALITIES

For complex numbers, we observe the important triangle inequality,

$$|z_1 + z_2| \leq |z_1| + |z_2| \tag{A.20}$$

That is, the magnitude of the sum of two complex numbers is at most equal to the sum of the magnitudes of the numbers. This inequality follows by noting that the points 0, z_1, and $z_1 + z_2$ are the vertices of the triangle shown in Figure A.4 with sides $|z_1|$, $|z_2|$, and $|z_1 + z_2|$, and the fact that one side in the triangle cannot exceed the sum of the other two sides. Other useful inequalities are

$$|\text{Re}\{z\}| \leq |z| \qquad \text{and} \qquad |\text{Im}\{z\}| \leq |z| \tag{A.21}$$

These follow by noting that for any $z = x + jy$, we have

$$|z| = \sqrt{x^2 + y^2} \geq |x|$$

and, similarly,

$$|z| \geq |y|$$

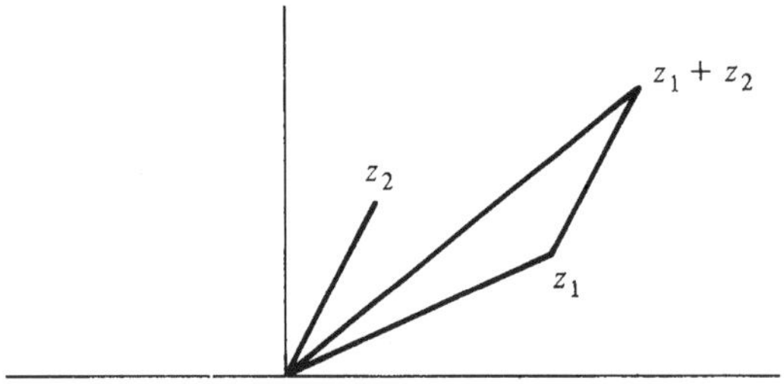

Figure A.4 Triangle inequality.

Appendix B

Mathematical Relations

Some of the mathematical relations encountered in electrical engineering are listed in this section for convenient reference. However, this appendix is not intended as a substitute for more comprehensive handbooks.

B.1 TRIGONOMETRIC IDENTITIES

$$\exp[\pm j\theta] = \cos\theta \pm j\sin\theta$$

$$\cos\theta = \tfrac{1}{2}(\exp[j\theta] + \exp[-j\theta]) = \sin\left(\theta + \frac{\pi}{2}\right)$$

$$\sin\theta = \frac{1}{2j}(\exp[j\theta] - \exp[-j\theta]) = \cos\left(\theta - \frac{\pi}{2}\right)$$

$$\sin^2\theta + \cos^2\theta = 1$$

$$\cos^2\theta - \sin^2\theta = \cos 2\theta$$

$$\cos^2\theta = \tfrac{1}{2}(1 + \cos 2\theta)$$

$$\sin^2\theta = \tfrac{1}{2}(1 - \cos 2\theta)$$

$$\cos^3\theta = \tfrac{1}{4}(3\cos\theta + \cos 3\theta)$$

$$\sin^3\theta = \tfrac{1}{4}(3\sin\theta - \sin 3\theta)$$

$$\sin(\alpha \pm \beta) = \sin\alpha\cos\beta \pm \sin\beta\cos\alpha$$

$$\cos(\alpha \pm \beta) = \cos\alpha\cos\beta \mp \sin\alpha\sin\beta$$

$$\tan(\alpha \pm \beta) = \frac{\tan\alpha \pm \tan\beta}{1 \mp \tan\alpha \tan\beta}$$

$$\sin\alpha \sin\beta = \tfrac{1}{2}[\cos(\alpha - \beta) - \cos(\alpha + \beta)]$$

$$\cos\alpha \cos\beta = \tfrac{1}{2}[\cos(\alpha - \beta) + \cos(\alpha + \beta)]$$

$$\sin\alpha \cos\beta = \tfrac{1}{2}[\sin(\alpha - \beta) + \sin(\alpha + \beta)]$$

$$\sin\alpha + \sin\beta = 2\sin\frac{\alpha + \beta}{2}\cos\frac{\alpha - \beta}{2}$$

$$\cos\alpha + \cos\beta = 2\cos\frac{\alpha + \beta}{2}\cos\frac{\alpha - \beta}{2}$$

$$\cos\alpha - \cos\beta = 2\sin\frac{\alpha + \beta}{2}\sin\frac{\beta - \alpha}{2}$$

$$A\cos\alpha + B\sin\alpha = \sqrt{A^2 + B^2}\cos\left(\alpha - \tan^{-1}\frac{B}{A}\right)$$

$$\sinh\alpha = \tfrac{1}{2}(\exp[\alpha] - \exp[-\alpha])$$

$$\cosh\alpha = \tfrac{1}{2}(\exp[\alpha] + \exp[-\alpha])$$

$$\tanh\alpha = \frac{\sinh\alpha}{\cosh\alpha}$$

$$\cosh^2\alpha - \sinh^2\alpha = 1$$

$$\cosh\alpha + \sinh\alpha = \exp[\alpha]$$

$$\cosh\alpha - \sinh\alpha = \exp[-\alpha]$$

$$\sinh(\alpha \pm \beta) = \sinh\alpha \cosh\beta \pm \cosh\alpha \sinh\beta$$

$$\cosh(\alpha \pm \beta) = \cosh\alpha \cosh\beta \pm \sinh\alpha \sinh\beta$$

$$\tanh(\alpha \pm \beta) = \frac{\tanh\alpha \pm \tanh\beta}{1 \pm \tanh\alpha \tanh\beta}$$

$$\sinh^2\alpha = \tfrac{1}{2}(\cosh 2\alpha - 1)$$

$$\cosh^2\alpha = \tfrac{1}{2}(\cosh 2\alpha + 1)$$

B.2 EXPONENTIAL AND LOGARITHMIC FUNCTIONS

$$\exp[\alpha]\exp[\beta] = \exp[\alpha + \beta]$$

$$\frac{\exp[\alpha]}{\exp[\beta]} = \exp[\alpha - \beta]$$

$$(\exp[\alpha])^{\beta} = \exp[\alpha\beta]$$

$\ln \alpha\beta = \ln \alpha + \ln \beta$

$\ln \dfrac{\alpha}{\beta} = \ln \alpha - \ln \beta$

$\ln \alpha^{\beta} = \beta \ln \alpha$

$\ln \alpha$ is the inverse of $\exp[\alpha]$; that is,

$$\exp[\ln \alpha] = \alpha \qquad \text{and} \qquad \exp[-\ln \alpha] = \exp\left[\ln\left(\frac{1}{\alpha}\right)\right] = 1/\alpha$$

$\log \alpha = M \ln \alpha, \qquad M = \log e \simeq 0.4343$

$\ln \alpha = \dfrac{1}{M} \log \alpha, \qquad \dfrac{1}{M} \simeq 2.3026$

$\log \alpha$ is the inverse of 10^{α}; that is,

$$10^{\log \alpha} = \alpha \qquad \text{and} \qquad 10^{-\log \alpha} = \frac{1}{\alpha}$$

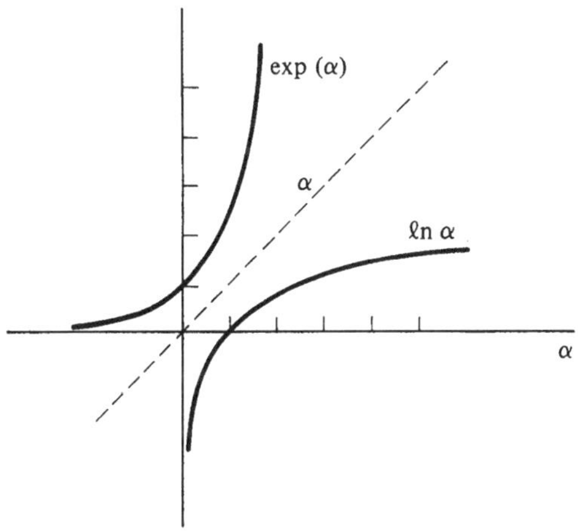

Figure B.1 Natural logarithmic and exponential functions.

B.3 SPECIAL FUNCTIONS

B.3.1 Gamma Functions

$\Gamma(\alpha) = \displaystyle\int_0^{\infty} t^{\alpha-1} \exp[-t]\,dt$

$\Gamma(\alpha + 1) = \alpha\Gamma(\alpha)$

$\Gamma(k + 1) = k!, \qquad k = 0, 1, 2, \ldots$

$\Gamma(\tfrac{1}{2}) = \sqrt{\pi}$

B.3.2 Incomplete Gamma Functions

$$I(\alpha, \beta) = \int_0^{\beta} t^{\alpha-1} \exp[-t]\,dt$$

$$Q(\alpha, \beta) = \int_{\beta}^{\infty} t^{\alpha-1} \exp[-t]\,dt$$

$$\Gamma(\alpha) = I(\alpha, \beta) + Q(\alpha, \beta)$$

B.3.3 Beta Functions

$$\beta(\mu, \nu) = \int_0^1 t^{\mu-1}(1-t)^{\nu-1}\,dt, \quad \mu > 0, \quad \nu > 0$$

$$\beta(\mu, \nu) = \frac{\Gamma(\mu)\Gamma(\nu)}{\Gamma(\mu + \nu)}$$

B.4 POWER-SERIES EXPANSION

$$(1+x)^n = 1 + nx + \frac{n(n-1)}{2!}x^2 + \cdots + \binom{n}{k}x^k + \cdots + x^n$$

$$\exp[x] = 1 + x + \frac{1}{2!}x^2 + \cdots + \frac{1}{n!}x^n + \cdots$$

$$\ln(1+x) = x - \frac{x^2}{2} + \frac{x^3}{3} - \cdots (-1)^{n+1}\frac{x^n}{n} + \cdots, \qquad |x| < 1$$

$$\sin x = x - \frac{x^3}{3!} + \frac{x^5}{5!} + \cdots + (-1)^n \frac{x^{2n+1}}{(2n+1)!} + \cdots$$

$$\cos x = 1 - \frac{x^2}{2!} + \frac{x^4}{4!} - \frac{x^6}{6!} + \cdots + (-1)^n \frac{x^{2n}}{(2n)!} + \cdots$$

$$\tan x = x + \frac{x^3}{3} + \frac{2}{15}x^5 + \cdots +$$

$$a^x = 1 + x\ln a + \frac{(x\ln a)^2}{2!} + \cdots + \frac{(x\ln a)^n}{n!} + \cdots$$

$$\sinh x = x + \frac{x^3}{3!} + \frac{x^5}{5!} + \cdots + \frac{x^{2n+1}}{(2n+1)!} + \cdots$$

$$\cosh x = 1 + \frac{x^2}{2!} + \frac{x^4}{4!} + \cdots + \frac{x^{2n}}{(2n)!} + \cdots$$

$$(1+x)^a = 1 + ax + \frac{a(a-1)}{2!}x^2 + \frac{a(a-1)(a-2)}{3!}x^3 + \cdots, \qquad |x| < 1$$

where a is negative or a fraction.

$$(1 + x)^{-1} = 1 - x + x^2 - x^3 + \cdots = \sum_{k=0}^{\infty} (-1)^{k-1} x^{k-1}$$

$$(1 + x)^{1/2} = 1 + \tfrac{1}{2}x - \tfrac{1}{2}\tfrac{1}{4}x^2 + \tfrac{1}{2}\tfrac{1}{4}\tfrac{3}{6}x^3 - \tfrac{1}{2}\tfrac{1}{4}\tfrac{3}{6}\tfrac{5}{8}x^4 + \cdots$$

B.5 SUMS OF POWERS OF NATURAL NUMBERS

$$\sum_{k=1}^{N} k = \frac{N(N+1)}{2}$$

$$\sum_{k=1}^{N} k^2 = \frac{N(N+1)(2N+1)}{6}$$

$$\sum_{k=1}^{N} k^3 = \frac{N^2(N+1)^2}{4}$$

$$\sum_{k=1}^{N} k^4 = \tfrac{1}{30}N(N+1)(2N+1)(3N^2+3N-1)$$

$$\sum_{k=1}^{N} k^5 = \tfrac{1}{12}N^2(N+1)^2(2N^2+2N-1)$$

$$\sum_{k=1}^{N} k^6 = \tfrac{1}{42}N(N+1)(2N+1)(3N^4+6N^3-3N+1)$$

$$\sum_{k=1}^{N} k^7 = \tfrac{1}{24}N^2(N+1)^2(3N^4+6N^3-N^2-4N+2)$$

$$\sum_{k=1}^{N} (2k-1) = N^2$$

$$\sum_{k=1}^{N} (2k-1)^2 = \tfrac{1}{3}N(4N^2-1)$$

$$\sum_{k=1}^{N} (2k-1)^3 = N^2(2N^2-1)$$

$$\sum_{k=1}^{N} k(k+1)^2 = \tfrac{1}{12}N(N+1)(N+2)(3N+5)$$

$$\sum_{k=1}^{N} k(k!) = (N+1)!-1$$

B.5.1 Sums of Binomial Coefficients

$$\sum_{k=0}^{m} \binom{n+k}{n} = \binom{n+m+1}{n+1}$$

$$1 + \binom{n}{2} + \binom{n}{4} + \cdots = 2^{n-1}$$

$$\binom{n}{1} + \binom{n}{3} + \binom{n}{5} + \cdots = 2^{n-1}$$

$$\sum_{k=0}^{N} (k+1)\binom{N}{k} = 2^{(N-1)}(n+2)$$

$$\sum_{k=0}^{N} (-1)^k \binom{N}{k} k^N = (-1)^N N! \qquad N \geq 0$$

$$\sum_{k=0}^{N} \binom{N}{k}^2 = \binom{2N}{N}$$

B.5.2 Series of Exponentials

$$\sum_{n=0}^{N} a^n = \frac{a^{N+1}-1}{a-1}, \qquad a \neq 1$$

$$\sum_{n=0}^{N} \exp\left[j\frac{2\pi kn}{N}\right] = \begin{cases} 0, & 1 \leq k \leq N-1 \\ N, & k = 0, N \end{cases}$$

$$\sum_{n=0}^{\infty} a^n = \frac{1}{1-a}, \qquad |a| < 1$$

$$\sum_{n=0}^{\infty} na^n = \frac{1}{(1-a)^2}, \qquad |a| < 1$$

$$\sum_{n=0}^{\infty} n^2 a^n = \frac{a^2+a}{(1-a)^3}, \qquad |a| < 1$$

B.6 DEFINITE INTEGRALS

$$\int_{-\infty}^{\infty} \exp[-\alpha x^2]\,dx = \sqrt{\frac{\pi}{\alpha}}$$

$$\int_{0}^{\infty} \exp[-\alpha x^2]\,dx = \frac{1}{2}\sqrt{\frac{\pi}{\alpha}}$$

$$\int_{0}^{\infty} x \exp[-\alpha x^2]\,dx = \frac{1}{2\alpha}$$

$$\int_0^\infty x^n \exp[-\alpha x]\,dx = \frac{n!}{\alpha^{n+1}}, \qquad \alpha > 0$$

$$\int_0^\infty x^2 \exp[-\alpha x^2]\,dx = \frac{1}{4\alpha^3}\sqrt{\pi}$$

$$\int_0^\infty \exp[-\alpha x]\cos\beta x\,dx = \frac{\alpha}{\beta^2+\alpha^2}, \qquad \alpha > 0$$

$$\int_0^\infty \exp[-\alpha x]\sin\beta x\,dx = \frac{\beta}{\beta^2+\alpha^2}, \qquad \alpha > 0$$

$$\int_0^\infty \exp[-\alpha x^2]\cos\beta x\,dx = \frac{1}{2\alpha}\sqrt{\pi}\exp\left[-\left(\frac{\beta}{2\alpha}\right)^2\right]$$

$$\int_0^\infty \frac{\sin\alpha x}{x}\,dx = \frac{\pi}{2}\operatorname{sgn}\alpha$$

$$\int_0^\infty \frac{1-\cos\alpha x}{x^2}\,dx = \frac{\alpha\pi}{2}, \qquad \alpha \ge 0$$

$$\int_0^\infty \frac{1-\cos\alpha x}{x(x-\beta)}\,dx = \pi\,\frac{\sin\alpha\beta}{\beta}, \qquad \alpha > 0, \qquad \beta \text{ real}, \qquad b \neq 0$$

$$\int_0^\infty \frac{\cos\alpha x - \cos\beta x}{x}\,dx = \ln\frac{\beta}{\alpha}, \qquad \alpha > 0, \qquad \beta > 0$$

$$\int_0^\infty \frac{\alpha\sin\beta x - \beta\sin\alpha x}{x^2}\,dx = \alpha\beta\ln\frac{\alpha}{\beta}, \qquad \alpha > 0, \qquad \beta > 0$$

$$\int_0^\infty \frac{\cos\alpha x - \cos\beta x}{x^2}\,dx = \frac{(\beta-\alpha)\pi}{2}, \qquad \alpha \ge 0, \qquad \beta \ge 0$$

$$\int_0^\infty \frac{\alpha\,dx}{\alpha^2+x^2} = \frac{\pi}{2}$$

$$\int_0^\pi \frac{\sin(2n-1)x}{\sin x}\,dx = \pi$$

$$\int_0^\pi \frac{\sin 2nx}{\sin x}\,dx = 0$$

$$\int_0^{\pi/2} \frac{\sin(2n-1)x}{\sin x}\,dx = \frac{\pi}{2}$$

$$\int_0^{\pi/2} \frac{\sin 2nx}{\sin x}\,dx = 1 - \frac{1}{3} + \frac{1}{5} - \cdots + \frac{(-1)^{n-1}}{2n-1}$$

$$\int_0^\pi \frac{\cos(2n+1)x}{\cos x}\,dx = (-1)^n\pi$$

$$\int_0^{\pi/2} \frac{\sin 2nx \cos x}{\sin x}\, dx = \frac{\pi}{2}$$

$$\int_0^{2\pi} (1 - \cos x)^n \sin nx \, dx = 0$$

$$\int_0^{2\pi} (1 - \cos x)^n \cos nx \, dx = (-1)^n \frac{\pi}{2^{n-1}}$$

$$\int_0^{\pi} \frac{\sin nx \cos mx}{\sin x}\, dx = \begin{cases} 0, & n \leq m \\ \pi, & n > m, \quad m + n \text{ odd} \\ 0, & n > m, \quad m + n \text{ even} \end{cases}$$

B.7 INDEFINITE INTEGRALS

$$\int u \, dv = uv - \int v \, du$$

$$\int x^n \, dx = \frac{1}{n+1} x^{n+1} + C, \qquad n \neq -1$$

$$\int \exp[x] \, dx = \exp[x] + C$$

$$\int x \exp[ax] \, dx = \frac{1}{a^2}(ax - 1)\exp[ax] + C$$

$$\int x^n \exp[ax] \, dx = \frac{1}{a} x^n \exp[ax] - \frac{n}{a} \int x^{n-1} \exp[ax] \, dx$$

$$\int \frac{dx}{x} = \ln|x| + C$$

$$\int \ln x \, dx = x \ln|x| - x + C$$

$$\int x^n \ln x \, dx = \frac{x^{n+1}}{(n+1)^2} [(n+1) \ln|x| - 1] + C$$

$$\int \frac{1}{x \ln x} dx = \ln|\ln x| + C$$

$$\int \cos x \, dx = \sin x + C$$

$$\int \sin x \, dx = -\cos x + C$$

$$\int \sec^2 x \, dx = \tan x + C$$

$$\int \csc^2 x \, dx = -\cot x + C$$

$$\int \tan x \, dx = \ln |\sec x| + C$$

$$\int \cot x \, dx = \ln |\sin x| + C$$

$$\int \sec x \, dx = \ln |\sec x + \tan x| + C$$

$$\int \csc x \, dx = \ln |\csc x - \cot x| + C$$

$$\int \sec x \tan x \, dx = \sec x + C$$

$$\int \csc x \cot x \, dx = -\csc x + C$$

$$\int \sin^2 x \, dx = \tfrac{1}{2}x - \tfrac{1}{4}\sin 2x + C$$

$$\int \cos^2 x \, dx = \tfrac{1}{2}x + \tfrac{1}{4}\sin 2x + C$$

$$\int \tan^2 x \, dx = \tan x - x + C$$

$$\int \cot^2 x \, dx = -\cot x - x + C$$

$$\int \sin^3 x \, dx = -\tfrac{1}{3}(2 + \sin^2 x) \cos x + C$$

$$\int \cos^3 x \, dx = \tfrac{1}{3}(2 + \cos^2 x) \sin x + C$$

$$\int \sin^n x \, dx = -\frac{1}{n}\sin^{n-1} x \cos x + \frac{n-1}{n}\int \sin^{n-2} x \, dx$$

$$\int \cos^n x \, dx = \frac{1}{n}\cos^{n-1} x \sin x + \frac{n-1}{n}\int \cos^{n-2} x \, dx$$

$$\int x \sin x \, dx = \sin x - x \cos x + C$$

$$\int x\cos x\,dx = \cos x + x\sin x + C$$

$$\int x^n \sin x\,dx = -x^n\cos x + n\int x^{n-1}\cos x\,dx$$

$$\int x^n \cos x\,dx = x^n \sin x - n\int x^{n-1}\sin x\,dx$$

$$\int \sinh x\,dx = \cosh x + C$$

$$\int \cosh x\,dx = \sinh x + C$$

$$\int \tanh x\,dx = \ln\cosh x + C$$

$$\int \coth x\,dx = \ln|\tanh x| + C$$

$$\int \operatorname{sech} x\,dx = \tan^{-1}|\sinh x| + C$$

$$\int \operatorname{csch} x\,dx = \ln|\tanh x| + C$$

$$\int \operatorname{sech}^2 x\,dx = \tanh x + C$$

$$\int \operatorname{csch}^2 x\,dx = -\coth x + C$$

$$\int \operatorname{sech} x\tanh x\,dx = -\operatorname{sech} x + C$$

$$\int \operatorname{csch} x\coth x\,dx = -\operatorname{csch} x + C$$

$$\int \frac{x\,dx}{a+bx} = \frac{1}{b^2}(a + bx - a\ln|x|) + C$$

$$\int \frac{x^2\,dx}{a+bx} = \frac{1}{2b^3}[(a+bx)^2 - 4a(a+bx) + 2a^2\ln|a+bx|] + C$$

$$\int \frac{dx}{x(a+bx)} = \frac{1}{a}\ln\left|\frac{x}{a+bx}\right| + C$$

$$\int \frac{dx}{\sqrt{x^2-a^2}} = \ln|x + \sqrt{x^2-a^2}| + C$$

$$\int \frac{dx}{x^2\sqrt{x^2 - a^2}} = \frac{\sqrt{x^2 - a^2}}{a^2 x} + C$$

$$\int \frac{dx}{(x^2 - a^2)^{3/2}} = -\frac{x}{a^2\sqrt{x^2 - a^2}} + C$$

$$\int \frac{dx}{a^2 + x^2} = \frac{1}{a}\tan^{-1}\frac{x}{a} + C$$

$$\int \frac{dx}{\sqrt{a^2 + x^2}} = \ln|x| + \sqrt{a^2 + x^2} + C$$

$$\int \frac{dx}{x\sqrt{a^2 + x^2}} = \frac{1}{a}\ln\left|\frac{\sqrt{a^2 + x^2} + a}{x}\right| + C$$

$$\int \frac{dx}{x^2\sqrt{a^2 + x^2}} = -\frac{\sqrt{a^2 + x^2}}{a^2 x} + C$$

$$\int \frac{dx}{(a^2 + x^2)^{3/2}} = \frac{x}{a^2\sqrt{a^2 + x^2}} + C$$

$$\int \frac{dx}{\sqrt{a^2 - x^2}} = \sin^{-1}\frac{x}{a} + C$$

$$\int \frac{dx}{\sqrt{2ax - x^2}} = \cos^{-1}\frac{a - x}{a} + C$$

$$\int \frac{x\,dx}{\sqrt{2ax - x^2}} = -\sqrt{2ax - x^2} + a\cos^{-1}\frac{a - x}{a} + C$$

$$\int \frac{dx}{x\sqrt{2ax - x^2}} = -\frac{\sqrt{2ax - x^2}}{ax} + C$$

$$\int \frac{dx}{x\sqrt{x^2 - a^2}} = \frac{1}{a}\sec^{-1}\frac{x}{a} + C$$

$$\int \sqrt{2ax - x^2}\,dx = \frac{x - a}{2}\sqrt{2ax - x^2} + \frac{a^2}{2}\cos^{-1}\frac{a - x}{a} + C$$

$$\int x\sqrt{2ax - x^2}\,dx = \frac{2x^2 - ax - 3a^2}{2}\sqrt{2ax - x^2} + \frac{a^2}{2}\cos^{-1}\frac{a - x}{a} + C$$

Appendix C

Elementary Matrix Theory

This appendix presents the minimum amount of matrix theory needed to comprehend the material in Chapters 2 and 6 in the text. It is recommended that even those well versed in matrix theory read the material herein to become familiar with the notation. For those not so well versed, the presentation is terse and oriented toward them. For a more comprehensive presentation of matrix theory, we suggest the study of textbooks solely concerned with the subject.

C.1 BASIC DEFINITION

A matrix, denoted by a capital boldface letter such as **A** or **Φ** or by the notation $[a_{ij}]$, is a rectangular array of elements. Such arrays ocur in various branches of applied mathematics. Matrices are useful because they enable us to consider an array of many numbers as a single object and to perform calculations on these objects in a compact form. Matrix elements can be real numbers, complex numbers, polynomials, or functions. A matrix that contains only one row is called a row matrix, and a matrix that contains only one column is called a column matrix. Square matrices have the same number of rows and columns. A matrix **A** is of order $m \times n$ (read m by n) if it has m rows and n columns.

The complex conjugate of a matrix **A** is obtained by conjugating every element in **A** and is denoted by $\mathbf{A}^*$. A matrix is real if all elements of the matrix are real. Clearly, for real matrices, $\mathbf{A}^* = \mathbf{A}$. Two matrices are equal if their corresponding elements are equal. $\mathbf{A} = \mathbf{B}$ means $[a_{ij}] = [b_{ij}]$, for all i and j. The matrices should be of the same order.

C.2 BASIC OPERATIONS

C.2.1 Matrix Addition

A matrix $\mathbf{C} = \mathbf{A} + \mathbf{B}$ is formed by adding corresponding elements; that is,

$$[c_{ij}] = [a_{ij}] + [b_{ij}] \tag{C.1}$$

Matrix subtraction is analogously defined. The matrices must be of the same order. Matrix addition is commutative and associative.

C.2.2 Differentiation and Integration

The derivative or integral of a matrix is obtained by differentiating or integrating each element of the matrix.

C.2.3 Matrix Multiplication

Matrix multiplication is an extension of the dot product of vectors. Recall that the dot product of the two N-dimensional vectors $\mathbf{u}$ and $\mathbf{v}$ is defined as

$$\mathbf{u} \cdot \mathbf{v} = \sum_{i=1}^{N} u_i v_i$$

Elements $[c_{ij}]$ of the product matrix $\mathbf{C} = \mathbf{AB}$ are found by taking the dot product of the ith row of the matrix $\mathbf{A}$ and the jth column of the matrix $\mathbf{B}$, so that

$$[c_{ij}] = \sum_{k=1}^{N} a_{ik} b_{kj} \tag{C.2}$$

The process of matrix multiplication is, therefore, conveniently referred to as the multiplication of rows into columns, as demonstrated in Figure C.1.

This definition requires that the number of columns of $\mathbf{A}$ be the same as the number of rows of $\mathbf{B}$. In that case, the matrices $\mathbf{A}$ and $\mathbf{B}$ are said to be compatible. Otherwise the product is undefined. Matrix multiplication is associative $[(\mathbf{AB})\mathbf{C} = \mathbf{A}(\mathbf{BC})]$, but not, in general, commutative $(\mathbf{AB} \neq \mathbf{BA})$. As an example, let

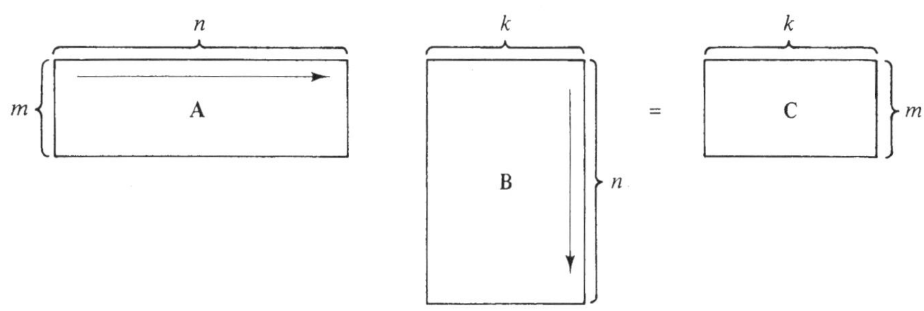

Figure C.1 Matrix multiplication.

$$\mathbf{A} = \begin{bmatrix} 3 & -2 \\ 1 & 5 \end{bmatrix} \quad \text{and} \quad \mathbf{B} = \begin{bmatrix} -4 & 1 \\ 1 & 6 \end{bmatrix}$$

Then, by Equation (C.2),

$$\mathbf{AB} = \begin{bmatrix} (3)(-4) + (-2)(1) & (3)(1) + (-2)(6) \\ (1)(-4) + (5)(1) & (1)(1) + (5)(6) \end{bmatrix} = \begin{bmatrix} -14 & -9 \\ 1 & 31 \end{bmatrix}$$

and

$$\mathbf{BA} = \begin{bmatrix} (-4)(3) + (1)(1) & (-4)(-2) + (1)(5) \\ (1)(3) + (6)(1) & (1)(-2) + (6)(5) \end{bmatrix} = \begin{bmatrix} -11 & -3 \\ 9 & 28 \end{bmatrix}$$

Matrix multiplication has the following properties:

$$(k\mathbf{A})\mathbf{B} = k(\mathbf{AB}) = \mathbf{A}(k\mathbf{B}) \tag{C.3a}$$

$$\mathbf{A}(\mathbf{BC}) = (\mathbf{AB})\mathbf{C} \tag{C.3b}$$

$$(\mathbf{A} + \mathbf{B})\mathbf{C} = \mathbf{AC} + \mathbf{BC} \tag{C.3c}$$

$$\mathbf{C}(\mathbf{A} + \mathbf{B}) = \mathbf{CA} + \mathbf{CB} \tag{C.3d}$$

$$\mathbf{AB} \neq \mathbf{BA}, \quad \text{in general} \tag{C.3e}$$

$$\mathbf{AB} = \mathbf{0} \quad \text{does not necessarily imply } \mathbf{A} = \mathbf{0} \quad \text{or} \quad \mathbf{B} = \mathbf{0} \tag{C.3f}$$

These properties hold, provided that **A**, **B**, and **C** are such that the expressions on the left are defined (k is any number). An example of (C.3f) is

$$\begin{bmatrix} 3 & 3 \\ 4 & 4 \end{bmatrix} \begin{bmatrix} -1 & 1 \\ 1 & -1 \end{bmatrix} = \begin{bmatrix} 0 & 0 \\ 0 & 0 \end{bmatrix}$$

The properties expressed by Equations (C.3e) and (C.3f) are quite unusual because they have no counterparts in the standard multiplication of numbers and should, therefore, be carefully observed. As with vectors, there is no matrix division.

C.3 SPECIAL MATRICES

Zero Matrix. The zero matrix, denoted by **0**, is a matrix whose elements are all zero.

Diagonal Matrix. The diagonal matrix, denoted by **D**, is a square matrix whose off-diagonal elements are all zeros.

Unit Matrix. The unit matrix, denoted by **I**, is a diagonal matrix whose diagonal elements are all ones. (*Note:* **AI** = **AI** = **A**, where **A** is any compatible matrix.)

Upper Triangular Matrix. The upper triangular matrix has all zeros below the main diagonal.

Lower Triangular Matrix. The lower triangular matrix has all zeros above the main diagonal.

In upper or lower triangular matrices, the diagonal elements need not be zero. An upper triangular matrix added to or multiplied by an upper triangular matrix results in an upper triangular matrix, and similarly for lower triangular matrices.

For example, matrices

$$\mathbf{T}_1 = \begin{bmatrix} 1 & 4 & 2 \\ 0 & 3 & -2 \\ 0 & 0 & 5 \end{bmatrix} \quad \text{and} \quad \mathbf{T}_2 = \begin{bmatrix} 1 & 0 & 0 \\ 2 & 4 & 0 \\ -3 & 0 & 7 \end{bmatrix}$$

are upper and lower triangular matrices, respectively.

Transpose Matrix. The transpose matrix, denoted by $\mathbf{A}^T$, is the matrix resulting from an interchange of rows and columns of a given matrix $\mathbf{A}$. If $\mathbf{A} = [a_{ij}]$, then $\mathbf{A}^T = [a_{ji}]$, so that the element in the ith row and the jth column of $\mathbf{A}$ becomes the element in the jth row and ith column of $\mathbf{A}^T$.

Complex Conjugate Transpose Matrix. The complex conjugate transpose matrix, denoted by $\mathbf{A}^\dagger$, is the matrix whose elements are complex conjugates of the elements of $\mathbf{A}^T$. Note that

$$(\mathbf{AB})^T = \mathbf{B}^T\mathbf{A}^T \quad \text{and} \quad (\mathbf{AB})^\dagger = \mathbf{B}^\dagger\mathbf{a}^\dagger$$

The following definitions apply to square matrices:

Symmetric Matrix. Matrix $\mathbf{A}$ is symmetric if

$$\mathbf{A} = \mathbf{A}^T$$

Hermitian Matrix. Matrix $\mathbf{A}$ is Hermitian if

$$\mathbf{A} = \mathbf{A}^\dagger$$

Skew-Symmetric Matrix. Matrix $\mathbf{A}$ is skew symmetric if

$$\mathbf{A} = -\mathbf{A}^T$$

For example, the matrices

$$\mathbf{A} = \begin{bmatrix} -1 & 2 & 4 \\ 2 & 5 & -3 \\ 4 & -3 & 6 \end{bmatrix} \quad \text{and} \quad \mathbf{B} = \begin{bmatrix} 0 & -3 & 4 \\ 3 & 0 & -7 \\ -4 & 7 & 0 \end{bmatrix}$$

are symmetric and skew-symmetric matrices, respectively.

Normal Matrix. Matrix $\mathbf{A}$ is normal if

$$\mathbf{A}^\dagger\mathbf{A} = \mathbf{A}\mathbf{A}^\dagger$$

Unitary Matrix. Matrix $\mathbf{A}$ is unitary if

$$\mathbf{A}^\dagger\mathbf{A} = \mathbf{I}$$

A real unitary matrix is called an orthogonal matrix.

C.4 THE INVERSE OF A MATRIX

In this section, we consider exclusively square matrices. The inverse of an $n \times n$ matrix $\mathbf{A}$ is denoted by $\mathbf{A}^{-1}$ and is an $n \times n$ matrix such that

$$\mathbf{A}\mathbf{A}^{-1} = \mathbf{A}^{-1}\mathbf{A} = \mathbf{I}$$

where $\mathbf{I}$ is the $n \times n$ unit matrix. If the determinant of $\mathbf{A}$ is zero, then $\mathbf{A}$ has no inverse and is called singular; on the other hand, if the determinant is nonzero, the inverse exists, and $\mathbf{A}$ is called a nonsingular matrix.

In general, finding the inverse of a matrix is a tedious process. For some special cases, the inverse is easily determined. For a 2×2 matrix

$$\mathbf{A} = \begin{bmatrix} a_{11} & a_{12} \\ a_{21} & a_{22} \end{bmatrix}$$

we have

$$\mathbf{A}^{-1} = \frac{1}{(a_{11}a_{22} - a_{12}a_{21})} \begin{bmatrix} a_{22} & -a_{12} \\ -a_{21} & a_{11} \end{bmatrix} \tag{C.4}$$

provided that $a_{11}a_{22} \neq a_{12}a_{21}$. For a diagonal matrix, we have

$$A = \begin{bmatrix} a_{11} & \cdot & \cdots & 0 \\ 0 & a_{22} & \cdots & 0 \\ \cdot & \cdot & \cdots & \cdot \\ \cdot & \cdot & \cdots & \cdot \\ \cdot & \cdot & \cdots & \cdot \\ 0 & \cdot & \cdots & a_{nn} \end{bmatrix}, \qquad A^{-1} = \begin{bmatrix} \frac{1}{a_{11}} & \cdot & \cdots & 0 \\ 0 & \frac{1}{a_{22}} & \cdots & 0 \\ \cdot & \cdot & \cdots & \cdot \\ \cdot & \cdot & \cdots & \cdot \\ \cdot & \cdot & \cdots & \cdot \\ 0 & \cdot & \cdots & \frac{1}{a_{nn}} \end{bmatrix} \tag{C.5}$$

provided that $a_{ij} \neq 0$ for any i.

The inverse of the inverse is the given matrix $\mathbf{A}$; that is,

$$(\mathbf{A}^{-1})^{-1} = \mathbf{A} \tag{C.6}$$

The inverse of a product $\mathbf{AC}$ can be obtained by inverting each factor and multiplying the results in reverse order:

$$(\mathbf{AC})^{-1} = \mathbf{C}^{-1}\mathbf{A}^{-1} \tag{C.7}$$

For higher order matrices, the inverse is computed using Cramer's rule:

$$\mathbf{A}^{-1} = \frac{1}{\det \mathbf{A}} \operatorname{adj} \mathbf{A} \tag{C.8}$$

Here, $\det \mathbf{A}$ is the determinant of $\mathbf{A}$, and $\operatorname{adj} \mathbf{A}$ is the adjoint matrix of $\mathbf{A}$. The following is a summary of the steps needed to calculate the inverse of an $n \times n$ square matrix $\mathbf{A}$:

1. Calculate the matrix of minors. (A minor of the element a_{ij}, denoted by det $\mathbf{M}_{ij}$, is the determinant of the matrix formed by deleting the ith row and the jth column of the matrix **A**.)
2. Calculate the matrix of cofactors. (A cofactor of the element a_{ij}, denoted by c_{ij}, is related to the minor by $c_{ij} = (-1)^{i+j} \det \mathbf{M}_{ij}$.)
3. Calculate the adjoint matrix of **A** by transposing the matrix of cofactors of **A**:

$$\text{adj } \mathbf{A} = [c_{ij}]^T$$

4. Calculate the determinant of **A** using

$$\det \mathbf{A} = \sum_{i=1}^{n} a_{ij} c_{ij}, \qquad \text{for any column } j$$

or

$$\det \mathbf{A} = \sum_{j=1}^{n} a_{ij} c_{ij}, \qquad \text{for any row } i$$

5. Use Equation (C.8) to calculate $\mathbf{A}^{-1}$.

C.5 EIGENVALUES AND EIGENVECTORS

The eigenvalues of an $n \times n$ matrix A are the solutions to the equation

$$\mathbf{Ax} = \lambda \mathbf{x} \tag{C.9}$$

Equation (C.9) can be written as $(\mathbf{A} - \lambda\mathbf{I})\mathbf{x} = \mathbf{0}$. Nontrivial solution vectors **x** exist only if $\det(\mathbf{A} - \lambda\mathbf{I}) = 0$. This is an algebraic equation of degree n in λ and is called the characteristic equation of the matrix. There are n roots for this equation, although some may be repeated. An eigenvalue of the matrix **A** is said to be distinct if it is not a repeated root of the characteristic equation. The polynomial $g(\lambda) = \det[\mathbf{A} - \lambda\mathbf{I}]$ is called the characteristic polynomial of **A**. Associated with each eigenvalue λ_i, is a nonzero vector $\mathbf{x}_i$ of the eigenvalue equation $\mathbf{x}_i = \lambda_i \mathbf{x}_i$. This solution vector is called an eigenvector. For example, the eigenvalues of the matrix

$$\mathbf{A} = \begin{bmatrix} 3 & 4 \\ 1 & 3 \end{bmatrix}$$

are obtained by solving the equation

$$\det \begin{bmatrix} 3-\lambda & 4 \\ 1 & 3-\lambda \end{bmatrix} = 0$$

or

$$(3-\lambda)(3-\lambda) - 4 = 0$$

This second-degree equation has two real roots, $\lambda_1 = 1$ and $\lambda_2 = 5$. There are two cigenvectors. The eigenvector associated with $\lambda_1 = 1$ is the solution to

$$\begin{bmatrix} 3 & 4 \\ 1 & 3 \end{bmatrix} \begin{bmatrix} x_1 \\ x_2 \end{bmatrix} = \begin{bmatrix} x_1 \\ x_2 \end{bmatrix}$$

or

$$\begin{bmatrix} 2 & 4 \\ 1 & 2 \end{bmatrix} \begin{bmatrix} x_1 \\ x_2 \end{bmatrix} = \begin{bmatrix} 0 \\ 0 \end{bmatrix}$$

Then $2x_1 + 4x_2 = 0$ and $x_1 + 2x_2 = 0$, from which it follows that $x_1 = -2x_2$. By choosing $x_2 = 1$, we find that the eigenvector is

$$\mathbf{x}_1 = \begin{bmatrix} -2 \\ 1 \end{bmatrix}$$

The eigenvector associated with $\lambda_2 = 5$ is the solution to

$$\begin{bmatrix} 3 & 4 \\ 1 & 3 \end{bmatrix} \begin{bmatrix} x_1 \\ x_2 \end{bmatrix} = 5 \begin{bmatrix} x_1 \\ x_2 \end{bmatrix}$$

or

$$\begin{bmatrix} -2 & 4 \\ 1 & -2 \end{bmatrix} \begin{bmatrix} x_1 \\ x_2 \end{bmatrix} = 0$$

which has the solution $x_2 = 2x_1$. Choosing $x_1 = 1$ gives

$$\mathbf{x}_2 = \begin{bmatrix} 1 \\ 2 \end{bmatrix}$$

C.6 FUNCTIONS OF A MATRIX

Any analytic scalar function $f(t)$ of a scalar t can be uniquely expressed in a convergent Maclaurin series as

$$f(t) = \sum_{k=0}^{\infty} \left\{ \frac{d^k}{dt^k} f(t) \right\}_{t=0} \frac{t^k}{k!}$$

The same type of expansion can be used to define functions of matrices. Thus, the function $f(\mathbf{A})$ of the $n \times n$ matrix $\mathbf{A}$ can be expanded as

$$f(\mathbf{A}) = \sum_{k=0}^{\infty} \left\{ \frac{d^k}{dt^k} f(t) \right\}_{t=0} \frac{\mathbf{A}^k}{k!} \tag{C.10}$$

For example,

$$\begin{aligned} \sin \mathbf{A} &= (\sin 0)\mathbf{I} + (\cos 0)\mathbf{A} + (-\sin 0)\frac{\mathbf{A}^2}{2!} \\ &\quad + \cdots + (-\cos 0)^n \frac{\mathbf{A}^{2n+1}}{(2n+1)!} + \cdots \\ &= \mathbf{A} - \frac{\mathbf{A}^3}{3!} + \frac{\mathbf{A}^5}{5!} - \cdots (-1)^n \frac{\mathbf{A}^{2n}+1}{(2n+1)!} + \cdots \end{aligned}$$

and

$$\exp[\mathbf{A}t] = \exp[0]\mathbf{I} + \exp[0]\mathbf{A}t + \exp[0]\frac{\mathbf{A}^2t^2}{2!} + \cdots + \exp[0]\frac{\mathbf{A}^nt^n}{n!} + \cdots$$
$$= \mathbf{I} + \mathbf{A}t + \frac{\mathbf{A}^2t^2}{2!} + \cdots + \frac{\mathbf{A}^nt^n}{n!} + \cdots$$

The Cayley-Hamilton (C-H) theorem states that any matrix satisfies its own characteristic equation. That is, given an arbitrary $n \times n$ matrix $\mathbf{A}$ with characteristic polynomial $g(\lambda) = \det(\mathbf{A} - \lambda\mathbf{I})$, it follows that $g(\mathbf{A}) = \mathbf{0}$. As an example, if

$$\mathbf{A} = \begin{bmatrix} 3 & 4 \\ 1 & 3 \end{bmatrix}$$

so that

$$\det[\mathbf{A} - \lambda\mathbf{I}] = g(\lambda) = \lambda^2 - 6\lambda + 5$$

then, by the Cayley-Hamilton theorem, we have

$$g(\mathbf{A}) = \mathbf{A}^2 - 6\mathbf{A} + 5\mathbf{I} = 0$$

or

$$\mathbf{A}^2 = 6\mathbf{A} - 5\mathbf{I} \tag{C.11}$$

In general, the Cayley-Hamilton theorem enables us to express any power of a matrix in terms of a linear combination of $\mathbf{A}^k$ for $k = 0, 1, 2, \ldots, n - 1$. For example, $\mathbf{A}^3$ can be found from Equation (C.11) by multiplying both sides by $\mathbf{A}$ to obtain

$$\mathbf{A}^3 = 6\mathbf{A}^2 - 5\mathbf{A}$$
$$= 6[6\mathbf{A} - 5\mathbf{I}] - 5\mathbf{A}$$
$$= 31\mathbf{A} - 30\mathbf{I}$$

Similarly, higher powers of $\mathbf{A}$ can be obtained by this method. Multiplying Equation (C.11) by $\mathbf{A}^{-1}$, we obtain

$$\mathbf{A}^{-1} = \frac{6\mathbf{I} - \mathbf{A}}{5}$$

assuming that $\mathbf{A}^{-1}$ exists. As a consequence of the C-H theorem, it follows that any function $f(\mathbf{A})$ can be expressed as

$$f(\mathbf{A}) = \sum_{k=0}^{n-1} \gamma_k \mathbf{A}^k$$

The calculation of $\gamma_0, \gamma_1, \ldots, \gamma_{n-1}$ can be carried out by the iterative method used in the calculation of $\mathbf{A}^n$ and $\mathbf{A}^{n+1}$. It can be shown that if the eigenvalues of $\mathbf{A}$ are distinct, then the set of coefficients $\gamma_0, \gamma_1, \ldots, \gamma_{n-1}$ satisfies the following equations:

$$
\begin{aligned}
f(\lambda_1) &= \gamma_0 + \gamma_1\lambda_1 + \cdots + \gamma_{n-1}\lambda_1^{n-1} \\
f(\lambda_2) &= \gamma_0 + \gamma_1\lambda_2 + \cdots + \gamma_{n-1}\lambda_2^{n-1} \\
&\vdots \\
f(\lambda_n) &= \gamma_0 + \gamma_1\lambda_n + \cdots + \gamma_{n-1}\lambda_n^{n-1}
\end{aligned}
$$

As an example, let us calculate $\exp[\mathbf{A}t]$, where

$$\mathbf{A} = \begin{bmatrix} 3 & 4 \\ 1 & 3 \end{bmatrix}$$

The eigenvalues of $\mathbf{A}$ are $\lambda_1 = 1$ and $\lambda_2 = 5$, with $f(\mathbf{A}) = \exp[\mathbf{A}t]$. Then

$$
\begin{aligned}
\exp[\mathbf{A}t] &= \sum_{k=0}^{1} \gamma_k(t)\mathbf{A}^k t^k \\
&= \gamma_0(t)\mathbf{I} + \gamma_1(t)\mathbf{A}t
\end{aligned}
$$

where $\gamma_0(t)$ and $\gamma_1(t)$ are the solutions to

$$
\begin{aligned}
\exp[t] &= \gamma_0(t) + \gamma_1(t) \\
\exp[5t] &= \gamma_0(t) + 5\gamma_1(t)
\end{aligned}
$$

so that

$$
\begin{aligned}
\gamma_1(t) &= \tfrac{1}{4}(\exp[5t] - \exp[t]) \\
\gamma_0(t) &= \tfrac{1}{4}(5\exp[t] - \exp[5t])
\end{aligned}
$$

Therefore,

$$
\begin{aligned}
\exp[\mathbf{A}t] &= (\tfrac{5}{4}\exp[t] - \tfrac{1}{4}\exp[5t])\begin{bmatrix} 1 & 0 \\ 0 & 1 \end{bmatrix} + (\tfrac{1}{4}\exp[5t] - \exp[t])\begin{bmatrix} 3 & 4 \\ 1 & 3 \end{bmatrix} \\
&= \begin{bmatrix} \dfrac{1}{2}\exp[5t] - \exp[t] & \exp[5t] - \exp[t] \\ \dfrac{1}{4}(\exp[5t] - \exp[t]) & \dfrac{1}{2}\exp[5t] - \exp[t] \end{bmatrix}
\end{aligned}
$$

If the eigenvalues are not distinct, then we have fewer equations than unknowns. By differentiating the equation corresponding to the repeated eigenvalue with resepect to λ, we obtain a new equation that can be used to solve for $\gamma_0(t), \gamma_1(t), \ldots, \gamma_{n-1}(t)$. For example, consider

$$\mathbf{A} = \begin{bmatrix} -1 & 0 & 0 \\ 0 & -4 & 4 \\ 0 & -1 & 0 \end{bmatrix}$$

This matrix has eigenvalues $\lambda_1 = -1$ and $\lambda_2 = \lambda_3 = -2$. The coefficients $\gamma_0(t)$, $\gamma_1(t)$, and $\gamma_2(t)$ are obtained as the solution to the following set of equations:

$$\begin{aligned}
\exp[-t] &= \gamma_0(t) - \gamma_1(t) + \gamma_2(t) \\
\exp[-2t] &= \gamma_0(t) - 2\gamma_1(t) + 4\gamma_2(t) \\
t\exp[-2t] &= \gamma_1(t) - 4\gamma_2(t)
\end{aligned}$$

Solving for γ_i yields

$$\begin{aligned}
\gamma_0(t) &= 4\exp[-t] - 3\exp[-2t] - 2t\exp[-2t] \\
\gamma_1(t) &= 4\exp[-t] - 4\exp[-2t] - 3t\exp[-2t] \\
\gamma_2(t) &= \exp[-t] - \exp[-2t] - t\exp[-2t]
\end{aligned}$$

Thus,

$$\begin{aligned}
\exp[\mathbf{A}t] &= \gamma_0(t)\begin{bmatrix} 1 & 0 & 0 \\ 0 & 1 & 0 \\ 0 & 0 & 0 \end{bmatrix} + \gamma_1(t)\begin{bmatrix} -1 & 0 & 0 \\ 0 & -4 & 4 \\ 0 & -1 & 0 \end{bmatrix} + \gamma_2(t)\begin{bmatrix} 1 & 0 & 0 \\ 0 & 12 & -16 \\ 0 & 4 & 4 \end{bmatrix} \\
&= \begin{bmatrix} \exp[-t] & 0 & 0 \\ 0 & \exp[-2t] - 2t\exp[-2t] & 4t\exp[-2t] \\ 0 & -t\exp[-2t] & -4\exp[-t] + 4\exp[-2t] + 4t\exp[-2t] \end{bmatrix}
\end{aligned}$$

Appendix D

Partial Fractions

Expansion in partial fractions is a technique used to reduce proper rational functions[1] of the form $N(s)/D(s)$ into sums of simple terms. Each term by itself is a proper rational function with denominator of degree 2 or less. More specifically, if $N(s)$ and $D(s)$ are polynomials, and the degree of $N(s)$ is less than the degree of $D(s)$, then it follows from the theorem of algebra that

$$\frac{N(s)}{D(s)} = T_1 + T_2 + \cdots + T_k \tag{D.1}$$

where each T_i has one of the forms

$$\frac{A}{(s+b)^{\mu}} \quad \text{or} \quad \frac{Bs+C}{(s^2+ps+q)^{\nu}}$$

where the polynomial $s^2 + ps + q$ is irreducible, and μ and ν are nonnegative integers. The sum in the right-hand side of Equation (D.1) is called the partial-fraction decomposition of $N(s)/D(s)$, and each T_i is called a partial fraction. By using long division, improper rational functions can be written as a sum of a polynomial of degree $M - N$ and a proper rational function, where M is the degree of the polynomial $N(s)$ and N is the degree of the polynomial $D(s)$. For example, given

$$\frac{s^4 + 3s^3 - 5s^2 - 1}{s^3 + 2s^2 - s + 1}$$

we obtain, by long division,

[1] A proper rational function is a ratio of two polynomials, with the degree of the numerator less than the degree of the denominator.

$$\frac{s^4 + 3s^3 - 5s^2 - 1}{s^3 + 2s^2 - s + 1} = s + 1 - \frac{6s^2 - 2s - 2}{s^3 + 2s^2 - s + 1}$$

The partial-fraction decomposition is then found for $(6s^2 - 2s - 2)/(s^3 + 2s^2 - s + 1)$.

Partial fractions are very useful in integration and also in finding the inverse of many transforms, such as Laplace, Fourier, and Z-transforms. All these operators share one property in common: linearity.

The first step in the partial-fraction technique is to express $D(s)$ as a product of factors $s + b$ or irreducible quadratic factors $s^2 + ps + q$. Repeated factors are then collected, so that $D(s)$ is a product of different factors of the form $(s + b)^{\mu}$ or $(s^2 + ps + q)^{\nu}$, where μ and ν are nonnegative integers. The form of the partial fractions depends on the type of factors we have for $D(s)$. There are four different cases.

D.1 CASE 1: NONREPEATED LINEAR FACTORS

To every nonrepeated factor $s + b$ of $D(s)$, there corresponds a partial fraction $A/(s + b)$. In general, the rational function can be written as

$$\frac{N(s)}{D(s)} = \frac{A}{s + b} + R(s)$$

where

$$A = \left(\frac{(s + b)N(s)}{D(s)}\right)_{s=-b} \tag{D.2}$$

Example D.1

Consider the rational function

$$\frac{37 - 11s}{s^3 - 4s^2 + s + 6}$$

The denominator has the factored form $(s + 1)(s - 2)(s - 3)$. All these factors are linear nonrepeated factors. Thus, for the factor $s + 1$, there corresponds a partial fraction of the form $A/(s + 1)$. Similarly, for the factors $s - 1$ and $s - 3$, there correspond partial fractions $B/(s - 2)$ and $C/(s - 3)$, respectively. The decomposition of Equation (D.1) then has the form

$$\frac{37 - 11s}{s^3 - 4s^2 + s + 6} = \frac{A}{s + 1} + \frac{B}{s - 2} + \frac{C}{s - 3}$$

The values of A, B, and C are obtained using Equation (D.2):

$$A = \left(\frac{37 - 11s}{(s - 2)(s - 3)}\right)_{s=-1} = 4$$

$$B = \left(\frac{37 - 11s}{(s + 1)(s - 3)}\right)_{s=2} = -5$$

$$C = \left(\frac{37 - 11s}{(s+1)(s-2)}\right)_{s=3} = 1$$

The partial-fraction decomposition is, therefore,

$$\frac{37 - 11s}{s^3 - 4s^2 + s + 6} = \frac{4}{s+1} - \frac{5}{s-2} + \frac{1}{s-3}$$

Example D.2

Let us find the partial-fraction decomposition of

$$\frac{2s + 1}{s^3 + 3s^2 - 4s}$$

We factor the polynomial as

$$D(s) = s^3 + 3s^2 - 4s = s(s+4)(s-1)$$

and then use the partial-fraction form

$$\frac{2s + 1}{s^3 + 3s^2 - 4s} = \frac{A}{s} + \frac{B}{s+4} + \frac{C}{s-1}$$

Using Equation (D.2), we find that the coefficients are

$$A = \left(\frac{2s + 1}{(s+4)(s-1)}\right)_{s=0} = -\frac{1}{4}$$

$$B = \left(\frac{2s + 1}{s(s-1)}\right)_{s=-4} = \frac{7}{20}$$

$$C = \left(\frac{2s + 1}{s(s+4)}\right)_{s=1} = \frac{3}{5}$$

The partial-fraction decomposition is, therefore,

$$\frac{2s + 1}{s^3 + 3s^2 - 4s} = -\frac{1}{4s} + \frac{7}{20(s+4)} + \frac{3}{5(s-1)}$$

D.2 CASE II: REPEATED LINEAR FACTORS

To each repeated factor $(s + b)^{\mu}$, there corresponds the partial fraction

$$\frac{A_1}{s+b} + \frac{A_2}{(s+b)^2} + \dots + \frac{A_{\mu}}{(s+b)^{\mu}}$$

The coefficients A_k can be determined by the formula

$$A_{\mu} = \left(\frac{(s+b)^{\mu} N(s)}{D(s)}\right)_{s=-b} \tag{D.3}$$

$$A_k = \left(\frac{1}{(\mu - k)!} \frac{d^{\mu-k}}{ds^{\mu-k}} \frac{(s+b)^{\mu} N(s)}{D(s)} \right)_{s=-b,} \quad k = 1, 2, \dots, \mu - 1 \tag{D.4}$$

Example D.3

Consider the rational function

$$\frac{2s^2 - 25s - 33}{s^3 - 3s^2 - 9s - 5}$$

The denominator has the factored form $D(s) = (s+1)^2(s-5)$. For the factor $s - 5$, there corresponds a partial fraction of the form $B/(s-5)$. The factor $(s+1)^2$ is a linear repeated factor to which there corresponds a partial fraction of the form $A_2/(s+1)^2 + A_1/(s+1)$ The decomposition of Equation (D.1) then has the form

$$\frac{2s^2 - 25s - 33}{s^3 - 3s^2 - 9s - 5} = \frac{B}{s-5} + \frac{A_1}{s+1} + \frac{A_2}{(s+1)^2} \tag{D.5}$$

The values of B, A_1, and A_2 are obtained using Equations (D.2), (D.3), and (D.4) as follows:

$$B = \left(\frac{2s^2 - 25s - 33}{s^2 + 2s + 1} \right)_{s=-5} = -3$$

$$A_2 = \left(\frac{2s^2 - 25s - 33}{s-5} \right)_{s=1} = -1$$

$$A_1 = \frac{1}{(2-1)!} \left(\frac{d}{ds} \frac{2s^2 - 25s - 33}{s-5} \right)_{s=1}$$

$$= \left(\frac{2s^2 - 20s + 158}{(s-5)^2} \right)_{s=1} = 5$$

Hence, the rational function in Equation (D.5) can be written as

$$\frac{2s^2 - 25s - 33}{s^3 - 3s^2 - 9s - 5} = -\frac{3}{s-5} + \frac{5}{s+1} - \frac{1}{(s+1)^2}$$

Example D.4

Let us find the partial-fraction decomposition of

$$\frac{3s^3 - 18s^2 + 9s - 4}{s^4 - 5s^3 + 6s^2 + 20s + 8}$$

The denominator can be factored as $(s+1)(s-2)^3$. Since we have a repeated factor of order 3, the corresponding partial fraction is

$$\frac{3s^3 - 18s^2 + 9s - 4}{s^4 - 5s^3 + 6s^2 + 20s + 8} = \frac{B}{s+1} + \frac{A_1}{s-2} + \frac{A_2}{(s-2)^2} + \frac{A_3}{(s-2)^3} \tag{D.6}$$

The coefficient B can be obtained using Equation (D.2):

$$B = \left(\frac{3s^3 - 18s^2 + 9s - 4}{(s-2)s^3} \right\}_{s=-1} = 2$$

The coefficients A_i, $i = 1, 2, 3$, are obtained using Equations (D.3) and (D.4). First,

$$A_3 = \left(\frac{3s^3 - 18s^2 + 9s - 4}{s + 1}\right)_{s=2} = 2$$

$$A_2 = \left(\frac{d}{ds}\frac{3s^3 - 18s^2 + 29s - 4}{s + 1}\right)_{s=2}$$

$$= \left(\frac{6s^3 - 9s^2 - 36s + 33}{(s + 1)^2}\right)_{s=2} = -3$$

Similarly, A_1 can be found using Equation (D.4).

In many cases, it is much easier to use the following technique, especially after finding all but one coefficient: Multiplying both sides of Equation (D.6) by $(s + 1)(s - 2)^3$ gives

$$3s^3 - 18s^2 + 9s - 4 = B(s - 2)^3 + A_1(s + 1)(s - 2)^2 + A_2(s + 1)(s - 2) + A_3(s + 1)$$

If we compare the coefficient of s^3 on both sides, we obtain

$$3 = B + A_1$$

Since $B = 2$, it follows that $A_1 = 1$. The resulting partial-fraction decomposition is then

$$\frac{3s^3 - 18s^2 + 9s - 4}{s^4 - 5s^3 + 6s^2 + 20s + 8} = \frac{2}{s + 1} + \frac{1}{s - 2} - \frac{3}{(s - 2)^2} + \frac{2}{(s - 2)^3}$$

D.3 CASE III: NONREPEATED IRREDUCIBLE SECOND-DEGREE FACTORS

In the case of a nonrepeated irreducible second-degree polynomial, we set up fractions of the form

$$\frac{As + B}{(s^2 + ps + q)} \tag{D.7}$$

The best way to find the coefficients is to equate the coefficients of different powers of s, as is demonstrated in the following example.

Example D.5

Consider the rational function

$$\frac{s^2 - s - 21}{2s^3 - s^2 + 8s - 4}$$

We factor the polynomial as $D(s) = (s^2 + 4)(2s - 1)$ and use the partial-fraction form:

$$\frac{s^2 - s - 21}{2s^3 - s^2 + 8s - 4} = \frac{As + B}{s^2 + 4} + \frac{C}{2s - 1}$$

Multiplying by the lowest common denominator gives

$$s^2 - s - 21 = (As + B)(2s - 1) + C(s^2 + 4) \tag{D.8}$$

The values of A, B, and C can be found by comparing the coefficients of different powers of s or by substituting values for s that make various factors zero. For example, substituting $s = 1/2$, we obtain $(1/4) - (1/2) - 21 = (17/4)C$, which has the solution $C = -5$. The remaining coefficients can be found by comparing different powers of s. Rearranging the right-hand side of Equation (D.8) gives

$$s^2 - s - 21 = (2A + C)s^2 + (-A + B)s - B + 4C$$

Comparing the coefficient of s^2 on both sides, we see that $2A + C = 1$. Knowing C results in $A = 3$. Similarly, comparing the constant terms yields $-B + 4C = -21$, or $B = 1$. Thus, the partial-fraction decomposition of the rational function is

$$\frac{s^2 - s - 21}{2s^3 - s^2 + 8s - 4} = \frac{3s + 1}{s^2 + 4} - \frac{5}{2s - 1}$$

D.4 CASE IV: REPEATED IRREDUCIBLE SECOND-DEGREE FACTORS

For repeated irreducible second-degree factors, we have factors of the form

$$\frac{A_1 s + B_1}{s^2 + ps + q} + \frac{A_2 s + B_2}{(s^2 + ps + q)^2} + \dots + \frac{A_\nu s + B_\nu}{(s^2 + ps + q)^\nu} \tag{D.9}$$

Again, the best way to find the coefficients is to equate the different powers of s.

Example D.6

As an example of repeated irreducible second-degree factors, consider

$$\frac{s^4 - 6s + 7}{(s^2 - 4s + 5)^2} \tag{D.10}$$

Note that the denominator can be written as $[(s - 2)^2 + 1]^2$. Therefore, applying Equation (D.9) with $\nu = 2$, we can write the partial fractions for Equation (D.10) as

$$\frac{s^4 - 6s + 7}{(s^2 - 4s + 5)^2} = \frac{A_2 s + B_2}{(s - 2)^2 + 1} + \frac{A_2 s + B_2}{[(s - 2)^2 + 1]^2} \tag{D.11}$$

Multiplying both sides of Equation (D.11) by $[(s - 2)^2 + 1]^2$ and rearranging terms, we obtain

$$s^2 - 6s + 7 = A_1 s^3 + (B_1 - 4A_1)s^2 + (5A_1 + A_2)s + B_2 - 5B_1 \tag{D.12}$$

The constants A_1, B_1, A_2, and B_2 can be determined by comparing the coefficients of s in the left- and right-hand sides of Equation (D.12). The coefficient of s^3 yields $A_1 = 0$, and the coefficient of s^2 yields

$$1 = B_1 - 4A_1, \quad \text{or} \quad B_1 = 1$$

Comparing the coefficient of s on both sides, we obtain

$$-6 = 5A_1 + A_2 \quad \text{or} \quad A_2 = -6$$

Comparing the constant term yields

$$7 = 5B_1 + B_2 \quad \text{or} \quad B_2 = 2$$

Finally, the partial-fraction decomposition of Equation (D.10) is

$$\frac{s^4 - 6s + 7}{(s^2 - 4s + 5)^2} = \frac{1}{[(s-2)^2 + 1]^2} - \frac{6s - 2}{(s-2)^2 + 1}$$

Bibliography

1. Brigham, E. Oram. *The Fast Fourier Transform and Its Applications*. Englewood Cliffs, NJ: Prentice-Hall, 1988.
2. Gabel, Robert A., and Richard A. Roberts. *Signals and Linear Systems*, 3d ed. New York: Wiley, 1987.
3. Johnson, Johnny R. *Introduction to Digital Signal Processing*. Englewood Cliffs, NJ: Prentice-Hall, 1989.
4. Lathi, B. P. *Signals and Systems*. Berkeley-Cambridge Press, Carmichael, CA, 1987.
5. McGillem, Claire D., and George R. Cooper. *Continuous and Discrete Signal and System Analysis*, 2d ed. New York: Holt, Reinhart and Winston, 1984.
6. O'Flynn, Michael, and Eugene Moriarity. *Linear Systems: Time Domain and Transform Analysis*. New York: Harper and Row, 1987.
7. Oppenheim, Alan V., and Ronald W. Schafer. *Discrete-Time Signal Processing*. Englewood Cliffs, NJ: Prentice-Hall, 1989.
8. Oppenheim, Alan V., Alan S. Wilsky, and S. Hamid Nawab. *Signals and Systems*. 2d ed. Englewood Cliffs, NJ: Prentice-Hall, 1997.
9. Papoulis, Athanasios. *The Fourier Integral and Its Applications*. New York: McGraw-Hill, 1962.
10. Philip, Charles L., and John Parr. *Signals, Systems and Transforms*. Englewood Cliffs, NJ: Prentice-Hall, 1995.
11. Poularikas, Alexander D., and Samuel Seely. *Elements of Signals and Systems*. Boston: PWS-Kent, 1988.
12. Proakis, John G., and Dimitris G. Manolakis. *Introduction to Digital Signal Processing*. New York: Macmillan, 1988.
13. Scott, Donald E. *An Introduction to Circuit Analysis: A Systems Approach*. New York: McGraw Hill, 1987.

14. Siebert, William M. *Circuits, Signals, and Systems*. New York: McGraw-Hill, 1986.
15. Strum, Robert D., and Donald E. Kirk. *First Principles of Discrete Systems and Digital Signal Processing*. Reading, MA: Addison-Wesley, 1988.
16. Swisher, George M. *Introduction to Linear Systems Analysis*. Beaverton, OR: Matrix, 1976.
17. Ziemer, Roger E., William H. Tranter, and D. Ronald Fannin. *Signals and Systems: Continuous and Discrete*, 2d ed. New York: Macmillan, 1989.

Index

D

E

F